Basic Vocational-Technical Mathematics

Fifth Edition

C. Thomas Olivo
Thomas P. Olivo

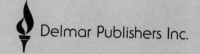

Delmar Publishers Inc.

About the Authors

C. Thomas Olivo is recognized nationally and internationally as one of the most experienced authors in the vocational-technical field. He has made, and continues to make, significant leadership contributions through his work in occupational and task analyses techniques, curriculum planning, instructional materials development, occupational competency assessment of teachers and students, school and industrial organization and management, international training, and human resource development.

Dr. Olivo entered teaching from industry and rose rapidly into successively responsible leadership positions including New York State supervisor and bureau chief of vocational-industrial-technical education and chief of vocational and practical arts curriculum development. He pioneered programs and services in post-secondary vocational-technical education as director of a state technical institute and as state director of vocational-technical education. He later served as tenured graduate professor at Temple University, Philadelphia, Pennsylvania.

Thomas P. Olivo has been a successful teacher in vocational-technical programs in secondary schools with additional service in relation to post-secondary institutions. He served as assistant professor in vocational-industrial teacher education at Long Beach State College. He is now executive director of Industrial/Vocational Consultants.

Mr. Olivo has worked in manufacturing and construction occupations and has made significant contributions to curriculum planning founded on task, job, and occupational analyses and competency-based instructional materials and testing programs. His service includes quality assessment of vocational-technical programs as a former state supervisor and curriculum planning, development, and instructional materials assessment through his earlier work in the Curriculum Laboratory at Clemson University.

This new edition is rededicated to Hilda G. Olivo, an outstanding teacher, and to Connie, Tom, and Judi.

For information, address Delmar Publishers Inc.
2 Computer Drive West, Box 15-015
Albany, New York 12212-5015

Previously published under the title *Basic Mathematics Simplified*

Copyright © 1985 by Delmar Publishers Inc.

Printed in the United States of America
Published simultaneously in Canada
by Nelson Canada,
A Division of International Thomson Limited

10 9 8 7 6 5

Delmar staff
Associate editor: Jonathan Plant
Production editor: Virginia Lawless Thompson

Cover photograph by Tom Carroll/FPG International

Library of Congress Cataloging in Publication Data

Olivo, C. Thomas.
  Basic vocational-technical mathematics.

  Rev. ed. of: Basic mathematics simplified. 4th ed. c1977.
  Includes index.
  1. Mathematics—1961-   . I. Olivo, Thomas P.
II. Olivo, C. Thomas. Basic mathematics simplified.
III. Title.
QA39.2.044   1985   510   84-23251
ISBN 0-8273-2226-7
ISBN 0-8273-2225-9 (pbk.)

CONTENTS

PREFACE

Scope of the Book

Basic Vocational-Technical Mathematics, fifth edition, is a major revision of the textbook previously titled *Basic Mathematics Simplified*. As this title change suggests, the fifth edition has been rewritten to meet the mathematical needs of today's vocational and technical students.

The content of *Basic Vocational-Technical Mathematics* has been thoroughly updated and expanded, based on broader demands for mathematics competence in business and industry. Computer-assisted design (CAD) and manufacturing (CAM) and use of electronic calculators for problem solving are some developments addressed in the fifth edition.

This new edition retains the solid approach to organizing and presenting mathematical topics that made *Basic Mathematics Simplified* a consistent best seller. The book's seven major parts are divided into sections, each of which concludes with an achievement review test. Sections are further subdivided into units, which include objectives and review and self-test problems. This organization allows the student to deal with mathematics in small, comprehensible portions. Worked-out examples are used frequently to demonstrate the steps involved in a particular procedure, and a wealth of applied examples and problems is included. These elements combine to make a book of high interest to vocational-technical students, and one they can easily understand.

Basic Vocational-Technical Mathematics is also an ideal core textbook to be used in conjunction with Delmar's *Practical Problems in Mathematics* series. These workbooks are designed to expand the mathematical skills of students by presenting problems from specific occupational areas.

New Features

The major changes in this edition include

- Expansion and updating of material in each part of the book. Examples and problems have been revised to reflect current costs, terms, and usage.
- Review and self-test items have been enriched with new applications, especially in the electricity/electronics area.
- Learning objectives are included at the beginning of each unit.
- The textbook has been redesigned and given a second color, making it more attractive and easier to follow.
- Part One (Fundamentals of Basic Mathematics) includes more coverage of precision instruments, multiple line graphs, and statistical measurement.
- Part Two (Fundamentals of SI Metric Measurement) now covers direct-reading metric micrometers, vernier calipers, and gage blocks. Historical information on metrics has been shortened in favor of a more practical approach.
- Part Three (Fundamentals of Electronic Calculators) places more emphasis on using calculators to solve applied mathematics problems. The general information given here is followed by new calculator applications in other parts of the book.
- Part Four (Fundamentals of Applied Algebra) includes new units on signed numbers, graphic solution of signed numbers, and the use of scientific notation. New worked-out examples have been added throughout this part, as have calculator applications.
- Part Five (Fundamentals of Applied Geometry) offers a more concise treatment of the subject by

consolidating information previously found in other parts, such as formulas applied to plane and solid forms. Additional examples, problems, and calculator applications are now included.

- Part Six (Fundamentals of Applied Trigonometry) includes strong coverage of vector and force measurements and calculator applications.
- Part Seven (High Technology Applications of Mathematics) is new to this edition. Cartesian coordinate and binary systems are explained, with applications to numerical control (NC), computer numerical control (CNC), computer-aided design (CAD) and manufacturing (CAM).
- The appendix and glossary have been carefully revised and updated.
- Additional test items are now provided in the *Instructor's Guide*.

Acknowledgments

Many people provided valuable guidance to this project as reviewers, including Manley Blevins, Gloucester High School; Doris Bratcher, Northwest Iowa Technical College; Frank Caldwell, York Technical College; Eddie Cameron, Gaston College; Joseph Dercoli, Newcastle School of Trades; R. R. Dolson, West Shore Community College; Thomas Karabees, Trident Technical College; William Nealan, Johnstown A.V.T.S.; Robert Peach, Greater Lowell Voc-Tech High School; Sidney Rabsatt, DeKalb Community College; Joanne Revelt, Erie County Technical School; J. Doug Richey, Henderson County Junior College; Chuck Saylor, Mid-Florida Technical Institute; Toby Shook, Asheville-Buncombe Technical College; and A. Lance Thompson, Modesto Junior College.

The authors express appreciation to Mr. Philip L. Starrett, manager, advertising and sales promotion, the L. S. Starrett Company; Mr. Robert Bledsoe, manager, and Crissey DeButts of the press relations department, Texas Instruments, Inc.; and Mr. Frank Cummings, manager, marketing division, Sharp Electronics Corporation, for providing technical assistance and furnishing industrial photographs.

NOTICE TO THE READER

PART ONE

Fundamentals of Basic Mathematics

SECTION 1

WHOLE NUMBERS

Unit 1 Addition of Whole Numbers

OBJECTIVES OF THE UNIT

After satisfactorily studying this unit, the student/trainee will be able to
- *Understand the meaning and use of Arabic numbers.*
- *Solve problems requiring the addition of whole numbers.*
- *Check the accuracy of each computation.*
- *Interpret and express whole numbers in the Roman numeral system.*

A. THE CONCEPT OF WHOLE NUMBERS

The term *whole numbers* refers to complete units where there are no fractional parts left over. Numbers such as 20 and 50; quantities like 144 machine screws, 10 spools, and 57 outlets; and measurements like 75 feet, 125 millimeters, or $875 represent whole numbers because the values do not contain a fraction.

The four basic operations in arithmetic include addition, subtraction, multiplication, and division. Addition is, by far, the most widely used of the four operations. However, even before the basic principles of addition may be applied, the Arabic system of numbers must be understood.

B. THE BASIC ARABIC NUMBER SYSTEM

The Arabic number system is the one that is widely used in this country and in many other parts of the world. This system includes ten digits: 0 1 2 3 4 5 6 7 8 9. These digits may be combined to express any desired number.

The ten numerals make up what is sometimes called the *system of tens* or the *decimal system*. An important fact about this system is that the location or position of a numeral in the written number expresses its value.

2

The term *digit* is used to identify a particular position. For instance, the first digit (column) in the extreme right position of a number is referred to as the *units digit* or *units column*. The digit in the next position to the left is in the *tens column;* the digit in the third position is in the *hundreds column*. Other examples of the place names of commonly used digits are shown in figure 1-1. Importantly, the combination of numerals that appear in particular digit positions in a written number expresses its value.

A whole number like 231 is a simple way of saying 200 + 30 + 1. The 231 means that every numeral in the *units* column has its value multiplied by 1; every numeral in the *tens* column, by 10; and every numeral in the *hundreds* column, by 100. The whole number is a shorthand way of representing the sum of the individual place values of the numerals.

Digit Place Names (Whole Numbers)						
Millions	Hundred-Thousands	Ten-Thousands	Thousands	Hundreds	Tens	Units
●	●	●	●	●	●	●

FIGURE 1-1

Numbers may be written in words as three hundred thirty-six; in an expanded form 300 + 30 + 6; or as the numeral, 336.

C. ADDING WHOLE NUMBERS

Addition is simply the process of adding all the numbers in each column in a problem. The answer is called the *sum*.

RULE FOR ADDING WHOLE NUMBERS

- Write one number under another so the unit numerals are in the units columns, tens are in the tens column, and hundreds are in the hundreds column.
- Add all the numbers in the units column.
- Write the result by placing the unit number in the units column.
  Note. Where the sum of a column is greater than 9, put the second number on the left in the tens column.
- Add all the numbers in the tens column. Write the last numeral in this sum in the tens column.
  Note. If this sum is greater than 9, put the second numeral on the left in the hundreds column.
- Continue to add each column of numbers. Find the sum of all columns by adding the numbers in the separate answers in each column.

E XAMPLE: Add 2765 + 972 + 857 + 1724.

Step 1   Arrange numbers in columns.

Step 2   Add all the numbers in *units* column.

$$5 + 2 + 7 + 4 = 18$$

Step 3   Add all numbers in *tens* column.

$$6 + 7 + 5 + 2 = 20$$

Step 4   Add all numbers in *hundreds* column.

$$7 + 9 + 8 + 7 = 31$$

Step 5   Add all numbers in *thousands* column.

$$2 + 1 = 3$$

Step 6   Add the sums in each column.

	2	7	6	5	
		9	7	2	
		8	5	7	
	1	7	2	4	
			1	8	
		2	0		
	3	1			
	3				
	6	3	1	8	Ans

RULE FOR ADDING WHOLE NUMBERS (SIMPLIFIED FORM)

- Write the numbers one under another with each digit in the proper column.
- Add all the numbers in the units column.
- Write the last numeral on the right in the units column.
- *Carry* the remaining numeral (mentally) on the left to the tens column and add it with the rest of the numbers in this column.
- Continue the same process with the remaining columns.

E XAMPLE: Add 2765 + 972 + 857 + 1724.

Step 1   Arrange numbers in columns.

Step 2   Add numbers in *units* column.

$$5 + 2 + 7 + 4 = 18$$

Place 8 in the *units* column. Carry the 1 (mentally) to the *tens* column.

Note. Add all columns downward from top to bottom.

Step 3   Add the numbers in the *tens* column and add the 1 carried over (mentally).

$$(6 + 7 + 5 + 2) + (1) = 21$$

Add Downward

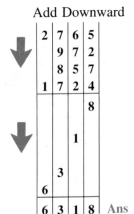

Place 1 in the *tens* column and carry the 2 over to
the *hundreds* column.

Step 4 Perform the same addition steps with the *hundreds*
and *thousands* columns.

RULE FOR CHECKING ACCURACY

* Find the sum of the numbers in the units column by adding
  upward (or in the reverse direction).
* Continue to add the numbers in the tens, then hundreds, then
  thousands columns.
  Note. If the sum of the numbers in the units column is more
  than 9, carry the number of tens to the tens column and add to
  the numbers in that column. Follow this same practice with the
  other columns.
* Compare the result (sum) with the sum obtained by adding in
  the reverse order.
  Note. If there is an error, repeat the addition step in the reverse
  order.

6	3	1	8	Ans
2	7	6	5	
	9	7	2	
	8	5	7	
1	7	2	4	

Add Upward

D. THE ROMAN NUMERAL SYSTEM

The Roman numeral system emerged toward the end of the sixteenth century as a second
numerical system. This system uses seven basic symbols. These are combined to produce any
required number or sequence of numerals.

This system is applied in the medical and allied health industry. Normally, the range of
numbers used in such occupations is from 1 to 100. There are other limited uses of Roman
numerals in architecture and design, different businesses, and for general consumer applications.

Many Roman numeral equivalents require more symbols and are more difficult to write
and apply than Arabic numerals. Seven basic Roman numerals with their values (together with
other combinations from 1 to 1,000) are given in figure 1-2.

Note from figure 1-2 that Roman numerals I, II, and III correspond with Arabic numerals 1,
2, and 3. The value for four is written as IV (V − I; 5 − 1). Six is VI (V + I; 5 + 1); seven
is VII (V + II; 5 + 2); and eight is VIII (V + III; 5 + 3). Nine is IX (X − I; 10 − 1).

Forty is written XL. This is the equivalent of L − X (50 − 10). Ninety is XC (C − X;
100 − 10); four hundred is CD (D − C; 500 − 100), and nine hundred is CM (M − C; 1,000
− 100). Intermediate numerals are stated as combinations of I, V, X, L, C, D and M. Although
capital letters are usually used, symbols for the Roman numerals also may be written in lowercase i, v, x, l, c, d, and m.

Basic Roman Numerals and Equivalent Values		Common Roman Numeral Combinations and Equivalent Values					
Roman Numeral	Value	Roman Numeral	Value	Roman Numeral	Value	Roman Numeral	Value
I	1	I	1	X	10	C	100
V	5	II	2	XX	20	CC	200
X	10	III	3	XXX	30	CCC	300
L	50	IV	4	XL	40	CD	400
C	100	V	5	L	50	D	500
D	500	VI	6	LX	60	DC	600
M	1,000	VII	7	LXX	70	DCC	700
		VIII	8	LXXX	80	DCCC	800
		IX	9	XC	90	CM	900
						M	1,000

FIGURE 1-2

RULE FOR WRITING ROMAN NUMERALS

- Break the numeral down into appropriate digits: units, tens, hundreds, thousands.
- Determine the Roman numeral equivalent of the largest digit value.
- Move to the next smaller digit to the right. Again, determine the Roman numeral equivalent.
- Repeat these steps for the numerical value in each digit position.
- Combine the separate Roman numerals in each digit position.
- Write the combination of numerals in a continuous sequence.

The answer is the equivalent Roman numeral value.

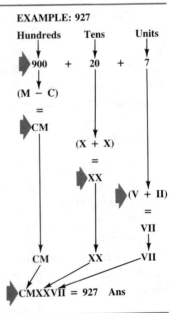

EXAMPLE: 927

Hundreds	Tens	Units
900	+ 20	+ 7
(M − C)	(X + X)	(V + II)
=	=	=
CM	XX	VII

CMXXVII = 927   Ans

The rule is reversed to determine the equivalent numerical value in the Arabic system. The Roman numeral is broken down into its respective digits. The Arabic numeral equivalent of each digit value is determined. The separate numerals in each digit position are combined. The resulting Arabic numeral is equivalent to the Roman numeral.

Review and Self-Test

ASSIGNMENT UNIT 1

A. The Concept of Whole Numbers (General Applications)

1. Write each number in Arabic figures.
   a. Fifty-nine
   b. One hundred twenty
   c. One thousand eight
   d. Twelve thousand nine hundred eighty-seven

2. Write each number in words.
   a. 12   b. 27   c. 140   d. 926   e. 1,700   f. 7,937

3. Write the numbers in each combination in simplified form.
   a. 50 + 7
   b. 80 + 10 + 9
   c. 100 + 3 + 2
   d. 400 + 50 + 7
   e. 1,000 + 600 + 50
   f. 8,000 + 50 + 7

4. Write each number in expanded form.
   a. 61   b. 103   c. 422   d. 1,006   e. 4,027 f. 6,931

5. State the equivalent Roman numeral for each Arabic numeral for values A through F.

A	B	C	D	E	F
6	24	49	94	444	1,976

6. Determine the equivalent Arabic numeral value for Roman numerals A through F.

A	B	C	D	E	F
XIX	XLIX	XCI	CXLIX	CDXCIX	MDCCCXC

B. Adding Whole Numbers (General Applications)

1. Add each column of numbers vertically. Check all answers.

A	B	C	D	E	F
17	35	67	96	900	808
20	35	24	48	65	15

G	H	I	J	K	L
		72	83	110	9,765
925	474	94	94	87	4,372
96	888		40	215	8,988
		53	12	468	1,009

2. Find the sum of the number combinations (a through f) by adding horizontally. Check each answer.

   a. 12 + 37   c. 77 + 69   e. 548 + 941 + 960

   b. 56 + 25   d. 962 + 829 + 13   f. 827 + 633 + 569

C. Adding Whole Numbers (Practical Problems)

1. Determine the overall length Ⓐ of the tapered pin.

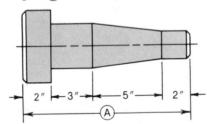

2. The tail lamps of an automobile draw 2 amperes of current; the head lamps, 11 amperes; the car heater, 9 amperes; and the ignition coil, 5 amperes. What is the total number of amperes drawn from the battery?
3. The following amounts of fabrics are required for decorating: 23 yards, 19 yards, 15 yards, and 8 yards. Determine the total yardage of fabric that is needed.
4. Six buildings are to be wired by electricians. The number of outlets that must be installed is 65, 75, 69, 81, 57, and 76, respectively. Find the total number of outlets that must be roughed-in.

5. Determine dimensions Ⓐ, Ⓑ, and Ⓒ for the structural steel beam.

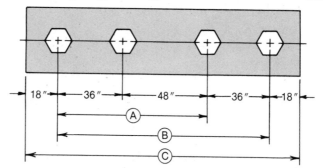

6. A carpenter laid 675 shingles in a half day; 1,425 the second day; and 1,054 the third day. How many shingles were laid in the 2½ days? Label the answer and check.

7. A plumber made the pipe connections as illustrated. What was the total length of pipe used? (The end-to-end measurement of each pipe is given in inches.)

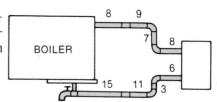

8. Three shipments of 1-inch native pine are received by a contractor: 7,556; 8,750; and 9,898 board feet, respectively. What is the total number of board feet of lumber delivered?

9. The monthly production of motors for refrigerators was as follows: January 29,220; February 32,416; March 37,240; April 39,374; May 45,666; June 52,487; July 36,458; August 35,000; September 32,250; October 51,750; November 62,475; December 50,525. Determine the total output of motors for the year.

10. Calculate dimensions Ⓐ, Ⓑ, Ⓒ, and Ⓓ for the aluminum template.

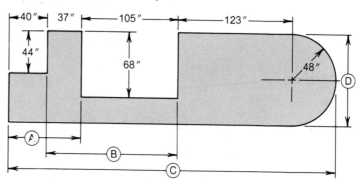

11. Find the total length of wire needed by a technician to connect the following lengths of wires: 15 feet, 32 feet, 96 feet, 142 feet, 68 feet, and 84 feet.

12. Determine the total resistance (R_T) of the series electrical circuit by adding the separate resistances ($R_1 + R_2 + R_3 + R_4 + R_5$).

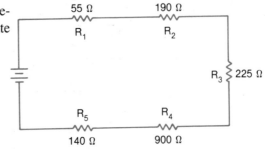

13. The kilowatt hours of electrical energy consumed monthly for six months were 1,412; 1,839, 27,000; 29,787; 32,496; and 1,934. Determine the total kilowatts of energy that were used.

14. A construction project required A, B, C, and D quantities of diesel fuel for four groups of equipment. (a) Find the number of gallons of diesel fuel used by each group. (b) Determine the total amount of fuel used by the four groups.

Equipment (A)	Equipment (B)	Equipment (C)	Equipment (D)
1,210	4,605	3,202	1,369
989	5,925	3,998	2,787
1,868	9,879	4,687	4,695
		3,896	5,489

15. Examine the main floor plan of the house. Determine the total number of square feet in the master bedroom, living and kitchen/dining rooms, and deck.

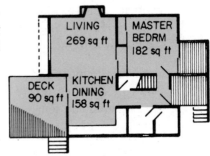

Unit 2 Subtraction of Whole Numbers

Subtraction is the process of determining the difference between two numbers or quantities. The number from which another number is to be taken (subtracted) is known as the *minuend*. The number to be subtracted is the *subtrahend*. The result of the process is called the *difference* or *remainder*.

RULE FOR SUBTRACTING WHOLE NUMBERS

* Write the larger of the two numbers as the minuend.
* Write the number to be subtracted *(subtrahend)* underneath the minuend, so that the digits are in their respective columns: units with units and tens with tens.
* Start with the units column and subtract the number in the subtrahend from the minuend.
* Continue the same process with the tens and hundreds columns.

EXAMPLE: Subtract 346 from 988.

Step 1   Write the larger number as the *minuend*.
Step 2   Place the digits in the *subtrahend* in proper columns.
Step 3   Start with the units column and take 6 away from 8. Record the difference (2) in the units column.
Step 4   Continue in the same manner with the tens and hundreds columns.
Step 5   The answer (642) is the difference between the two numbers.

Many times the digits in one or more columns of the subtrahend are larger than the corresponding digits in the minuend. In such cases, numbers are *exchanged* from the minuend. The principle of *exchanging* is based on redistributing numbers. For instance, if 254 is to be subtracted from 723, the digits in both the units and tens places of the subtrahend are larger than those in the minuend.

RULE FOR EXCHANGING NUMBERS IN SUBTRACTION

- Consider 723 as being equal to $700 + 20 + 3$.
- Exchange one ten from the tens column for 10 units. Add these 10 units to the units column. The 723 is now equal to $700 + 10 + 13$.
- Subtract the 4 in the subtrahend from 13.
- Exchange 100 from the hundreds column and add it to the 10 in the tens column. Thus, the 723 is now equal to $600 + 110 + 13$.
  Note. The 5 in the tens column of the subtrahend means 5 times 10 or 50.
- Subtract the 50 in the minuend from the 110 in the subtrahend.
- Subtract the 2 in the hundreds column from the 6 in the same column of the minuend; $600 - 200 = 400$.
- Simplify the result. $400 + 60 + 9 = 469$ **Ans**

To summarize, when 254 is subtracted from 723, consider the

$$
\begin{aligned}
\text{minuend} &= 600 + 110 + 13 \\
\text{subtrahend} &= \underline{200 + \ \ 50 + \ \ 4} \\
\text{difference} &= 400 + \ \ 60 + \ \ 9 \quad \text{or} \quad 469 \quad \textbf{Ans}
\end{aligned}
$$

With continued practice, the steps in exchanging become automatic. At this point, there is no need to write the steps in detail.

RULE FOR CHECKING THE SUBTRACTION OF WHOLE NUMBERS

- Arrange the numerals in each digit in the subtrahend and those in the answer in columns.
- Add the subtrahend and the difference. When the difference is correct, the sum of the difference and the subtrahend is equal to the minuend.
- Recheck if the answers are not equal. First check the addition, which is the easiest step. If the answers still do not agree, rework the original problem.

Review and Self-Test

ASSIGNMENT UNIT 2

A. Subtracting Whole Numbers and Checking (General Applications)

1. Subtract each pair of numbers. Check each answer.

A	B	C	D	E	F
78	87	45	98	286	364
34	26	29	59	142	158

G	H	I	J	K	L
753	946	473	707	1,642	2,537
225	168	289	198	456	1,659

2. Perform each operation as indicated. Check each answer.

A	246 − 134	E	3,015 − 2,127
B	727 − 415	F	6,007 − 5,188
C	965 − 847	G	4,112 + 705 + 1,293 − 2,097
D	1,752 − 1,263	H	15,625 + 16,596 + 8,989 − 7,349

3. Determine the difference between each set of numbers and check each answer.
   a. 19,264 and 11,156   c. 10,065 feet and 9,047 feet
   b. 8,537 and 6,759   d. 20,003 miles and 13,365 miles
4. Subtract and check each answer.
   a. 51,219 from 63,422   c. 7,603 acres from 9,502 acres
   b. 9,655 from 13,004   d. 17,092 square miles from 25,001 square miles

B. Subtracting Whole Numbers (Practical Problems)

Note. Check all answers.
1. A contractor has 5,500 board feet of oak flooring. If 2,625 board feet are used on one house, how much flooring is left?
2. A customer's service bill for electricity shows that a total of 1,235 kilowatt hours was used. Of this total, 367 kilowatt hours were used for lighting service and the balance for domestic hot water. How many kilowatt hours were used for hot water?
3. Determine the number of miles traveled for each of five weeks from the odometer readings shown.

Week	1	2	3	4	5
Reading (Start)	32,119	32,899	33,988	35,976	37,065
Reading (End)	32,899	33,988	35,976	37,065	39,001

4. A container (drum) holds 55 gallons of turpentine. In a one-month period, these quantities were used: 5 gallons, 10 gallons, 8 gallons, 7 gallons, 16 gallons, and 8 gallons. How much turpentine is left?

5. Subtract the following electrical quantities.

a.
$$297 \text{ millihenries}$$
$$-148 \text{ millihenries}$$

b.
$$632 \text{ watts}$$
$$-419 \text{ watts}$$

c.
$$54\ 500 \text{ ohms}$$
$$-29\ 750 \text{ ohms}$$

6. The cubic yards of gas energy used in five projects for two processes (A) and (B) are given in the table.

Projects	1	2	3	4	5
Process (A)	99	365	1,277	3,018	41,605
Process (B)	87	246	1,098	1,129	32,719

a. Find the increased number of cubic yards of energy required for process A over process B for projects 1 through 5.
b. Determine the total cubic yards of energy required for process A.
c. Determine the total cubic yards for process B.
d. Find the difference between the amounts of gas energy required for process A and process B for the total of projects 1 through 5.
e. Show how the answer to item d. may be checked by a second method.

7. A time and motion study of three workers shows the production reported in the table for processes A through E.

Workers	Production for Processes				
	A	B	C	D	E
1	22	131	2,027	1,169	2,235
2	36	257	3,249	2,479	1,357
3	15	215	2,116	3,368	1,799

a. Identify the fastest production worker for each process A through E.
b. Make a table and indicate the difference in production between (1) the fastest worker and the second fastest worker and (2) the fastest worker and slowest worker.

8. Complete the weekly garment sales chart. Compute the missing quantities for garment types A through D.

Type	Garments Received	Daily Sales - Week Ending June 20						Quantity on Hand June 22
		M	T	W	Th	F	S	
A	178	21	17	9	14	25	31	
B	2,347	345	196	187	294	468	613	
C		93	79	67	105	198	94	217
D		216	237	259	316	419	177	597

9. a. Determine the difference in calories of food quantities A and B for each item (1, 2, 3, and 4).
   b. Find the total food values for A and B.
   c. Indicate the total difference in calories between quantities A and B.

Item	Quantity Food Values in Calories		Difference in Calories
	A	B	
1	46	228	
2	2,280	1,682	
3	7,357	5,469	
4	9,874	10,042	
Total			

10. Compute dimensions (A), (B), (C), and (D). Check each answer.

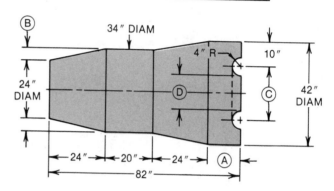

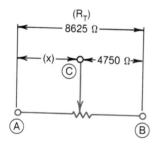

11. Find the ohms of resistance *(X)* between lugs (A) and (B) and (C) of the variable resistor. X = maximum resistance (R_T) − ohms of resistance $(B - C)$.

12. Calculate dimensions Ⓐ, Ⓑ, Ⓒ, and Ⓓ for machining the Link Plate. All dimensions are in inches.

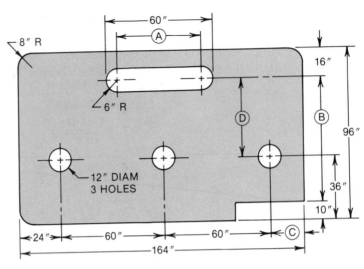

Unit 3 Multiplication of Whole Numbers

OBJECTIVES OF THE UNIT

After satisfactorily studying this unit, the student/trainee will be able to
- *Interpret the arithmetical process for multiplying whole numbers.*
- *Multiply whole numbers in general and practical problems.*
- *Check answers by reviewing each step or by reworking each problem.*

Multiplication is a simplified method of adding a quantity a given number of times. For example, instead of writing the number 27 nine times and adding a long column of numbers, 27 may be multiplied by 9. The multiplication process saves time and simplifies the problem. The answer or result obtained by multiplication is called the *product*.

A. ARRANGING NUMBERS FOR MULTIPLICATION

When one number is multiplied by another number, the product is the same, regardless of how the numbers are arranged. For instance, the product of 22 × 11 is the same as 11 × 22. In either case, the number to be multiplied is called the *multiplicand* and the other number, the *multiplier*. The multiplication process is simplified when the smaller number is the multiplier and the larger is the multiplicand. Each digit in the multiplier and multiplicand (and those digits in the number obtained by multiplying) should be placed in its proper column. This simple practice saves time and makes greater accuracy possible.

B. EXPLANATION OF THE MULTIPLICATION PROCESS

In the multiplication process, every number in the multiplicand is multiplied by every number of the multiplier. For example, when the number 47 is multiplied by 26, the numbers in the units column of both multiplicand and multiplier are multiplied first.

Because the product of 7 times 6 is greater than 9, the second digit is mentally *carried over* to the tens column. In this instance, the 42 means 2 units and 4 tens. The 2 is written in the units column and the 4 is carried over and added to the result in the tens column.

Continue by multiplying the number in the tens column of the multiplicand (4) by the same multiplier (6). To this product of 6×4 add the 4 that is carried over.

Write the 8 in the tens column. Since no other numbers in the multiplicand are to be multiplied by 6, the 2 is written in the hundreds column.

The same multiplication processes are carried out until every number in the multiplicand is multiplied by every number in the multiplier. When the number in the multiplier is in the tens place, the first digit in the product is written in this column. Thus, the next step is to multiply each number in the multiplicand by the number in the tens column of the multiplier.

Actually, the 7 in the multiplicand is multiplied by 20. If this were done, 0 would be written in the units column, 4 in the tens column, and 1 in the hundreds column.

Next, the 4 in the multiplicand represents 4 tens or 40. If the 40 is multiplied by the same 20 in the multiplier, the product is 800. The last step is to add $282 + 140 + 800 = 1222$.

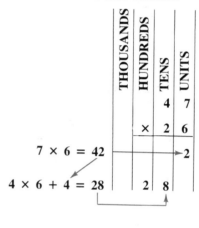

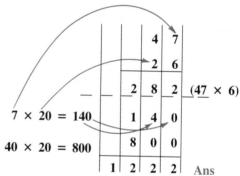

Shortening the Multiplication Process

The multiplication process is shortened and simplified by multiplying by the actual number in the tens column of the multiplier rather than by ten times that number. Then enter the first right digit of the product in the tens column and carry over the remainder to the hundreds column. When multiplying by a number in the hundreds column, put the first right digit in the hundreds column.

C. MULTIPLYING WHOLE NUMBERS

Regardless of the number of digits in either the multiplicand or multiplier, the multiplication process is the same.

RULE FOR MULTIPLYING WHOLE NUMBERS

- Write the larger of the two numbers as the multiplicand; write the smaller as the multiplier.
  Note. Place the numerals in both multiplicand and multiplier under each other in columns: tens in tens column, hundreds in hundreds column.
- Multiply the numbers in the units column of both multiplicand and multiplier. Write the units result in the units column and carry over the tens.
- Multiply the number in the tens column of the multiplicand by the number in the units column of the multiplier. Add the tens remainder to this product.
- Write the first digit on the right in the result in the tens column.
- Carry over any numerals representing hundreds and add to the next result.
  Note. If no other numbers are to be multiplied, write each digit in the result in the proper column.
- Continue to multiply every number in the multiplicand by every number in the multiplier.
- Add the results in each column.

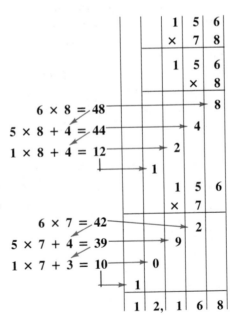

EXAMPLE: Multiply 156 by 78.

Step 1 Write numbers in columns.

Step 2 Multiply every number in the multiplicand by 8.

$6 \times 8 = 48$
$5 \times 8 + 4 = 44$
$1 \times 8 + 4 = 12$

Step 3 Multiply every number in the multiplicand by the number in the tens place of the multiplier (7).

$6 \times 7 = 42$
$5 \times 7 + 4 = 39$
$1 \times 7 + 3 = 10$

Step 4 Add the numbers in each column. The result, 12,168, is the product of 156×78.

D. CHECKING THE MULTIPLICATION PROCESS

Two methods of checking multiplication problems are commonly used.

Method I (Checking Each Step)
- Multiply the numbers in the units column of both multiplicand and multiplier.
- Continue until all numbers in the multiplicand are multiplied by all the numbers in the multiplier.
- Check the product against the original product. If the products are not the same, check the steps again.

Method II (Reworking)
- Write the multiplicand as the multiplier; write the multiplier as the multiplicand.
- Multiply each number. Check the final product against the original product.

Regardless of which method is used, the checking process takes almost as much time and effort as the original problem. One of the principal places for error is in failing to remember the numeral that is to be carried over from one column to another. This difficulty may be overcome by writing the number to be carried lightly over the number to be multiplied next. For instance, if 97 is to be multiplied by 7, the problem may be worked out as shown in the example.

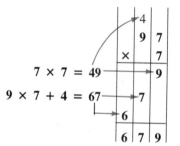

$7 \times 7 = 49$

$9 \times 7 + 4 = 67$

The multiplication process is simplified further by learning all the combinations of numbers in a multiplication table from 1 to 10. The product of any two of these numbers may then be given quickly without computation.

ASSIGNMENT UNIT 3

A. Multiplying Whole Numbers (General Applications)

1. Multiply each set of numbers mentally.

A	B	C	D	E
14	22	78	240	121
6	7	9	8	7

2. Multiply each set of numbers and check each product.

A	B	C	D	E
212 × 6	343 × 5	508 × 9	689 × 6	987 × 7
F	G	H	I	J
411 ×14	627 ×26	303 ×97	879 ×78	687 ×90
K	L	M	N	O
8,165 × 72	6,057 ×324	5,009 ×620	7,987 ×869	97,009 ×308

B. Multiplying Whole Numbers (Practical Problems)

1. A crew of 17 worked on a construction job 153 days of 8 hours each without a lost-time accident. How many accident-free hours did the crew work?
2. A bricklayer lays an average of 145 bricks an hour. At this rate, how many bricks can be laid in 37 hours?
3. A mason purchased 223 cubic yards of ready-mixed concrete at $36.00 a cubic yard, 51 barrels of lime at $24.00 a barrel, and 38 cubic yards of sand at $9.00 per cubic yard. What was the total cost of materials?
4. The cost of a certain size brass elbow is $7.00 per pound. Determine the cost of 147 pounds of elbows.
5. Compute center distance measurements Ⓐ, Ⓑ, Ⓒ, and Ⓓ for the Fixture Plate.

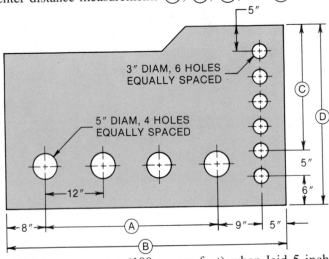

6. It takes 760 shingles per square (100 square feet) when laid 5 inches to weather. How many shingles will be needed to cover 37 squares with the same weathering?

7. A bundle of white cedar shingles contains 250 shingles. How many shingles are in 378 bundles?

8. Two magnets are wound. The first has 57 layers of 98 turns each; the second magnet has 38 layers of 179 turns each. Give the total number of turns in both coils.

9. An electrical contractor purchased 196 conduit bodies at 89 cents each and 87 of another type for $1.35 each. Give the total cost of these materials.

10. A car has three 34-candlepower bulbs, two 26-candlepower, four 9-candlepower, and three 15-candlepower. Find the total candlepower.

11. A truck averages 42 miles an hour for 6 hours daily for 19 days. Another truck averages 37 miles an hour for 6 hours daily for 23 days. A third truck averages 39 miles an hour for 6 hours daily for 22 days. Determine the total mileage of the three trucks.

12. A power plant consumes an average of 17 gallons of fuel per hour. Determine the consumption (a) for a 24-hour day, (b) each week, and (c) for a 31-day month.

13. The hourly production for four different parts (A, B, C, and D) produced on three machines (1, 2, and 3) is recorded in the chart.

Machine	Part A	Part B	Part C	Part D
1	4	10	100	234
2	6	11	120	247
3	9	17	132	379
Unit Cost	3	5	12	27

a. How many of each part are produced on each machine in an 8-hour shift?

b. What is the weekly production of each part on each machine for two shifts, each working a 36-hour week?

c. Determine the weekly cost for each part for the two shifts using the unit cost per part.

14. Compute the amount of material that is needed in the total order for
a. Hem and seam allowance
b. Cover length
c. Completed covers

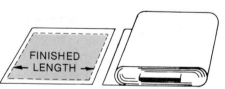

COVER SPECIFICATIONS

MATERIAL: PERCALE MUSLIN
FINISHED LENGTH: 43″
HEM SEAM ALLOWANCE: 4″
QUANTITY: 725 COVERS

FINISHED
LENGTH

HEM AND SEAM ALLOWANCE

15. Each cylinder of an engine has a displacement of 37 cubic inches. Compute the total displacement of a 6-cylinder engine.

16. Find the voltage in simple circuits Ⓐ and Ⓑ. The voltage (V) equals the current (measured in amperes, A) multiplied by the resistance (R) (measured in ohms, Ω).

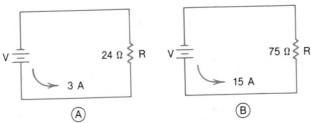

Unit 4 Division of Whole Numbers

OBJECTIVES OF THE UNIT

After satisfactorily studying this unit, the student/trainee will be able to
- *Interpret the arithmetical process of dividing whole numbers.*
- *Divide whole numbers in general and practical problems.*
- *Check answers to problems involving the division of whole numbers.*

Division is a simplified method of subtracting a quantity a given number of times. For example, if 23 is subtracted from 92, the subtraction process is repeated four times until there is no remainder.

- Subtract 23 from 92 $92 - 23 = 69$
- Subtract 23 from 69 $69 - 23 = 46$
- Subtract 23 from 46 $46 - 23 = 23$
- Subtract 23 from 23 $23 - 23 = 0$

This method of repeated subtraction is long and becomes more involved as the numbers get larger. In its place, a shorter method known as *division* is used to save time and effort.

In any division problem, the number to be divided is called the *dividend*. The number by which the dividend is divided is the *divisor*. The number that indicates how many times the divisor may be subtracted from the dividend is the *quotient*. When the divisor cannot be subtracted from the dividend an even number of times, the number left over is referred to as the *remainder*. The two signs or symbols that are commonly used to denote the process of division are ($\div$) and ($\overline{)}$).

$$\begin{array}{r} 4 \quad \text{◀ Quotient} \\ \text{Divisor ▶} \quad 23\overline{)92} \quad \text{◀ Dividend} \\ 92 \\ \hline \end{array}$$

A. THE DIVISION PROCESS

Division, like multiplication, is based on the fact that the dividend may be written in expanded form with any combination of smaller numbers. For instance, the number 525 may be thought of as consisting of 500 + 25, or 350 + 175, or any other combination that adds up to 525. The combination depends largely on the divisor. The first step is to break the dividend into a combination of numbers into which the divisor will divide evenly.

If 525 is to be divided by 35, the division process is simplified when the 525 is considered as 350 + 175.

$$35\overline{)525} = 35\overline{)350} + 35\overline{)175} = (10 + 5) = 15 \text{ Ans}$$

In this case, 525 is divided an even number of times by 35. If the number were to be divided by 25, instead of 35, the 525 may be considered in expanded form to be equal to 500 + 25.

$$25\overline{)525} = 25\overline{)500} + 25\overline{)25} = (20 + 1) = 21 \text{ Ans}$$

In these two examples, the dividend in expanded form is made up of different combinations of numbers depending on the divisor.

B. DIVIDING WHOLE NUMBERS

Although this is the principle on which division is based, the actual process is simplified by following a few basic steps.

RULE FOR DIVIDING WHOLE NUMBERS

* Write the number to be divided as the dividend within the division frame; write the divisor on the outside.
* Determine how many times the numerals in the first few digits of the dividend may be divided by the divisor.
* Multiply the divisor by the *trial quotient*. The numeral in the units digit of the divisor is multiplied first, then the tens. The product is placed under the dividend.
  Note. If the *trial quotient* is larger than it should be, the product will be greater than the dividend. When this happens, change the quotient to the next lower number.
* Subtract the product from the dividend.
* Bring down the numeral in the next place in the dividend. If the remainder cannot be divided by the divisor, bring down the next digit in the dividend.
* Repeat the division process until all the digits in the dividend are used.
  Note. When the divisor does not divide evenly into the dividend, the number resulting from the last subtraction is the *remainder.*
* Express the quotient in terms of the quantities that are being divided.

Divisor $\overline{)\text{Dividend}}$

Trial
Quotient
▼
$\bullet\bullet\bullet\overline{)\bullet\ \bullet\ \bullet\ \bullet\ \bullet}$
$\underline{\bullet\ \bullet\ \bullet}$
▲
Place Product
Here

EXAMPLE: *Case 1.* (Without a remainder.) Divide 1984 by 64.

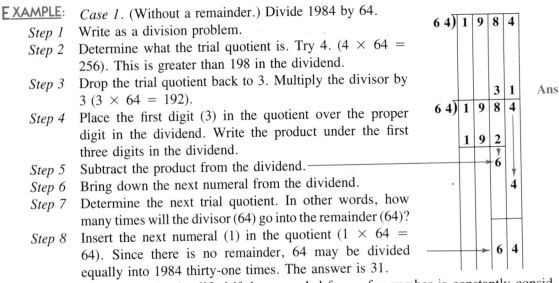

Step 1   Write as a division problem.
Step 2   Determine what the trial quotient is. Try 4. (4 × 64 = 256). This is greater than 198 in the dividend.
Step 3   Drop the trial quotient back to 3. Multiply the divisor by 3 (3 × 64 = 192).
Step 4   Place the first digit (3) in the quotient over the proper digit in the dividend. Write the product under the first three digits in the dividend.
Step 5   Subtract the product from the dividend.
Step 6   Bring down the next numeral from the dividend.
Step 7   Determine the next trial quotient. In other words, how many times will the divisor (64) go into the remainder (64)?
Step 8   Insert the next numeral (1) in the quotient (1 × 64 = 64). Since there is no remainder, 64 may be divided equally into 1984 thirty-one times. The answer is 31.

The division process is simplified if the expanded form of a number is constantly considered. In this example, the first 3 in the quotient is actually 30. Multiplying the divisor (64) by 30 = 1920. Subtracting 1920 from the original 1984 leaves 64. The second digit in the units column is 1, so the quotient is equal to 30 + 1 or 31.

EXAMPLE: *Case 2.* (With a remainder.) Divide 3900 by 47.

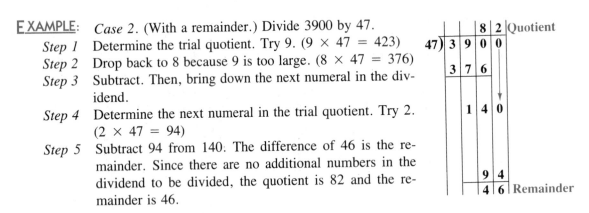

Step 1   Determine the trial quotient. Try 9. (9 × 47 = 423)
Step 2   Drop back to 8 because 9 is too large. (8 × 47 = 376)
Step 3   Subtract. Then, bring down the next numeral in the dividend.
Step 4   Determine the next numeral in the trial quotient. Try 2. (2 × 47 = 94)
Step 5   Subtract 94 from 140. The difference of 46 is the remainder. Since there are no additional numbers in the dividend to be divided, the quotient is 82 and the remainder is 46.

C. CHECKING THE DIVISION OF WHOLE NUMBERS

RULE FOR CHECKING DIVISION

- Multiply the quotient by the divisor.
- Add the remainder to the product. The sum is equal to the dividend when the division is correct.

4. a. Compute the center-center measurements for slots Ⓐ and holes Ⓑ and Ⓒ.

b. Check each measurement by the multiplication process.

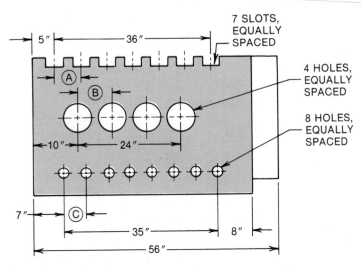

5. A tank holds 4,851 cubic inches of coolant. How many gallons of liquid are needed to fill the tank? (Each gallon contains 231 cubic inches.)

6. How many columns spaced 8'-0" on centers are required for a girder 72'-0" long? (Both ends of the girder are supported on foundation walls.)

7. Determine the rise (A) of each step from the drawing of the stringer.

8. Determine the run (B) of each step.

9. A total load of 23,256 watts is distributed equally over 18 branch circuits. Find the load per circuit in watts.

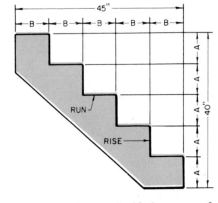

10. What is the average number of feet of wire per outlet used on a job that takes 1,896 feet of *BX* for 82 outlets?

11. How much money per person can a catering service allot if there will be 350 guests at a total cost of $2,100?

12. A florist has 912 flowers on hand. How many bouquets can be made if there are 6 flowers in each bouquet?

13. A hospital patient is allowed a total of 29,500 calories during a 20-day period. What is the patient's daily allotment?

14. A 24-foot structural steel I-beam weighs 2,136 pounds. Determine the weight per linear foot.

15. The total current in a parallel circuit is 728 milliamperes. There are four branches (A, B, C, and D) into which the current splits equally. Determine and label the current in one branch when A = B = C = D.

Note. If the two quantities are not equal, check the steps again in the checking process. If necessary, rework the original steps in division.

EXAMPLE: Check the correctness of $47\overline{)3900}$. The given answer
is 82 with a remainder of 46.

$82 \times 47 = 3,854$

Step 1 Multiply the quotient by the divisor.

Step 2 Add the remainder (46).

$+ \ 46$

Step 3 Check the sum (3,900) with the original dividend (3,900).
Since both agree, the answer is correct.

$3,900$ Check

Review and Self-Test

ASSIGNMENT UNIT 4

A. Dividing Whole Numbers and Checking (General Applications)

1. Divide each pair of numbers. Check each quotient.

A	B	C	D	E
$6\overline{)126}$	$4\overline{)120}$	$7\overline{)147}$	$3\overline{)135}$	$7\overline{)182}$
F	G	H	I	J
$11\overline{)110}$	$12\overline{)132}$	$27\overline{)810}$	$53\overline{)742}$	$92\overline{)2024}$

2. Perform the operation indicated. Where there is a remainder, mark it (Rem). Check each answer.

   a. Divide 1,250 by 25   c. Divide 1, 782 by 162   e. 9,002 ÷ 45
   b. Divide 1,638 by 39   d. 14,091 ÷ 33   f. 80,208 ÷ 121

B. Dividing Whole Numbers (Practical Problems)

1. A mason plastered an area of 425 square yards in 5 days. What was the average number of square yards plastered each day?
2. A contractor agreed to furnish and pour 27 cubic yards of concrete for $918.00. What is the cost per cubic yard?
3. A plumber sets 14 water closets and uses 42 pounds of caulking compound. How much caulking compound is used on an average for each closet?

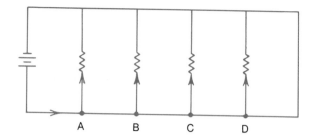

Unit 5 Achievement Review on Whole Numbers

OBJECTIVES OF THE UNIT

This achievement review serves as an overall test for Section 1. The unit is designed to measure the student's/trainee's ability to
- *Write whole numbers in simplified or expanded form.*
- *Solve general and practical problems involving addition, subtraction, multiplication, and division of whole numbers.*
- *Find the solutions to problems involving any combination of addition, subtraction, multiplication, and division of whole numbers.*
- *Check each answer.*

A. THE CONCEPT OF WHOLE NUMBERS

1. Write each number in expanded form.
   a. 305   b. 620   c. 735   d. 1,025   e. 7,572
2. Write each total in simplified form.
   a. 70 + 5   c. 700 + 20 + 5   e. 6,000 + 200 + 60
   b. 100 + 10   d. 2,000 + 15   f. 10,000 + 700 + 50 + 6
3. Each word or phrase in column I expresses one of the four basic mathematical processes listed in column II. Match each process with the correct term or phrase. Write the A, S, M, or D symbol where it applies in column I.

Column I				Column II
a.	Times	f.	Product of	Addition(A)
b.	Minus	g.	Difference between	Subtraction (S)
c.	Plus	h.	Added to	Multiplication (M)
d.	Divided by	i.	Sum of	Division (D)
e.	Increased by	j.	Quotient of	

B. ADDITION OF WHOLE NUMBERS

1. Building materials for the repair of a barn include masonry $968, electrical $356, hardware $134, painting $336, and lumber $1,376. Find the total cost of the materials.
2. A school has sixteen electrical circuits with capacities of 2,360; 1,648; 1,235; 660; 978; 2,296; 1,586; 1,975; 462; 855; 343; 592; 2,325; 847; 2,015; and 1,238 watts, respectively. Determine the total number of watts consumed when all the circuits are used to capacity.
3. Add each column of numbers vertically. Then add each row across.

201	9,716	7,061	5,012
76	253	2,562	9,109
18	2,009	4,928	6,007
35	5,265	7,465	8,160
257	374	968	5,789

4. Calculate dimensions Ⓐ through Ⓔ from the Die Plate drawing.

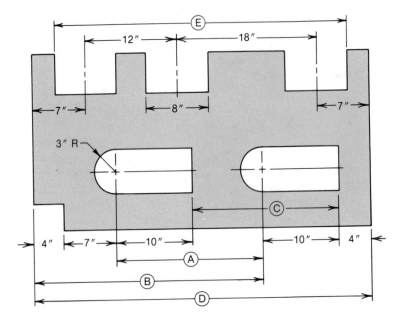

C. SUBTRACTION OF WHOLE NUMBERS

1. Subtract the quantities A through F. State each answer in terms of the unit of measure specified in each case.

A	B	C
12,695 square miles − 2,797 square miles	9,085 tons −6,187 tons	3,012 yards −1,029 yards

D	E	F
4,855 board feet −2,068 board feet	6,537 watts −4,659 watts	28,007 barrels −19,018 barrels

2. Calculate dimensions Ⓐ, Ⓑ, and Ⓒ for the Turned Shaft.

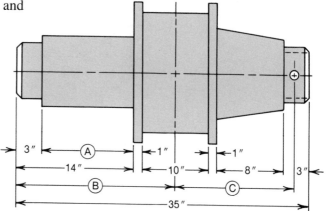

3. Determine the distance Ⓐ from the outside wall to the center of the side door opening from the dimensions given on the floor plan of the two-car garage.
4. Find the width of the wall Ⓑ.
5. What is dimension Ⓒ at the rear of the garage?

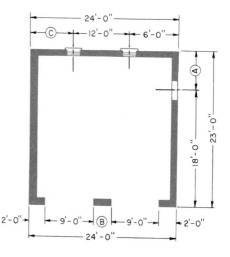

D. MULTIPLICATION OF WHOLE NUMBERS

1. Find the total cost of these materials: 65 barrels of granulated lime at a cost of $29 per barrel, 52 cubic yards of ready-mixed concrete at $42 per cubic yard, and 25 cubic yards of sand at $12 per cubic yard.
2. A coil magnet has 56 layers of magnet wire wound around a core. If there are 165 turns of wire per layer, how many turns of wire are there in the coil?
3. Determine the total number of watts in an electrical lighting circuit with this load: twelve 150-watt lamps, three 100-watt lamps, nine 60-watt lamps, and eleven 15-watt lamps.
4. A book contains 277 pages. The type matter on each page measures 5 inches wide by 7

inches long and there are 6 lines of type per inch. Determine how many 5-inch lines must be typeset.

5. Compute center distances Ⓐ, Ⓑ, Ⓒ, and Ⓓ.

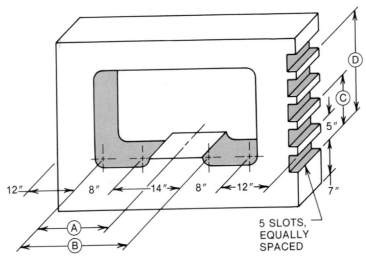

12"   8"   14"   8"   12"   7"

5 SLOTS,
EQUALLY
SPACED

E. DIVISION OF WHOLE NUMBERS

1. A total load of 22,931 watts is distributed equally over 23 branch circuits. Give the load per circuit in watts.

2. A 192-page book is to be printed. There are 1,500 copies required, using a paper stock that weighs 178 pounds per thousand sheets and costs $2.00 per pound. From each sheet 32 pages can be printed. Determine the cost of paper stock.

3. In six consecutive months the following quantities of stamped metal parts were heat treated: 462,925; 378,916; 417,829; 382,885; 415,297; and 450,628 pounds. Determine the six-month hourly average per man if the department employs 16 men on a nine-hour shift, 12 on an eight-hour shift, and 4 on a seven-hour shift. In the six-month period, each man worked 130 days.

4. Find the average hourly production of parts A through E from the monthly (156 hours) production schedule.

Parts Manufactured				
A	B	C	D	E
Average Monthly (156 Hours) Production				
1,248	2,964	21,372	202,332	5,584,332

5. A pattern layout requires 51 inches of cloth. Find (a) how many garments can be cut from the material, (b) the amount of material left over, and (c) the number of yards in 324 inches. (1 yard = 36 inches)

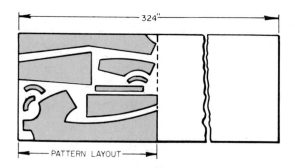

F. COMBINATIONS OF ADDITION, SUBTRACTION, MULTIPLICATION, OR
DIVISION OF WHOLE NUMBERS

1. Establish whether there was a profit or loss on sales A through G. Find the amount of profit (P) or loss (L).

Item	Sales	Cost of Item	Overhead Expense	Profit (P) or Loss (L)
A	$98,000	$75,000	$22,500	
B	15,750	11,429	3,000	
C	9,342	6,210	3,350	
D	56,735	38,750	17,000	
E	126,960	97,873	18,250	
F	7,942	4,975	2,635	
G	11,697	7,539	2,973	

2. Calculate dimensions Ⓐ through Ⓔ for machining the Serrated Plate. Use the basic mathematical processes needed to find each dimension.

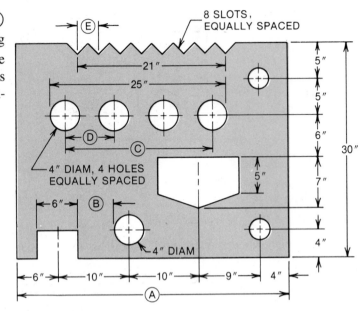

3. Stock control records show an inventory of 9,241 integrated electronic circuits at the close of a four-month period. The following quantities were shipped each month: February, 928; March, 1,641; April, 3,077; and May, 2,104.

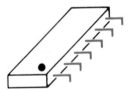

Provide the following information.
a. The total number shipped.
b. The average of the number of circuits shipped each month during the period.
c. The number of integrated circuits remaining in stock at the close of the period.
d. The anticipated shipments for a twelve-month period based on the average per month for the first four-month period.

COMMON FRACTIONS

Unit 6 The Concept of Common Fractions

OBJECTIVES OF THE UNIT

After satisfactorily studying this unit, the student/trainee will be able to
- *Understand the terms and functions of common fractions in relation to shop, laboratory, and everyday problems.*
- *Transfer common fraction values to equivalent measurements on steel tapes, rules, and other measuring tools.*
- *Reduce proper and improper fractions to their lowest terms.*

Craftspersons are required constantly to take measurements, do layout work, and perform hand and machine operations. These jobs require the use of mathematics to compute missing or needed dimensions.

The use of whole numbers alone is not sufficient to obtain this information. All computations involve either whole numbers or fractions or a combination of both. Numbers, whether they are whole or just parts of a whole, must be added, subtracted, multiplied, and divided.

A. INTERPRETING COMMON FRACTIONS

A *fraction* is a part of a whole quantity. For example, a triangle, square, or circle is divided into two equal parts (*figure 6-1*). One part is involved in an operation, as shown by the shaded portion of each illustration. The fractional part of the whole triangle, square, or circle is one-half, as shown at (A), (B), and (C).

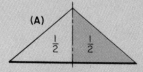
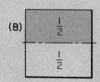
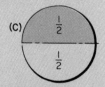

FIGURE 6-1

On shop prints and sketches and in mathematical computations the one-half is usually written $\frac{1}{2}$ and is called a *common fraction*. This fraction shows the number of equal parts of a unit that are taken. If the fraction $\frac{3}{4}$ appears on a drawing, it means the unit one ● is divided into four equal parts ⊕ and that three of the four parts are taken ⟶ ◕ .

Common fractions are used daily when taking measurements with line-graduated measuring tools and instruments. Rules, levels, tapes, and other measuring tools are commonly graduated in fourths, eighths, sixteenths, and thirty-seconds of an inch. For greater precision, steel rules are graduated in sixty-fourths of an inch. In the printing trades, measurements are expressed as fine as seventy-seconds of an inch.

Where the measurements are given as fractions, they indicate that the inch has been divided into an equal number of parts. The object must measure a stated number of these parts.

The steel rule, a common measuring tool, may be used to show how the inch looks when it is divided into four, eight, sixteen, thirty-two, and sixty-four equal parts.

1. The inch contains four-fourths. Each equal part is expressed as $\frac{1}{4}$ of the whole.

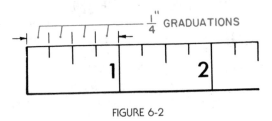

FIGURE 6-2

2. The inch contains eight-eighths. Each equal part is expressed as $\frac{1}{8}$ of the whole.

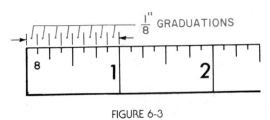

FIGURE 6-3

3. The inch contains sixteen-sixteenths. Each equal part is expressed as $\frac{1}{16}$ of the whole.

FIGURE 6-4

4. The inch contains thirty-two-thirty-seconds. Each equal part is expressed as $\frac{1}{32}$ of the whole.

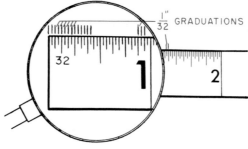

FIGURE 6-5

5. The inch contains sixty-four-sixty-fourths. Each equal part is expressed as $\frac{1}{64}$ of the whole.

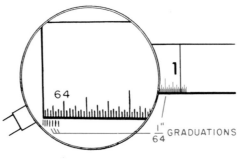

FIGURE 6-6

B. DEFINING PARTS OF FRACTIONS

If on the eighth scale of the steel rule three of the eight equal parts (into which the inch is divided) are needed for a measurement, the fraction would be shown as $\frac{3}{8}$. The terms of this fraction are a *numerator*, which appears over a horizontal line, and a *denominator*, which appears under the line.

$$\text{Numerator} \blacktriangleright \frac{3}{8} \blacktriangleleft \text{Denominator}$$

RULE
_____

• The denominator (which is always written below the line) indicates the equal number of parts into which the unit is divided.

RULE
_____

- The numerator (which is always written above the line) indicates the number of equal parts of the denominator that is taken.

_____

C. REDUCING FRACTIONS

A required measurement that is the result of adding, subtracting, multiplying, or dividing fractions is not always expressed as simply as possible. Measurements can be taken or read with greater facility when the fraction is given in its *lowest terms*.

Reducing Common Fractions

The mathematical expression *proper fraction* is used to indicate a fraction whose numerator is smaller than its denominator. Quantities such as

$$\frac{1}{4}, \frac{3}{8}, \frac{29}{64}, \text{ and } \frac{5}{8}$$

are examples of proper fractions.

RULE
_____

- The value of a fraction is not changed when both the numerator and denominator are multiplied or divided by the same number. This number, which can be used to divide both the numerator and denominator of a fraction without a remainder, is called a *common factor*.

_____

RULE FOR REDUCING A COMMON FRACTION TO ITS LOWEST TERMS
_____

- Divide the numerator and denominator by the same number.
  Note. When both the numerator and the denominator cannot be divided further by the same number, the fraction is expressed in its lowest terms.

_____

EXAMPLE: Reduce $\frac{8}{16}$ to its lowest terms.

   Step 1   Select a number (common factor) that will divide evenly into both the numerator and denominator.

$$\frac{8}{16} = \frac{4}{8} \qquad \frac{4}{8} = \frac{2}{4} = \frac{1}{2}$$

   Step 2   Continue this division until the numerator and denominator can no longer be evenly divided by the same number.

   Note. When it is apparent that both numerator and denominator can be divided by a larger number, for example, 8 in the case of $\frac{8}{16}$, the intermediate steps are omitted.

Reducing Improper Fractions

   A fraction whose numerator is greater than its denominator is called an *improper fraction.* Examples of improper fractions are

$$\frac{3}{2}, \frac{25}{16}, \text{ and } \frac{71}{32}.$$

RULE FOR REDUCING AN IMPROPER FRACTION TO ITS LOWEST TERMS
_____

   • Divide the numerator (above the line) by the denominator (below the line).
   • Reduce the resulting fraction to its lowest terms.

_____

EXAMPLE: Reduce $\frac{20}{16}$ to its lowest terms.

   Step 1   Divide 20 by 16. $\frac{20}{16} = 1\frac{4}{16}$

   Step 2   Reduce $\frac{4}{16}$ to lowest terms. $\frac{4}{16} = \frac{1}{4}$

   Step 3   Answer $\frac{20}{16} = 1\frac{1}{4}$ Ans

Review and Self-Test

ASSIGNMENT UNIT 6

A. Interpretation of Common Fractions (General Applications)

1. The squares and circles are divided equally into 6, 8, 16, 32, or 64 parts. Visualize (by actual counting of spaces, if necessary) the part of the square or circle represented by fractions (b) to (z).

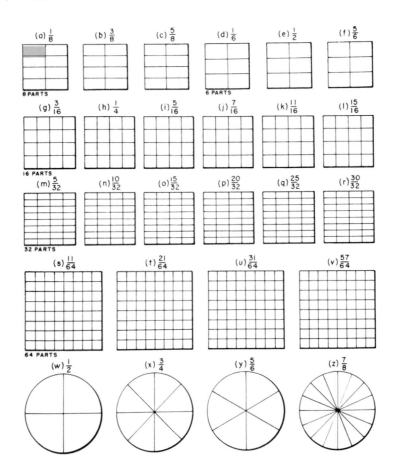

B. Reading Fractional Values (Practical Problems)

1. Rule a table as illustrated. Include the letters in the boxes, the arrows, and all words. Arrange the following twist drills in the table in sizes ranging from the smallest to the largest.

2. Locate the $\frac{1}{4}''$ and $\frac{3}{4}''$ graduations on a steel tape or rule. *Note:* If a rule is not available for problems 2 and 3, place a transparent sheet over the illustrations and trace the rules.

a. $\frac{1}{4}''$ 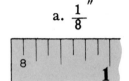   b. $\frac{3}{4}''$

3. Locate common fractions (a) through (d) on the eighth scale of a rule.

a. $\frac{1}{8}''$ b. $\frac{3}{8}''$ c. $\frac{5}{8}''$ d. $\frac{7}{8}''$

C. Reduction of Fractions (Practical Problems)

1. Reduce common fractions (a) through (f) to their lowest terms. Then locate each fraction on a rule.

EXAMPLE:

$\frac{6}{16}'' = \boxed{\frac{3}{8}} =$

a. $\frac{14}{32}''$ d. $\frac{44}{64}''$

b. $\frac{48}{64}''$    e. $\frac{10}{64}''$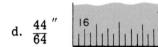

c. $\frac{10}{16}''$    f. $\frac{18}{128}''$

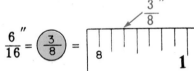

2. Reduce each improper fraction to its lowest terms and locate each measurement on a rule.

   a. $\frac{5''}{2}$ b. $\frac{25''}{4}$ c. $\frac{35''}{8}$ d. $\frac{42''}{32}$ e. $\frac{85''}{64}$

3. Locate each common fraction on the sixteenth scale of a rule for problems (a) through (f).

   a. $\frac{15''}{16}$ b. $\frac{13''}{16}$ c. $\frac{7''}{16}$ d. $\frac{9''}{16}$ e. $\frac{3''}{16}$ f. $\frac{11''}{16}$

4. Give each measurement as shown at Ⓐ, Ⓑ, Ⓒ, and Ⓓ on the eighth, sixteenth, and thirty-second scales.

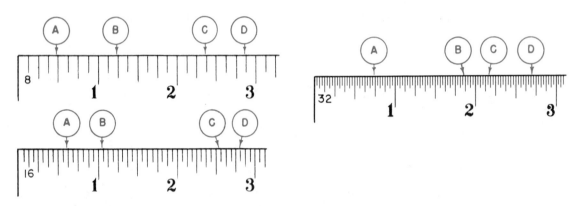

5. Draw lines to the lengths given in (a) through (f).

   a. $3''$ b. $4\frac{1''}{2}$ c. $5\frac{1''}{4}$ d. $3\frac{1''}{8}$ e. $4\frac{3''}{4}$ f. $4\frac{7''}{8}$

6. Give the measurements of the stepped parts shown at Ⓐ and the extensions Ⓑ, Ⓒ, Ⓓ, and Ⓔ.

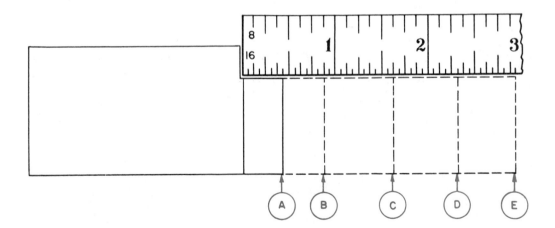

7. Reduce each dimension Ⓐ through Ⓙ to its lowest terms as a common fraction or as an improper fraction.

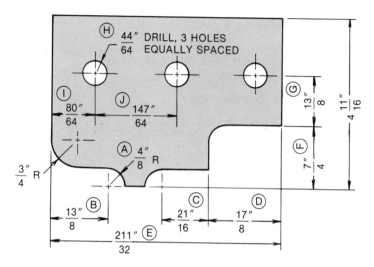

Unit 7 Addition of Fractions

OBJECTIVES OF THE UNIT

After satisfactorily studying this unit, the student/trainee will be able to
- *Reduce common fractions to lowest common denominators.*
- *Add common fractions.*
- *Add combinations of whole numbers, mixed numbers, and common fractions.*

The mechanic must often determine overall sizes by adding dimensions that are given on a drawing or written into specifications. The answers to these problems require the addition of whole numbers, common fractions, and combinations of whole numbers and fractions. Such combinations are referred to as *mixed numbers*.

To add combinations of whole numbers and fractions, the denominators of each fraction must be the same number. The smallest number that can be divided by all the denominators is called the *lowest common denominator*.

A. DETERMINING THE LOWEST COMMON DENOMINATOR (LCD)

The simplest method of determining the lowest common denominator (LCD) is used when it is evident that all of the denominators divide evenly into a given number. This method is covered in this unit.

On the drawing of the special pin (*figure 7.1*) all of the given sizes must be added to compute the overall dimension Ⓐ. Before these measurements (2, $1\frac{31}{32}$, $\frac{3}{8}$, and $\frac{3}{64}$) can be added, the lowest common denominator must be determined.

The lowest number into which each of the denominators (32, 8, and 64) will divide evenly is 64. The 64 is called the *lowest common denominator*.

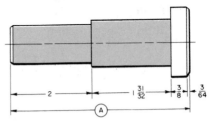

FIGURE 7-1

RULE FOR REDUCING FRACTIONS TO THE LOWEST COMMON DENOMINATOR

- Divide the number selected as the lowest common demoninator by the denominator of each given fraction.
- Multiply both the numerator and denominator by this quotient.

EXAMPLE: The lowest common denominator (LCD) for the pin dimension is 64.

Step 1 Divide the LCD by the denominator of first fraction.

Step 2 Multiply the numerator (31) and denominator (32) by the quotient (2).

Step 3 Continue the same process with the other fractions.

Denominator
$$2 \leftarrow \text{Quotient}$$
$$32\overline{)64}$$

$$\frac{31 \times 2}{32 \times 2} = \frac{62}{64}$$

$$8\overline{)64}^{\,8}$$
$$\frac{3 \times 8}{8 \times 8} = \frac{24}{64}$$

$$64\overline{)64}^{\,1}$$
$$\frac{3 \times 1}{64 \times 1} = \frac{3}{64}$$

B. ADDING COMMON FRACTIONS

RULE FOR ADDING FRACTIONS

- Change to fractions having a least common denominator.
- Add the numerators.
- Write the sum over the common denominator.
- Reduce the result to its lowest terms.

EXAMPLE: Add $\frac{31}{32}$, $\frac{3}{8}$, and $\frac{3}{64}$.

Step 1 Write fractions in vertical column.

Step 2 Change fractions to same denominator (64).

$$\frac{31 \times 2}{32 \times 2} = \frac{62}{64}$$

$$\frac{3 \times 8}{8 \times 8} = \frac{24}{64}$$

$$\frac{3 \times 1}{64 \times 1} = \frac{3}{64}$$

Step 3   Add numerators.

Step 4   Place result (89) over lowest common denominator (64).

Step 5   Reduce to lowest terms.

$62 + 24 + 3 = 89$

$\frac{89}{64}$

$\frac{89}{64} = 1\frac{25}{64}$ **Ans**

C. ADDING WHOLE NUMBERS, COMMON FRACTIONS, AND MIXED NUMBERS

A mixed number consists of two parts: (1) a whole number and (2) a fraction. Numbers such as $1\frac{1}{2}$, $256\frac{3}{8}$, and $1927\frac{3}{5}$ are called *mixed numbers*.

RULE FOR ADDING WHOLE NUMBERS, MIXED NUMBERS, AND COMMON FRACTIONS

- Add the whole numbers.
- Add the fractions.
- Add the two sums.
- Reduce the result to lowest terms.

EXAMPLE: Determine overall dimension Ⓐ of the shaft.

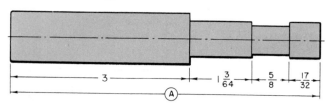

FIGURE 7-2

Step 1   Write all dimensions in vertical columns.

Step 2   Change fractions to same denominator (64).

$3 \qquad = 3$

$1\frac{3}{64} \times \frac{1}{1} = 1\frac{3}{64}$

$\frac{5}{8} \times \frac{8}{8} = \frac{40}{64}$

$\frac{17}{32} \times \frac{2}{2} = \frac{34}{64}$

Step 3   Add numerators.

Step 4   Place result (77) over lowest common denominator (64).

Step 5   Reduce $\frac{77}{64}$ to lowest terms.

Step 6   Add column of whole numbers.

Step 7   Add sum of whole numbers (4) to sum of common fractions $\left(1\frac{13}{64}\right)$.

$3 + 40 + 34 = 77$

$\frac{77}{64}$

$1\frac{13}{64}$

$(3 + 1 = 4)$

$4 + 1\frac{13}{64} = 5\frac{13}{64}''$ overall dimension of shaft

Review and Self-Test

ASSIGNMENT UNIT 7

A. Addition of Common Fractions (General Applications)

1. $\frac{1}{6} + \frac{5}{6}$ 4. $\frac{1}{3} + \frac{1}{6}$ 7. $\frac{5}{8} + \frac{3}{4} + \frac{3}{8}$

2. $\frac{1}{8} + \frac{5}{8}$ 5. $\frac{1}{2} + \frac{3}{8}$ 8. $\frac{1}{32} + \frac{7}{8} + \frac{3}{16} + \frac{5}{32}$

3. $\frac{1}{4} + \frac{4}{8}$ 6. $\frac{1}{6} + \frac{1}{6} + \frac{5}{6}$ 9. $\frac{61}{64} + \frac{13}{16} + \frac{5}{8} + \frac{23}{64}$

B. Addition of Common Fractions and Mixed Numbers (General Applications)

1. $121 + 7\frac{5}{12}$ 4. $1\frac{17}{64} + 1\frac{13}{64} + \frac{9}{32}$

2. $10\frac{9}{16} + 4\frac{1}{8}$ 5. $4\frac{3}{16} + 10\frac{21}{64} + 1\frac{5}{16} + \frac{3}{32}$

3. $23\frac{5}{8} + 10\frac{5}{8}$ 6. $3\frac{5}{32} + 2\frac{13}{64} + 1\frac{13}{32} + 3\frac{1}{16} + \frac{1}{4}$

C. Addition of Whole Numbers, Mixed Numbers, and Common Fractions (Practical Problems)

1. Determine dimensions of Ⓐ, Ⓑ, Ⓒ, Ⓓ, and Ⓔ (in inches).

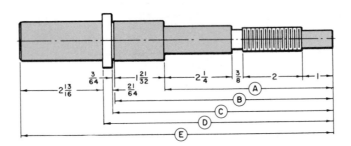

2. Calculate dimensions Ⓐ through Ⓔ for the stand.

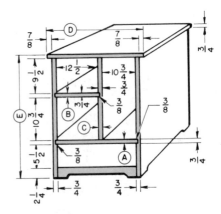

3. Determine the overall lengths of springs A through D.

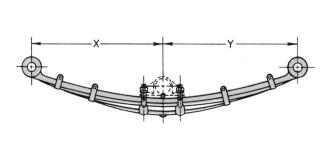

Spring	Distance		Overall Length
	X	Y	
A	$17\frac{1}{2}''$	$15\frac{3}{4}''$	
B	$16\frac{3}{4}''$	$15\frac{7}{8}''$	
C	$17\frac{3}{32}''$	$16\frac{1}{8}''$	
D	$18\frac{21}{32}''$	$17\frac{3}{4}''$	

4. Compute the total horsepower for (a) motors 1, 2, and 3; (b) motors 4, 5, and 6; and (c) all 6 motors. The rated horsepower is indicated.

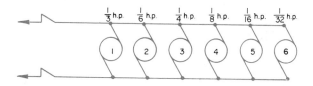

5. Find the total length of $2\frac{1}{2}''$ diameter plastic tubing needed on the job. The required lengths are $35\frac{1}{4}''$, $12\frac{1}{2}''$, $26\frac{3}{8}''$, $8\frac{5}{16}''$, and $19\frac{21}{32}''$. Reduce the answer to lowest terms.

Unit 8 Subtraction of Fractions

OBJECTIVES OF THE UNIT
After satisfactorily studying this unit, the student/trainee will be able to
* *Solve general and practical problems that require the subtraction of proper fractions.*
* *Carry on the addition and subtraction of fractions in the same problem.*

The value of a missing dimension must often be determined by subtracting whole numbers, fractions, and mixed numbers. Fractions cannot be subtracted unless they first have the same common denominator.

A. SUBTRACTING PROPER FRACTIONS

RULE FOR SUBTRACTING FRACTIONS

- Express all fractions using the lowest (least) common denominator.
- Subtract the numerators.
- Write the difference over the lowest (least) common denominator.
- Express the resulting fraction in lowest terms.

EXAMPLE: Subtract $\frac{7}{32}$ from $\frac{11}{16}$.

Step 1   Determine the lowest (least) common denominator.

Step 2   Write the fractions in terms of lowest (least) common denominator. $\frac{7}{32} = \frac{7}{32}$ $\frac{11}{16} = \frac{22}{32}$

Step 3   Subtract numerators. ⟶ **22 − 7 = 15**

Step 4   Place the numerator result over the lowest (least) common denominator. $\frac{15}{32}$ **Ans**

B. SUBTRACTING A FRACTION FROM A WHOLE NUMBER

RULE FOR SUBTRACTING A FRACTION FROM A WHOLE NUMBER

- Take one unit from the whole number. Change it to a fraction having the same denominator as the fraction that is to be subtracted.
- Subtract the numerators of the original fraction from the one unit that was changed to its fractional value.
- Express the resulting fraction in lowest terms.
- Place the whole number next to the fraction.

EXAMPLE: Subtract $\frac{21}{32}$ from 9.

Step 1   Take one unit from the whole number.

Step 2   Change the one unit to a fractional equivalent having the same denominator as the fraction to be subtracted.

Step 3   Arrange the fractions in one column; the whole number in another.

Step 4   Subtract the numerators.

Step 5   Place the whole number (8) next to the fraction $\frac{11}{32}$ to get the answer.

$$9 - 1 = 8$$
$$1 = \frac{32}{32}$$

$$8\frac{32}{32}$$
$$-\frac{21}{32}$$

$$\frac{32}{32} - \frac{21}{32} = \frac{11}{32}$$

$$8\frac{11}{32} \quad \textbf{Ans}$$

C. SUBTRACTING A MIXED NUMBER FROM A WHOLE NUMBER

RULE FOR SUBTRACTING A MIXED NUMBER FROM A WHOLE NUMBER

- Borrow one unit from the whole number and express it as a fraction that has the same denominator as the mixed number.
- Subtract the fraction part of the mixed number from the fraction part of the whole number.
- Subtract the whole numbers and reduce the resulting mixed number to its lowest terms.

EXAMPLE: Subtract $1\frac{35}{64}$ from 6.

Step 1 $6 \;\; = 5\frac{64}{64}$

Step 2 $1\frac{35}{64} = 1\frac{35}{64}$

Step 3 $\qquad\qquad 4\frac{29}{64}$ Ans

D. SUBTRACTING MIXED NUMBERS FROM MIXED NUMBERS

RULE FOR SUBTRACTING MIXED NUMBERS

- Express the fractional part of each mixed number using the least common denominator.
- Borrow one unit, when necessary, to make up a fraction larger than the one to be subtracted.
- Subtract the fractions first and the whole numbers next. Express the result in its lowest terms.

EXAMPLE: Subtract $2\frac{7}{16}$ from $5\frac{11}{32}$.

Step 1 $5\frac{11}{32} = 4 + \frac{32}{32} + \frac{11}{32} = 4\frac{43}{32}$

Step 2 $2\frac{7}{16} = \qquad\qquad\quad = 2\frac{14}{32}$

Step 3 $\qquad\qquad\qquad\qquad\quad 2\frac{29}{32}$ Ans

E. COMBINING ADDITION AND SUBTRACTION OF FRACTIONS

RULE FOR ADDING AND SUBTRACTING FRACTIONS IN THE SAME PROBLEM

- Change all fractions to the least (lowest) common denominator.
- Add or subtract the numerators as required.
- Express the result in lowest terms.

EXAMPLE: Add $1\frac{9}{16} + 3\frac{5}{8} + 2\frac{1}{4}$ and from the sum subtract $2\frac{13}{16}$.

$$1\frac{9}{16} + 3\frac{5}{8} + 2\frac{1}{4} - 2\frac{13}{16} =$$

Step 1 $1\frac{9}{16} + 3\frac{10}{16} + 2\frac{4}{16} - 2\frac{13}{16} =$

Step 2

$$1\frac{9}{16}$$
$$+3\frac{10}{16}$$
$$+2\frac{4}{16}$$

Step 3

$$6\frac{23}{16}$$
$$-2\frac{13}{16}$$

Step 4

$$4\frac{10}{16} = 4\frac{5}{8} \quad \text{Ans}$$

Review and Self-Test

ASSIGNMENT UNIT 8

A. Subtraction of Proper Fractions (General Applications)

Subtract.

1. $\begin{array}{r} \frac{5}{8} \\ -\frac{4}{8} \\ \hline \end{array}$
2. $\begin{array}{r} \frac{5}{6} \\ -\frac{1}{6} \\ \hline \end{array}$
3. $\begin{array}{r} \frac{13}{16} \\ -\frac{5}{16} \\ \hline \end{array}$
4. $\begin{array}{r} \frac{9}{32} \\ -\frac{7}{32} \\ \hline \end{array}$
5. $\begin{array}{r} \frac{1}{2} \\ -\frac{1}{4} \\ \hline \end{array}$

6. $\frac{1}{8}$ from $\frac{5}{8}$

7. $\frac{5}{32}$ from $\frac{9}{16}$

8. $\frac{9}{64}$ from $\frac{23}{32}$

9. $\frac{3}{8} + \frac{1}{8}$ from $\frac{9}{16}$

10. $\frac{37}{64} + \frac{21}{64}$ from $\frac{63}{64}$

B. Subtraction of Fractions from a Whole Number (General Applications)

Subtract.

1. $\quad 4$ $\quad -\ \frac{3}{4}$	2. $\quad 7$ $\quad -\ \frac{15}{16}$	3. $\quad 32$ $\quad -\ \frac{13}{32}$	4. $\quad 175$ $\quad -\ \frac{4}{5}$	5. $\quad 72$ $\quad -\ \frac{61}{64}$

C. Subtraction of a Mixed Number from a Whole Number (General Applications)

Subtract.

1. $\quad 2$ $\quad -1\frac{1}{3}$	2. $\quad 3$ $\quad -1\frac{3}{8}$	3. $\quad 27$ $\quad -1\frac{5}{16}$	4. $\quad 142$ $\quad 6\frac{21}{32}$	5. $\quad 372$ $\quad -21\frac{5}{64}$

6. $1\frac{21}{32}$ from 3

7. $3\frac{57}{64}$ from 4

8. $3\frac{5}{32} + 1\frac{9}{32}$ from 5

9. $2\frac{53}{64} + 1\frac{1}{64} + 3\frac{1}{4}$ from 8

10. $3\frac{7}{16} + 2\frac{25}{64} + 16\frac{17}{32}$ from 42

D. Subtraction of Mixed Numbers from Mixed Numbers (General Applications)

Subtract.

1. $\quad 1\frac{3}{5}$ $\quad -1\frac{1}{5}$	2. $\quad 7\frac{5}{6}$ $\quad -2\frac{1}{6}$	3. $\quad 18\frac{7}{8}$ $\quad -\ 9\frac{3}{8}$	4. $\quad 35\frac{5}{8}$ $\quad -\ 8\frac{1}{2}$	5. $\quad 172\frac{21}{64}$ $\quad -\ 22\frac{5}{32}$

6. $1\frac{19}{32}$ from $4\frac{29}{32}$

7. $2\frac{31}{32}$ from $5\frac{15}{32}$

8. $4\frac{7}{64}$ from $8\frac{1}{32}$

9. $2\frac{9}{16} + 1\frac{1}{8}$ from $5\frac{51}{64}$

10. $7\frac{1}{64} + 2\frac{31}{32} + 1\frac{1}{4}$ from $12\frac{3}{32}$

E. Addition or Subtraction Processes (Practical Problems)

1. Determine the inside diameter of each size bushing. All dimensions are in inches.

Bushing	Outside Diameter	Single Wall Thickness (T)
A	1	$\frac{1}{16}$
B	$2\frac{1}{8}$	$\frac{3}{16}$
C	$2\frac{15}{32}$	$\frac{9}{64}$
D	$3\frac{1}{64}$	$\frac{15}{32}$
E	$1\frac{9}{64}$	$\frac{9}{32}$

T →

OUTSIDE DIAMETER

2. Determine dimensions Ⓐ, Ⓑ, Ⓒ, Ⓓ, and Ⓔ. All dimensions are in inches.

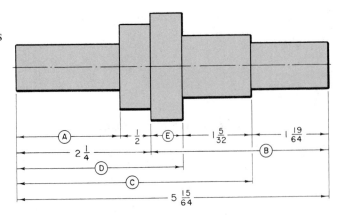

3. How much material will be left from an 18-inch length of strip rule after four pieces are cut off in the following lengths: $2\frac{1}{2}$, $1\frac{1}{4}$, $3\frac{1}{2}$, and $3\frac{3}{4}$ inches.

4. A motor base is to be set up $4\frac{1}{2}$ inches from the floor. Two wooden blocks must be used. If one block is $1\frac{5}{8}$ inches thick, how thick must the second block be?

5. A concrete sidewalk $4\frac{1}{4}$ inches thick consists of a base course and a finish course. The base course thickness is $3\frac{5}{8}$ inches. What is the thickness of the finish course?

6. Three pieces of 3-inch lead pipe are cut from a piece $35\frac{1}{2}$ inches long. The lengths are $7\frac{1}{4}$ inches, $11\frac{3}{8}$ inches, and $6\frac{1}{2}$ inches. If $\frac{3}{8}$ inch of stock is wasted in cutting, how much pipe is left?

7. Four pieces are cut from a 12-foot 2 × 4. The pieces measure 3 feet $9\frac{1}{2}$ inches, 2 feet $6\frac{1}{4}$ inches, 1 foot $4\frac{1}{2}$ inches, and 2 feet $3\frac{3}{8}$ inches. If a total of $\frac{1}{2}$ inch is allowed for the four cuts, how much of the 2 × 4 is left?

8. A piece of radiator hose is $32\frac{1}{2}$ inches long. Short pieces of the following lengths are cut from it: $6\frac{1}{2}$ inches, $5\frac{1}{4}$ inches, $8\frac{13}{16}$ inches, and $10\frac{9}{16}$ inches. How much hose is left?

9. Determine dimensions Ⓐ, Ⓑ, Ⓒ, and Ⓓ.

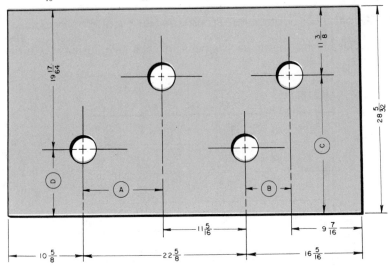

10. Compute dimensions Ⓐ, Ⓑ, Ⓒ, and Ⓓ.

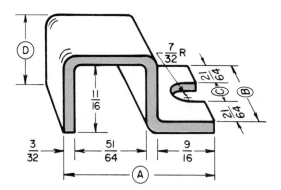

11. The front and rear axles of a light truck are to be aligned. Center distance A measures $16'-3\frac{5''}{8}$; center distance B measures $16'-4\frac{7''}{16}$. Determine how much the axles need to be changed to be aligned.

12. Find the voltage Ⓒ in the series circuit according to the voltages indicated for Ⓐ, Ⓑ, and Ⓓ. The source voltage (E_S) is $25\frac{3}{4}$ volts.

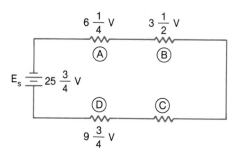

13. Compute the length (L) of the conduit needed according to the dimensions given on the drawing. Express the length in feet and inches.

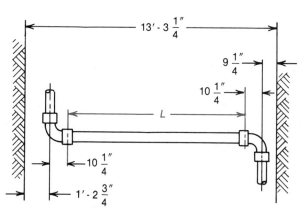

Unit 9 Multiplication of Fractions

OBJECTIVES OF THE UNIT
After satisfactorily studying this unit, the student/trainee will be able to
* *Solve general and practical problems requiring the multiplication of two or more fractions.*
* *Multiply fractions, whole numbers, and mixed numbers.*
* *Simplify the multiplication process by the cancellation method.*

The multiplication of fractions, like the multiplication of whole numbers, is a simplified method of addition. The multiplication of common fractions, whole numbers, and mixed numbers, typical of those used daily, are covered in this unit.

A. MULTIPLYING COMMON FRACTIONS

RULE FOR MULTIPLYING TWO OR MORE FRACTIONS

* Multiply the numerators.
* Multiply the denominators.
* Write the product of the numerators over the product of the denominators.
* Express the resulting fraction in lowest terms.

EXAMPLE: Multiply $\frac{7}{8}$ by $\frac{3}{4}$.
Step 1 Multiply the numerators.
Step 2 Multiply the denominators.
Step 3 Express as a fraction.

$(7 \times 3) = 21$
$(8 \times 4) = 32$
$\frac{21}{32}$ Ans

B. MULTIPLYING A COMMON FRACTION BY A MIXED NUMBER

RULE FOR MULTIPLYING A COMMON FRACTION BY A MIXED NUMBER

* Express the mixed number as an improper fraction where the numerator is larger than the denominator.
* Multiply the numerator of the improper fraction by the numerator of the common fraction.
* Multiply the denominators.

EXAMPLE: Multiply $4\frac{1}{2}$ by $\frac{1}{8}$.

Step 1 Express the mixed number $\left(4\frac{1}{2}\right)$ as an improper fraction $\left(\frac{9}{2}\right)$.

Step 2 Multiply the numerators.

Step 3 Multiply the denominators.

$$4\frac{1}{2} = \frac{4 \times 2 + 1}{2} = \frac{9}{2}$$

$$(9 \times 1) = 9$$

$$(2 \times 8) = 16$$

$$4\frac{1}{2} \times \frac{1}{8} = \frac{9}{2} \times \frac{1}{8} = \frac{9}{16} \quad \text{Ans}$$

C. MULTIPLYING FRACTIONS AND WHOLE AND MIXED NUMBERS

RULE FOR MULTIPLYING FRACTIONS, WHOLE NUMBERS, AND MIXED NUMBERS IN ANY COMBINATION

- Express all mixed numbers as improper fractions.
- Place all whole numbers over a denominator of 1.
- Multiply all numerators.
- Multiply all denominators.
- Express resulting product in lowest terms.

EXAMPLE: Multiply $8\frac{3}{8} \times 1\frac{1}{4} \times 2$.

Step 1 Express $\left(8\frac{3}{8}\right)$ as an improper fraction $\left(\frac{67}{8}\right)$.

Express $\left(1\frac{1}{4}\right)$ as an improper fraction $\left(\frac{5}{4}\right)$.

$$\frac{8 \times 8 + 3}{8} = \frac{67}{8}$$

$$\frac{1 \times 4 + 1}{4} = \frac{5}{4}$$

Step 2 Place 2 over denominator of 1.

Step 3 Multiply all numerators.

Step 4 Multiply all denominators.

Step 5 Express $\left(\frac{670}{32}\right)$ in lowest terms.

$$\frac{2}{1}$$

$$(67 \times 5 \times 2) = 670 \searrow 670$$

$$(8 \times 4 \times 1) = 32 \longrightarrow 32$$

$$\frac{670}{32} = 20\frac{30}{32} = 20\frac{15}{16} \quad \text{Ans}$$

D. CANCELING TO SIMPLIFY THE MULTIPLICATION PROCESS

The multiplication of fractions can be simplified by removing the *common factor*. The common factor is any number by which the numerator and denominator may be evenly divided. The process is called *cancellation*.

RULE FOR EVENLY DIVIDING A NUMERATOR AND DENOMINATOR

- Select a number *(common factor)* by which the numerator and denominator may be evenly divided.
- Divide by the common factor.
- Reduce the result to the lowest terms.

EXAMPLE: Multiply $72 \times 3\frac{5}{8}$.

Step 1   Express the mixed number $\left(3\frac{5}{8}\right)$ as an improper fraction $\left(\frac{29}{8}\right)$.

Step 2   Select a number that is common to any numerator and any denominator (8 in this case).

Step 3   Divide the numerator and denominator by this factor (8).

$$\frac{\overset{9}{\cancel{72}}}{1} \times \frac{29}{\cancel{8}} = 261 \quad \textbf{Ans}$$

E. SHORTCUTS IN MULTIPLYING BY ONE-HALF $\left(\frac{1}{2}\right)$

EXAMPLE: *Case 1.* Find $\frac{1}{2}$ of $\frac{7}{8}$.

Step 1   Multiply the denominator (8) by 2.

Step 2   Use the numerator (7) as it is. The answer is $\frac{7}{16}$.

EXAMPLE: *Case 2.* Find $\frac{1}{2}$ of $2\frac{3}{4}$.

Step 1   Take one-half of 2.

Step 2   Multiply the denominator (4) of the fraction $\left(\frac{3}{4}\right)$ by 2.

Step 3   Use the same numerator (3).

Step 4   Combine the whole number 1 and the fraction $\frac{3}{8}$. The answer is a mixed number $\left(1\frac{3}{8}\right)$.

$$\frac{1}{2} \times 2 = 1$$
$$4 \times 2 = 8$$
$$\frac{3}{8}$$
$$1 + \frac{3}{8} = 1\frac{3}{8} \quad \textbf{Ans}$$

EXAMPLE: *Case 3.* Find $\frac{1}{2}$ of $5\frac{13}{16}$.

Step 1   Take one-half of 4 (the largest number in the mixed number that will divide exactly).

Step 2   Express the remainder $\left(1\frac{13}{16}\right)$ as an improper fraction.

Step 3   Place the numerator (29) over twice the denominator.

Step 4   Combine the whole number (2) with the fraction $\left(\frac{29}{32}\right)$.

$$\frac{1}{2} \times 4 = 2$$
$$1\frac{13}{16} = \frac{29}{16}$$
$$\frac{29}{2 \times 16} = \frac{29}{32}$$
$$2 + \frac{29}{32} = 2\frac{29}{32} \quad \textbf{Ans}$$

Special Note. When more than one operation is called for within a problem, multiplication or division operations are completed first. Other operations (addition and subtraction) are then performed. These are done in order from left to right.

Review and Self-Test

ASSIGNMENT UNIT 9

A. Multiplication of Proper Fractions (General Applications)

Multiply.

1. $\frac{1}{4}$ by $\frac{1}{2}$

2. $\frac{5}{9}$ by $\frac{1}{8}$

3. $\frac{1}{6}$ by $\frac{7}{12}$

4. $\quad \frac{7}{8}$
   $\underline{\text{by } \frac{1}{4}}$

5. $\quad \frac{5}{6}$
   $\underline{\text{by } \frac{5}{12}}$

6. $\left(\frac{21}{32} + \frac{3}{32}\right)$
   $\underline{\text{by } \frac{1}{8}}$

7. $\left(\frac{5}{8} + \frac{3}{16}\right)$
   $\underline{\text{by } \frac{1}{2}}$

8. $\left(\frac{19}{32} + \frac{1}{4}\right)$ by $\frac{3}{8}$

9. $\left(\frac{1}{64} + \frac{17}{32} - \frac{5}{64}\right)$ by $\frac{13}{16}$

10. $\left(\frac{1}{8} + \frac{19}{64} - \frac{3}{16}\right)$ by $\frac{13}{32}$

B. Multiplication of Common Fractions and Mixed Numbers (General Applications)

Multiply.

1. $\frac{1}{6}$ by $1\frac{1}{6}$

2. $1\frac{1}{4}$ by $\frac{1}{2}$

3. $2\frac{3}{8}$ by $\frac{1}{4}$

4. $\frac{5}{8}$ by $6\frac{1}{4}$

5. $\frac{5}{6}$ by $3\frac{7}{12}$

6. $2\frac{9}{16}$
   $\underline{\times \frac{3}{8}}$

7. $\quad \frac{15}{32}$
   $\underline{\times 2\frac{1}{8}}$

8. $3\frac{1}{2} + \left(1\frac{3}{4} \times \frac{7}{8}\right)$

9. $1\frac{9}{16} + \left(\frac{5}{32} \times \frac{5}{8}\right)$

10. $\left(17\frac{1}{4} + 3\frac{1}{32} - 2\frac{1}{8}\right) \times \frac{27}{64}$

C. Multiplication of Mixed Numbers (General Applications)

Multiply.

1. $1\frac{1}{3}$ by $2\frac{1}{6}$

2. $1\frac{9}{16}$ by $10\frac{3}{4}$

3. $6\frac{5}{8}$ by $2\frac{7}{32}$

4. $3\frac{5}{6} \times 1\frac{3}{4} \times 3\frac{1}{8} \times 6\frac{1}{2}$

5. $\left(2\frac{53}{64} - 1\frac{9}{32}\right) \times 2\frac{1}{2} \times 3\frac{3}{4}$

D. Shortcuts in Multiplying by One-Half $\left(\frac{1}{2}\right)$ (General Applications)

Find by the shortcut method.

1. $\frac{1}{2}$ of $\frac{1}{4}$

2. $\frac{1}{2}$ of $\frac{7}{16}$

3. $\frac{1}{2}$ of $2\frac{3}{32}$

4. $\frac{1}{2}$ of $3\frac{5}{6}$

E. Multiplication of Whole Numbers, Fractions, and Mixed Numbers (Practical Problems)

1. What lengths of bar stock will be needed to machine the quantity of parts A, B, C, D, and E?

Parts	Quantity	Length	Allowance for Each Saw Cut
A	10	$\frac{1}{2}''$	$\frac{1}{16}''$
B	12	$1\frac{5}{32}''$	$\frac{3}{32}''$
C	64	$2\frac{21}{64}''$	$\frac{1}{8}''$
D	100	$1\frac{3}{32}''$	$\frac{7}{64}''$
E	75	$1\frac{9}{16}''$	$\frac{1}{8}''$

2. Determine the cost of materials A, B, C, D, and E. Round off answers to two places.

Materials	Length (ft)	Weight (lbs/ft)	Unit Cost (lb)
A	2	$\frac{1}{2}$	\$ .72
B	$10\frac{1}{4}$	$\frac{1}{4}$	\$ .92
C	$7\frac{5}{12}$	$1\frac{1}{2}$	\$1.32$\frac{1}{2}$
D	$9\frac{1}{2}$	$3\frac{3}{4}$	\$ .66$\frac{3}{8}$
E	$11\frac{3}{4}$	$2\frac{1}{16}$	\$ .48$\frac{9}{16}$

3. Compute dimensions Ⓐ, Ⓑ, Ⓒ, and Ⓓ.

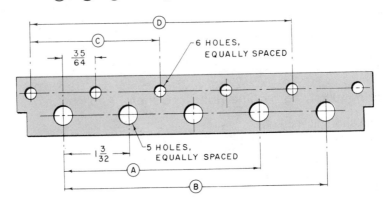

4. Find the special incentive earnings for weeks A, B, C, D, and E and the total earnings from data given in the table. Overtime incentive earnings amounting to time and a half are paid for all hours over 40 per week.

Week	Hours per Week	Rate per Hour
A	40	$.87\frac{1}{2}$
B	44	$.87\frac{1}{2}$
C	48	$.87\frac{1}{2}$
D	$43\frac{1}{4}$	$.99\frac{1}{2}$
E	$43\frac{3}{4}$	$.99\frac{1}{2}$

5. a. Find the daily production on parts A and B produced by methods 1, 2, and 3 for a $7\frac{1}{2}$ hour day.
   b. Determine the unit costs for all conditions.

Method	Part A		Part B	
	Hourly Production	Unit Cost	Hourly Production	Unit Cost
1	10	2	21	$2\frac{4}{5}$
2	14	$2\frac{1}{2}$	27	$3\frac{1}{8}$
3	15	$2\frac{3}{4}$	$29\frac{1}{2}$	$4\frac{1}{8}$

6. Four different foods and quantities are given in the table. Compute the cost of each food.

Foods	Quantity	Cost Per Unit	Total Quantity
A	$127\frac{1}{2}$ lbs	$2.16/lb	
B	$29\frac{3}{4}$ doz	4.63/doz	
C	$63\frac{1}{4}$ qts	.87/qt	
D	$17\frac{1}{2}$ yds	7.47/yd	

7. Compute the total length (feet/inches) of *BX* cable required to fill the bill of materials for rooms A, B, C, and D.

Room	# Pieces	Length
A	7	10'-6"
B	26	9"
C	8	6'-3"
D	7	9'-9"

Unit 10 Division of Fractions

OBJECTIVES OF THE UNIT

After satisfactorily studying this unit, the student/trainee will be able to
* *Solve general, shop, and laboratory problems involving the division of whole numbers and mixed numbers.*
* *Determine the dimensional and other measurement quantities that involve multiplication and division processes.*

The division of fractions refers to the process of determining how many times one number is contained in another. While the division of fractions is not used as often as the other mathematical processes, the principles are applied constantly.

A. DIVIDING FRACTIONS

RULE FOR DIVIDING FRACTIONS

* Turn the dividing fraction around so the denominator becomes the numerator and the numerator becomes the denominator. This step is often expressed as "invert the divisor."
* Change the division sign to a multiplication sign and multiply.

EXAMPLE: Divide $\frac{7}{8}$ by $\frac{3}{4}$.

Step 1 Invert the divisor $\left(\frac{3}{4}\right)$ to $\left(\frac{4}{3}\right)$.

Step 2 Cancel the factor (4) common to both numerator and denominator.

$$\frac{7}{\overset{}{\underset{2}{\cancel{8}}}} \times \frac{\overset{1}{\cancel{4}}}{3} =$$

Step 3 Multiply remaining fractions.

$$\frac{7}{2} \times \frac{1}{3} = \frac{7}{6}$$

Step 4 Express in lowest terms.

$$\frac{7}{6} = 1\frac{1}{6} \quad \text{Ans}$$

B. DIVIDING FRACTIONS AND WHOLE NUMBERS

RULE FOR DIVIDING A FRACTION AND A WHOLE NUMBER

* Express the whole number as a fraction whose denominator is 1.
* Invert the divisor.
* Proceed as in the multiplication of fractions.

EXAMPLE: *Case 1.* Divide 20 by $\frac{7}{8}$.

   Step 1 Express the whole number (20) as a fraction having
             an equivalent value.

   Step 2 Invert the divisor $\left(\frac{7}{8}\right)$ to $\left(\frac{8}{7}\right)$.

   Step 3 Multiply and simplify.

$$20 = \frac{20}{1}$$

$$\frac{20}{1} \times \frac{8}{7} = \frac{20 \times 8}{1 \times 7} = \frac{160}{7} = 22\frac{6}{7} \quad \textbf{Ans}$$

EXAMPLE: *Case 2.* Divide $\frac{15}{32}$ by 6.

   Step 1 Express the divisor (6) as a fraction having an
             equivalent value $\left(\frac{6}{1}\right)$.

   Step 2 Invert the divisor $\left(\frac{6}{1}\right)$ to become $\left(\frac{1}{6}\right)$.

   Step 3 Cancel factor (3).

   Step 4 Multiply.

$$\frac{\overset{5}{\cancel{15}}}{32} \times \frac{1}{\underset{2}{\cancel{6}}} =$$

$$\frac{5}{32} \times \frac{1}{2} = \frac{5}{64} \quad \textbf{Ans}$$

C. DIVIDING MIXED NUMBERS

RULE FOR DIVIDING MIXED NUMBERS

- Express the mixed numbers as improper fractions.
- Invert the divisor.
- Multiply the fractions.

EXAMPLE: Divide $1\frac{9}{16}$ by $3\frac{1}{8}$.

   Step 1 Express the $1\frac{9}{16}$ as $\frac{25}{26}$; the divisor $\left(3\frac{1}{8}\right)$
             as $\frac{25}{8}$.

   Step 2 Invert the divisor $\left(\frac{25}{8}\right)$ to $\frac{8}{25}$.

   Step 3 Cancel like factors (25) and (8).

   Step 4 Multiply.

$$1\frac{9}{16} \div 3\frac{1}{8} =$$

$$\frac{25}{16} \div \frac{25}{8} =$$

$$\frac{25}{16} \times \frac{8}{25} =$$

$$\frac{\overset{1}{\cancel{25}}}{\underset{2}{\cancel{16}}} \times \frac{\overset{1}{\cancel{8}}}{\underset{1}{\cancel{25}}} =$$

$$\frac{1}{2} \times \frac{1}{1} = \frac{1}{2} \quad \textbf{Ans}$$

D. COMBINING THE MULTIPLICATION AND DIVISION OF FRACTIONS AND WHOLE AND MIXED NUMBERS

RULE FOR SOLVING PROBLEMS REQUIRING THE MULTIPLICATION AND DIVISION OF FRACTIONS

- Express all mixed numbers as improper fractions.
- Invert the divisor or divisors and change the division sign or signs to multiplication sign(s).
- Cancel like factors from numerator and denominator.
- Multiply remaining fractions.
- Express product in lowest terms.

EXAMPLE: $9\frac{1}{4} \div \frac{7}{8} \div \frac{1}{32} \times 1\frac{17}{32} =$

Step 1 Express all mixed numbers as improper fractions.

$$9\frac{1}{4} = \frac{37}{4} \quad 1\frac{17}{32} = \frac{49}{32}$$

Step 2 Invert the divisors $\left(\frac{7}{8}\right)$ and $\left(\frac{1}{32}\right)$ to $\left(\frac{8}{7}\right)$ and $\left(\frac{32}{1}\right)$ and change division signs to multiplication signs.

$$9\frac{1}{4} \div \frac{7}{8} \div \frac{1}{32} \times 1\frac{17}{32} =$$

$$\frac{37}{4} \times \frac{8}{7} \times \frac{32}{1} \times \frac{49}{32} =$$

Step 3 Cancel like factors (7), (4), and (32).

$$\frac{37}{\cancel{4}} \times \frac{\cancel{8}^2}{7} \times \frac{\cancel{32}^1}{1} \times \frac{\cancel{49}^7}{\cancel{32}} =$$

Step 4 Multiply remaining fractions and express result in lowest terms.

$$\frac{37}{1} \times \frac{2}{1} \times \frac{1}{1} \times \frac{7}{1} = \frac{518}{1} = \mathbf{518}$$

Ans

Review and Self-Test

ASSIGNMENT UNIT 10

A. Division of Common Fractions (General Applications)

Divide.

1. $\frac{3}{4}$ by $\frac{1}{4}$

2. $\frac{1}{2}$ by $\frac{1}{6}$

3. $\frac{1}{4}$ by $\frac{3}{8}$

4. $\frac{5}{8}$ by $\frac{1}{2}$

5. $\frac{3}{16}$ by $\frac{1}{4}$

6. $\frac{5}{16}$ by $\frac{11}{16}$

7. $\frac{7}{12}$ by $\frac{5}{6}$

8. $\frac{9}{64}$ by $\frac{5}{32}$

9. $\left(\frac{13}{16} \times \frac{3}{32}\right)$ by $\frac{53}{64}$

10. $\left(\frac{21}{32} \times \frac{15}{64} \times \frac{1}{4}\right)$ by $\frac{7}{16}$

B. Division of Fractions and Whole Numbers (General Applications)

Divide.

1. 3 by $\frac{1}{3}$
2. 4 by $\frac{1}{2}$
3. 5 by $\frac{5}{8}$
4. $\frac{7}{16}$ by 7
5. $\frac{15}{32}$ by 30

6. $\frac{55}{64}$ by 11
7. 15 by $\frac{9}{16}$
8. 8 by $\frac{7}{16}$
9. $\left(\frac{5}{8} - \frac{1}{4} + \frac{1}{16} + \frac{1}{8}\right)$ by $\frac{3}{16}$
10. $\left(\frac{9}{32} - \frac{5}{64} \times \frac{1}{16} \times \frac{3}{32}\right)$ by $\frac{5}{8}$

C. Division of Mixed Numbers (General Applications)

1. $1\frac{1}{2} \div 1\frac{1}{2}$
2. $3\frac{1}{2} \div 4\frac{1}{4}$
3. $5\frac{3}{8} \div 3\frac{1}{4}$
4. $2\frac{3}{4} \div 4\frac{3}{8}$

5. $12\frac{1}{4} \div 6\frac{5}{16}$
6. $13\frac{3}{16} \div 11\frac{3}{4}$
7. $2\frac{15}{16} \div 6\frac{3}{8}$
8. $12\frac{3}{32} \div 1\frac{7}{64}$

D. Multiplication and Division of Fractions, Whole Numbers, and Mixed Numbers

Reduce all answers to lowest terms.

1. $\frac{1}{2} \times \frac{1}{4} \div \frac{1}{2}$
2. $1\frac{1}{6} \times \frac{5}{12} \div \frac{1}{6}$
3. $12 \times 6\frac{3}{8} \div \frac{7}{8}$
4. $21\frac{1}{4} \times 9\frac{5}{8} \div \frac{3}{4}$

5. $2\frac{9}{32} \div 1\frac{3}{16} \times 2\frac{1}{4}$
6. $16\frac{21}{64} \times 12\frac{3}{8} \div 2\frac{1}{4} \div \frac{7}{8}$
7. $1\frac{1}{4} \times \frac{9}{16} \times \frac{5}{8} \div 3\frac{1}{32}$
8. $\left(17\frac{1}{2} + \frac{3}{8} + \frac{1}{4} - \frac{5}{8}\right) \div 12\frac{1}{8}$

E. Division or Multiplication of Fractions, Whole Numbers, and Mixed Numbers (Practical Problems)

1. How many leads 15 picas long $\left(2\frac{1}{2}\text{ inches}\right)$ can be cut from a 20-inch strip? No waste allowance is made for trim.

2. How many pieces of stock $\frac{7}{8}$ of an inch long can be cut from a 30-inch bar of drill rod if $\frac{1}{16}$ of an inch is allowed on each piece for cutting?

3. How many pieces $10\frac{5}{16}$ inches long may be cut from a 12-foot length of a 2 × 4? Allow $\frac{3}{16}$ of an inch between cuts for waste.

4. How many billets of cold-drawn steel $3\frac{9}{32}$ inches long can be cut from a bar 48 inches long? Allow $\frac{1}{16}$ of an inch for the saw cut and another $\frac{1}{16}$ of an inch for facing.

5. Determine the number of pieces that can be blanked from a 50-yard roll of brass when each stamping is $4\frac{1}{2}$ inches long and each piece requires an additional $\frac{5}{32}$ of an inch for positioning.

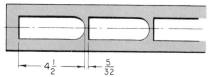

6. A drafting scale of $\frac{1}{4}$ of an inch to 1 foot is used on a drawing of a house. Compute the length of rooms A, B, C, and D.

Room	Scale Measurement (length)	Actual Room Length
A	$9\frac{1}{4}$ in.	
B	$10\frac{1}{4}$ in.	
C	$13\frac{1}{8}$ in.	
D	$14\frac{5}{16}$ in.	

7. The I-beam lintel of a brick doorway opening is 4'-4" long. It weighs $32\frac{1}{2}$ pounds. Find the number of pounds per foot in the I-beam.

8. Determine dimensions Ⓐ and Ⓑ.

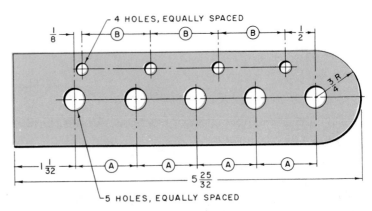

9. In milling flat surfaces, the work is fed against a revolving plain milling cutter with a specified feed.
   a. Find the length of time it will take to mill pieces A, B, C, D, and E for the lengths of work and feed indicated in the table.
   b. Determine the total time for milling.

Cut	Length of Work	Feed per Minute
		(in inches)
A	4	$\frac{3}{4}$
B	11	$1\frac{3}{8}$
C	$11\frac{1}{2}$	$2\frac{7}{8}$
D	$13\frac{9}{16}$	$3\frac{3}{4}$
E	$6\frac{3}{32}$	$8\frac{1}{8}$

10. The average hourly production of a plastic part, the number of work hours per item per station, and the required production are given.
    a. Find the daily production for items 1, 2, and 3.
    b. Determine the number of days required to produce the units needed of the three items.

Item	Average Hourly Production	Daily Work Hours per Station	Required Units
1	20	14	3220
2	18	15	3720
3	$16\frac{1}{2}$	$20\frac{1}{2}$	5158

11. Calculate the current of a heating element in a simple circuit in amperes. The resistance of the element is $32\frac{3}{4}$ ohms, operating in a 48-volt circuit.
    Note. The current equals the voltage divided by the resistance. Round the answer to the nearest $\frac{1}{2}$ ampere.

Unit 11 Achievement Review on Common Fractions

OBJECTIVES OF THE UNIT

This achievement review serves as an overall test for Section 2. The unit is designed to measure the student's/trainee's ability to

- *Visualize fractional parts of an object and reduce answers for measurements and other computed quantities to lowest terms.*
- *Apply the appropriate mathematical processes to solve problems that require the addition, subtraction, multiplication, or division of fractions, whole numbers, or mixed numbers.*
- *Check each answer.*

A. THE CONCEPT OF COMMON FRACTIONS: REPRESENTATION AND REDUCTION

1. What common fraction is represented by the shaded area of each square?

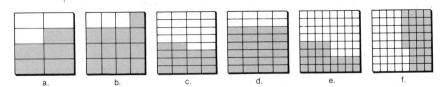

a.   b.   c.   d.   e.   f.

2. Reduce the fractional dimensions of Ⓐ, Ⓑ, Ⓒ, Ⓓ, Ⓔ, and Ⓕ, given in the drawing, to lowest terms. Locate each dimension on a rule.

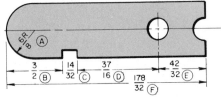

B. ADDITION OF FRACTIONS

1. Add the following fractions.

   a. $\frac{1}{8} + \frac{1}{4} + \frac{3}{8}$ d. $1\frac{1}{16} + \frac{3}{64} + 4\frac{17}{32}$

   b. $\frac{3}{32} + \frac{5}{8} + \frac{3}{4}$ e. $3\frac{23}{32} + 2\frac{1}{8} + \frac{57}{64} + \frac{1}{4} + \frac{3}{16}$

   c. $\frac{27}{64} + \frac{3}{16} + \frac{3}{32}$ f. $\frac{1}{9} + \frac{1}{6} + 2\frac{7}{8} + 3\frac{27}{32}$

2. Determine the dimensions Ⓐ, Ⓑ, Ⓒ, Ⓓ, Ⓔ, and Ⓕ for the stripper plate.

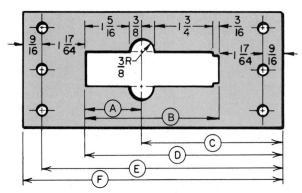

3. A pattern layout for a jacket is illustrated. Determine the length of material needed for one pattern.

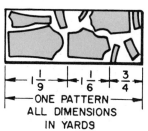

C. SUBTRACTION OF FRACTIONS

1. Subtract the following fractions.

   a. $\frac{9}{16} - \frac{3}{16}$ d. $122 - 3\frac{3}{8}$ g. $19 - 2\frac{25}{64}$

   b. $\frac{19}{64} - \frac{7}{32}$ e. $17 - \frac{55}{64}$ h. $13\frac{9}{16} - 7\frac{9}{32}$

   c. $3 - \frac{27}{32}$ f. $203 - 6\frac{9}{16}$ i. $5\frac{37}{64} - 2\frac{1}{4} - 1\frac{7}{64}$

2. Determine dimensions Ⓐ, Ⓑ, Ⓒ, Ⓓ, Ⓔ, and Ⓕ. Reduce all answers to lowest terms.

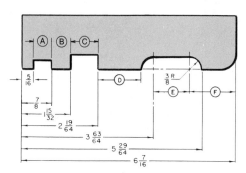

D. MULTIPLICATION OF FRACTIONS

1. Determine dimensions Ⓐ, Ⓑ, Ⓒ, Ⓓ, Ⓔ, and Ⓕ. Reduce all answers to lowest terms.

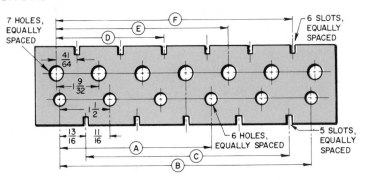

2. Determine the length of corduroy fabric needed for jackets A, B, and C. State any fractional answer in terms of the nearest next whole number.

Jacket	Material per Jacket (including selvages)	Quantity	Length of Material Needed
A	$2\frac{7}{8}$ yds.	70	
B	$3\frac{1}{4}$ yds.	120	
C	$3\frac{5}{8}$ yds.	325	

E. DIVISION OF FRACTIONS

1. Divide the following fractions and reduce all answers to lowest terms.

   a. $\frac{1}{8} \div \frac{3}{4}$ c. $11 \div \frac{5}{6}$ e. $3\frac{3}{32} \div 1\frac{5}{64}$

   b. $\frac{21}{64} \div \frac{7}{16}$ d. $\frac{21}{32} \div 7$ f. $2\frac{3}{16} \div 1\frac{3}{64} \div 4 \div 1\frac{3}{64}$

2. Determine dimensions Ⓐ, Ⓑ, Ⓒ, Ⓓ, Ⓔ, and Ⓕ. Reduce all answers to lowest terms.

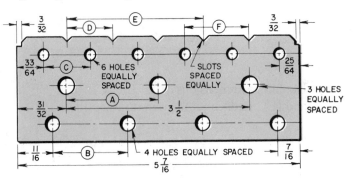

F. COMBINED PROCESSES WITH COMMON FRACTIONS

1. The resistance of each of three resistors in series is given in ohms (Ω) on the diagram. The current through the circuit is $1\frac{1}{2}$ amperes (A). Determine the following values.

   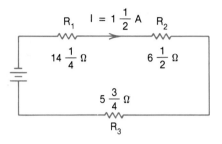

   a. The total resistance (R_T).
   b. The source voltage (E_S).
   Note. The total resistance (R_T) equals the sum of the three separate resistors (R_1, R_2, and R_3). The source voltage (E_S) equals the current (I) multiplied by the total resistance (R_T).

2. The current through a diode is $\frac{1}{24}$ ampere. The voltage drop across the diode is $\frac{3}{10}$ volt. Find the resistance reduced to lowest terms.
   Note. Resistance is equal to the voltage (E) divided by the current (I).

$$E = \frac{3}{10} \text{ V}$$

$$I = \frac{1}{24} \text{ A}$$

DECIMAL FRACTIONS

Unit 12 The Concept of Decimal Fractions

OBJECTIVES OF THE UNIT
After satisfactorily studying this unit, the student/trainee will be able to
* *Understand how fractional measurements and other quantities are written in the decimal system.*
* *Read and write equivalent values for whole numbers and fractional parts in the decimal system.*
* *Round off decimals.*

Machine, hand, and assembly operations must often be performed to a greater degree of accuracy than a fractional part of an inch. Where this accuracy is required, as in mating or interchangeable parts, precise dimensions are given on specifications, drawings, and sketches. These dimensions are given in thousandths, ten-thousandths, and, for extremely accurate work, in hundred-thousandths and millionths of an inch.

A. INTERPRETING THE DECIMAL SYSTEM

This system, which is based on ten (10), is known as the decimal system. The decimal system has been adopted universally throughout many industries because of the ease and accuracy with which dimensions may be measured and computed. Steel rules, micrometers, indicators, and other precision instruments are available for taking measurements based on the decimal system.

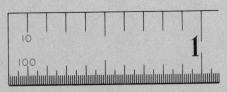

FIGURE 12-1

Describing a Decimal Fraction

A decimal fraction is a fraction. The denominator is 10, 100, 1,000, 10,000, or any other value that is obtained by multiplying 10 by itself a specified number of times. Instead of looking like a common fraction, the decimal fraction is written on one line with a period in front of it.

This is possible because the denominator is always one (1) followed by zeros. By placing a period before the number that appears in the numerator, the denominator may be omitted. This period is called a decimal point. For example, the common fraction $\frac{5}{10}$ is written as the decimal .5; $\frac{5}{100}$ is written as .05; and $\frac{5}{1000}$ is written as .005.

Writing Decimal Fractions

Any whole number with a decimal point in front of it is a decimal fraction. The numerator is the number to the right of the decimal point. The denominator is always one (1) with as many zeros after it as there are places in the number to the right of the decimal point.

For example, the fraction $\frac{9}{10}$ may be written as the decimal fraction .9. This means that 9 is the numerator; the denominator is 1 with as many zeros as there are places (or digits) in the number to the right of the decimal point. In this case, there is one place to the right of the decimal point so the denominator is 10. The decimal fraction .9 is, therefore, the same as $\frac{9}{10}$.
To illustrate further

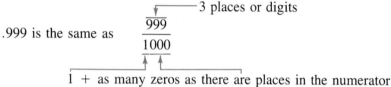

$$.999 \text{ is the same as} \quad \frac{999}{1000}$$

1 + as many zeros as there are places in the numerator

B. EXPRESSING DECIMAL VALUES

Writing Whole Numbers and Decimal Fractions

With whole numbers and fractions, the whole number is placed to the left of the decimal point. The decimal fraction appears to the right. Three examples are given to show how different quantities may be expressed.

EXAMPLE:

$5\frac{7}{10}$ is written   Whole Number ——→ 5.7 ←—— Decimal Fractions (Tenths)

$55\frac{7}{100}$ is written   Whole Number ——→ 55.07 ←—— Hundredths, Tenths

$555\frac{77}{1000}$ is written   Whole Number ——→ 555.077 ←—— Thousandths, Hundredths, Tenths

Indicating Degree of Accuracy

On drawings and in computations, whole numbers are sometimes expressed in the decimal system with zeros following the decimal point to indicate the degree of precision to which certain dimensions must be held.

EXAMPLE: The quantity 2″ is written 2.00″ if the dimension must be accurate to the second decimal place. If accuracy to the thousandth part of an inch is required, the 2″ is written 2.000″.

A decimal fraction such as .46 is often written (0.46). The zero that is placed before the decimal point emphasizes the fact that the decimal fraction is less than one.

C. READING DECIMALS

A decimal is read like a whole number except that the name of the last column or place to the right of the decimal point is added.

EXAMPLE:

- 0.63 is read sixty-three *hundredths*.
- 0.136 is read one hundred thirty-six *thousandths*.
- 0.5625 is read five thousand six hundred twenty-five *ten-thousandths*.
- 3.5 is read three and five *tenths*.
- 2.15625 is read two and fifteen thousand six hundred twenty-five *hundred-thousandths*.
- 0.0625 is read six hundred twenty-five *ten-thousandths*. This quantity is also commonly expressed as sixty-two and a half *thousandths*.

Simplified Method of Reading Decimals

Dimensions involving whole numbers and decimals are frequently expressed in an abbreviated form.

EXAMPLE:

- A dimension like (7.625) is spoken of as ''seven point six two five.''
- A dimension like (21.3125) is spoken of as ''twenty-one point three one two five.''

The use of decimal fractions provides an easy method of solving problems. Accurate computations may be made in addition, subtraction, multiplication, and division of fractions having a denominator of 10, 100, 1000, and the like. The units that follow deal with each one of the four fundamental mathematical operations as applied to decimal fractions.

D. ROUNDING OFF DECIMALS

The degree of precision to which a part is to be machined or finished sometimes determines how accurately the answer to a problem is computed. Many drawings indicate an accuracy in terms of thousandths or ten-thousandths of an inch. However, in computing dimensions, the answers may be accurate to four, five, or more decimal places.

The process of expressing a decimal to the number of decimal places needed for a predetermined degree of accuracy is called *rounding off decimals*.

RULE FOR ROUNDING OFF DECIMALS

- Check the drawing, sketch, or specifications to determine the required degree of accuracy.
- Look at that digit in the decimal place that indicates the required degree of accuracy.
- Increase that digit by 1 if the digit that follows immediately is 5 or more.
- Leave that digit as it is if the digit that follows is less than 5. Drop all other digits that follow.

EXAMPLE: The sum of a column of decimals is .739752. A part must be machined to an accuracy of only three places. Round off the decimal to three places.

Step 1	Write the computed decimal (6 places).	.739752
Step 2	Locate the digit that shows the number of .001″. The third digit (9) does this.	.739
Step 3	Look at the fourth-place digit to determine whether or not the third digit should remain the same or be increased.	.7397
Step 4	Increase the 9 by 1 because the fourth-place digit (7) is greater than (5). The correct answer is .740.	.740 Ans

All of the intermediate steps in the example are given to serve as a guide in rounding off decimals. With actual practice, it is possible to round off a decimal to any desired degree of accuracy by just looking at it.

Review and Self-Test

ASSIGNMENT UNIT 12

A. Writing Equivalent Decimal Values (General Applications)

1. Examine each circle and square. Determine visually what fractional part of each circle or square is represented by the shaded portions (A through I).

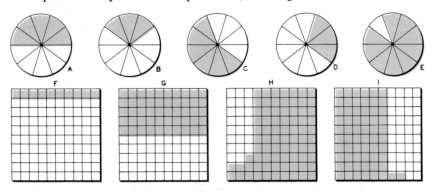

2. Express decimals (a, b, c, d, e, and f) in words.

    a. .3   c. 1.25   e. 5.375

    b. .07   d. 0.3125   f. 27.01563

3. Write the value of each quantity (a, b, c, d, e, and f) as a decimal.

    a. Seven tenths

    b. Sixteen hundredths

    c. Fifteen thousandths

    d. Eleven ten-thousandths

    e. Two thousand one hundred fifty-two ten-thousandths

    f. Three point one eight seven five

4. Write the following fractions as decimal fractions.

    a. $\frac{1}{10}$ e. $\frac{73}{100}$ i. $\frac{3}{10,000}$ m. $\frac{1,000}{10,000}$

    b. $\frac{3}{10}$ f. $\frac{1}{1,000}$ j. $\frac{19}{10,000}$ n. $\frac{793}{100,000}$

    c. $\frac{9}{100}$ g. $\frac{93}{1,000}$ k. $\frac{205}{10,000}$ o. $\frac{1,027}{100,000}$

    d. $\frac{29}{100}$ h. $\frac{157}{1,000}$ l. $\frac{1,923}{10,000}$ p. $\frac{30,019}{100,000}$

5. Write the following mixed numbers as decimals.

    a. $1\frac{1}{10}$ c. $25\frac{91}{100}$ e. $2525\frac{21}{10,000}$

    b. $3\frac{9}{100}$ d. $272\frac{67}{1,000}$ f. $362\frac{2,007}{10,000}$

B. Reading Decimal Measurements and Rounding Off Values (Practical Problems)

1. Secure a rule with a tenth and hundredth scale on it. Locate dimensions (a, b, c, d, and e) on the tenth scale.

    a. $\frac{1}{10}$ d. 1.2

    b. $\frac{3}{10}$ e. 1.5

    c. $\frac{5}{10}$

2. Locate dimensions (a, b, c, d, and e) on the hundredth scale.

    a. $\frac{10}{100}$ d. 1.20

    b. $\frac{33}{100}$ e. 1.32

    c. $\frac{77}{100}$

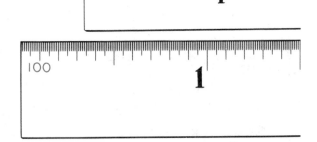

3. Express dimensions Ⓐ through Ⓔ on the drawing as decimals.

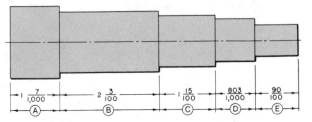

4. Read dimensions Ⓐ through Ⓖ on the rules as illustrated. Express as decimal fractions and decimals.

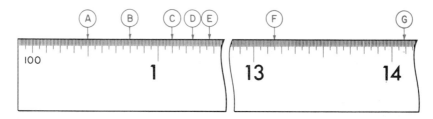

5. Round off decimal fractions (a) through (i) to two decimal places.

   a. .756   d. 29.409   g. 221.7557

   b. 1.952   e. 2.5644   h. 0.89673

   c. 7.324   f. 18.2707   i. 20.99974

6. Express each electrical measurement as a common fraction or mixed number. Reduce each answer to lowest terms.

   a. A copper wire diameter of 0.090″

   b. A resistance of 22.35 ohms

   c. A voltage of 105.625 volts

   d. An average unit cost of $.875

7. Convert each quantity to its equivalent decimal value, rounded to three decimal places.

   a. A wire resistance of $9\frac{3}{32}$ ohms

   b. A wire length of $6\frac{3}{8}''$

   c. An installation time of 4 hours 20 minutes

   d. A power rating of $3\frac{15}{16}$ watts

Unit 13 Addition of Decimals

OBJECTIVE OF THE UNIT

After satisfactorily studying this unit, the student/trainee will be able to
- *Add decimals in general applications and in practical measurement problems.*

On many drawings and sketches, dimensions must be computed that require the addition of two or more decimals.

A typical example is illustrated (*figure* 13-1). The distance Ⓐ on the gage may be determined by adding the decimal dimensions 2.20″, 2.76″, and .50″. The addition of these decimals is the same as the addition of regular whole numbers. The exception is that the location of the decimal point must be given.

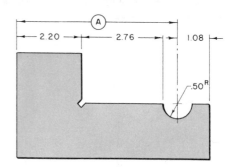

FIGURE 13-1

A. ADDING DECIMALS

RULE FOR ADDING DECIMALS

* Write the given numbers one under the other so that all of the decimal points are aligned in a vertical column.
* Add each column of numbers the same as for regular whole numbers.
* Locate the decimal point in the answer by placing it in the same column in which it appears with each number.

EXAMPLE: Add .875 + 1.2 + 375.007 + 71.1357 + 735.

.875
1.2
375.007
71.1357
735.

Step 1   Write the numbers under each other so that all of the decimal points are aligned in a vertical line.

.8750
1.2000
375.0070
71.1357
735.0000

Note. Zeros are sometimes added to the numbers so that they all have an equal number of places after the decimal point. This practice may be followed to eliminate errors.

Step 2   Add each column.

Step 3   Locate the decimal point in the answer in the same column in which it appears with the numbers being added.

1183.2177   Ans

Review and Self-Test

ASSIGNMENT UNIT 13

A. Addition of Decimals (General Applications)

1. Add.

a.	.5	b.	9.3	c.	76.8	d.	195.7
	.6		17.7		119.32		83.02
	8.3		72.4		24.6		9.006

e. .4 + .7 + .4

f. 269.1 + 201.3

g. 0.57 + 29.35 + 1.6

h. 0.872 + 1.54 + 725.093

i. 2.9834 + 0.7256 + 329.7 + 21.0006

j. 0.00850 + 0.93006 + 3225.06 + 0.0875

2. Add decimals (a) through (e). Then round off each sum correct to three decimal places.

a. 25.0097
    0.9237
    <u>1.125</u>

c. 11.61254 + 0.735 + 1.3 + 625.003125

d. .7 + 1.707 + 22.0625 + 3.09375 + 0.625

e. 7.251 + 0.98475 + .03125 + 25.0 + 5.105

b.   .7895
       .6842
    12.7
   <u>231.0924</u>

B. Addition of Decimal Measurements (Practical Problems)

1. Determine dimensions Ⓐ, Ⓑ, Ⓒ, Ⓓ, and Ⓔ for the Die Plate.

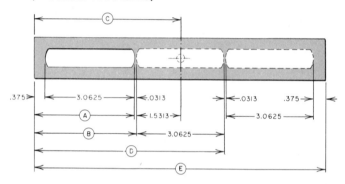

2. Determine baseline dimensions Ⓐ, Ⓑ, Ⓒ, Ⓓ, and Ⓔ.

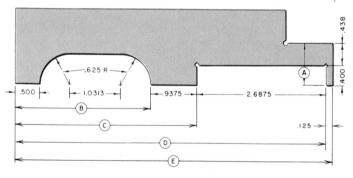

3. Determine overall dimension Ⓐ of the Template. Round off the answer to three decimal places.

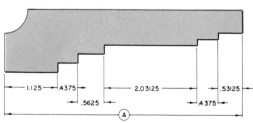

4. Find dimensions Ⓐ through Ⓔ. Start each computation with the dimensions given on the drawing. Round off answers to two decimal places.

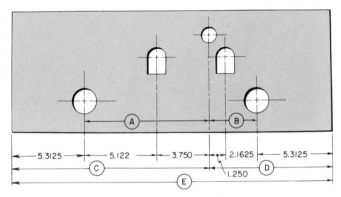

5. A counter top is 1.725″ thick. It is covered with laminated plastic .0625″ thick. Give the total thickness of the top.

6. Find the thickness of the plywood panel as illustrated.

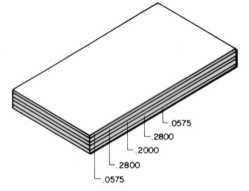

7. The total current in the circuit is measured by an ammeter Ⓐ. The total current is distributed in the four appliances. State what the reading is on the ammeter to the closest hundredth ampere.

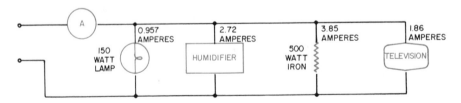

Unit 14 Subtraction of Decimals

OBJECTIVE OF THE UNIT

After satisfactorily studying this unit, the student/trainee will be able to
* *Subtract decimals in practical measurement problems.*

Subtracting one decimal dimension from another is common practice in the factory, store, hospital, and home. The mathematical operation is the same as the subtraction of whole numbers with one exception. The position of the decimal point in the answer is similar to the addition of decimals.

The drawing of the tapered plug gage (*figure 14-1*) shows a typical method of dimensioning a taper. The difference in diameters can readily be determined by the subtraction of decimals. The smaller diameter (1.5936″) is subtracted from the larger diameter (1.875″). The answer is referred to as the *difference*.

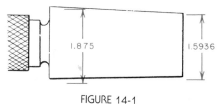

FIGURE 14-1

A. SUBTRACTING DECIMALS

RULE FOR SUBTRACTING DECIMALS

* Write the given numbers so that the decimal points are under each other.
* Subtract each column of numbers the same as for regular whole numbers.
* Locate the decimal point in the answer by placing it under the column where it appeared in the problem.

EXAMPLE: Determine the amount of taper of the tapered plug gage. Give the dimension to the nearest thousandth.

Step 1   Write the two dimensions (1.875″ and 1.5936″) so that the smaller is under the larger. The decimal points are in the same column.

$$\begin{array}{r} 1.875 \\ \underline{1.5936} \end{array}$$

Note. Zeros may be added to one of the numbers (1.875). It has fewer digits after the decimal point than the other number. The addition of zeros does not change the value.

$$\begin{array}{r} 1.8750 \\ \underline{1.5936} \end{array}$$

Step 2   Subtract the numbers by starting with the last digit on the right.

Step 3   Locate the decimal point in the answer.

$$\begin{array}{r} 1.8750 \\ \underline{1.5936} \\ 0.2814 \end{array}$$

Step 4   Round off the answer to the required number of places. 0.2814″ rounded off to three places.

= **0.281**   Ans

Review and Self-Test

ASSIGNMENT UNIT 14

A. Subtracting Decimals (Practical Problems)

1. Find the difference in diameters (taper) of the Taper Plugs A, B, C, D, and E.

Plug	Diameter	
	Large End	**Small End**
A	1.750″	1.500″
B	3.0935″	2.837″
C	4.5312″	3.687″
D	10.1563″	8.3125″
E	7.0781″	5.7812″

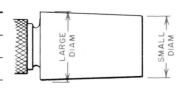

2. Find dimensions Ⓐ, Ⓑ, Ⓒ, Ⓓ, and Ⓔ for the plate.

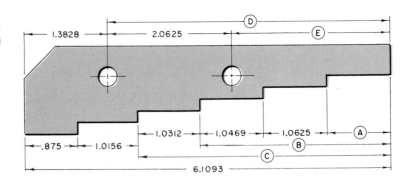

3. Determine dimensions Ⓐ, Ⓑ, Ⓒ, Ⓓ, and Ⓔ. Round off each answer to three decimal places.

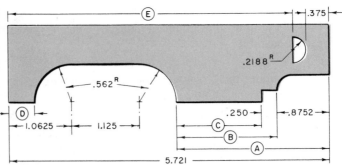

4. Determine dimensions Ⓐ, Ⓑ, Ⓒ, Ⓓ, and Ⓔ correct to two decimal places.

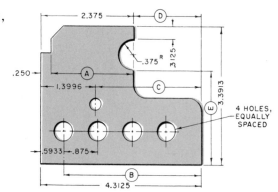

5. Find dimensions Ⓐ, Ⓑ, Ⓒ, Ⓓ, and Ⓔ correct to three decimal places.

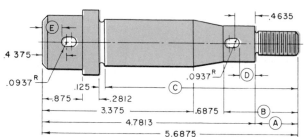

6. The wire diameter for sizes 10 through 16 of copper conductors is given in the table. Indicate the difference in diameter between sizes.
   a. 10 and 16
   b. 10 and 11
   c. 13 and 14
   d. 15 and 16

Wire Size	
National Gage Number	Diameter in Inches
10	0.10190
11	0.09074
12	0.08081
13	0.07196
14	0.06408
15	0.05707
16	0.05082

7. The opening between two parts is three and a half thousandths (0.0035″). A shim 0.00225 is available. Find the additional thickness to fill the opening.
8. A diet permits a daily intake of 1.67 quarts of liquid a day; 0.375 quart and 0.750 quart of liquid are used for two meals. Determine the remaining amount of liquid that may be taken during the day. Round off the answer to two decimal places.

9. Refer to the diagram for the quantity of current (*I*) flowing through each branch circuit. The total supply current (*I_T*) in the parallel circuit is 2.625 amperes (A). The current in I_A is 0.938 A and the current in I_B is 0.625 A. Find the current in I_C. *Note.* $I_C = I_T - I_A - I_B$.

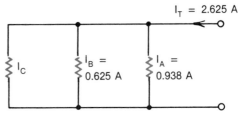

Unit 15 Multiplication of Decimals

OBJECTIVES OF THE UNIT

After satisfactorily studying this unit, the student/trainee will be able to
- *Solve general and practical measurement problems involving the multiplication of decimals.*
- *Convert percent values to decimals and decimals to percent values.*
- *Check solutions obtained by the multiplication of decimals.*

Multiplying decimals is a convenient and simplified method of adding them. One method of multiplying is to take one of the numbers to be multiplied, list it the number of times indicated by the multiplier and then add. This procedure is cumbersome, however, and it is easier with less chance for error if the two numbers are multiplied.

The multiplication process is identical to the multiplication process used for whole numbers. The exception is in pointing off the decimal places in the answer.

A. MULTIPLYING DECIMALS

RULE FOR MULTIPLYING DECIMALS

- Multiply the same as for whole numbers.
- Total the number of decimal places to the right of the decimal point in both of the numbers being multiplied.
- Locate the decimal point in the answer.
  Note. Start at the extreme right digit in the answer. Count off as many places to the left as there are in both the multiplier and multiplicand.

EXAMPLE: Multiply 27.935×7.07.

Step 1 Multiply (27.935) by (7.07).

Step 2 Add the number of digits to the right of the decimal point in both numbers that have been multiplied.

Step 3 Start at the right in the product. Count off the five decimal places to the left. Place the decimal point here. The result is the product of multiplying the two decimals.

27.935 ◀(3 decimal places)

7.07 ◀(2 decimal places)

195545

195545

197.50045   Ans

▲

(5 decimal places)

B. CONVERTING PERCENT VALUES TO DECIMALS

The term percent means comparison in terms of so many hundredths of a quantity. In most percentage problems, multiplication is the mathematical process involved. The percent value is often converted to the decimal system for ease in computing and accuracy in pointing off the required number of decimal places. Instead of writing out the word percent each time, the symbol (%) is used.

RULE FOR CHANGING A PERCENT TO A DECIMAL

* Remove the percent (%) sign.
* Place a decimal point two digits to the left of the number for the given percent.
  Note. If the percent is a mixed number, change the fraction to a decimal and place this value after the whole number.
* Use the decimal value for the given percent the same as any other decimals to perform the required mathematical operations.

EXAMPLE: Change 12% to a decimal.

Step 1 Remove percent (%) sign.   **12**

Step 2 Place decimal point two digits to the left.   **.12**

Step 3 $12\% = .12$ **.12**   Ans

C. CONVERTING DECIMALS TO PERCENT

RULE FOR CHANGING A DECIMAL TO A PERCENT

* Move the decimal point two places to the right.
* Place the percent (%) sign after this number.

EXAMPLE: Change .055 to a percent.

Step 1	Move decimal point two places to the right.	**05.5**
Step 2	Place percent (%) sign after this number.	**5.5%**
Step 3	.055 = 5.5%	**5.5%** **Ans**

Different ways of representing common and decimal fraction and percent values of the same quantities are shown in figure 15-1.

Fraction	Decimal Equivalent	Percent Value	Expressed in Words
$\frac{1}{100}$	1.0	1%	One one-hundredth
$\frac{10}{100}$	.10	10%	Ten hundredths
$\frac{100}{100}$	1.00	100%	One hundred hundredths
$\frac{150}{100}$	1.50	150%	One hundred fifty hundredths
$\frac{175}{1000}$	.175	17.5%	One hundred seventy-five thousandths

FIGURE 15-1

D. CHECKING

Problems in multiplication may be checked by one of two methods (the same as for whole numbers).

- The multiplier and multiplicand may be interchanged.
- The multiplication process may be repeated, using the same multiplier and multiplicand.

In either case, one of the most important steps is to check the location of the decimal point in the product.

Review and Self-Test

ASSIGNMENT UNIT 15

A. Multiplication of Decimals

1. Multiply each whole number and decimal fraction. Check each answer.

a. 9 by .8	d. .37 by 100	g. 11.7×1.82	j. 1.0313×2.937
b. 16 by 1.5	e. 1.3×98	h. 11.31×6.14	k. 125.002×2.14
c. 12 by .72	f. $9.5 \times .76$	i. 92.07×7.392	l. 10.063×2.030

2. Multiply these decimal fractions. Check each answer. Then, round off the answers to four decimal places.

   a. 10.0625 × 6.437   b. 1.0937 × 3.0313   c. 1.5 × 3.7 × 5.12

B. Conversion of Percent Values

1. Determine the weight of metals A, B, C, and D, which were alloyed to cast a bronze plate weighing 79.5 pounds. The composition of the bronze is indicated by percent of each alloying metal.

A	B	C	D
Copper	Tin	Zinc	Lead
81%	6.5%	7.25%	5.25%

C. Multiplication of Decimal Values (Practical Problems)

1. Determine the cost of bar stock needed for parts A, B, C, D, and E. Give the answer in terms of dollars and cents for each part and the total cost.

Parts	Weight (Pounds per Foot)	Required Number of Feet	Cost Per Pound
A	2.5	3.5	.75
B	2.25	7.5	.63
C	1.25	7.75	.93
D	7.5	27.125	.527
E	.0125	10.25	4.375

2. Find the lengths of insulating strip required to blank out each of the quantity of plates specified in A, B, C, D, and E. Round off all answers to one decimal place. Also determine the total length required.

Part	Quantity	Length of Plate	Allowance for Blanking
A	100	1.5	.25
B	100	2.25	.25
C	75	2.375	.25
D	75	2.625	.125
E	75	2.8906	.0937

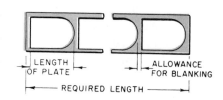

LENGTH OF PLATE   ALLOWANCE FOR BLANKING

REQUIRED LENGTH

3. Determine the distance in inches that a tool travels for each of five cuts (A, B, C, D, and E). Each distance should be rounded off to one decimal place. Also determine the total distance.

Cuts	RPM	Feed per Rev. (in inches)	Time (in minutes)
A	900	.005	1.5
B	424	.008	2.25
C	368	.015	6.75
D	336	.062	5.75
E	128	.062	25.25

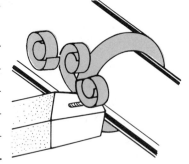

4. The table gives the thickness, number of laminations, and meter readings of resistance for each lamination for each of three different core materials.
   a. Determine the thickness of cores 1, 2, and 3.
   b. Find the total resistance of the laminations in the three cores.
   Note. Round off all answers to two decimal places.

Core Material	Thickness per Lamination	Number of Laminations	Resistance per Lamination
1	.003"	200	.1 amp
2	.012"	21	.125 amp
3	.1352"	15	.3157 amp

5. It takes an average of 3.4 hours to lay 100 square feet (one square) of finish flooring. Compute the time required to lay finish flooring in rooms A, B, and C.

Room	Floor Area (Squares)	Flooring Time
A	12.2	
B	13.75	
C	17.87	

6. The unit price, weight per container, and quantity required for four foods are given. Determine the cost of each item correct to two decimal places.

Food	Cost ¢ per Ounce	Ounces per Container	Quantity	Cost
A	.086	12	24 pkgs	
B	.078	18	12 pkgs	
C	.062	15.75	24 cans	
D	.034	15.75	48 cans	

7. An electric (kilowatt hour) meter registers 7 0 2 3 on June 1. On July 1 the reading is 7 8 6 9 . The difference represents electrical energy used. The cost of electrical energy is \$0.07252 per kilowatt hour (kW·h).
   a. Find the cost of the electrical energy.
   b. The bill is reduced by a fuel adjustment of \$0.00303 per kW·h. Indicate the amount the bill is reduced.
   c. A sales tax of 0.975% is levied against each adjusted utility bill. Compute the sales tax on the adjusted bill.
   d. Determine the total charges.

8. The inductive reactance (in ohms) in the ac circuit equals $2\pi \cdot I \cdot f$. Substitute the value of 0.032 Hz for *(f)* and 60/s for *(I)*.
   Calculate the inductive reactance in ohms. Round the answer to two decimal places. Use $\pi = 3.1416$.

9. Determine the power in watts for the values given in the circuit diagram, correct to two decimal places.
   Note. The power *(W)* equals the voltage (V) multiplied by the current in amperes (A).

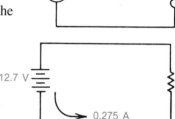

Unit 16 Division of Decimals

OBJECTIVES OF THE UNIT

After satisfactorily studying this unit, the student/trainee will be able to
- *Solve general and practical measurement problems involving the division of decimals.*
- *Translate and express common fractions as decimal fractions and mixed numbers as decimals.*
- *Use decimal equivalent charts and shop methods to select equivalent decimal or fractional measurements.*

Division is the simplified process of computing the number of times one number is contained in another. The division of decimals, like all other mathematical operations for decimals, is essentially the same as for whole numbers. An additional consideration is the location of the decimal point in the answer.

A. DIVIDING DECIMALS

RULE FOR DIVIDING DECIMALS

- Place the number to be divided (called *dividend*) inside the division box.
- Place the *divisor* outside.
- Move the decimal point in the divisor to the extreme right. The divisor then becomes a whole number.
- Move the decimal point the same number of places to the right in the dividend.
  Note. Zeros are added in the dividend if it has fewer digits than the divisor.
- Mark the position of the decimal point in the *quotient*. The position is directly above the decimal point in the dividend.
- Divide as for whole numbers. Place each figure in the quotient directly above the digit involved in the dividend.
- Add zeros after the decimal point in the dividend if the dividend cannot be divided exactly by the divisor.
- Continue the division until the quotient has as many places as are required for the answer.

___

EXAMPLE: *Case 1.* Divide 25.5 by 12.75.

Step 1 Move the decimal point in the divisor to the right (2 places).

Step 2 Move the decimal point in the dividend to the right the same number of places (2). Since there is only one digit after the decimal, add a zero to the dividend.

Step 3 Place the decimal point in the quotient.

Step 4 Divide as for whole numbers.

$$
\begin{array}{r}
2. \\
12.75.\overline{)25.50.}
\end{array}
$$

EXAMPLE: *Case 2.* Divide 123.573 by 137.4.

The answer must be correct to three decimal places. *Note.* The division process is usually carried out to one more than the required number of places in the answer. The last digit may then be rounded off for greater accuracy. In this case, .8993 is rounded off to .899 **Ans**

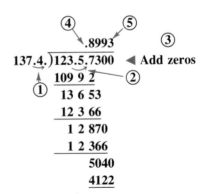

$$
\begin{array}{r}
.8993 \\
137.4.\overline{)123.5.7300} \quad \blacktriangleleft \text{ Add zeros} \\
109\ 9\ 2 \\
\hline
13\ 6\ 53 \\
12\ 3\ 66 \\
\hline
1\ 2\ 870 \\
1\ 2\ 366 \\
\hline
5040 \\
4122 \\
\hline
\end{array}
$$

B. EXPRESSING COMMON FRACTIONS AS DECIMAL FRACTIONS

Dimensions used in machine, bench, or assembly operations are given in terms of common fractions or decimal fractions. Yet, when the actual measurements are taken, it is often necessary to either express a common fraction as a decimal fraction or a decimal fraction as a common fraction. Converting from one system to the other is comparatively easy. It involves, in most cases, the division of numbers with three or less digits.

RULE FOR EXPRESSING A COMMON FRACTION AS A DECIMAL FRACTION

* Divide the numerator by the denominator.

EXAMPLE: Express $\frac{5}{16}$ as a decimal fraction.

Step 1 Divide the numerator (5) by the denominator (16).
Step 2 Place a decimal point after the 5.
Step 3 Locate the decimal point in the quotient.
Step 4 Add as many zeros as are needed to obtain a quotient that can be rounded off to the required number of decimal places.
Step 5 Divide. The resulting answer (.3125) is the decimal fraction equivalent of the common fraction $\left(\frac{5}{16}\right)$.

$$\begin{array}{r} .3125 \text{ Ans} \\ 16\overline{)5.0000} \\ \underline{4\ 8} \\ 20 \\ \underline{16} \\ 40 \\ \underline{32} \\ 80 \\ \underline{80} \end{array}$$

C. EXPRESSING MIXED NUMBERS AS DECIMALS

RULE FOR EXPRESSING A MIXED NUMBER AS ITS DECIMAL EQUIVALENT

* Express the mixed number as an improper fraction.
* Divide the numerator by the denominator.
* Carry out the division to the number of decimal places required for the degree of accuracy involved.

EXAMPLE: Determine the decimal equivalent of $1\frac{3}{64}$ correct to three decimal places. The answer is 1.047.

Step 1 Divide the numbers to four decimal places.
Step 2 Round off the decimal equivalent (1.0468) to three decimal places. The answer is 1.047.

$$1\frac{3}{64} = \frac{67}{64} = \begin{array}{r} 1.0468 = 1.047 \text{ Ans} \\ 64\overline{)67.0000} \\ \underline{64} \\ 3\ 00 \\ \underline{2\ 56} \\ 440 \\ \underline{384} \\ 560 \\ \underline{512} \\ 48 \end{array}$$

D. EXPRESSING DECIMAL FRACTIONS AS COMMON FRACTIONS

RULE FOR EXPRESSING A DECIMAL FRACTION AS A COMMON FRACTION

- Write the number after the decimal point as the numerator of a common fraction.
- Write the denominator as 1 with the same number of zeros after it as there are digits to the right of the decimal point.
- Express resulting fraction in lowest terms.

EXAMPLE: Express .09375 as a common fraction.

Step 1 Write number after the decimal point as the numerator. $9375 = 9,375$

Step 2 Determine denominator. $1 + 5 \text{ zeros} = 100,000$

Step 3 Express fraction in lowest terms. $\dfrac{9,375}{100,000} = \dfrac{3}{32}$ Ans

E. SIMPLIFIED METHODS OF DETERMINING DECIMAL OR FRACTIONAL EQUIVALENTS

Tables of Decimal and Fractional Equivalents (*figure* 16-1) are usually found in all trades, drafting rooms, production shops, and toolrooms. These tables are printed in many forms: as enlarged wall charts, as reference tables in handbooks and kits, and as handy-guide plastic cards. The tables are used extensively for determining either the decimal or fractional equivalent of a given value. The simplified use of the charts is preferred to the longhand method of dividing. The charts ensure accuracy and speed up the process of computation.

Selecting Decimal or Fractional Values

The fractions on these charts are given in steps of $\frac{1}{64}''$ in one column with the corresponding decimal equivalents in another. Most tables are carried out to six places. Decimal fractions may be rounded off to any desired degree of accuracy. The decimal equivalent is found by locating the given fraction in the left-hand column. The equivalent value is located in the decimal column. Reverse the practice for finding fractional equivalents of decimals.

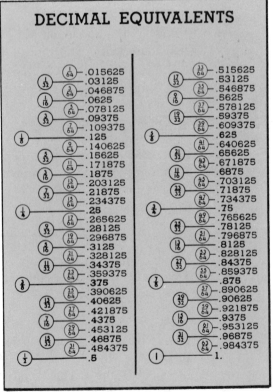

FIGURE 16-1

Shop Method of Determining Value

Even without the charts, most good mechanics can quickly determine the decimal equivalents of all fractions by increments of $\frac{1}{64}''$. In addition to knowing the decimal equivalents of eighths and quarters, many craftspeople also learn that the decimal equivalent of $\frac{1}{64}''$ is .015625, that of $\frac{1}{32}''$ is .03125'', and $\frac{1}{16}''$ is .0625''.

By thinking of the required fractions as being equal to or so much smaller or larger than a known decimal value, it is comparatively easy to determine decimal equivalents of fractions. For example, assume that the decimal equivalent of $\frac{29}{32}$ is needed. The mechanic usually thinks of that fraction as a "thirty-second over seven-eighths." It would be equal to .875 plus $\frac{1}{32}$ (.03125) or (.90625). This same line of reasoning provides a shortcut in determining decimal equivalents of most fractions.

Review and Self-Test

ASSIGNMENT UNIT 16

A. Division of Decimals (General Applications)

1. Divide.
   a. 11.8 by 100
   b. 23.7 by 1,000
   c. 10 by 2.5
   d. 104.26 ÷ 26
   e. 7.5 ÷ 22.5
   f. 26.0313 ÷ 10.25

2. Divide and round off each answer correct to the number of decimal places indicated.
   a. .875 ÷ 6.25 (one place)
   b. 2.234 ÷ 24.63 (three places)
   c. .4375 ÷ 156.25 (three places)
   d. 145.26 ÷ 13.750 (two places)

B. Reduction of Common Fractions and Mixed Numbers to Decimals (General Applications)

1. Compute the three-place decimal fraction equivalent of each of the following.
   a. $\frac{3}{4}$
   b. $\frac{5}{8}$
   c. $\frac{9}{16}$
   d. $\frac{1}{6}$
   e. $\frac{7}{72}$
   f. $\frac{13}{32}$
   g. $\frac{29}{64}$

C. Determination of Decimal and Fractional Sizes by Table

1. Find the decimal value of each fractional size drill by using a decimal equivalents table.
   a. $\frac{1}{2}''$
   b. $\frac{3}{4}''$
   c. $\frac{3}{8}''$
   d. $\frac{5}{16}''$
   e. $\frac{7}{32}''$
   f. $\frac{19}{64}''$

2. Find the fractional equivalent to each decimal by using a decimal equivalents table.
   a. .250''   b. .875''   c. .5625''   d. .6875''   e. .71875''   f. .046875''

D. Division of Decimals (Practical Problems)

1. Determine the depth of cut for each tooth on broaches A, B, C, D, and E.

Part	Depth to be Broached (in inches)	Teeth in Broach
A	.1	10
B	.126	42
C	.255	150
D	.063	30
E	.0924	42

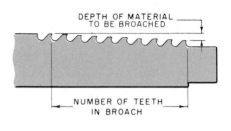

DEPTH OF MATERIAL TO BE BROACHED

NUMBER OF TEETH IN BROACH

2. Determine how many spacing collars are needed for sawing operations A, B, C, D, and E. (All dimensions are given in inches.)

Operation	Thickness of Collars	Spacing Required
A	.5	4.5
B	.25	3.5
C	.125	2.375
D	.1875	1.875
E	.09375	1.125

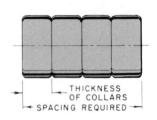

THICKNESS OF COLLARS
SPACING REQUIRED

3. In a production operation, pins A, B, C, D, and E are cut off to the lengths indicated. Determine the whole number of pieces that can be cut from the workable lengths of stock given. (All dimensions are given in inches.)

Pins	Workable Length	Length of Pin	Allowance for Cutoff and Facing
A	10	$\frac{1}{2}$	.120
B	33	$\frac{3}{4}$	.120
C	68	$1\frac{1}{4}$	.125
D	68	$2\frac{1}{8}$	.125
E	11	$\frac{17}{32}$	.094

LENGTH OF STOCK THAT CAN BE USED (WORKABLE LENGTH)
ADDITIONAL STOCK FOR CHUCKING

4. A power sewing machine is set at eight stitches per inch of seam.
   a. Determine the length of one stitch.
   b. Calculate the number of stitches in a seam that is 20.25″ long.

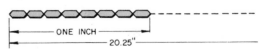

ONE INCH
20.25″

5. Express the decimal quantities of food products A, B, C, and D.

Food Product	Contents	
	Decimal Value	Fractional Equivalent
A	14.437 oz.	
B	25.4 g	
C	175.375 qts.	
D	9.56 lbs.	

6. Six wall receptacles are to be equally spaced starting four feet from each wall to the center line of the first receptacle. The distance between walls is 117'-0". Determine the center-to-center distance between each receptacle.

Unit 17 Achievement Review on Decimal Fractions

OBJECTIVES OF THE UNIT

This achievement review serves as an overall test for Section 3. The unit is designed to measure the student's/trainee's ability to
- *Express fractional dimensions and other quantity measurements as decimal fractions.*
- *Apply the appropriate mathematical processes to solve problems that require the addition, subtraction, multiplication, or division of decimals.*

A. THE CONCEPT OF DECIMAL FRACTIONS

1. Write dimensions Ⓐ, Ⓑ, Ⓒ, Ⓓ, and Ⓔ as decimals. Locate each dimension on a steel rule.

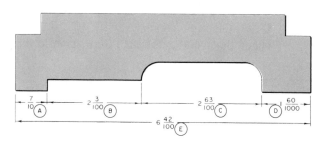

B. ADDITION OF DECIMALS

1. Determine distances Ⓐ, Ⓑ, Ⓒ, Ⓓ, and Ⓔ.

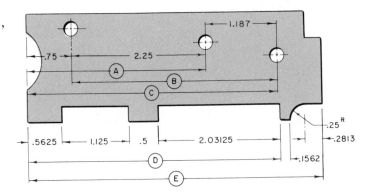

2. Prepare a table similar to the one illustrated and fill in the appropriate columns.
   a. List the values obtained for A, B, C, D, and E in the problem B.1.
   b. Round off each measurement to three decimal places.
   c. Express each decimal as a mixed number.

	Value Obtained	Rounded to Three Places	Expressed as Mixed Number
A			
B			
C			
D			
E			

3. The elevation at the top of the girder is 39.72 feet. Determine the elevation on the surface of the terrazzo floor.

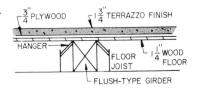

C. SUBTRACTION OF DECIMALS

1. Determine dimensions Ⓐ, Ⓑ, Ⓒ, Ⓓ, and Ⓔ. Recheck the accuracy of each dimension to three places.

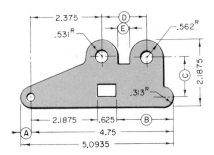

2. Determine dimensions Ⓐ, Ⓑ, Ⓒ, Ⓓ, and Ⓔ correct to three decimal places.

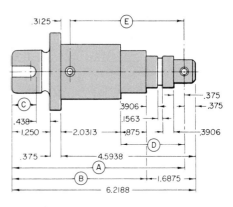

D. MULTIPLICATION OF DECIMALS

1. Determine the overall dimension of the saw and spacing collar combination for each of five setups given in the table (A, B, C, D, and E). Dimensions must be correct to three decimal places.

Setup	Collars		Saws	
	No.	Width	No.	Width
A	4	.5	3	.1
B	8	.25	7	.1
C	7	.25	6	.06
D	10	.187	9	.03
E	12	.1875	11	.0313

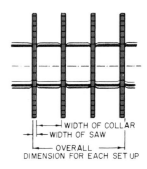

2. A bar of Babbitt weighing 142.5 pounds is composed of 79.5% tin, 11.25% copper, and 9.25% antimony.
   a. Determine the amount of tin, copper, and antimony in the bar.
   b. Determine the cost of the copper and tin if these metals sell for $1.744 and $2.628 per pound, respectively.

E. ADDITION, MULTIPLICATION, AND DIVISION OF DECIMALS

1. Determine the whole number of parts A, B, C, D, and E that can be machined from the workable lengths of stock given. (Each dimension is in inches.) *Note.* Round off the length of each part to two decimal places before determining the number of parts.

Parts	Length of Part	Allowance for		Spoilage	Workable Length
		Cutting Off	Facing		
A	$\frac{1}{2}$	.100	.010	1%	28
B	$\frac{3}{4}$	.100	.010	2.5%	34
C	1.25	.100	.005	4.5%	$69\frac{1}{2}$
D	2.125	.120	.015	$3\frac{1}{4}$%	$69\frac{1}{2}$
E	3.187	.125	.015	$5\frac{1}{2}$%	$69\frac{1}{2}$

2. How many pieces can be stamped from each length of stock for parts A, B, C, D, and E?

Part	Length of Stock (in feet)	Length of Stamping	Allowance on One End
			(in inches)
A	16	$1\frac{1}{10}$	$\frac{2}{10}$
B	$12\frac{1}{2}$	2.25	$\frac{2}{10}$
C	$12\frac{3}{4}$	3.13	.25
D	$25\frac{1}{4}$	3.187	.25
E	$240\frac{3}{4}$	5.125	.375

3. Compute the micrometer readings for the finished diameters of cylinders A, B, and C.

Cylinder		Oversize Bore	Finished Diameter
	Diameter		
A	$3\frac{1}{2}''$	+0.040"	
B	$4\frac{1}{4}''$	+0.0625"	
C	$5\frac{1}{8}''$	+0.0781"	

4. A carbon brush has a 0.016" copper plating on all sides. Find the width and thickness of the carbon brush as illustrated.

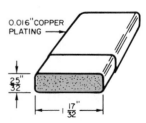

0.016" COPPER PLATING
$\frac{25}{32}''$
$\frac{17}{32}''$

5. Determine the power consumption *(P)* in watts of the transistor shown in the diagram. The quantity of current *(I)* is 0.125 amperes (A); voltage *(E)* across the power transistor is 8.25 V. *Note.* Power equals current multiplied by the voltage.

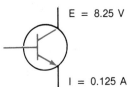

E = 8.25 V

I = 0.125 A

6. Find the voltage *(E)* in a circuit where a transistor draws a current *(I)* of 1.02 amperes and has a power consumption *(P)* of 0.75 watts.
   Note. The voltage equals power divided by current.

7. A new piston is 3.875″ diameter. The rate of wear (reduction of diameter) is 0.0015″ for each 7,500 miles of use.
   a. Determine the total wear at (1) 41,250 miles and again at (2) 76,875 miles. Round the answer to four decimal places.
   b. Use each answer to find the piston diameter at the end of (1) 41,250 miles and (2) 76,875 miles.

SECTION **4**

MEASUREMENT: DIRECT AND COMPUTED (CUSTOMARY UNITS)

Unit 18 Principles of Linear Measure

OBJECTIVES OF THE UNIT

After satisfactorily studying this unit, the student/trainee will be able to

- *Understand how common fractional and decimal units of linear measurement are applied to rules, tapes, and other line-graduated measuring instruments.*
- *Interpret standard and vernier micrometer readings and relate these to precision measurements in the ranges of 0.001" and 0.0001".*
- *Set up combinations of working gage blocks for extremely accurate linear measurements of ± 0.000 004".*

The exchange of goods and services among nations depends on communications. Though languages differ, it is necessary that universally accepted international standards for measurements, technical terms, and data be established. The current worldwide standards movement is toward metrication, using the International System of Units (SI). "Metrication" refers to any program or process of conversion to SI metrics. This means that the continuing development of SI metric units by the International Standards Organization must in turn be adopted by the nations of the world.

The United States is one of the few nations that continues to use a nonmetric system of measurement. Even after the legislative adoption of SI metrics, the system will not be mandated nor will it be the sole system of measurement. Conversion is intended to be voluntary, based on industry-by-industry decision.

The measurements with which each individual must be familiar in order to solve common mathematical problems in business, industry, agriculture, health occupations, and in the home and for other daily activities, include

- Linear measure
- Circular measure
- Area measure
- Volume measure

These measurements are covered in this section by customary units. These units are used to define the quantity of each measurement. The term "customary" refers to the American system of units based on the British system. "Conventional metric" units and the newer SI metrics follow in Part 2. Standardization, in terms of these three major systems, affects the life of each individual.

Direct and Indirect Measurement

Some measurements are taken directly. When measuring tools, weights, instruments, and other line-graduated rules are used and the quantity is read directly, the term "direct measurement" applies. There are other instances where it is impractical or impossible to take direct measurements. In such cases, dimensions are computed, resulting in an *indirect or computed measurement*.

A. CONDITIONS AFFECTING DEGREE OF PRECISION

The British, American, metric, SI metric, and other special systems of measurement all provide for varying degrees of precision. For some applications, a value rounded off to the closest $\frac{1}{100}$ is adequate. In other cases, a dimensional tolerance, accurate to three or more decimal places, is required. Besides dimensional accuracy, precision also is affected by surface finish and temperature. The movement of parts and mechanisms requires differing conditions of finish and measurement precision.

Precision applies equally to indirect measurements. These may relate in the fields of science to light, heat, and sound and nuclear and other energy sources. Precision relates to all industries: foods, health, construction, banking, manufacturing, and services. These, and all other industries, use measurements.

Measuring Tools and Instruments

The greater the degree of precision, the more precise the measuring instruments must be. Higher precision usually involves higher production expenses because of additional processing costs. Where a rough direct measurement within $\frac{1}{16}''$ is required, the dimension may be measured easily with a ruler. If a drawing shows a tolerance of plus or minus $\frac{1}{32}''$ or $\frac{1}{64}''$, a line-graduated steel rule is a practical measuring tool to use.

A micrometer is needed for tolerances within a plus or minus "one-thousandth" range ($\pm 0.001''$). An operation like precision grinding to a tolerance of plus or minus one-ten-thousandth ($0.0001''$) requires a micrometer having vernier graduations reading in ten-thousandths of an inch. Still other parts require that direct measurements be taken to within limits of two-millionths ($0.000002''$) of an inch. Gage blocks are used industrially in combination with other measuring instruments to make direct measurements to this limit.

Common rules, measuring instruments, and accessories are described in this section. The different measurement applications range from a precision of $\frac{1}{16}''$ to two-millionths ($0.000002''$). These experiences are intended to develop skill in applying mathematical principles to direct and computed measurements.

Linear Measurements

Linear measure is the measurement of straight-line distances between two points, lines, or surfaces. In this section, linear measurements are treated in terms of British linear units, which are still the most widely used. The yard is the standard unit of length. The smallest unit of measure in the British system is the inch.

In 1856 England presented the United States with two bronze bars as a standard representation of the yard. The American system of linear measure is based on the British system. The bronze bars are kept for historical significance and not accuracy. The standard for all linear measurements was authorized by law in 1893 as the National Standard of Length.

Table of Linear Measure		
12 inches	=	1 foot
3 feet	=	1 yard
$5\frac{1}{2}$ feet	=	1 rod

Smallest unit of measure = one inch

FIGURE 18-1

A. APPLICATIONS OF RULES, TAPES, AND LINE-GRADUATED TOOLS TO LINEAR MEASUREMENTS

Linear measurements may be made by craftspersons and technicians with solid rules, flexible steel tapes, and other line-graduated measuring instruments. Containers and other vessels are sometimes graduated to permit linear measurements to be taken directly.

Consumers generally use a ruler, yardstick, steel tape, or tape measure. When the tape measure is made of fabric, measurements should be checked for accuracy against a more precise

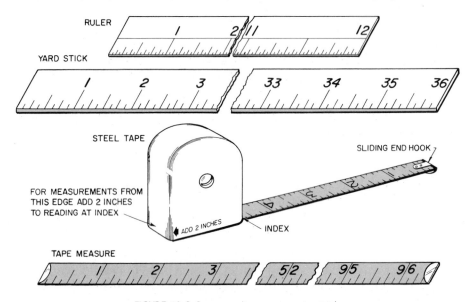

FIGURE 18-2 Consumer linear measuring tools

measuring tool. Line graduations of sixteenths and eights are common on consumer measuring tools.

Standard Unit of Linear Measure

The most commonly used unit of measure is the inch. As a standard unit of linear measure, the inch is subdivided into smaller fractional parts representing either common or decimal fraction equivalents.

The fractional divisions of an inch that are most commonly used on rules represent halves, quarters, eighths, sixteenths, thirty-seconds, and sixty-fourths of an inch.

The decimal system is used when smaller units of measure are required. It is common practice in the shop and laboratory to express fractional parts of an inch in decimals, which are called "decimal equivalents." For example, the decimal equivalent of one-fourth $\frac{1}{4}''$ would be two hundred fifty thousandths (.250").

Measurements up to $\frac{1}{100}''$ may be made directly with a steel rule graduated in fiftieths and hundredths of an inch. The use of such rules reduces the possibility of error that results from changing common fractions to decimals.

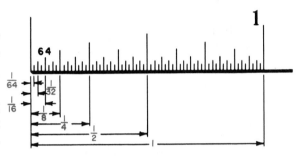

FIGURE 18-3 Enlarged view of fractional parts of an inch

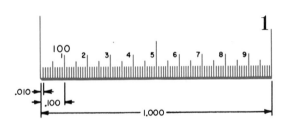

FIGURE 18-4 Enlarged view of decimal fractions

B. APPLICATION OF THE CALIPER TO LINEAR MEASUREMENTS

The outside and inside caliper, while not a graduated measuring tool, is used in combination with the rule to measure linear distances. The caliper is used to transfer linear measurements from the work to the steel rule.

Ordinarily, the smallest measurement that can be taken with a caliper and rule is $\frac{1}{64}''$ in the case of common fractions and $\frac{1}{100}''$ for decimal fractions. Where measurements in terms of thousandths are to be taken, the caliper size is measured with a micrometer.

C. PRINCIPLES OF MICROMETER MEASUREMENT

The standard micrometer is used to measure parts requiring an accuracy of one thousandth. These readings are obtained by turning a graduated thimble on a graduated barrel. The movement of this thimble is at the rate of $\frac{1}{40}''$ per turn. The $\frac{1}{40}''$ is determined by the pitch of the screw threads, which are concealed. As the thimble turns, the spindle moves closer to or further away from the anvil. The anvil is a stationary part of the frame.

The barrel of the micrometer has 40 vertical graduations to indicate this movement of (.025''). Each fourth division on the barrel is marked for ease in reading.

The thimble of the micrometer is divided into 25 equal parts. As each line crosses the horizontal line on the barrel, the space between the anvil and spindle is greater or smaller by $\frac{1}{25}$ of a revolution or (.001''). Each fifth division of the spindle is numbered 5, 10, 15, 20, and 0 or 25. These numbers are in terms of thousandths: .005'', .010'', .015'', .020'', and .025''.

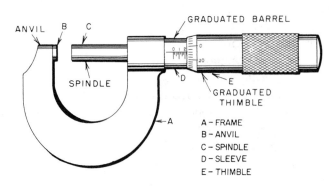

A – FRAME
B – ANVIL
C – SPINDLE
D – SLEEVE
E – THIMBLE

FIGURE 18-5

ONE REVOLUTION OF THIMBLE MOVES THE SPINDLE $\frac{1''}{40}$ OR .025''

FIGURE 18-6

GRADUATIONS ON BARREL OF MICROMETER

ONE DIVISON $= \frac{1''}{40} = .025''$

FIGURE 18-7

THIMBLE DIVIDED INTO 25 EQUAL PARTS

FIGURE 18-8

RULE FOR READING A MICROMETER

- Note the last vertical line that is visible on the barrel.
- Determine its reading with respect to a numbered graduation.
- Add to this reading the number of the line on the sleeve that crosses the horizontal line on the barrel.

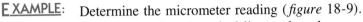

EXAMPLE: Determine the micrometer reading (*figure* 18-9).

    Step 1   Read last numbered vertical line on barrel.

    Step 2   Add .025″ for each additional line that shows (reading on barrel).

    Step 3   Add number of graduation on sleeve (reading on sleeve).

    Step 4   Read the required reading as the sum of all additions.

.400
.025
.017
442″ Ans

FIGURE 18-9

Some micrometer manufacturers recommend that the number of vertical graduations on the barrel be multiplied by 25 to get the reading. Although this may be necessary at first, with practice the value of the graduations can be determined readily. It is also possible to measure accurately to the half-thousandth by splitting the distance between lines on the thimble.

EXAMPLE: The decimal equivalent of $\frac{9″}{16}$ is .5625″. The measurement would indicate a reading of .550″ on the barrel plus .0125″ on the thimble.

BARREL

.013

.012

READING MIDWAY
BETWEEN .013 AND
.012 = .0125

.550

FIGURE 18-10

The micrometer head includes the graduated barrel, graduated sleeve and spindle, and modified frame. The micrometer head has many applications in addition to the micrometer caliper. On internal work the micrometer principle is applied in a measuring tool called an inside micrometer. Examples of other applications are the depth micrometer and height gage attachment for V-blocks.

Micrometer heads of slightly different construction are used widely for accurate machining, measuring, and inspection processes. Regardless of the construction or size, the micrometer principle remains the same. The total measurement is equal to the sum of the reading on the barrel and sleeve.

D. PRINCIPLES OF VERNIER MICROMETER MEASUREMENT

In appearance and construction the vernier micrometer, with which measurements can be taken to .0001″, is identical to the standard micrometer. The only difference is that the vernier micrometer has additional graduations running lengthwise on the barrel.

Readings in one ten-thousandths of an inch are obtained by the vernier principle. Pierre Vernier applied the vernier principle in 1631. The vernier principle is comparatively simple. There are ten graduations on the top of the barrel. These occupy the same space as nine divisions on the thimble.

The difference between the width of one of

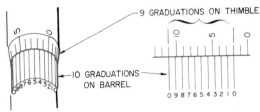

9 GRADUATIONS ON THIMBLE

10 GRADUATIONS ON BARREL

0 9 8 7 6 5 4 3 2 1 0

FIGURE 18-11

the nine spaces on the thimble and one of the ten spaces on the barrel is one-tenth of one space.

Since each space on the thimble represents one-thousandth of an inch, the difference between the graduation on the thimble and barrel is one-tenth of one-thousandth or one ten-thousandth ($\frac{1}{10}$ of $\frac{1}{1,000} = \frac{1}{10,000} = .0001''$).

RULE FOR READING A VERNIER MICROMETER

- Read as a standard micrometer graduated in thousandths.
- Add to this reading the number of the line on the vernier scale of the barrel that coincides with a line on the sleeve. This number gives the ten-thousandths to be added.

EXAMPLE: The micrometer reading in the illustration in thousandths is (.281''). Add to this reading the number of the line on the vernier scale that coincides with a line on the thimble. In this case, the first line coincides. This indicates one ten-thousandth. The second line would be two-tenths of one-thousandth, the third line three-tenths. These tenth readings continue until the zero line on the barrel coincides with a line on the thimble. In this position the fourth-place number in the reading is zero.

To read the vernier micrometer in figure 18-12, correct to four decimal places:

Step 1   Take regular reading.

Step 2   Determine which lines on vernier scale coincide.   **.281**

Step 3   Add regular reading to vernier reading.   **.0001**

.2811''   Ans

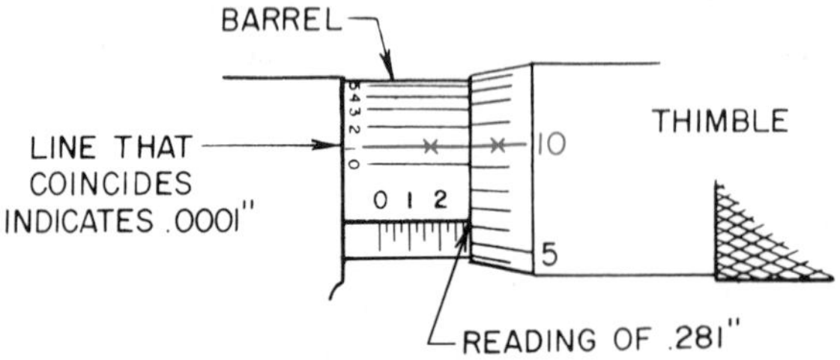

FIGURE 18-12

E. PRINCIPLES OF VERNIER CALIPER MEASUREMENT

The vernier caliper differs from the vernier micrometer in construction and principle of operation. The reading on the vernier caliper is not obtained by any relationship between the pitch of a screw and the movement of a thimble. Instead, the vernier caliper legs are slid into position. They are accurately adjusted for measurement by means of a fine screw that moves a sliding leg on a beam. The measurement is then determined by adding the reading on the beam and a graduation on the vernier scale.

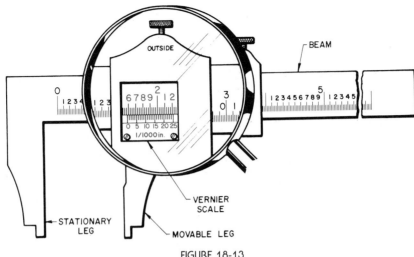

FIGURE 18-13

By varying the design of the stationary or solid leg, the beam can be used for other measuring needs as a height gage and depth gage. In many instances the beam is fitted to the table of a machine and the scale to a stationary part. Accurate linear measurements for machine operations may thus be taken with greater ease and less chance of error than by using the graduated collar on machines.

Each inch on the beam of a vernier caliper is divided into 40 equal parts. The distance between each graduation is $\frac{1}{40}''$ or .025".

The vernier scale has 25 divisions that correspond to 24 divisions on the beam (*figures* 18-14 and 18-15). The difference between one division on the scale and one on the beam is $\frac{1}{25}$ of .025" or .001".

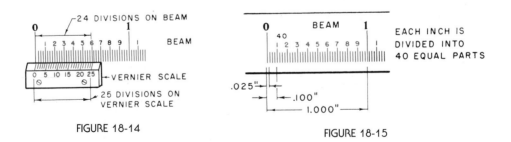

FIGURE 18-14

FIGURE 18-15

If the movable or sliding leg of the caliper (to which the vernier is attached) is moved to the right until the first line on the beam and vernier coincide, the leg will open .001" (*figure* 18-16).

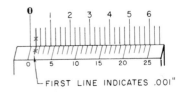

FIGURE 18-16

If this movement continues until the jaw is opened (.025″), the zero line on the vernier scale and the first line on the beam will coincide (*figure* 18-17).

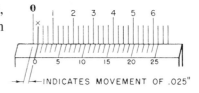

INDICATES MOVEMENT OF .025″

FIGURE 18-17

RULE FOR MEASURING WITH THE VERNIER CALIPER

- Read the graduation on the beam to the left of the zero on the vernier scale.
- Determine the number of the line on the vernier that coincides with a line on the beam.
- Add the reading on the beam to that of the vernier.

EXAMPLE: *Case 1.*

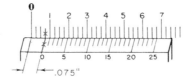

.075″

FIGURE 18-18

Step 1	Read graduation on beam to left of zero on vernier scale.	.075
Step 2	Determine which line on vernier scale coincides with line on beam.	.000
Step 3	Add beam and vernier scale readings.	.075″ Ans

EXAMPLE: *Case 2.*

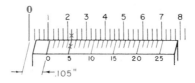

.105″

FIGURE 18-19

Step 1	Read graduation on beam to left of zero on vernier scale.	.100
Step 2	Determine number of line on vernier scale that coincides with line on beam.	.005
Step 3	Add beam and vernier scale readings.	.105″ Ans

EXAMPLE: *Case 3.*

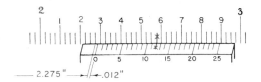

FIGURE 18-20

Step 1	Read graduation on the beam to left of zero on vernier scale (two inches plus .275).	**2.275**
Step 2	Determine number of line on vernier scale that coincides with line of beam.	**.012**
Step 3	Add beam and vernier scale readings.	**2.287″** Ans

F. PRINCIPLES OF PRECISION GAGE BLOCK MEASUREMENT

Gage blocks are hardened rectangular- or square-shaped blocks of steel. Gage blocks have a high quality surface finish and are dimensionally accurate within millionths of an inch.

The dimensional accuracy of a gage block relates to its *size, degree of flatness, parallelism* between the reference faces, and the *quality of* the *surface finish*. In general, the highest quality newer gage blocks are dimensionally accurate to within ±0.000 002″. This high precision level is required for calibrating instruments and other laboratory and scientific work under tempera-ture-controlled conditions.

Gage blocks that are used for regular layout and precise linear measurements in the shop are dimensionally accurate to within ±0.000 004″. Blocks earlier produced are accurate to within ±0.000 008″.

Gage blocks are available in different combinations of sizes and numbers in a *set*. Frac-tional gage blocks range from $\frac{1}{64}$″ to 12.000″ by sixty-fourths. Other gage block sets are produced in increments (steps) of 0.001″, 0.0001″, or 0.000 025″, depending on the required degree of accuracy and range of linear measurements. Each set provides for a tremendous range. For example, it is possible to make more than 80,000 different linear combinations with a 35-gage block set.

In practice, the combination of blocks to use for a specified dimension is determined by the addition and subtraction of decimals.

RULE FOR DETERMINING GAGE BLOCK COMBINATIONS

- Determine from the required dimension what number is in the ten-thousandths col-umn.
- Select the gage block from the series .1001 to .1009 whose fourth-place number is the same as the required dimension.
- Select any gage block from the .001 series whose third-place (thousandths) dimen-sion ends in the same number of thousandths as the required dimension.

Note. Make sure that the sum of the two blocks is not greater than the required dimension.
- Select any gage block from the .01 series whose second-place dimension (hundredths) ends in the same number of hundredths as the right dimension.
- Continue to add blocks until the sum of the gages equals the required dimension.

EXAMPLE: Determine the combination of blocks needed for a measurement of 5.9325″ using the 35-piece set indicated below.

Series	Thicknesses of English Measurement Gage Blocks in Inches								
.0001 Series	.1001	.1002	.1003	.1004	.1005	.1006	.1007	.1008	.1009
.001 Series	.101	.102	.103	.104	.105	.106	.107	.108	.109
.010 Series	.110	.120	.130	.140	.150	.160	.170	.180	.190
.100 Series	.100	.200	.300	.400	.500				
1.000 Series	1.000	2.000	3.000						

FIGURE 18-21

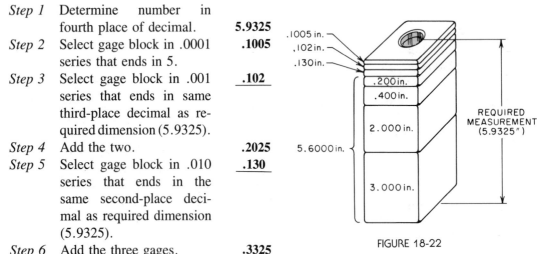

FIGURE 18-22

Step 1 Determine number in fourth place of decimal. **5.9325**

Step 2 Select gage block in .0001 series that ends in 5. **.1005**

Step 3 Select gage block in .001 series that ends in same third-place decimal as required dimension (5.9325). **.102**

Step 4 Add the two. **.2025**

Step 5 Select gage block in .010 series that ends in the same second-place decimal as required dimension (5.9325). **.130**

Step 6 Add the three gages. **.3325**

Step 7 Subtract the decimal value of the gages (.3325″) from the required dimension (5.9325″). = 5.600″

Step 8 Select the smallest number of blocks in the .100 and 1.000 series to obtain this number. 5.600″ = 3.000″ + 2.000″ + .400″ + .200″

Step 9 Check the gage block combination for accuracy. Also, determine whether or not a combination having fewer blocks may be used.

Review and Self-Test

ASSIGNMENT UNIT 18

A. Direct Measurement (Practical Problems)

1. Determine the reading of each measurement.

a.

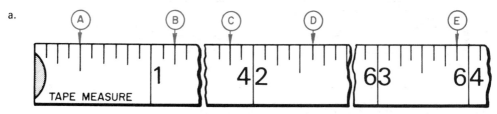

b.

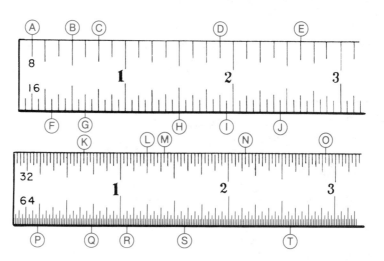

2. Determine the reading of each measurement indicated on the rule, which is graduated in 10ths, 50ths, and 100ths of an inch.

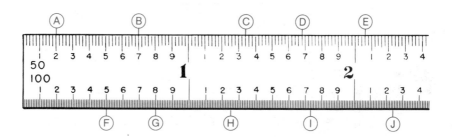

3. Measure the length of lines (a) through (j) to the degree of accuracy indicated in each case.

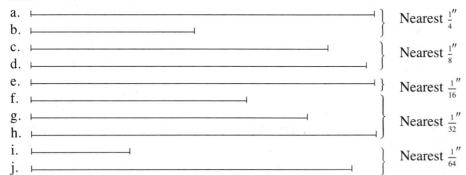

a. $\qquad\}$ Nearest $\frac{1}{4}''$
b. $\qquad$

c. $\qquad\}$ Nearest $\frac{1}{8}''$
d. $\qquad$

e. $\qquad\}$ Nearest $\frac{1}{16}''$
f. $\qquad$

g. $\qquad\}$ Nearest $\frac{1}{32}''$
h. $\qquad$

i. $\qquad\}$ Nearest $\frac{1}{64}''$
j. $\qquad$

4. Measure lengths Ⓐ, Ⓑ, Ⓒ, and Ⓓ and check the sum of these against the overall dimension Ⓔ.

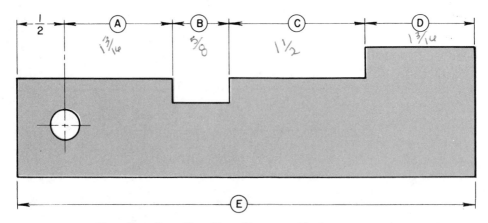

5. Measure lengths Ⓐ, Ⓑ, Ⓒ, Ⓓ, Ⓔ, Ⓕ, and Ⓖ and check the sum against the overall dimension Ⓗ.

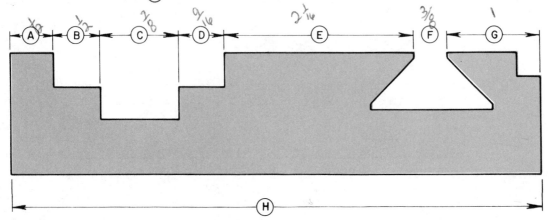

6. Measure the outside diameters of the metal bars (A), (B), (C), (D), and (E) with a caliper. Transfer the measurement to a rule and record the diameter to the nearest $\frac{1}{50}''$.

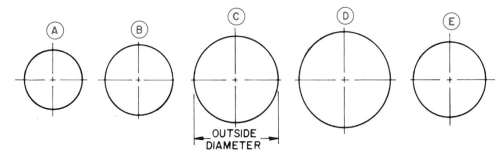

7. Measure the length of lines (a) through (d) to the degree of accuracy indicated in each case.

a. ⊢————————————————————⊣ } Nearest $\frac{1}{10}''$
b. ⊢——————⊣
c. ⊢———————————————⊣ } Nearest $\frac{1}{50}''$
d. ⊢———————⊣

8. Measure the diameter of bored holes (A), (B), (C), (D), (E), and (F) with an inside caliper. Then transfer the measurement to a rule and record the diameter to the nearest $\frac{1}{32}''$.

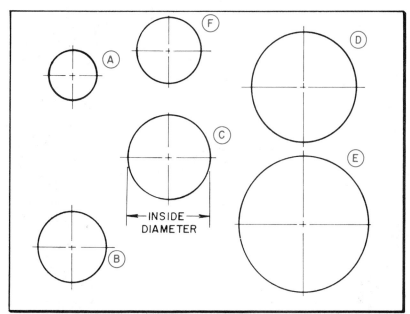

B. Indirect Measurement (Practical Problems)

1. Determine dimensions (A), (B), (C), (D), (E), and (F).

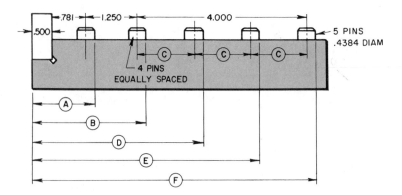

2. Find dimensions (A) through (J) to which the shaft must be rough turned. Allow $\frac{1}{16}''$ on all diameters and $\frac{1}{32}''$ on all faces for finish machining.

3. Calculate dimensions (A) through (J) to which the part must be finish turned before grinding. Allow .010″ on all diameters and .008″ on all faces for grinding.

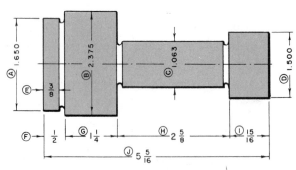

4. Determine (a) the outside perimeter of the house and (b) the inside perimeter of each of the six rooms. The perimeter equals the sum of the linear measurements of each side of an object.

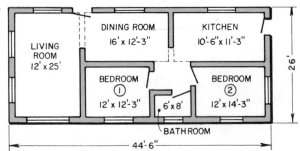

C. Direct Measurement: Standard Micrometer (Practical Problems)

1. Determine the linear dimension indicated at A, B, C, D, E, and F on the standard micrometer.

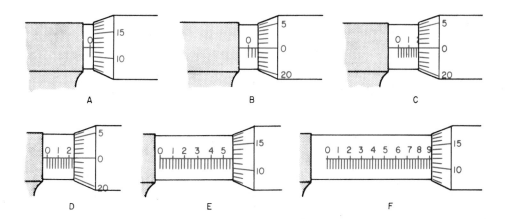

D. Direct Measurement: Vernier Micrometer

1. Determine the linear dimension on the vernier micrometer settings at A, B, and C.

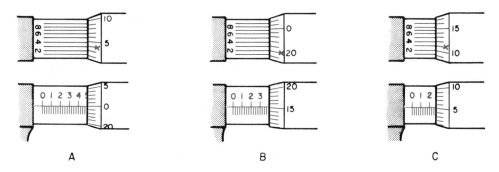

E. Direct Measurement: Vernier Caliper (Practical Problems)

1. Determine the vernier caliper readings A, B, C, and D.

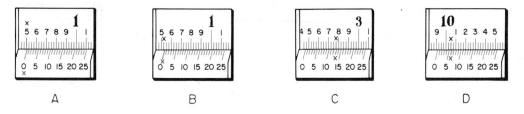

F. Direct and Computed Measurement: Gage Blocks (Practical Problems)

1. The set of 81 gage blocks may be used to make 120,000 accurate combinations of measurements in steps of .0001″ from .200 ″ to over 24″.

Series	Thicknesses of Gage Blocks (in inches)									
.0001 Series	.1001	.1002	.1003	.1004	.1005	.1006	.1007	.1008	.1009	
	.101	.102	.103	.104	.105	.106	.107	.108	.109	
	.110	.111	.112	.113	.114	.115	.116	.117	.118	.119
.001 Series	.120	.121	.122	.123	.124	.125	.126	.127	.128	.129
	.130	.131	.132	.133	.134	.135	.136	.137	.138	.139
	.140	.141	.142	.143	.144	.145	.146	.147	.148	.149
.050 Series	.050	.100	.150	.200	.250	.300	.350	.400	.450	
	.500	.550	.600	.650	.700	.750	.800	.850	.900	.950
1.000 Series	1.000	2.000	3.000	4.000						

Determine the sizes of gage blocks that, when combined, will give the required dimensions A through J.

A	.500
B	.750
C	.875
D	.265

E	.6001
F	.7507
G	1.2493

H	2.2008
I	11.0049
J	11.885

Unit 19 Principles of Angular and Circular Measure

OBJECTIVES OF THE UNIT

After satisfactorily studying this unit, the student/trainee will be able to

- Interpret parts and features of circles and angles in relation to computing circular and angular measurements.
- Express angular measurements in terms of degrees, minutes, and seconds.
- Compute the circular length of an arc.
- Measure and lay out angles by using a flat semicircular protractor or a more accurate bevel protractor.

The measurement of circles, curved surfaces, cylinders, and angles takes place daily in the home, in business, and in industry. Measuring these rounded surfaces calls for an understanding of the fundamental principles relating to circular and angular measurement.

Some problems require the application of a principle to compute a missing value or dimension. Under actual conditions, measurements are made directly on the job, using tools and instruments that will measure to the required degree of accuracy.

This unit gives step-by-step procedures for computing answers (where basic principles alone are applied) and for direct measurement with tools and instruments.

A. DEVELOPING A CONCEPT OF CIRCULAR AND ANGULAR MEASURE

Defining the Circle and Its Parts

A circle is defined as a closed, curved line on a flat surface. Every point on the closed, curved line is the same distance from a fixed given point called a *center*. The distance around the circle, or the periphery of the circle, is the *circumference*. This circumference is measured either in terms of the standard units of linear measure or in degrees.

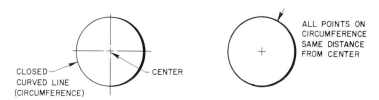

FIGURE 19-1

The *diameter* is the straight line through the center of the circle. Its ends terminate in the circumference. The diameter divides the circle into two equal half circles called *semicircles*. The *radius* is a straight line starting at the center of the circle and terminating in the circumference. The symbol *(R)* or *(r)* is usually used to indicate radius. The term *radial* is in common usage. A line or surface on a circular object is radial if, when extended, it cuts through the center.

In all circles, the circumference is related to the diameter in a specific way. It is approximately $3\frac{1}{7}$ or 3.1416 times the diameter. This value is a constant relationship between the diameter and circumference of the same circle. For convenience it is called ''pi'' or π.

To find the circumference of a circle, multiply the diameter by the value of π.

On many objects, only a *segment* of a circle is used. The segment refers to the area of that portion of the circle that is being considered. The segment consists of an *arc* on the circumference whose ends are joined by a straight line in the circle.

An *arc* is a curved part of the circumference. The straight line that extends inside the circle to join both ends of the arc is known as a *chord*.

 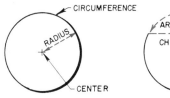

FIGURE 19-2

Two other terms are widely used: *concentric circles* and *eccentric circles*. When two or more circles have a common center, they are *concentric*. Circles that are on the same flat surface and are used in the same part but do not originate from a common center are called *eccentric* circles. On concentric circles the inner circle is sometimes designated by the letters *(ID)* for inside diameter; the outer circle is designated by the letters *(OD)* for outside diameter.

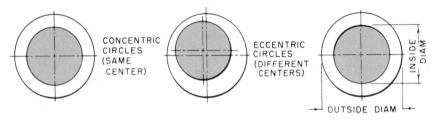

FIGURE 19-3

Defining the Angle and Its Parts

Most measurements of round surfaces are in linear measure. However, many dimensions are expressed in angular measure. Angular measure indicates the size of an opening formed by two lines or surfaces (intersecting lines). The lines or surfaces open as they extend from a common starting point called a *vertex*. The angles formed are measured in *degrees, minutes,* and *seconds*.

The size of an angle depends upon the space between the sides and not on the length of the sides. In speaking of an angle, it may be referred to as ∠ABC, or ∠1, or ∠A and the angle is marked or lettered accordingly.

A circle is divided into 360 equal parts that are called degrees. An angle of one degree is formed by drawing two lines from the center to two consecutive points that cut $\frac{1}{360}$ of the circumference.

Degrees are designated by placing a small symbol to the right and slightly above the number: thus, 4 degrees is 4°. Sometimes degrees are given in tenths. For example, four and three-tenths degrees is written 4.3°.

FIGURE 19-4

Each degree is divided into 60 equal parts, each of which is called a *minute*. A degree therefore equals 60 minutes. Minutes are indicated by placing a single line in the same relative position as the degree sign. Thus, 15 minutes is written 15′.

FIGURE 19-5

For greater accuracy, measurements are made in divisions finer than the minute. Each minute is divided into 60 equal parts, each of which is called a *second*. Seconds are written with two lines to the right and slightly above the number. An angle ending in 25 seconds is written 25″.

To summarize, angles are measured in degrees (°), minutes (′), and seconds (″). There are 360 degrees in a circle, 60 minutes in a degree, and 60 seconds in a minute (*figure* 19-6).

The circle is very often divided into four equal parts, sometimes called quadrants, each of which represents an angle of 90°. The term *right angle* is used to denote such an angle. When the circle is divided into two equal parts, the 180° for each part is called a *straight angle* (*figure* 19-7).

Table of Angular Measure	
1 Circumference (or circle)	= 360 degrees
1 Degree	= 60 minutes
1 Minute	= 60 seconds

FIGURE 19-6

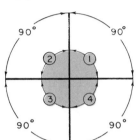

THE CIRCLE DIVIDED INTO 4 EQUAL PARTS OF 90° OR 4 RIGHT ANGLES OR QUADRANTS ①, ②, ③ AND ④

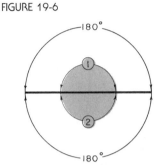

THE CIRCLE DIVIDED INTO 2 EQUAL PARTS OF 180° OR 2 STRAIGHT ANGLES ① AND ②

FIGURE 19-7

B. COMPUTING ANGULAR MEASURE

Problems in angular measure involve the four basic mathematical processes of addition, subtraction, multiplication, and division of angles and parts of angles. To do these operations, it is necessary to change degrees, minutes, and seconds to whichever one of the three units will expedite the mathematical processes involved. Angles given in fractional or decimal parts of a degree, minute, or second may be added, subtracted, multiplied, or divided the same as any fraction or decimal.

RULE FOR EXPRESSING DEGREES IN MINUTES

- Multiply the number of degrees by 60.
- Express the product in terms of minutes.

EXAMPLE: Express 15 degrees (15°) in minutes.
Step 1 Multiply 15 degrees (15°) by 60. $15 \times 60 = 900$
Step 2 Express answer in minutes (′). $15° = 900′$ Ans

RULE FOR EXPRESSING MINUTES IN SECONDS

- Multiply the number of minutes by 60.
- Express the product in terms of seconds.

EXAMPLE: Change 32 minutes (32′) to seconds.

Step 1   Multiply 32 minutes (32′) by 60.   **32 × 60 = 1920**

Step 2   Express answer in seconds (″).   **32′ = 1920″   Ans**

RULE FOR EXPRESSING AN ANGLE IN DEGREES AND MINUTES AS SECONDS

- Multiply the given number of degrees by 60.
  Note. The product denotes minutes.
- Multiply the product by 60.
  Note. The product denotes seconds in the given number of degrees.
- Multiply the given number of minutes by 60.
- Add the number of seconds in the given number of degrees to the number of seconds in the given number of minutes.
- Express the answer in terms of seconds.

EXAMPLE: *Case 1*. Change 3°10′ to seconds.   **3 × 60 = 180′**

Step 1   Multiply 3° by 60.   **180 × 60 = 10,800″**

Step 2   Multiply 180′ by 60.   **10′ × 60 = 600″**

Step 3   Express ten minutes 10′ as seconds.   **3° = 10,800″**

Step 4   Add values of 3° and 10′ in seconds.   **+ 10′ = ___600″**

                                                     11,400″   Ans

EXAMPLE: *Case 2*. Express 2.75° as minutes.

Step 1   Multiply 2.75° by 60.   **2.75 × 60 = 165.00**

Step 2   Locate the decimal point.   **2.75° = 165′   Ans**

EXAMPLE: *Case 3*. Express 7.12° as minutes and seconds.   **7.12 × 60 = 427.20**

Step 1   Multiply 7.12° by 60.   **.20 × 60 = 12.00**

Step 2   Point off decimal places.   **7.12° = 427′12″   Ans**

        Note. The whole number of minutes is 427. The remainder (.20′) must be changed to seconds.

Step 3   Multiply the decimal part of the minutes (.20) by 60 and locate the decimal point.

Step 4   Combine the number of minutes and seconds.

RULE FOR EXPRESSING MINUTES AS DEGREES

- Divide the number of minutes by 60.
- Express the quotient in terms of degrees.
  Note. Seconds are changed to minutes in the same way.

E XAMPLE: Express 45′ as degrees.
  Step 1 Divide the minutes (45′) by 60.
  Step 2 Express the quotient as degrees $\left(\frac{30}{4}\right)$. $\frac{45}{60} = \frac{30}{4}$ Ans
    Note. This quotient may also be written as (.75°).

C. COMPUTING CIRCULAR LENGTH

There are many cases where the length of an arc must be computed. This length is equal to the length of a portion of the circumference that is included in a given angle.

RULE FOR MEASURING THE LENGTH OF AN ARC (CIRCULAR LENGTH)

- Determine the number of degrees in the arc.
- Determine the diameter of the circle.
- Place the number of degrees in the angle over the number of degrees in the circle (360°) to determine what part the included angle is of the whole circle.
- Multiply this quantity by the circumference of the circle.
  Note. The circumference is equal to its diameter multiplied by 3.1416. The result is in the same dimension as the diameter.

E XAMPLE: What is the length of an arc included in an angle of 45° when the diameter of the circle is 2 inches?
  Step 1 Determine size of angle (45°) and diameter of circle (2″).
  Step 2 Place number of degrees in angle (45°) over the number of degrees in circle (360°). $\frac{45}{360} = \frac{1}{8}$
  Step 3 Compute circumference of (2″) circle.   **Circumference =**
  Step 4 Multiply the circumference (6.2832) by the fractional part of circumference in the included angle $\left(\frac{1}{8}\right)$.   **2 × 3.1416 = 6.2832**

    6.2832 × $\frac{1}{8}$ = .7854
  Step 5 The product is the required length of the arc in inches.   .7854″ Ans

D. APPLICATION OF THE PROTRACTOR TO ANGULAR MEASUREMENTS

Three general types of instruments, called *protractors,* are used to measure and lay out angles. The simplest type of protractor used for comparatively rough work is flat, semicircular in shape, and graduated in degrees. Each tenth division is usually marked for ease in reading. The semicircular protractor finds wide application in laying out and measuring simple angles.

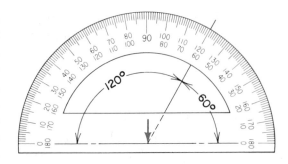

FIGURE 19-8

In the shop and on the job, the bevel protractor is the one most widely used. This protractor can be adjusted accurately to lay out or measure any angle from 0° to 180° by degrees. It is possible to estimate fairly accurately to a half degree or 30'. The number of degrees is read directly from the graduations on a movable turret. The swinging blade protractor is sometimes more convenient to use because of its simplicity of construction. Here again, the degrees from 0° to 180° are read at a center mark on one of the blades.

The third type of angle measuring tool is the universal vernier bevel protractor. This instrument is equipped with a vernier scale for very accurate measurement and layout of angles in terms of degrees and minutes.

Review and Self-Test

ASSIGNMENT UNIT 19

A. Determining Equivalent Values of Angles (General Applications)

1. How many degrees are there in the following parts of a circle?
   a. $\frac{1}{6}$ b. $\frac{2}{9}$ c. $\frac{3}{8}$ d. $\frac{5}{36}$ e. $\frac{7}{24}$
2. What part of a circle is each of the following angles?
   a. 40°   b. 108°   c. 9°   d. 54°   e. 96°
3. Change each angular measurement (a) through (h) to its equivalent value in the unit of measure indicated.

   a. 30' to degrees
   b. 45' to degrees
   c. 75' to degrees
   d. 5° to minutes

   e. $7\frac{1}{2}°$ to minutes
   f. 1.4° to minutes
   g. 5' to seconds
   h. $5\frac{1}{2}'$ to seconds

B. Computation of Circular and Angular Measurements (General Applications)

1. Multiply and reduce each result to degrees, minutes, and seconds in whichever combination is needed.

   a. $5° \times 6$
   b. $17° \times 15$
   c. $12\frac{1}{2}° \times 8$

   d. $12.75' \times 6$
   e. $9° \ 12' \times 9$
   f. $23° \ 16' \ 20'' \times 4$

2. Divide each angular measurement.

   a. $180° \div 9$
   b. $135° \div 10$
   c. $120° \div 9$

   d. $90° \ 30' \div 6$
   e. $75.8° \div 10$
   f. $144° \ 24' \ 48'' \div 12$

C. Measurement of Angles: Direct and Computed (Practical Problems)

1. Read angles (A) through (H) on the semicircular protractor.

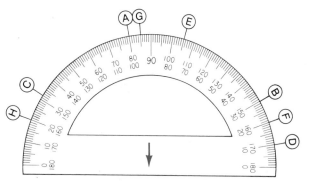

2. Measure angles (A) through (F) with a flat semicircular protractor.
   Note. It may be necessary to extend the sides of the angles to measure them.

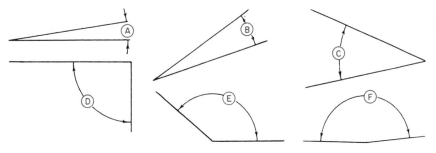

3. Lay out angles (a) through (d) with a semicircular protractor.
   a. $10°$ b. $25°$ c. $100°$ d. $120°$

4. The bevel protractor readings A, B, and C appear on one make of instrument; those at D, E, and F on another. Determine the reading for each setting.

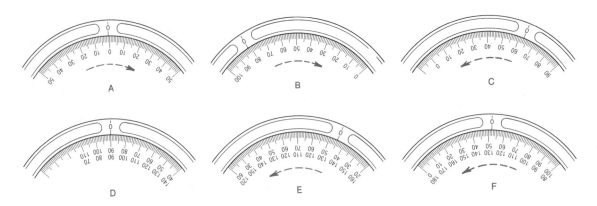

5. The table and the illustration relate to four engine strokes. The angular valve opening/ closing or movement of the crankshaft are given. Compute the missing angular (X°) measurements for strokes A, B, C, and D.

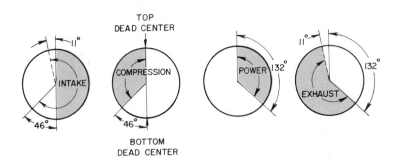

Stroke		Exhaust Valve		Angular Movement of Crankshaft
		Opening	Closing	
A	Intake	11° before top dead center	46° after top dead center	(X°)
B	Compression	top dead center	46° past bottom dead center	(X°)
C	Power	top dead center	(X°) before bottom dead center	132°
D	Exhaust	(X°) end of power stroke	11° before top dead center	132°

D. Computation of Circular Length (Practical Problems)

1. Compute the length of arcs A through D according to the given diameter for each part.

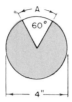  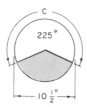 

2. The diameters of three roof plates are (a) 7'-0" (b) 8'-6", and (c) 29'-1", respectively. Use π = 3.1416. Give the length (circumference) of each circular roof plate to the nearest inch.

3. Find the semicircular length of forms A, B, and C. State each length in feet and inches.

Form	Diameter	Semicircular Length
A	3'-0"	
B	$4\frac{1}{3}'$	
C	5.25'	

Unit 20 Principles of Surface Measure

OBJECTIVES OF THE UNIT

After satisfactorily studying this unit, the student/trainee will be able to
- *Understand the concept of square and surface measure.*
- *Change one unit value of surface measure to another unit.*
- *Express larger units of surface measure in terms of smaller units.*
- *Solve general and practical problems relating to finding the areas of rectangles, parallelograms, trapezoids, triangles, circles and sectors, and cylinders.*

The term *surface measure* refers to the measurement of an object or part that has length and height. There are eight common objects or shapes for which area or surface measure must be computed. The list includes surfaces that are defined by pairs of lines like the square, rectangle, and parallelogram; by four lines like the trapezoid; by three lines like the triangle; then the circle and sector; and finally, the cylinder. The characteristics by which each of these shapes is recognized and defined are covered separately in this unit.

A. DEVELOPING A CONCEPT OF SQUARE OR SURFACE MEASURE

A surface is any figure that has length and height but no thickness. To measure a surface, its length and height must be in the same unit before they are multiplied. The result of this mathematical process is called the *area* of the surface. The area is expressed in square units of the same kind as the linear units. For example, if the length and height of an object are given in inches, the area will be in square inches. In this case, the surface contains a number of square inches. One square inch is the area a square figure measures that is one linear inch long and one linear inch high.

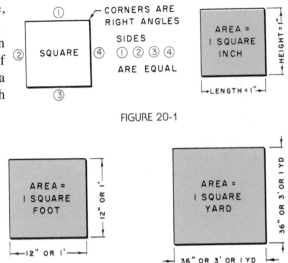

FIGURE 20-1

One square foot is the area a square figure measures that is 12 linear inches long and 12 linear inches high. One square yard is the area a square figure measures that is 36 linear inches long and 36 linear inches high.

FIGURE 20-2

In each case, the area of the square is found by multiplying the linear length by the linear height. Both dimensions must be in the same unit of measure.

By comparison, the area of a square one inch on a side is 1 square inch. The area of a square 12 inches on a side is 144 square inches. The area of a square 36 inches on a side is 1,296 square inches.

FIGURE 20-3

The area of the second square illustrated in figure 20-3 is 144 square inches. Since this value is obtained by multiplying the length and height (each of which is equal to one foot), the area may be given as 1 square foot. The square foot is, therefore, equivalent to 144 square inches.

By the same reasoning, in the third square in figure 20-3 the 36 inches = 3 feet, which equals 1 yard. The area of the square, therefore, is equal to 1,296 square inches, or 9 square feet, or 1 square yard.

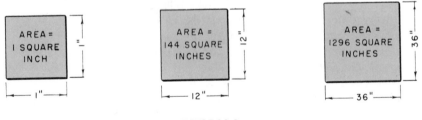

STANDARD UNIT OF SURFACE MEASURE

1 SQ IN.

1 SQ FT = 144 SQ IN.

1 SQ YD = 9 SQ FT = 1,296 SQ IN.

FIGURE 20-4

In place of continually writing the word "square" and the unit of measure, the symbol □ is used to indicate *square*. This symbol is followed by other symbols for inches (□″), feet (□′), or yards (□ yd). Another way of writing areas is to abbreviate the word square (sq) and the units of measure: (in.) for inches, (ft) for feet, and (yd) for yard. One square inch may be written 1 sq in. or 1 □″; one square foot, 1 sq ft or 1 □′; and one square yard, 1 sq yd or 1 □ yd.

The three common units of surface measure that find daily application are the square inch, the square foot, and the square yard. The value of each of these units is shown in figure 20-5.

Table of Surface Measure

Unit of Surface Measure	= 1 sq in.
144 sq in.	= 1 sq ft
9 sq ft	= 1 sq yd

FIGURE 20-5

B. CHANGING FROM ONE UNIT VALUE OF SURFACE MEASURE TO ANOTHER

A value expressed in square inches may be changed to square feet by dividing by 144 (144 sq in. = 1 sq ft). By the same process, an area given in square feet may be changed to square yards by dividing by 9 (9 sq ft = 1 sq yd). If an area is given in inches, the value may be changed to square yards or a fractional part by dividing by 1,296 (1,296 sq in. = 1 sq yd).

RULE FOR EXPRESSING A UNIT OF SURFACE MEASURE AS A LARGER UNIT

- Divide the given area by the number of square units contained in one of the required larger units.
- Express the quotient in terms of the required larger unit.

EXAMPLE: *Case 1.* Express 288 sq in. in sq ft.
Step 1 Divide given area by the number of sq in. in a sq ft (144). $\frac{288}{144} = 2$
Step 2 Express quotient (2) in terms of required unit (sq ft). **288 sq in. = 2 sq ft   Ans**

EXAMPLE: *Case 2.* Express 27 sq ft in sq yd.
Step 1 Divide given area by the number of sq ft in a sq yd (9). $\frac{27}{9} = 3$
Step 2 Express quotient (3) in terms of required unit (sq yd). **27 sq ft = 3 sq yd   Ans**

EXAMPLE: *Case 3.* Express 2,592 sq in. in sq yd.
Step 1 Divide the given area by the number of sq in. in a sq yd (1,296). $\frac{2592}{1296} = 2$
Step 2 Express quotient (2) in terms of required unit. **2,592 sq in. = 2 sq yd   Ans**

When the given area cannot be divided exactly, then the quotient may be expressed in more than one unit. For example, if 12 square feet is changed to square yards, the result may be given as $1\frac{1}{3}$ square yards or as 1 square yard and 3 square feet. In other words, the fraction $\left(\frac{1}{3}\right)$ is expressed in terms of a smaller unit.

RULE FOR EXPRESSING LARGER UNITS AS SMALLER UNITS OF SURFACE MEASURE

- Multiply the given unit by the number of smaller units contained in one of the required units.
- Express the product in terms of the required smaller unit.

EXAMPLE: *Case 1.* Express 5 sq ft in sq in.

Step 1   Multiply given unit (5) by the number of smaller units contained in one of the required units (144).

$5 \times 144 = 720$

Step 2   Express the product (720) in terms of the required unit (sq in.).

5 sq ft = 720 sq in.   Ans

EXAMPLE: *Case 2.* Express 3 sq yd in sq in.

Step 1   Multiply given unit (3) by the number of smaller units contained in one of the required units (1,296).

$3 \times 1,296 = 3,888$

Step 2   Express the product (3,888) in terms of the required unit (sq in.).

3 sq yds = 3,888 sq in.   Ans

An area given in terms of two or more units (2 square feet 9 square inches) may be changed in value to a smaller unit. Only that portion of the area that is not in terms of the required unit is multiplied by the number of smaller units contained in one of the given units. The remainder of the given area is added to this product.

EXAMPLE: Express 2 sq ft 9 sq in. in sq in.

Step 1   Multiply only that portion of the given area that is not in terms of the required unit. Multiply by the number of smaller units contained in one of the given units.

2 sq ft = 2 × 144 (sq in. in 1 sq ft) = 288

Step 2   Express the product in terms of the required unit (sq in.).

= 288 sq in.

Step 3   Add the remainder of the given unit (9 sq in.).

288 + 9 = 297 sq in.   Ans

C. APPLYING SURFACE MEASURE TO THE SQUARE AND RECTANGLE

The area of a square surface may be found by multiplying the length by the height. Since both of these dimensions are equal, the area is equal to the side multiplied by itself.

The term *rectangle* refers to a surface whose opposite sides are parallel. The adjacent sides are at right angles to each other. The area of this surface is the number of square units that it contains. In place of adding the number of square units contained in both the height and length, the process is simplified by multiplication.

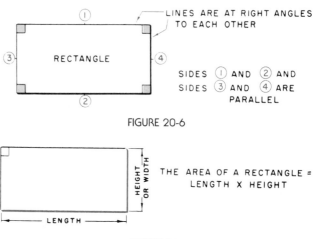

FIGURE 20-6

FIGURE 20-7

RULE FOR FINDING THE AREA OF A RECTANGLE
_____

- Express the dimensions for length and width (sometimes called *height*) in the same linear unit of measure.
- Multiply the length by the height.
- Express the product in units of surface measure.
- Express product in lowest terms, if required.

_____

EXAMPLE: *Case 1.* Determine the area of a rectangle 9″ long by 3″ high.

Step 1 Multiply length by height. $9 \times 3 = 27$

Step 2 Express product (27) in terms of units of surface measure.   **27 sq in.** Ans

EXAMPLE: *Case 2.* Determine the area of a rectangular surface 3′ long and 10″ wide.

Step 1 Express dimensions in same unit of linear measure. $3' = 36''$

Step 2 Multiply length (36″) by height (10″). $36 \times 10 = 360$ **sq in.**

Step 3 Express product (360) in units of surface measure.

Step 4 Express (360 sq in.) in lowest terms by dividing by (144). $\frac{360}{144} = $ **2 sq ft and 72 sq in.** Ans

D. APPLYING SURFACE MEASURE TO THE PARALLELOGRAM

A *parallelogram* has two pairs of sides that are parallel to each other. The sides are not necessarily at right angles as in the case of the rectangle. The parallelogram may be made into a rectangle, as illustrated by the shaded triangles in figure 20-8, by cutting off the triangular surface (A) and placing it in position (B). It should be noted that the rectangle is a special kind of parallelogram in which the angles are right angles.

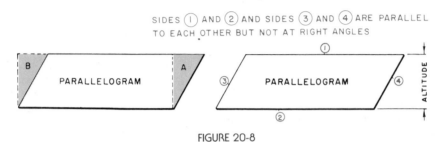

FIGURE 20-8

The rectangle thus formed has the same length base as the parallelogram. The height is the same as the altitude. Since the area of a rectangle is equal to the product of the length times the height, the area of a parallelogram is equal to the product of its base × altitude. The term *length* is often used instead of *base;* similarly, *height* is used for *altitude*. Regardless of which term is used, the process of finding the area is the same.

RULE FOR FINDING THE AREA OF A PARALLELOGRAM

- Express the dimensions for base and altitude in the same unit of linear measure, if needed.
- Multiply the base by the altitude.
- Express the product in units of surface measure.
- Express product, if needed, in lowest terms.

EXAMPLE: A parallelogram has a base 4 feet long and an altitude of 18 inches. Find its area.

Step 1 Express dimensions in same unit of measure.	$18 \text{ in.} = 1\frac{1}{2} \text{ ft}$
Step 2 Multiply the base and altitude.	$4 \times 1\frac{1}{2} = 6$
Step 3 Express product in units of surface measure.	**Area = 6 sq ft** Ans

E. APPLYING SURFACE MEASURE TO A TRAPEZOID

A *trapezoid* is a four-sided figure. Two of the trapezoid sides, called bases, are parallel (*figure* 20-9).

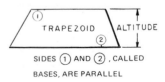

FIGURE 20-9

RULE FOR FINDING THE AREA OF A TRAPEZOID

- Express the dimensions for the bases and altitude in the same unit of linear measure, if needed.
- Add the lengths of the two bases.
- Multiply the sum by one-half of the altitude.
- Express the product, if needed, in lowest terms.

EXAMPLE: Find the area of the trapezoid (*figure* 20-10).

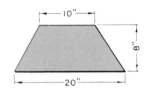

FIGURE 20-10

Step 1 Add the lengths of the two bases. $\qquad$ **10 + 20 = 30**

Step 2 Multiply the sum by $\frac{1}{2}$ of the altitude (8). $\quad$ **30 × $\frac{1}{2}$ × 8 =**

Area = 120 sq in. Ans

F. APPLYING SURFACE MEASURE TO A TRIANGLE

Two triangles of the same size and shape, when placed so that the longest side is their common side, form a parallelogram (*figure* 20-11).

The opposite pairs of sides of this figure are parallel. The parallelogram thus formed has the same size base and altitude as the original triangle.

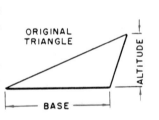
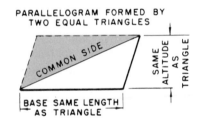

FIGURE 20-11

The area of the parallelogram is equal to the product of its base times altitude. Since the parallelogram is made up of two equal triangles, the area of each triangle is equal to one-half the area of the parallelogram. Thus, the area of a triangle may be computed directly by multiplying the base times $\frac{1}{2}$ the altitude.

RULE FOR FINDING THE AREA OF A TRIANGLE

- Multiply the base by $\frac{1}{2}$ the altitude.
- Express the product in units of surface measure.

EXAMPLE: Compute the area of a triangle having a 16-inch base
and an altitude of 6 inches.
>   *Step 1*   Multiply the base (16) by $\frac{1}{2}$ the altitude (6). $16 \times \frac{1}{2} \times 6 = 48$
>   *Step 2*   Express the product in units of surface measure.   **Area = 48 sq in.**   Ans

G. APPLYING SURFACE MEASURE TO A CIRCLE AND A SECTOR

The Circle

If a circle is divided into an equal number of parts and the sectors thus formed are stretched out along two parallel lines and these two rows of sectors are brought together, a figure approaching a rectangle is formed. As the number of sectors into which the circle is divided is increased, the length of the rectangle approaches one-half the circumference of the circle.

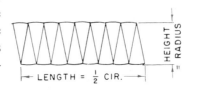
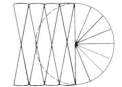

FIGURE 20-12

The area of the rectangle is equal to its *length* times *height*. The length is equal to $\frac{1}{2}$ the circumference of the circle. The *height* is equal to the radius of the circle. By substituting these values, the area of the rectangle formed is equal to its length $\left(\frac{1}{2}\text{ the circumference}\right)$ times height (radius of circle).

$$\textbf{Area of Circle} = \frac{3.1416 \times \text{Diam of Circle}}{2} \times \text{Radius of Circle}$$

Since the diameter of the circle is twice the radius, the area of a circle in terms of its diameter

$$= \frac{3.1416 \times \text{Diam}}{2} \times \frac{\text{Diam}}{2} \; or \; \frac{3.1416 \times \text{Diam} \times \text{Diam}}{4}$$

$$= .7854 \times \text{Diam} \times \text{Diam}$$

The area of a circle may be expressed in simpler terms. However, the two forms as given will be used until these values are later expressed as formulas. Where the radius of a circle is given instead of the diameter

$$\text{Area} = \frac{3.1416 \times 2\,(\text{Radius})}{2} \times \text{Radius} = 3.1416 \times \text{Radius} \times \text{Radius}$$

In place of continually using 3.1416, the Greek letter pi (π) is used in writing a problem. The actual value is applied when working out a problem.

RULE FOR FINDING THE AREA WHEN THE DIAMETER IS GIVEN

- Multiply .7854 × diam × diam.
- Express the product (area of the circle) in units of surface measure.

EXAMPLE: Find the area of a circle whose diameter is 4 inches, correct to two decimal places.

Step 1 Multiply .7854 × Diam × Diam.   .7854 × 4 × 4 = 12.5664

Step 2 Round off product to two decimal places.   = 12.57

Step 3 Express the product in units of surface measure.   Area = 12.57 sq in.   Ans

RULE FOR FINDING THE AREA WHEN THE RADIUS IS GIVEN

- Multiply 3.1416 × radius × radius.
- Express the product (area of the circle) in units of surface measure.

EXAMPLE: Find the area of a circle whose radius is 2.8″, correct to two decimal places.

Step 1 Multiply π × radius (2.8) × radius (2.8).   3.1416 × 2.8 × 2.8 = 24.630144

Step 2 Round off product to two decimal places.   = 24.63

Step 3 Express in units of surface measure.   Area = 24.63 sq in.   Ans

The Sector

The *sector* of a circle is the surface or area between the center and circumference. The sector is included within a given angle.

The area of a sector is equal to the area of the circle divided by the fractional part of the whole circle occupied by the sector. The angle of the sector is usually expressed as an *included angle*.

The fractional part of a circle occupied by a sector equals the number of degrees in the included angle divided by the number of degrees in a circle (360°).

RULE FOR FINDING THE AREA OF A SECTOR

- Compute the area of the circle.
- Determine the fractional part of the circle that the sector occupies by dividing the angle of the sector by 360°.
- Multiply the area of the circle by this fraction.
- Express the result (area) in square measure.

EXAMPLE: Determine the area of the sector removed from the disc.

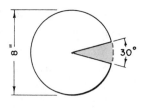

FIGURE 20-13

Step 1 Multiply (.7854 × Diam × Diam) to get the area of the circle.

.7854 × 8 × 8 = 50.2656

Step 2 Divide the number of degrees in the sector by 360°.

$$\frac{30}{360} = \frac{1}{12}$$

Step 3 Multiply the area of the circle (50.2656) by the fractional part occupied by the sector $\left(\frac{1}{12}\right)$.

$$\frac{\overset{4.1888}{\cancel{50.2656}}}{1} \times \frac{1}{\cancel{12}} = 4.1888$$

Step 4 Express the product in units of surface measure.

Area = 4.1888 sq in. Ans

H. APPLYING SURFACE MEASURE TO THE SURFACE OF A CYLINDER

Occasionally, the area of a figure known as a *right cylinder* or a *cylinder* of *revolution* must be computed. These cylinders consist of two round bases of equal diameter and the vertical outside surface (called *lateral surface*) around the bases.

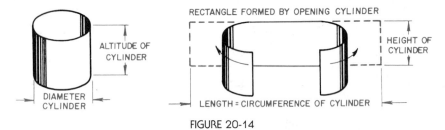

FIGURE 20-14

If the outside or lateral surface of the cylinder was unrolled, it would form a rectangle. The length of the rectangle is equal to the circumference of the cylinder. The rectangle height is the same as the altitude of the cylinder.

The area of this lateral surface = Circumference × Height
= (π × Diameter) × Altitude

RULE FOR FINDING THE AREA OF THE SURFACE OF A CYLINDER

- Find the circumference of the base.
- Multiply the circumference by the altitude of the cylinder.
- Express the area of the lateral surface in units of surface measure.

EXAMPLE: Determine the area of the lateral surface of a cylinder 2 inches in diameter and 5 inches high.

Step 1 Circumference = 3.1416 × Diam.   **3.1416 × 2 = 6.2832**
Step 2 Multiply circumference (6.28232) by altitude (5).   **6.2832 × 5 = 31.4160**
Step 3 Express area (31.416) in units of linear measure.   **31.416 sq in.**   Ans

The *total area of a cylinder* is equal to the area of the two bases plus the area of the lateral surface.

RULE FOR FINDING THE TOTAL AREA OF A CYLINDER

- Compute area of one base (.7854 × Diam × Diam).
- Multiply this area by 2 for two bases.
- Compute area of lateral surface (Circumference × Height).
- Add area of both bases to area of lateral surface.
- Express total area in units of surface measure.

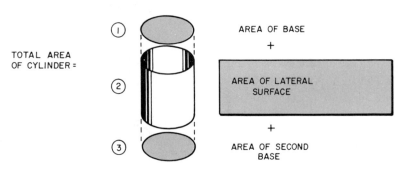

FIGURE 20-15

EXAMPLE: Determine the total area of a cylinder 4 inches in diameter and 10 inches in height, correct to two decimal places.

Step 1 Compute area of base (.7854 × Diam × Diam).   **.7854 × 4 × 4 = 12.5664**
Step 2 Multiply area of one base by 2.   **12.5664 × 2 = 25.1328**
Step 3 Compute area of lateral surface. Multiply the circumference of base times altitude.   **3.1416 × 4 × 10 = 125.6640**
Step 4 Add areas of two bases ⟶ **25.1328**
to area of lateral surface ⟶ **125.6640**
                    Total area = **150.7968**
Step 5 Round off to two decimal places.   **Total area = 150.80 sq in.**   Ans

Review and Self-Test

ASSIGNMENT UNIT 20

A. Areas of Squares and Rectangles (General and Practical Problems)

1. Determine the areas of squares A, B, and C and rectangles D, E, and F.

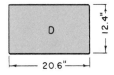
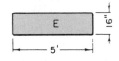
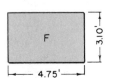

2. Find the cross-sectional areas of parts A and B and the area of the shaded portion of part C.

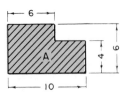
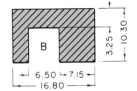
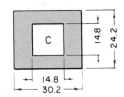

3. Find the total square foot area of the roof. The front portion is 46′ long × 18′ wide.

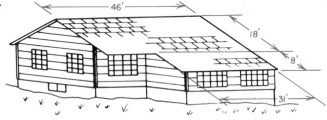

4. Assume the entire rectangular face of the carbon brush makes contact on a motor commutator. Calculate the contact surface areas of brushes A, B, and C. Round off answers to two decimal places.

Brush	Dimensions		Contact Area
	Length (L)	Width (W)	
A	$1\frac{1}{4}''$	$\frac{1}{2}''$	
B	$1\frac{1}{4}''$	$\frac{7}{8}''$	
C	$3\frac{3}{4}''$	$1\frac{1}{8}''$	

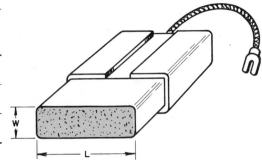

B. Areas of Parallelograms (Practical Problems)

1. Determine the areas of A, B, and C. All dimensions are in inches.

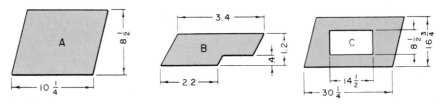

C. Areas of Trapezoids (Practical Problems)

1. Find the areas of stamped pieces A, B, and C that are shaped as trapezoids. Use the lengths of bases and altitudes given in each case. All dimensions are in inches.

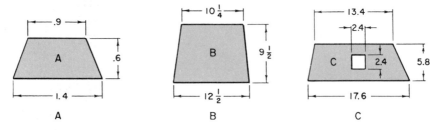

| A | B | C |

2. Determine the altitude of a trapezoid whose area is 122 square inches and whose bases are 12.2 inches and 18.3 inches.

3. Compute the land area of each of the two trapezoidal-shaped lots (A) and (B).

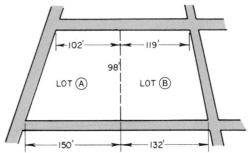

4. Compute the area in square feet in the end of the concrete retaining wall.

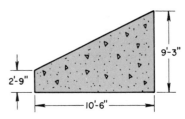

D. Areas of Triangles (Practical Problems)

1. Find the areas of triangles A, B, and C, and triangular stamping D. All dimensions are in inches.

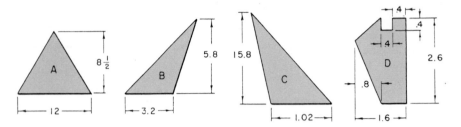

2. Compute the area of a triangular louver that is 4'-6" wide × 2'-6" high.

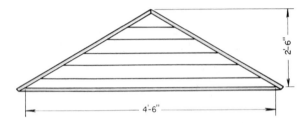

E. Areas of Circles (Practical Problems)

1. Determine the areas of circles A through F. The diameter or radius is given. Express the result in each case correct to two decimal places.

	A	B	C		D	E	F
Diameter	.8"	6"	5.25"	Radius	.5"	2"	3.8"

2. Find the area of stampings A and B. All dimensions are in inches.

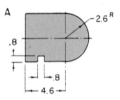

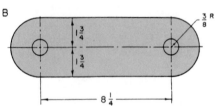

3. The brake lining of a magnetic disc motor brake has an outside diameter of 12". It is $2\frac{3}{4}''$ wide. Calculate the braking area of one disc.

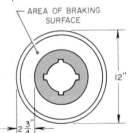

F. Areas of Sectors (Practical Problems)

1. Determine the area of the shaded areas A, B, and C, correct to three decimal places.

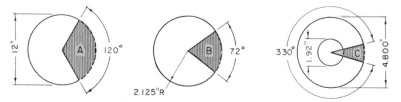

2. Compute the area of the garment pattern shown in the sketch. Give the answer in the nearest whole number.

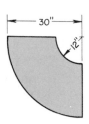

G. Areas of Lateral Surfaces and Cylinders (Practical Problems)

1. Determine the area of the lateral surface of cylinder A and the total area of cylinder B. The answers should be correct to two decimal places.

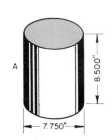

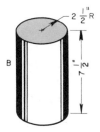

Unit 21 Principles of Volume Measure

OBJECTIVES OF THE UNIT

After satisfactorily studying this unit, the student/trainee will be able to
- *Understand the concept of volume measure.*
- *Express units of volume measure as larger, smaller, or combined units.*
- *Solve general and practical problems relating to the volumes of cubes, rectangular solids, cylinders, and irregular solid forms.*
- *Apply units of volume measure to solve problems involving liquid measure and interchange units of liquid and volume measure.*

Volume or cubic measure refers to the measurement of the space occupied by a body. Each body has three linear dimensions: length, height, and depth. The principles of volume measure are applied in this unit to three common shapes and the combinations of these three shapes: (1) the cube, (2) the rectangular solid, and (3) the cylinder.

A. DEVELOPING A CONCEPT OF VOLUME MEASURE

Volume measure is the product of three linear measurements. Each measurement must be in the same linear unit before being multiplied. The product is called the *volume* of the solid or body. Volume is expressed in cubic units of the same kind as the linear units. For example, if the length, height, and depth of a solid are given in inches, the volume will be in cubic inches.

The standard unit of volume or cubic measure is the cubic inch. The *cubic inch* is the space occupied by a body (called a cube). The cube is one linear inch long, one inch high, and one inch deep.

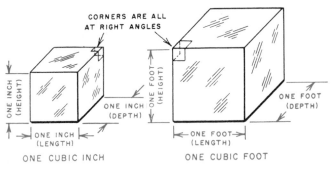

FIGURE 21-1

One cubic foot is the space occupied by a cubical body that is one linear foot long, one foot high, and one foot deep. If 12 inches is used as the length of each side in place of the linear foot, the volume is equal to 1,728 cubic inches (12 × 12 × 12).

One cubic yard is the space occupied by a cube that is one linear yard long, one yard wide, and one yard deep. Using the equivalent value of the yard, namely, three feet, the volume is equal to 27 cubic feet (3 × 3 × 3).

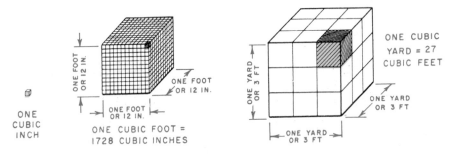

FIGURE 21-2

Cubic or volume measure is computed by multiplying three dimensions. By comparison, only two dimensions are multiplied in surface measure. The value of each of three standard units of volume measure may be compared. The number of cubic inches that the cubic foot and the cubic yard contain is illustrated (*figure* 21-3).

The word cubic is abbreviated (cu) for ease in writing. It is followed by the symbol for the linear unit of measure. The three units of volume measure that are used most often are the cubic inch (cu in.), the cubic foot (cu ft), and the cubic yard (cu yd).

Table of Cubic or Volume Measure
Standard unit of measure = 1 cu in.
1728 cu in. = 1 cu ft
27 cu ft = 1 cu yd

FIGURE 21-3

B. EXPRESSING UNITS OF VOLUME MEASURE

A volume in cubic inches may be expressed in cubic feet by dividing by 1,728 (1,728 cu in. = 1 cu ft). Volumes given in cubic feet may be expressed in cubic yards by dividing by 27 (27 cu ft = 1 cu yd).

RULE FOR EXPRESSING A UNIT OF VOLUME MEASURE AS A LARGER UNIT

- Divide the given volume by the number of cubic units contained in the required larger units.
- Express the quotient in terms of the required larger unit.

EXAMPLE: Express 5,184 cu in. as cu ft.

Step 1 Divide the given volume (5,184) by the number of cu in. in one cu ft. (1,728). $\frac{5184}{1728} = 3$

Step 2 Express the quotient (3) in terms of the required larger unit. **5,184 cu in. = 3 cu ft** Ans

Note. If the given volume cannot be divided exactly, the quotient may be expressed in more than one unit. A given number of cubic feet plus a stated number of cubic inches is an example.

RULE FOR EXPRESSING A LARGER UNIT OF VOLUME MEASURE AS A SMALLER UNIT

- Multiply the given unit by the number of smaller units contained in one of the required units.
- Express the product in terms of the required smaller unit.

EXAMPLE: Express 10 cu yd in cu ft.

Step 1 Multiply the given unit (10) by the number of cu ft in one cu yd (27 cu ft = 1 cu yd). **10 × 27 = 270**

Step 2 Express the product (270) in terms of the required unit (cu ft). **10 cu yd = 270 cu ft** Ans

RULE FOR EXPRESSING TWO OR MORE UNITS OF MEASURE

If a volume expressed in two or more units of measure is to be expressed in a smaller unit,
- Multiply those units of measure that are not in terms of the required unit by the number of smaller units equal to one given unit.
- Add the remaining units in the original given volume to this product.

EXAMPLE: Express 2 cu yd, 10 cu ft as cu ft.

Step 1 Multiply 2 cu yd by the number of cu ft in one cu yd.

Step 2 Add to the product (54 cu ft) the remainder of the given volume (10 cu ft).

$$2 \times 27 = \textbf{54 cu ft}$$
$$\underline{\textbf{10 cu ft}}$$
$$\textbf{64 cu ft} \quad \text{Ans}$$

C. APPLYING VOLUME MEASURE TO THE CUBE AND RECTANGULAR SOLID

In volume measure the three linear dimensions that express length, height, and depth or their equivalents are multiplied to determine the cubical contents of a regular solid. The product is in cubic inches if the dimensions are in inches, cubic feet when the dimensions are given in feet, and cubic yards when the dimensions are given in yards.

When the area of one surface is extended in a third direction, a solid is formed. The space the solid occupies is measured in terms of the number of cubic units that it contains.

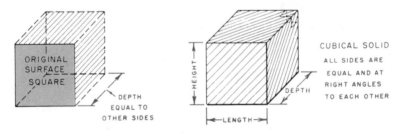

FIGURE 21-4

If the original surface is a square and its face is extended to add depth, the resulting figure is called a solid. When all corners of the solid are square and the linear length, height, and depth are equal, the object is called a *cube* or *cubical solid*. The volume of this cubical solid is equal to the product of its length times depth times height.

RULE FOR COMPUTING THE VOLUME OF A CUBE

- Express the dimensions for length, depth, and height in the same linear unit of measure, when needed.
- Multiply the length × depth × height.

- Express the product in terms of units of volume measure.
- Express the resulting product, if needed, in lowest terms.

EXAMPLE: *Case 1.* Find the volume of a cube, each side of which is 8 inches long.

Step 1	Multiply the length (8) × depth (8) × height (8).	**8 × 8 × 8 = 512**
Step 2	Express the product (512) in terms of volume measure.	**512 cu in.** Ans

EXAMPLE: *Case 2.* Determine the volume of a cube that measures 1'-9" on a side.

Method I

Step 1	Express 1'-9" as 21".	
Step 2	Multiply length × depth × height.	**21 × 21 × 21 = 9261**
Step 3	Express product (9,261) in terms of volume measure.	**9261 cu in.**
Step 4	Express as cu ft and cu in.	$\frac{9261}{1728}$ **= 5 cu ft, 621 cu in.**
		5 cu ft, 621 cu in. Ans

Method II

Step 1	Express 1'-9" as $1\frac{3}{4}''$.	$1\frac{3}{4} \times 1\frac{3}{4} \times 1\frac{3}{4} = 5\frac{23}{64}$ **cu ft**
Step 2	Multiply length × depth × height.	
Step 3	Express $\left(\frac{23}{64}\right)$ cu ft as cu in. by multiplying by (1,728).	$\frac{23}{\cancel{64}} \times \overset{27}{\cancel{1728}} = 621$ **cu in.**
Step 4	Add the number of cu in. to the cu ft to get volume of the cube.	**5 cu ft, 621 cu in.** Ans

A *rectangular solid* resembles a cube except that the faces or sides are rectangular in shape. The volume of a rectangular solid is equal to the product of the length × depth × height. Sometimes the volume is expressed as the product of the area of the base (length × depth) × the height.

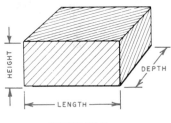

FIGURE 21-5

RULE FOR FINDING THE VOLUME OF A RECTANGULAR SOLID

- Express the dimensions of length, depth, and height in the same linear unit of measure if needed.
- Multiply the length × depth × height.
- Express the product in terms of units of volume measure and reduce to lowest terms, if needed.

EXAMPLE: Find the volume of the block in figure 21-6.

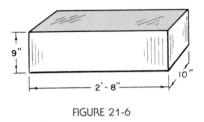

FIGURE 21-6

Step 1   Express all dimensions in the same linear unit.
Step 2   Multiply length × depth × height.
Step 3   Express product as units of volume measure.
Step 4   Express in lowest terms.

$10 \times 32 \times 9 = 2880$
2880 cu in.
$\frac{2880}{1728} = 1$ **cu ft, 1152 cu in.**   Ans

The volume or weight of a hollow rectangular solid is computed by using the same rules and mathematical processes as for the rectangular solid. Such problems often require a double computation. One computation is required for the outer surface and one for a cored or cut-away section.

D. APPLICATION OF VOLUME MEASURE TO CYLINDERS

The volume of a cylinder is the number of cubic units of a given kind that it contains. This number is found by multiplying the area of the base by the length or height of the cylinder.

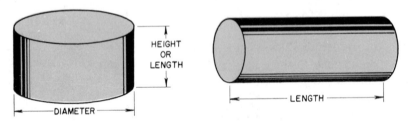

FIGURE 21-7

RULE FOR FINDING THE VOLUME OF A CYLINDER

• Compute the area of the base.
• Multiply this area by the height or length of the cylinder.
• Express the product (volume of cylinder) in units of volume measure.

EXAMPLE: Find the volume of a cylinder 3″ in diameter and 10″ long, correct to two decimal places.

Step 1   Compute the area of the base by multiplying **.7854 × 3 × 3 = 7.0686 sq in.** (.7854 × Diam × Diam).

Step 2   Multiply this area by the length of the cylinder (10″).   **7.0686 × 10 = 70.686**

Step 3   Express product (70.686), correct to two decimal   **70.69 cu in.   Ans** places, in units of volume measure.

E. APPLICATION OF VOLUME MEASURE TO IRREGULAR FORMS

In addition to regular solids like the cube, rectangle, and cylinder, many objects are a combination of these shapes in a modified form.

The volume of an irregular solid can be computed by dividing it into solids having regular shapes. The volume of each regular solid (or part of one) can be computed. The sum of the separate volumes equals the volume of the irregular solid.

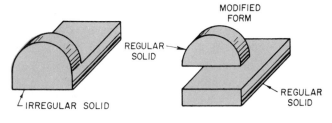

FIGURE 21-8

RULE FOR FINDING THE VOLUME OF AN IRREGULAR SOLID

- Divide the solid into regular forms.
- Compute the volume of each regular solid or part of one.
- Add the separate volumes.

EXAMPLE: Determine the volume of the brass casting in figure 21-9. Round the answer to two decimal places.

Step 1   Divide the irregular form into two regular solids (a cube and a cylinder).

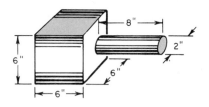

FIGURE 21-9

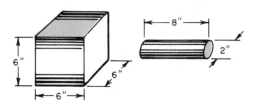

Step 2   Compute the volume of each regular solid.

a. Volume of cube = length × depth × height

$6 \times 6 \times 6 = 216$ cu in.

b. Volume of the cylinder = area of base × height

$(.7854 \times 2 \times 2) \times 8 = \underline{25.1328 \text{ cu in.}}$

Step 3   Add the separate volumes.

Total volume of casting = 241.13 cu in.   Ans

F. APPLICATION OF VOLUME MEASURE TO LIQUID MEASURE

Constant reference is made in the shop and laboratory to the measurement of liquids for cutting oils, oils for heat-treating metals, coolant solutions, marking fluids, cleaning agents, and lubricants. Also, an understanding of the measurement of liquids is essential in clinics and hospitals, business, merchandising, and the home.

Liquids are measured by cubical units of measure known as *liquid* measure. One common method of determining liquid capacity requires, first, computing the cubical contents of the object. Second, the resulting units of volume measure are changed to units of liquid measure.

The standard units of liquid measure are the gill, pint, quart, gallon, and barrel. These units are sometimes abbreviated for ease in writing. The pint is written (pt), quart (qt), gallon (gal), and barrel (bbl). A comparison of values for each of these units is found in figure 21-10.

The gallon as established by law contains 231 cubic inches of liquid. With this known value, it is possible to solve problems requiring the use of liquid measure by dividing the volume, expressed in cubic inches, by 231.

Table of Liquid Measure
4 gills = 1 pint (pt)
2 pints (pt) = 1 quart (qt)
4 quarts (qt) = 1 gallon (gal) = 231 cu in.
$31\frac{1}{2}$ gallons = 1 standard barrel (bbl)

FIGURE 21-10

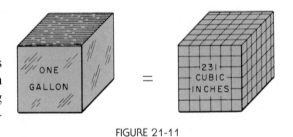

FIGURE 21-11

RULE FOR CHANGING UNITS OF VOLUME MEASURE TO UNITS OF LIQUID MEASURE

- Compute the volume of the object in terms of cubic inches.
- Divide this volume by 231 (231 cu in. = 1 gal).
- Express the quotient in terms of liquid measure (gal).

EXAMPLE: Determine the liquid capacity of a coolant tank whose volume is 1,155 cubic inches.

    Step 1   Divide volume in cubic inches (1,155) by the number of cubic inches (231) in one gallon.

$$\frac{1155}{231} = 5$$

    Step 2   Express the quotient (5) in terms of liquid measure.

Capacity of tank = 5 gallons   Ans

RULE FOR EXPRESSING LARGER UNITS OF LIQUID MEASURE IN SMALLER UNITS (BARRELS TO GALLONS TO QUARTS TO PINTS TO GILLS)

- Determine the number of smaller units of liquid measure in one larger unit.
- Multiply the given units by this number.
- Express the product in terms of the required unit of measure.

EXAMPLE: *Case 1.* Express $4\frac{1}{2}$ gallons in quarts.

    Step 1   Determine the number of smaller units (quarts) in one larger unit (gallons).

4 qt = 1 gal

    Step 2   Multiply the given units $\left(4\frac{1}{2}\right)$ by this number (4).

$4\frac{1}{2} \times 4 = 18$

    Step 3   Express the product (18) in the required units of liquid measure (qt).

$4\frac{1}{2}$ **gals = 18 qt**   Ans

EXAMPLE: *Case 2.* Express $3\frac{1}{8}$ quarts as pints and gills.

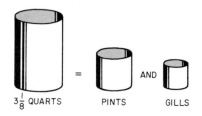

    Step 1   Determine the number of smaller units (pints and gills) in one larger unit (quart).

2 pt = 1 qt, 4 gills = 1 pt

    Step 2   Multiply the given unit $\left(3\frac{1}{8} \text{ qt}\right)$ by the number of smaller units (2 pt) in one of the given units (1 qt). *Note.* The fractional part of a pint $\left(\frac{1}{4}\right)$ may be changed to gills by multiplying by 4 the number of gills in one pint.

$3\frac{1}{8} \times 2 = 6\frac{1}{4}$ **pt**

$\frac{1}{4} \times 4 = 1$ **gill**

    Step 3   Combine both values (6 pints and 1 gill). The result is the equivalent of $3\frac{1}{8}$ quarts in terms of pints and gills.

$3\frac{1}{8}$ **qt = 6 pt, 1 gill**   Ans

RULE FOR EXPRESSING SMALLER UNITS OF LIQUID MEASURE IN LARGER UNITS (GILLS TO PINTS TO QUARTS TO GALLONS TO BARRELS)

- Determine the number of smaller units of liquid measure in one of the required larger units.
- Divide the number of given units by this number.
   Note. Where the result is a mixed number, the fractional part is sometimes changed to the next smaller unit. For example $3\frac{1}{2}$ gallons may be written 3 gallons, 2 quarts.

EXAMPLE: *Case 1.* Express 24 pints in gallons.

Step 1 Determine the number of smaller units (pt) in one of the larger required units (gal).
$8 \text{ pt} = 1 \text{ gal}$

Step 2 Divide the number of given units (24) by this number (8) to get gallons.
$\frac{24}{8} = 3$
$24 \text{ pt} = 3 \text{ gal}$ **Ans**

EXAMPLE: *Case 2.* Express 76 gills in gallons, quarts, and pints.

Step 1 Determine the number of smaller units of liquid measure (gills) in one of the required larger units (gallons and quarts).
8 gills = 1 qt
32 gills = 1 gal

Step 2 Divide the 76 gills by 32 to determine the number of gallons.
$\frac{76}{32} = $ **2 gal, 12 gills**

Step 3 Express the 12 gills as quarts by dividing by 8.
$\frac{12}{8} = $ **1 qt, 4 gills**

Step 4 Express the remaining 4 gills as pints by dividing by 4.
$\frac{4}{4} = $ **1 pt**

Step 5 Combine all values. The result in gallons, quarts, and pints is the equivalent of 76 gills.
76 gills = 2 gal, 1 qt, 1 pt **Ans**

Review and Self-Test

ASSIGNMENT UNIT 21

Note. Unless otherwise stated, all dimensions for all problems are expressed in inches.

A. Changing Values of Units of Volume Measure (General Applications)

1. Express each of the following volumes (a) through (l) in the unit of volume measure specified in each case.

   a. 2 cu ft in cu in.

   b. $1\frac{1}{2}$ cu ft in cu in.

c. $3\frac{5}{8}$ cu ft in cu in.

d. 10 cu ft, 19 cu in. in cu in.

e. 3456 cu in. in cu ft

f. 18.144 cu in. in cu ft

g. 8640 cu in. in cu ft

h. 1944 cu in. in cu ft and cu in.

i. 3 cu yd in cu ft

j. $4\frac{1}{3}$ cu yd in cu ft.

k. 5 cu yd, 7 cu ft in cu ft

l. 7 cu yd, 19 cu ft in cu ft

B. Applying Volume Measure to Cubes and Rectangular Solids (Practical Problems)

1. Determine the volume of cubes A, B, and C.

	A	B	C
Length	6	$8\frac{1}{2}$	1'-6"
Depth	6	$8\frac{1}{2}$	1'-6"
Height	6	$8\frac{1}{2}$	1'-6"

2. Compute the volume of rectangular solids A, B, and C, correct to two decimal places.

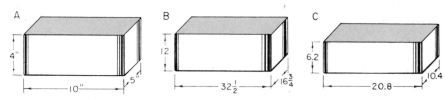

3. Determine the number of cubic yards of concrete mix that is needed to pour footings A, B, and C.

Footing	Number Required	Dimensions			Required Cubic Yards
		Width	**Length**	**Height**	
A	4	1'-6"	1'-6"	1'-6"	
B	2	1'-6"	1 yd.	9"	
C	6	1'-6"	4'-6"	1'-6"	

LALLY COLUMN

FOOTING

C. Applying Volume Measure to Rectangular Solids (Practical Problems)

1. Compute the number of cubic yards of earth that is to be removed for a basement. The basement dimensions are 8' deep × 36' wide × 48' long.

2. Determine the volume of concrete in the foundation wall. State the answer to the nearest cubic yard.

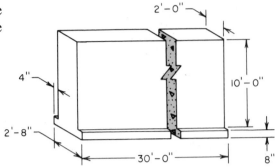

3. Compute the volume of hollow rectangular solids A and B. Round off the volume of B correct to one decimal place.

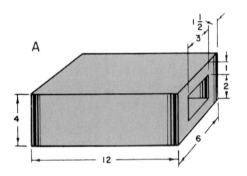

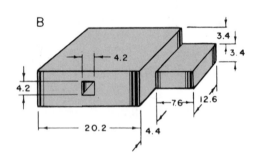

D. Applying Volume Measure to Cylinders (Practical Problems)

1. Determine the volume of cylinders A, B, and C correct to two decimal places.

	A	B	C
Diameter	4	12.5	
Radius			1.6
Length	10	24.5	6.4

2. Compute the liquid capacity of cisterns A, B, and C in gallons of water. Round off the value to one decimal place.

Cistern	Inside Diameter	Height
A	4'-0"	6'-0"
B	5'-6"	8'-0"
C	6'-6"	8'-6"

1 cu ft = $7\frac{1}{2}$ gal

3. Find the volume of cored cylinders A and B, correct to one decimal place.

	A	B
Outside Diam	5	$4\frac{1}{4}$
Inside Diam	2	$1\frac{1}{2}$
Length	10	12

4. Find the weight of cored brass casting A and cored bronze casting B.

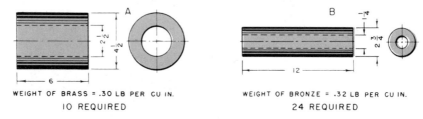

WEIGHT OF BRASS = .30 LB PER CU IN.
10 REQUIRED

WEIGHT OF BRONZE = .32 LB PER CU IN.
24 REQUIRED

5. Determine the cost of material needed to machine 250 special pins of $\frac{3}{4}$ inch round stock, $3\frac{1}{2}$ inches long. Allow $\frac{1}{8}$ inch for cutting off each pin and an additional 107 inches of the total length for waste and stock spoilage. The material weighs 0.28 pounds per cubic inch and costs \$0.42 per pound. Find the cost to the nearest dollar.

6. Determine the weight of the aluminum parts shown at A and B, correct to two decimal places. The weight of aluminum is .09 pounds per cubic inch.

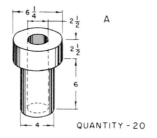

QUANTITY - 20

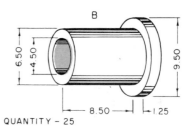

QUANTITY - 25

E. Applying Volume Measure to Irregular Shapes (Practical Problems)

1. Determine the weight of the rectangular cast iron blocks specified, correct to one decimal place.

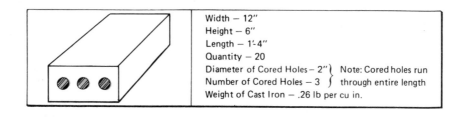

| Width — 12" |
| Height — 6" |
| Length — 1'-4" |
| Quantity — 20 |
| Diameter of Cored Holes – 2" } Note: Cored holes run |
| Number of Cored Holes – 3 } through entire length |
| Weight of Cast Iron — .26 lb per cu in. |

2. Determine the cost of 75 steel drop forgings conforming to the specifications given. The cost should be correct to two decimal places.

QTY - 75

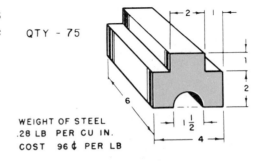

WEIGHT OF STEEL
.28 LB PER CU IN.
COST 96¢ PER LB

F. Expressing Units of Measure (General Applications and Practical Problems)

1. Express each of the following values in the unit or units of liquid or volume measure indicated in each case.

a.	4 gal	qt	i.	37 qt	gal and qt
b.	$6\frac{1}{2}$ gal	qt	j.	63 gal	bbl
c.	$3\frac{3}{4}$ gal	qt	k.	$96\frac{1}{2}$ gal	bbl and gal
d.	$6\frac{1}{2}$ qt	pt	l.	693 cu in.	gal and qt
e.	$5\frac{1}{4}$ qt	pt	m.	577.5 cu in.	gal and qt
f.	$8\frac{3}{4}$ qt	pt and gills	n.	5 gal	cu in.
g.	$3\frac{7}{8}$ qt	pt and gills	o.	4 gal, 3 qt	cu in.
h.	17 qt	gal and qt			

2. Determine the liquid capacity of the rectangular coolant tank A and the circular portable container B. Compute A to the nearest gallon and B to the nearest quart.

A

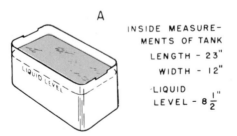

INSIDE MEASURE-
MENTS OF TANK
LENGTH - 23"
WIDTH - 12"
LIQUID
LEVEL - $8\frac{1}{2}$"

B

INSIDE
MEASUREMENTS
HEIGHT - 11"
DIAMETER - 14"

Unit 22 Achievement Review on Measurement (Customary Units)

OBJECTIVES OF THE UNIT

This achievement review serves as an overall test for Section 4. The unit is designed to measure the student's/trainee's ability to

- *Express measurements in appropriate units of linear, circular, surface, and volume measure.*
- *Perform basic mathematical processes of addition, subtraction, multiplication, and division to solve problems involving linear, circular, surface, or volume measurements.*

A. APPLICATION OF LINEAR MEASURE (GENERAL APPLICATIONS AND PRACTICAL PROBLEMS)

1. Express each measurement in the unit indicated in each case.

   a. 15 ft to yd

   b. 8 ft to in.

   c. 3 yd 2 ft to ft

   d. 7 ft 5 in. to in.

   e. 176 in. to ft

   f. 12.5 ft to in.

   g. $6\frac{1}{2}$ yd to ft

   h. 9 ft 8 in. to in.

   i. 4 yd 6 ft 3 in. to in.

2. Add each series of measurement and reduce to lowest terms.

   a. $10' + 7' + 5'$

   b. 6 yd + 9 ft + 6 ft

   c. $.5'' + .375'' + .125''$

   d. $6.500'' + 1'\text{-}3\frac{1}{4}$ ft + 8 in.

   e. $10\frac{1}{2}$ yd + $17\frac{1}{4}$ ft + 8 in.

   f. $4\frac{3}{4}$ yd + 19 ft 6 in.

3. Perform the arithmetical process required in each case. Give result in simplest form.

   a. $\begin{array}{r} 3 \text{ yd } 6 \text{ ft } 9 \text{ in.} \\ -1 \text{ yd } 4 \text{ ft } 5 \text{ in.} \\ \hline \end{array}$

   b. $\begin{array}{r} 9 \text{ yd } 2 \text{ ft } 4 \text{ in.} \\ -8 \text{ ft } 6 \text{ in.} \\ \hline \end{array}$

   c. $\begin{array}{r} 6 \text{ ft } 2 \text{ in.} \\ \times \ 28 \\ \hline \end{array}$

   d. $\begin{array}{r} 12' \text{ - } 3.5'' \\ \times \ 10 \\ \hline \end{array}$

   e. $\begin{array}{r} 280 \text{ in.} \\ \div \ 14 \text{ in.} \\ \hline \end{array}$

   f. $\begin{array}{r} 9'\text{-}10'' \\ \div \ 7'' \\ \hline \end{array}$

4. The diesel engine plate gage illustrated is to be machined. The part is to be rough machined, finish machined, and ground to the finished sizes given on the drawing.

   a. Allow $\frac{1}{32}$ " on all faces and determine the rough machining dimensions for Ⓐ through Ⓗ.

   b. Determine the size to which dimensions Ⓐ through Ⓗ are to be machined before grinding if .010″ is allowed on each dimension for the grinding operation.

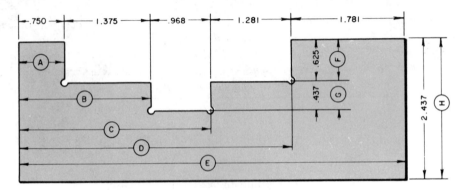

5. Determine the standard micrometer readings A, B, and C.

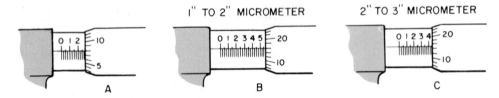

B. APPLICATION OF CIRCULAR MEASURE (PRACTICAL PROBLEMS)

1. Give the bevel protractor readings A and B.

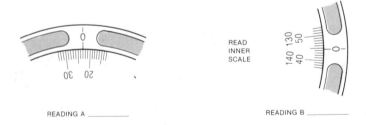

READING A _____

READING B _____

2. Determine angles Ⓐ through Ⓞ for the jig plate in order to machine the required slots and drill the holes.

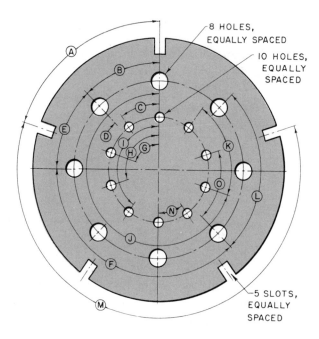

C. APPLICATION OF SURFACE MEASURE

1. Find the area of the plastic part.

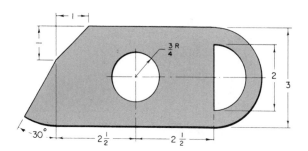

2. What is the cost of the 1-inch thick cored iron castings?

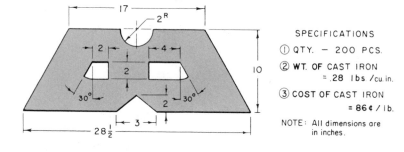

SPECIFICATIONS

① QTY. − 200 PCS.

② WT. OF CAST IRON
= .28 lbs /cu. in.

③ COST OF CAST IRON
= 86¢ / lb.

NOTE: All dimensions are in inches.

D. APPLICATION OF VOLUME MEASURE

1. Determine the amount of liquid held in a rectangular oil reservoir of a hydraulic machine for the liquid levels indicated in the table. Round off each answer to the nearest quart.

Gage Level	Height of Liquid
A	5"
B	$5\frac{1}{2}''$
C	6"
D	$6\frac{1}{2}''$
E	7"

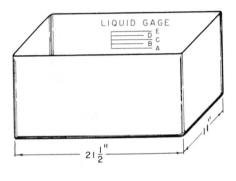

2. What is the weight of 2500 brass parts that are stamped from $\frac{1}{16}''$ sheet brass weighing .3 pound per cubic inch? Give the total weight to the nearest pound.

NOTE: All dimensions are in inches.

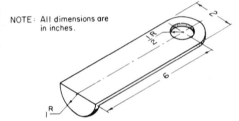

PERCENTAGE AND AVERAGES

Unit 23 The Concepts of Percent and Percentage

OBJECTIVES OF THE UNIT

After satisfactorily studying this unit, the student/trainee will be able to
- *Express numerical values in terms of percent.*
- *Determine any required percent of a measurement value.*
- *Change percent values to decimals and fractions.*

Percents are given in catalogs, magazines, newspapers, handbooks for technicians, and other publications. Percents show how many parts of a total are taken out. Percents are used to make comparisons and compute wages, taxes, discounts, and increases or decreases in production. An ever-increasing number of applications of percents are found in health, business distribution and merchandizing occupations, industry, and the home.

A. FORMS FOR EXPRESSING PERCENT

The word *percent* is a short way of saying "by the hundred" or "hundredths part of the whole." A percent refers to a given number of parts of the whole which is equal to 100 percent. "Fifteen percent" is the same as writing 15%.

Percent may be shown graphically by two illustrations (*figure 23-1*). The square is divided into 100 equal parts. Each of the small squares is one one-hundredth of the whole (100%) or $\frac{1}{100}$ of 100% = 1%.

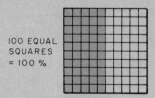

100 EQUAL SQUARES = 100 %

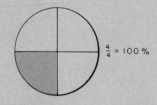

$\frac{4}{4}$ = 100 %

FIGURE 23-1

By the same reasoning, the 50 shaded squares are $\frac{50}{100}$ of the total = $\frac{50}{100}$ of 100% or 50%. In the circle, the shaded area is $\frac{1}{4}$ of the whole circle or $\frac{1}{4}$ of 100% or 25%.

Percent is another form of mathematical expression. A value like 50% may be represented as such. It also may be given in an equivalent form as the fraction $\left(\frac{1}{2}\right)$ or as the decimal (.5). The form in which a value is given depends on the requirements of the problem.

B. DETERMINING ONE HUNDRED PERCENT OF A NUMBER

One hundred percent of a number is the same as one hundred hundredths or 1. It represents the whole number or total. Therefore, 100% of a number is the number itself.

C. DETERMINING ONE PERCENT OF A NUMBER

RULE FOR FINDING ONE PERCENT OF A NUMBER

* Change one percent to an equivalent decimal and remove the percent sign.
  1% = one hundredth = .01.
* Multiply the given number by the decimal equivalent of 1% (.01).
* Label the answer in the required unit of measure.

EXAMPLE: Find 1% of 275 feet.

    275
  × .01

Step 1 Multiply the given number (275) by .01.   275

Step 2 Point off, starting at the right, the same number of decimal places in the answer as there are in the multiplier and multiplicand.   2.75

Step 3 Label answer with correct unit of measure.   2.75 feet   Ans

Since 1% is the same as .01, the process may by performed mentally by just placing a decimal in the original number. For instance, 1% of 120.33 is found by just moving the decimal point two more places to the left.

$$1\% \text{ of } 120.33 = .01 \times 120.33 = 1.2033 \quad \textbf{Ans}$$

D. DETERMINING ANY PERCENT OF A NUMBER

A percent may be converted to any equivalent mathematical form. At times, the fractional equivalent is preferred. In other cases, the decimal form is best suited to a specific problem.

RULE FOR FINDING ANY PERCENT OF A NUMBER

* Convert the percent to either a fractional or decimal equivalent.
* Multiply the given number by this equivalent.

- Point off the same number of decimal places in the product as there are in the multiplier and multiplicand.
- Label answer with appropriate unit of measure.

EXAMPLE: Find 16% of 1,218 square yards.

Step 1   Change 16% to a decimal.   16% = .16

Step 2   Multiply the given number (1,218) by the decimal (.16).

$$\begin{array}{r} 1218 \\ \times\ .16 \\ \hline 19488 \end{array}$$

Step 3   Point off two decimal places in the product.   194.88

Step 4   Label answer.   194.88 sq yd   Ans

These same steps are followed when the percent is a mixed number.

EXAMPLE: Find $6\frac{1}{4}$% of 782 horsepower (hp).

Step 1   Change $6\frac{1}{4}$% to a decimal. $6\frac{1}{4}$% = $.06\frac{1}{4}$ = .0625

Step 2   Multiply the given number (782) by the decimal (.0625).

$$\begin{array}{r} 782 \\ \times\ .0625 \\ \hline 488750 \end{array}$$

Step 3   Point off four decimal places in the product.   48.8750

Step 4   Label answer.   48.875 hp   Ans

E. CHANGING PERCENTS TO DECIMALS AND FRACTIONS

A percent may be changed to its decimal equivalent. Simply move a decimal point two places to the left and take away the percent sign. The decimal value may then be changed to a fraction to get the fractional equivalent of the original percent. A few relationships between percents, decimals, and fractions are shown in figure 23-2.

Percent	Equivalent		Parts of the Whole
	Decimal	Fraction	
1%	.01	$\frac{1}{100}$	One hundredth
100%	1.00	$\frac{100}{100}$ or $\frac{1}{1}$	One hundred hundredths
325%	3.25	$3\frac{1}{4}$	Three hundred twenty-five hundredths
.1%	.001	$\frac{1}{1000}$	One tenth of one percent

FIGURE 23-2

Review and Self-Test

ASSIGNMENT UNIT 23

A. The Concept of Percent (General Applications)

1. Give the percent of the whole that the shaded area of the rectangle (a), triangle (b), and circle (c) represents in each case.

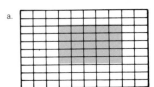
a.

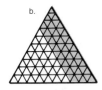
b.

c.

2. Shade and indicate by percent the area of the whole square (a), rectangle (b), and circle (c) that is represented in each instance by 100%, 50%, 25%, 10%, and 1%.

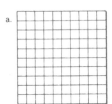
a.

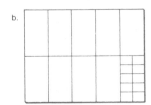
b.

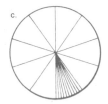
c.

3. Write each percent (A through E) using the symbol (%).

A	B	C	D	E
65 percent	6.5 percent	.6 percent	$6\frac{2}{3}$ percent	12.25 percent

4. Draw a table similar to the one illustrated and insert the decimal equivalent of percents A through O.

	Percent	Decimal		Percent	Decimal		Percent	Decimal
A	50%		F	125%		K	.1%	
B	30%		G	205%		L	.25%	
C	25%		H	312%		M	.05%	
D	9%		I	101%		N	$3\frac{1}{2}\%$	
E	1%		J	250%		O	$2\frac{1}{4}\%$	

5. Prepare a table and insert the percent equivalent of each decimal A through F.

A	B	C	D	E	F
1.50	.75	.125	2.05	.025	.004

B. Determining 100% and 1% of a Number (General Applications)

1. Find 100% of values A through E.

A	B	C	D	E
200	325	62.5	1.25 tons	.563 lb

2. Find 1% of the same values.

C. Determining Any Percent of a Number (General Applications)

1. Find the percent indicated in each case for quantities A through F.

A	4% of $500	D	.2% of 325 cu yd
B	8% of 120 sheets	E	3.25% of 800 kW
C	$1\frac{1}{2}$% of 840 acres	F	$37\frac{1}{2}$% of 972 lb

2. The absenteeism in a large business is $1\frac{1}{2}$% of all employee-hours worked. If the work-week is 40 hours and there are 1,750 employees, determine the total number of workdays (of 8 hours each) that are lost during each 260 workday year.
3. Find the total number of pounds of tin and lead required to make three different solder compositions (A, B, and C).

Composition	A	50% tin 50% lead	B	90% tin 10% lead	C	60% tin 40% lead
Pounds Solder Required		212 lb		48 lb		125 lb

4. The area of a floor is 320 square feet. In laying this floor, there is a lumber allowance of 25% for waste and matching. Determine the amount of flooring required.
5. The power output of a motor is 58 horsepower. If 7.6% of the power is used to overcome friction and other losses, what horsepower is available under a full load?
6. It takes 3.35 yards of material to make a suit. An additional 12% is required for waste and matching. Determine the total yardage that is needed to make 12 suits.

Unit 24 Application of Percentage, Base, and Rate

OBJECTIVES OF THE UNIT

After satisfactorily studying this unit, the student/trainee will be able to
• Apply base, rate, and percentage terms.
• Compute base, rate, or percentage values in practical occupational problems.

A. DESCRIPTION OF PERCENTAGE TERMS AND RULES

Percentage refers to the value of any percent of a given number. In every percentage problem three numbers are involved. The first number is known as the *base* because a definite percent is to be taken of it. The second number, the *rate,* refers to the percent that is to be taken of the base. The third number is the *percentage.* The relationship of the base, rate, and percentage to one another may be stated in the rule: the product of the base times the rate equals the percentage.

In simplified form,

$$\text{percentage} = \text{base} \times \text{rate}.$$

Sometimes, the letter B is used for base, R for rate, and P for percentage. Using letters instead of words, the rule may be given as $P = B \times R$. Wherever this rule is used, the rate must always be in decimal form.

B. DETERMINING PERCENTAGE

RULE FOR FINDING PERCENTAGE (GIVEN: BASE AND RATE)

- Write the rule for percentage in simplified form $(P = B \times R)$.
- Change the given percent to its decimal equivalent.
- Substitute the given values for B and R in the rule.
- Multiply and label answer.
- Check by substituting the answer for P. Rework to see that the quantities on both sides of the ($=$) sign are equal

---

EXAMPLE: Find $12\frac{1}{2}\%$ of $360''$.

Step 1	Write the rule.	$P = B \times R$
Step 2	Change $12\frac{1}{2}\%$ to its decimal equivalent.	$12\frac{1}{2}\% = .125$
Step 3	Substitute $360''$ for the base and .125 as the rate.	$P = 360 \times .125$
Step 4	Multiply and label the answer.	$P = 45''$ Ans

C. DETERMINING BASE OR RATE

It is possible to use the percentage rule to determine the base or rate when the two other terms in the rule are known.

RULE FOR FINDING BASE OR RATE (GIVEN: PERCENTAGE AND BASE OR RATE)

- Write the rule in simplified form. $P = B \times R$
- Substitute the two known values in the rule.
  Note. If a percent is given, be certain that it is changed to its decimal equivalent.
- Divide the percentage by the base to get the rate as a decimal or by the rate to find the base.
- Convert the decimal or fractional value of the rate to a percent if this is required.
- Label the answer with the appropriate unit of measure.

EXAMPLE: *Case 1.* Two hundred (200) is what percent of 500?

Step 1	Write the rule.	$P = B \times R$
Step 2	Substitute known values: the base is 500 and the percentage 200.	$200 = 500 \times R$
Step 3	Divide P by B to get R.	$\dfrac{\overset{2}{\cancel{200}}}{\underset{5}{\cancel{500}}} = R$
Step 4	Change $\frac{2}{5}$ of the whole to a decimal.	$\frac{2}{5} = .4$
Step 5	Change the decimal .4 to its percent equivalent by moving the decimal two places to the right and adding the (%) symbol.	$.4 = 40\%$ Ans
Step 6	Check the values on both sides of the (=) sign.	$200 = 500 \times .4$
		$200 = 200$ Proof

EXAMPLE: *Case 2.* If 56 castings are 20% of the total, what is the total number?

Step 1	Write the rule.	$P = B \times R$
Step 2	Substitute the known values: the percentage is 56, the rate is 20%.	
	Note. Change the 20% rate to its decimal equivalent .20 before substituting.	$56 = B \times .20$
		$\frac{56}{20} = B$
Step 3	Divide P by R to get B.	$280 = B$
Step 4	Label answer in the appropriate unit of measure.	**280 castings** Ans

With the rule, *percentage = base × rate*, it is possible to solve any percentage problems when two of the three quantities are given and the rate is expressed in decimal form.

Review and Self-Test

ASSIGNMENT UNIT 24

A. Determining Percentage

1. Find the percentage in each problem (A through E) for each value given for the base and rate.

	A	B	C	D	E
Base	2400	1875 gal	142.6 in.	3268.5 sq ft	$296\frac{1}{2}$ sheets
Rate	80%	45%	3.8%	$4\frac{1}{2}\%$	$6\frac{1}{4}\%$

2. A piece of meat weighs 25.6 pounds before it is cooked and 23.2 pounds after cooking. Determine the percent of weight lost in cooking.
3. The total receipts of a merchandising store for one week total $3,800. The expenses include 40% wages, 12% rent, 9% for heating and cooling, and 27% for taxes and other overhead. Establish (a) the amount of profit and (b) the percent of profit.
4. A car speedometer registers 52 miles per hour. The actual car speed is 55 miles per hour. State the percent of error correct to two decimal places.
5. The original cost of 1,290 aluminum castings is 97.5 cents each. Five percent (5%) is scrapped as poor castings and another $7\frac{1}{2}\%$ is spoiled in machining. (a) How many *whole castings* are used and (b) what is the new unit cost (rounded to the nearest two-place value) of each good casting?

B. Determining Base or Rate (Practical Problems)

1. Find the base in each problem (A through E) when the percentage and rate are given.

	A	B	C	D	E
Percentage	120 hp	2016 cables	137.8 lb	$78\frac{1}{2}$ bars	126.5 in.
Rate	90%	54%	7.2%	$6\frac{1}{2}\%$	$3\frac{1}{4}\%$

2. Find the rate in each problem (A through E) when the percentage and base are given.

	A	B	C	D	E
Percentage	30	360°	24" D	8.2	$92\frac{1}{2}''$
Base	30	30°	36" D	19.68	294"

3. What percent is wasted when 2.4 of every 120 sheets of metal are spoiled?

4. What percent of metal is allowed for cut-off on each 2-inch length of stock?

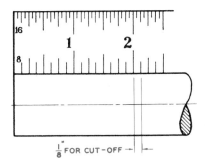

C. Determining Percentage or Base or Rate

1. One part of acid and four parts of water are mixed as an electrolyte for a storage battery. What percent is acid and what percent is water?

2. A generator rating is 42,500 kilowatts. If the output is 29,500 kilowatts, what percent of the rating is the generator delivering?

3. What is the operating spindle speed of a lathe spindle traveling at 346 RPM when 18% is lost through slippage and cutting force?

4. The cutting speed of a milling cutter is 85 feet per minute. Friction and other cutting losses amount to $12\frac{1}{2}\%$. Find the base cutting speed.

5. Two special bronze castings are composed of six different metals. The percent of each metal used in each casting and the casting weight are given. Determine the weight (in pounds and fractional parts) of each metal. Round off each decimal to one place.

Casting	Casting Weight	Composition (% by Weight)					
		Copper	Tin	Zinc	Phosphorus	Lead	Iron
A	400 lb	80	11	8.2	0.4	0.3	0.1
B	525 lb	83	8.7	7.38	0.34	0.52	0.06

Unit 25 Averages and Estimates

OBJECTIVES OF THE UNIT

After satisfactorily studying this unit, the student/trainee will be able to
- *Compute the average of several quantities.*
- *Find an unknown value when all quantities except the missing one and the average are given.*
- *Use estimating to check the mathematical accuracy of a computed value.*
- *Apply basic steps to estimating to compute quantities and values when the given data have a number of variable factors.*

Averages

Averages are used in linear, circular, angular, temperature, weight, and all other measurements. The *average* of given quantities and values is the starting point on which many other computations and factual data are based.

A. AVERAGE QUANTITIES

The average of two or more quantities that are in the same unit of measure is found by simple addition and division.

RULE FOR AVERAGING SEVERAL QUANTITIES

- Check the units of measure in each quantity to be sure they are the same.
- Arrange the quantities in a column and add.
- Divide by the number of quantities to get the average.

EXAMPLE: What is the average of the linear dimensions in figure 25-1?

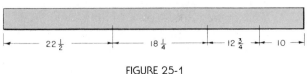

FIGURE 25-1

Step 1 Check each dimension and change, if necessary, to the same unit of measure.

Step 2 Arrange in a column and add.

Note. The mathematical processes may be simplified by expressing fractional values as decimals before averaging.

$$22\tfrac{1}{2} = 22.50$$
$$18\tfrac{1}{4} = 18.25$$
$$12\tfrac{3}{4} = 12.75$$
$$10\ \ = \underline{10.00}$$
$$63.50$$

Step 3 Divide the sum (63.50) by the number of quantities (4).

$$4\overline{)63.50}$$
$$15.875$$

Step 4 Express the decimal value (.875) as a fraction $\left(\tfrac{7}{8}\right)$, if necessary. The average of the four dimensions is $15\tfrac{7}{8}$.

$$15.875 = 15\tfrac{7}{8}\ \ \textbf{Ans}$$

B. DETERMINING VALUES FROM AVERAGES

An unknown value may be computed when the average and all quantities but the unknown are given.

RULE FOR FINDING AN UNKNOWN VALUE

- Multiply the average by the number of quantities.
- Add the given quantities.
- Subtract this sum from the product. The difference is the missing value.

- Check by adding all quantities and dividing by the number of quantities. The numbers are correct when the given average equals the computed average in the check.

EXAMPLE: The average of four temperature readings is 1672°. Three of the actual readings are 1525°, 1683°, and 1726°. Give the fourth reading.

Step 1   Multiply the average (1672) by the number of readings (4).

Step 2   Add the three given readings.

$$1525 + 1683 + 1726 = 4934$$

1672
× 4
6688

Step 3   Subtract this sum (4934) from the product of the average (6688). The difference of 1754° is the fourth temperature reading.

Step 4   Check the average of the four numbers with the given average.

6688
−4934
1754

$$\frac{1525 + 1683 + 1726 + 1754}{4} = \frac{6888}{4} = 1672$$

1754°   Ans

Estimating

Estimating has a twofold meaning. Estimating may refer to a shortcut mathematical process of determining a range against which an actual answer may be compared for accuracy. Estimating, in another instance, may require actual computations to determine cost, time, material, and other essential data. The estimate is as accurate as variations in materials and working conditions permit.

Estimating is valuable in checking computations and in arriving at quantities and values that vary as the basic conditions change.

C. ESTIMATING AS A MATHEMATICAL CHECK

Where estimating is used to check the accuracy of a solution to a problem, certain simple steps may be used.

RULE FOR ESTIMATING MATHEMATICAL ACCURACY

- Work the problem in a conventional way.
- Reread the original problem and determine what numbers may be rounded off.
- Use these numbers and perform the operations as required.
- Pay special attention to counting off decimal places in the answer, as most errors occur at this point.
- Check the computed value by comparing the original answer with the estimated value.
  Note. If both values are close, the answer is accurate in most instances.

E XAMPLE: What is $24\frac{1}{2}\%$ of 996?

Step 1   Compute answer in the conventional way.   **996**
Step 2   Estimate the answer.   **×.245**
          Round off $24\frac{1}{2}\%$ to 25% or $\frac{1}{4}$ **244.02**
          Round off 996 to 1000
          Take $\frac{1}{4}$ of 1000 = 250 **Ans**
Step 3   Compare the estimated value (250) with the computed value (244.02).
          Note. Since the 25% and the 1000 are larger than either of the original quantities, the computed value must be smaller than the estimate. If the answer were larger, the problem would require reworking.

D. ESTIMATING AS A BASE FOR ADDITIONAL COMPUTATION

Where there are a number of variable factors in given data, an estimate may provide the only sound basis for determining costs, appropriations, and employee-hours. On jobs where new materials are used, or all conditions are not known at any given time, the estimate furnishes about the best working information that is obtainable.

While estimates are determined in a number of ways, there are steps that are common to all estimating.

Step 1   Determine the data that must be computed.
Step 2   Analyze the available information to see what is given.
Step 3   Select averages (if available) or compute averages where needed.
Step 4   Perform the required mathematical operations and combine like quantities to get an accurate estimate.
          Note. Take sufficient space for each part of an estimate. Label all answers for greater accuracy and speed.
Step 5   Total all computations. Check by reworking to see that all items are included.
Step 6   Estimate by rounding off quantities and amounts. Then perform the mathematical operations to get an estimated answer.
Step 7   Check the *accurate estimate* against the *rounded-off estimate* to see that the final result is within an acceptable range.
Step 8   Rework the original accurate estimate if the variation is too great.
Step 9   Label all answers with the appropriate units of measure.

Review and Self-Test

ASSIGNMENT UNIT 25

A. Averaging Quantities and Estimating Accuracy (Practical Problems)

Solve each problem. Then check each solution by estimating.

1. Find the average length of the five rods.

2. What is the average weight of five castings that weigh $17\frac{1}{4}$ pounds, $12\frac{7}{8}$ pounds, $9\frac{1}{4}$ pounds, 4 pounds, and $8\frac{1}{2}$ pounds?

3. Micrometer measurements, taken at five places on a metal part, are recorded. Determine the average thickness of the part.

Measurements in Inches				
A	B	C	D	E
1.252	1.249	1.249	1.248	1.251

4. Five variations in temperature are recorded on a graph for a heat-treating operation. Find the average temperature.

Reading	A	B	C	D	E
Temperature	2272°	2346°	2147°	2286°	2304°

B. Determining Missing Values with Known Averages (Practical Problems)

1. The weekly production of a mechanism averages 1,235 units. The number of units produced in each of four days is 212, 224, 232, and 275. How many units must be produced the fifth day to meet the required average?

2. The space available on three floors of a loft building is 900 sq ft, 1475 sq ft, and 1350 sq ft. If 140 production machines averaging 32 sq ft of space apiece are to be installed, how much additional space is required?

Unit 26 Achievement Review on Percentage and Averages

OBJECTIVES OF THE UNIT

This achievement review serves as an overall test for Section 5. The unit is designed to measure the student's/trainee's ability to
- *Interpret percent equivalents of decimal and fractional measurements.*
- *Solve practical problems involving base, rate, or percentage quantities.*
- *Average quantities used in shop and laboratory problems and check by estimating.*
- *Compute missing values when average values are known.*

A. THE CONCEPT OF PERCENT

1. Give the percent of the whole that the shaded area represents in a, b, and c.

a. 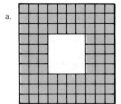 b. 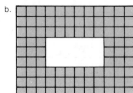 c.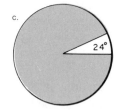

2. Prepare a table similar to the one shown and insert the missing decimal, fractional, or percent equivalent.

	Percent	Decimal	Fraction
A	16.5%		
B	3.5%		
C		.5	
D			$\frac{1}{40}$
E		.0025	

B. DETERMINING PERCENTAGE OR BASE OR RATE (PRACTICAL PROBLEMS)

1. A $191.10 charge for a set of forgings represents 22.5% of the total cost. Determine what the job costs.
2. Determine the percent by which the area of each square or round bar is increased or decreased for each size change given. (Use $\pi = 3.14$.)

	Square Stock			Round Stock	
	Original Dimension of Side	Increased or Decreased		Original Diameter	Increased or Decreased
A	2"	4"	E	2"	4"
B	2.5"	4"	F	2"	3"
C	.75"	.50"	G	1.500"	.750"
D	$1\frac{1}{4}$"	$\frac{3}{4}$"	H	$2\frac{1}{4}$"	$1\frac{1}{4}$"

3. The revolutions per minute of a milling machine spindle are reduced by friction and cutting pressure the percents indicated at A through E. In each instance, determine the rpm that the cutter actually turns.

	A	B	C	D	E
RPM	80	150	225	335	465
Speed Losses	10%	12.5%	$16\frac{1}{2}$%	$12\frac{1}{4}$%	$9\frac{1}{4}$%

4. A restaurant buys three sides of beef. The original weight and trim waste are given. Compute (a) the percent of trim losses and (b) the average percent of losses for the three sides.

Sides of Beef	Original Weight (lb)	Trim Losses	
		Weight (lb)	Percent
A	106.54	38.35	
B	148.91	49.14	
C	173.44	60.7	

C. AVERAGING QUANTITIES AND CHECKING BY ESTIMATING (PRACTICAL PROBLEMS)

Solve each problem and then check the answer by estimating.
1. Twenty-four castings weigh 272.5 pounds and cost $1.125 per pound. Determine the average weight of one casting and its cost.
2. Find the average weight of six sheets of metal that weigh $16\frac{1}{2}$ pounds, $12\frac{3}{4}$ pounds, 12 pounds, $12\frac{1}{4}$ pounds, $11\frac{1}{2}$ pounds, and $12\frac{1}{4}$ pounds.
3. Find the average length of rods measuring 2'-3", 3'-2", 2'-8", and 2'-6".

D. DETERMINING MISSING VALUES WITH KNOWN AVERAGES (PRACTICAL PROBLEMS)

1. At what temperature during the last hour must a piece of metal be held in a furnace to average 1128°? Hourly readings for the first four hours were 1062°, 1110°, 1174°, and 1158°.

2. During the first four days of a workweek, the total daily output reached 276, 320, 342, and 386 parts. The rejects each day of these totals were 5%, $4\frac{1}{2}$%, 6%, and 5%, respectively. The weekly quota to meet a contract is 325 perfect parts per day. How many parts must be produced the fifth day to meet the schedule? (Assume that the spoilage on the fifth day is the average percent of the four other days.)

FINANCE

Unit 27 Money and Time Calculations

OBJECTIVES OF THE UNIT

After satisfactorily studying this unit, the student/trainee will be able to
* *Make calculations that involve the addition, subtraction, multiplication,*
and division of money quantities.
* *Solve general and practical problems that deal with units of time and conversion*
from larger to smaller units and smaller to larger units.

A. CONCEPT OF THE MONEY SYSTEM

Money transactions are made daily in all branches of business, industry, and the home. The money system in the United States is based on the dollar as a unit. All fractional parts of the dollar are expressed in the decimal system. Thus, all amounts from one cent to ninety-nine cents may be written as decimals.

Two symbols, the dollar ($) and the cent (¢) signs are used with money. In writing a sum like ten dollars, the ($) sign is placed in front of the number, for example, $10 or $10.00. An amount like twenty-five cents may be written with the symbol (¢) as 25¢ or as the decimal ($.25). Neither the ($) sign nor the decimal point is used with the cent sign (¢).

The decimal point separates the dollar values from the fractional parts of the dollar. The value of a few common digits in money calculations is illustrated.

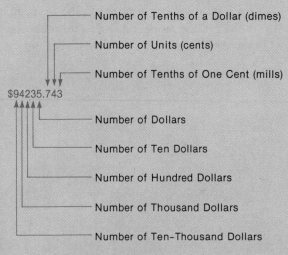

FIGURE 27-1

B. CALCULATIONS INVOLVING ADDITION, SUBTRACTION, MULTIPLICATION, AND DIVISION OF MONEY

Since the money system is a decimal one, it is possible to add, subtract, multiply, and divide the numbers representing different amounts in the same way as any other decimal. Care must be taken to keep the units in each number in columns, one under another, to point off the correct number of places in an answer, and to check each step.

C. CONCEPT OF THE TIME SYSTEM

Most workers are interested in time calculations as they enter into production and wages, both as a producer and a consumer. The table of time on which calculations are based is given in figure 27-2.

The units of time measure may be added, subtracted, multiplied, or divided and may be used as mixed numbers or decimals.

Table of Time	
60 seconds (sec)	= 1 minute (min)
60 minutes (min)	= 1 hour (hr)
24 hours (hr)	= 1 day (da)
7 days (da)	= 1 week (wk)
365 days (da)	= 1 year (yr)
52 weeks (wk)	= 1 year (yr)

FIGURE 27-2

RULE FOR EXPRESSING A LARGER UNIT OF TIME AS A SMALLER UNIT

- Express the larger unit as a decimal, where practical.
- Multiply the larger unit by the number of smaller units in one of the larger ones.
- Count off decimal places in the product, if decimals are used.
- Add to the product any smaller units contained in the original number.
- Label the answer with the smaller unit of time measure.

EXAMPLE: Express 7 hours, 20 minutes in minutes.

Step 1 Multiply the 7 hours by the number of minutes (60) in one hour.

$$\begin{array}{r} 7 \\ \times\,60 \\ \hline 420 \text{ min} \end{array}$$

Step 2 Add the remaining units and label the answer.

$$\begin{array}{r} +\,20 \\ \hline 440 \text{ min} \quad \text{Ans} \end{array}$$

This same problem may also be worked with the original time as the fraction $7\frac{1}{3}$ (7 hr 20 min = 7 20/60 or $7\frac{1}{3}$ hrs.) or as the decimal 7.33 multiplied by 60.

RULE FOR EXPRESSING A SMALLER UNIT OF TIME AS A LARGER UNIT

* Divide the smaller unit by the number of smaller units contained in one larger unit.
* Label the answer with the larger unit of time measure.

EXAMPLE: Express 4,200 seconds in minutes.

Step 1	Determine how many smaller units make one larger unit.	**60 sec = 1 min**
Step 2	Divide the given number of smaller units (4,200) by this number (60).	$\frac{4200}{60} = 70$
Step 3	Label the answer in the required unit of measure.	**70 min** Ans

Review and Self-Test

ASSIGNMENT UNIT 27

A. Concept of the Money System (General Applications)

1. Write the amounts A to G in words.

A	B	C	D	E	F	G
$1.12	$20.07	$2,692.75	$0.02	37¢	2.37\frac{1}{2}$	46$\frac{1}{4}$¢

2. Check each amount and indicate those that are incorrectly written.

A	B	C	D	E
.29¢	$0.17¢	$124.04¢	$6,292.16	57¢

B. Addition, Subtraction, Multiplication, and Division of Money (Practical Problems)

1. Find the cost of three lengths of steel as shown on the sketch.

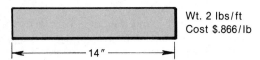

Wt. 2 lbs/ft
Cost $.866/lb

|←——— 14″ ———→|

2. Twenty castings cost $47.35. If the cost per pound is $1.65, how much does each casting weigh and cost? Round the weight and cost to two decimal places.
3. How much coolant is required and what is the cost to fill a machine reservoir $\frac{3}{4}$ full of a ready-mix preparation that sells for $5.50 per gallon? The reservoir measures $2'$ wide × $3'$ long × $8'$ deep.

4. The labor and material costs for a job total $675.28. The labor costs are $68\frac{1}{2}\%$ of this amount; the material is $31\frac{1}{2}\%$. Determine, correct to two decimal places, what the materials cost and what charges were made for labor.

C. Concept of the Time System (Practical Problems)

1. Express the units of time measure for A through F as indicated.

Change	A	B	C	D	E	F
From	6 hr	76 hr	12 min	$8\frac{1}{2}$ hr	100 sec	1400 min
To	min	days	sec	min	min	hr-min

2. Three plastic parts require the time per operation and the number of operations indicated in the table.
   a. Find the total time (days, hours, minutes, and seconds) required to produce the quantities of parts 1, 2, and 3, separately.
   b. Determine the total time required to produce the three parts. (Reduce to lowest terms.)

Part	Time per Operation	Number of Operations	Quantity Needed
1	13 seconds	3	10
2	3 minutes and 30 seconds	4	19
3	12 minutes and 20 seconds	5	134

Unit 28 Manufacturing Costs and Discounts

OBJECTIVES OF THE UNIT
After satisfactorily studying this unit, the student/trainee will be able to
- *Apply percentages in computing manufacturing costs.*
- *Solve practical cost problems involving single and multiple discounts.*

A. DETERMINING MANUFACTURING COSTS

Manufacturing costs consist of three basic elements: (1) raw materials costs, (2) labor costs to convert raw materials into marketable products, and (3) overhead costs such as taxes, rent, light, power, equipment depreciation and maintenance, and worker benefits.

In arriving at manufacturing costs, simple addition, subtraction, multiplication, and division are used. Also, percentages and averages are needed for comparison purposes. The manufacturing cost is equal to the cost of raw materials, direct labor, and overhead.

RULE FOR DETERMINING MANUFACTURING COSTS AND PERCENTAGES

- Total the cost of all materials.
- Total all labor charges.
- Determine what items of overhead to include.
- Check by subtracting from the manufacturing costs any one of the three cost figures. The difference is the sum of the other two costs.

EXAMPLE: Determine the manufacturing costs and the percent of each item. The raw materials cost $245, the labor charges are $1,535, and the overhead costs are $450.

Step 1 Add raw materials and labor and overhead costs.

$$\begin{array}{r} \$\ 245 \\ 1535 \\ \$\ 450 \\ \hline \end{array}$$

Manufacturing cost = **$2230**   Ans

Step 2 Divide *raw materials* cost by manufacturing cost to get percent of total.

$\dfrac{245}{2230} = 11\%$

Step 3 Divide the *labor cost* by the manufacturing cost for percent.

$\dfrac{1535}{2230} = 69\%$

Step 4 Divide *overhead costs* by manufacturing costs for percent.

$\dfrac{450}{2230} = 20\%$

Step 5 Check percent by adding to see if the total is 100%.

$$\begin{array}{r} 11\% \\ 69\% \\ 20\% \\ \hline 100\% \quad \text{Proof} \end{array}$$

B. APPLYING SINGLE AND MULTIPLE DISCOUNTS

The *list price* is equal to the manufacturing cost and profit. As an incentive to sell or to have a product paid for faster and because production costs usually decrease with greater quantities, the manufacturer offers discounts.

A *discount* is a reduction in the list price of a part or mechanism. Discounts may be *single* like a 10% discount or *multiple,* for instance, 10%, 10%, and 5%. With multiple discounts, each successive discount is taken on the remainder.

RULE FOR TAKING A SINGLE DISCOUNT

- Change the percent discount to a decimal.
- Multiply the list price by the discount.
- Point off the decimal places in the answer. The product then represents the discount.

EXAMPLE: Find the net cost of an instrument that lists at $630 and discounts at 12%.

Step 1	Change the 12% to its decimal equivalent.	12% = .12
Step 2	Multiply the list price (630) by the discount (.12).	$ 630
		× .12
		$75.60

Step 3	Subtract the discount ($75) from the list price ($630) to get the net price of $555.	$ 630
		− 75
		$ 555 Ans

RULE FOR TAKING MULTIPLE DISCOUNTS

- Take the first of the series of discounts as a single discount.
- Subtract the discount from the list price to get a net price.
- Multiply the net price by the decimal equivalent of the second discount.
- Subtract the second discount from the net price.
- Continue these same steps for each discount. The last difference represents the actual cost to the consumer.

EXAMPLE: Determine the cost of 4,200 pounds of metal at $1.45 per pound less 15%, 10%, and 5% discounts.

Step 1	Change the percents to decimal equivalents.	4200
Step 2	Multiply the 4,200 pounds by $1.45 to get the gross cost.	× 1.45
		$6090.00
Step 3	Take the first discount of 15% on $6,090.00	$6090.00
		× .15
		$ 913.50

Step 4 Subtract the first discount from $6,090.00	**$6090.00** **− 913.50** **$5176.50**
Step 5 Take the second discount of 10% on the remaining $5,176.50	**$5176.50** **× .10** **517.65**
Step 6 Subtract the second discount from the $5,176.50	**$5176.50** **− 517.65** **$4658.85**
Step 7 Repeat the last two steps with the third discount of 5%.	**$4658.85** **× .05** **232.94**
Note. The $4,425.91 is the actual cost of the metal.	**4658.85** **− 232.94** **$4425.91** Ans

Review and Self-Test

ASSIGNMENT UNIT 28

A. Determining Manufacturing Costs (Practical Problems)

1. Determine the manufacturing costs according to the charges given for jobs A to E.

	Material Costs	Labor Costs	Overhead
A	$ 75.00	$232.00	$ 74.00
B	96.00	384.50	142.50
C	84.46	295.23	216.24
D	165.29	347.63	210.98
E	220.06	400.97	$309.58

2. Compute the amounts spent for materials, labor, and overhead for parts A to E. Use the manufacturing costs and percents shown in the table.

	Mfg. Costs	Materials	Labor	Overhead
A	$1000	50%	30%	20%
B	2428	25%	60%	15%
C	1378	12%	70%	18%
D	2532	18.5%	72.5%	9%
E	1864	22%	$57\frac{1}{2}\%$	$20\frac{1}{2}\%$

3. The cost of manufacturing 1,000 parts is $792.00. Of this, 35% is spent for materials, 62% for labor, and the remainder for overhead.
   a. Determine the cost for materials.
   b. Find the cost for labor.
   c. Find the overhead costs.
   d. During the next quarterly period, the materials increased 10% in price, the labor costs rose 5%, and the overhead increased 5%. Determine the new cost for materials, labor, and overhead.
   e. What is the new cost for 1,000 parts?

B. Application of Single and Multiple Discounts (Practical Problems)

1. Carbon steel twist drills of a certain size list at $19.50 a dozen less a trade discount of 22%. Determine the cost of each drill to the nearest cent.

2. The materials cost of making a machine guard is 18%, the labor cost is 75%, and the overhead is 7%. Fifty guards are sold for $3,050 less a discount of 10%, 8%, and 2% for cash. Determine
   a. The cost of each guard.
   b. The materials, labor, and overhead costs for the lot, based on discounted price.

3. Three textile manufacturers offer the same grade of fabric at the same list price of $324. The distributor discounts are given in the table.
   a. Establish which distributor offers the best total discount.
   b. Compute the net cost of the fabric from the distributor that provides the best discount.

Distributor	Multiple Discounts
A	18% and 17%
B	20% and 15%
C	12%, 18%, and 5%

4. Compute (a) the total discount and (b) the net cost for equipment items A through D, using (1) the single discount and (2) the successive discount rates given.

| Equipment | List Price | Single Discount | | | Successive Discounts | | | | |
		Discount %	Total Discount	Net Cost	First	Second	Third	Total Discount	Net Cost
A	$ 750	18%			12%	6%	X		
B	$ 975	$16\frac{2}{3}\%$			10%	7%	X		
C	$1,440	$37\frac{1}{2}\%$			20%	10%	$7\frac{1}{2}\%$		
D	$3,945	$\frac{1}{3}$ off			15%	15%	5%		
		Grand Total					Grand Total		

5. Add the separate discounts and costs in problem 4 and determine (a) the grand total of the discounts and (b) the net cost of the equipment, applying the (1) single discount and (2) successive discounts.
6. Compare the single discount and the total successive discount rates for each equipment item with the net costs in problems 4 and 5. State the importance of computing and comparing the net costs of equipment items A through D.

Unit 29 Payrolls and Taxes

OBJECTIVES OF THE UNIT

After satisfactorily studying this unit, the student/trainee will be able to
- *Understand different taxes on wages and personal benefit deductions that affect take-home pay and other transactions.*
- *Apply the four basic mathematical processes to compute Social Security payments, withholding (income) taxes, and take-home pay.*

Each worker applies the four principles of addition, subtraction, multiplication, and division of decimals to keep accurate records. These records deal with time worked and checking the accuracy of computations made for social security and income taxes, disability, health, dental, and other benefits and deductions. As the number of items to be considered increases, payrolls become more complex.

A. COMPUTING PAYROLLS AND WAGES

One important payroll entry is the number of hours worked each week. In this connection, three terms are used: straight time, overtime, double time. *Straight time* is normally credited to a worker for the standard workweek.

Overtime is usually paid for hours worked in excess of the standard number of hours. The *overtime rate* is straight time or the regular hourly rate times $1\frac{1}{2}$. The *double time* rate applies to hours worked on Sundays, legal holidays, and for all time over a stated number of overtime hours.

RULE FOR COMPUTING WAGES (BASED ON HOURLY, OVERTIME, AND DOUBLE TIME RATES)

- Multiply the straight time each day by 1.
- Multiply the overtime by 1.5.
- Multiply the double time by 2.

EXAMPLE: Compute the total working time according to the daily hours entered in figure 29-1.

Day	Mon.	Tues.	Wed.	Thur.	Fri.	Sat.	Sun.
Time in Hours	8	9	9	9	9	4	8

FIGURE 29-1

Step 1 Multiply the straight time each day by 1 and add to get the total.

M	$8 \times 1 = 8$	
T	$8 \times 1 = 8$	
W	$8 \times 1 = 8$	
Th	$8 \times 1 = 8$	
F	$8 \times 1 = 8$	

Straight time $= 40$ hr

Step 2 Multiply the overtime hours by 1.5 and add to find the total.

T	$1 \times 1.5 = 1.5$	
W	$1 \times 1.5 = 1.5$	
Th	$1 \times 1.5 = 1.5$	
F	$1 \times 1.5 = 1.5$	
S	$4 \times 1.5 = 6$	

Overtime $= 12$ hr

Step 3 Multiply the double time hours by 2.

Sun. $8 \times 2 = 16$ hr

Step 4 Add the straight time (40), overtime (12), and double time (16).

$40 + 12 + 16 = 68$ hr Ans

Step 5 Multiply the total number of hours (68) by the hourly rate to get wages due.

An added incentive to improve quality and to get greater production is to increase the hourly rate by a bonus after a quota is reached. Bonus plans are given to individuals, to groups, and to departments. The bonus for additional production or quota is added either to the weekly wage or given over longer periods of time.

B. COMPUTING PAYROLL TAXES

The term *take-home pay* denotes the actual money that the worker receives after all deductions are made. While the deductions may include civic contributions, union dues, and other deductions, only income tax and Social Security are considered in this unit.

Social Security Payments

Social Security contributions are paid by employer and/or employee at a specified percent of a defined maximum amount of yearly income. Each employee (including special groups of self-employed persons) has a permanent, lifetime Social Security number. All payments by the individual/employer are credited to the employee's Social Security account. At this point, Social Security taxes are treated in terms of contributions or payments into the system.

RULE FOR CALCULATING SOCIAL SECURITY TAXES

- Multiply the weekly earnings by the prevailing tax rate. The product represents the weekly Social Security deduction.
  Note. Social Security deductions withheld by an employer stop each year when the prescribed base yearly income is reached.
- Add the employer's contribution.
  Note. The total represents the amount credited to the employee's Social Security account.

EXAMPLE: A worker earns $348 a week. Using a 7% tax rate, determine the amount withheld from the employee earnings for Social Security.

Step 1 Check the earned income for the year to determine whether it is within the prescribed base amount. Social Security taxes are deductible on this amount only.

Step 2 Multiply the weekly earning ($348) by the Social Security rate. The decimal equivalent (.07) of 7% is used.

Step 3 Point off two decimal places. The result ($24.36) is the Social Security tax withheld.

$$\begin{array}{r} \$\ 348 \\ \times\ .07 \\ \hline \$24.36 \end{array}\ \text{Ans}$$

Income Taxes

Practically all people who receive payment for work are required to pay an income tax. These taxes help to finance the cost of federal government operations. Each employer is required by federal law to withhold a percentage of salary from each employee. In turn, at regular periods, the employer turns over the money withheld to the director of Internal Revenue.

The amount withheld for income taxes depends on the number of dependents and on other allowable contributions and deductions. These are either itemized in a *long form* tax return or an average is taken on a *short form*. An example of an exemption allowance table is illustrated in figure 29-2. Withholding tax computations are based on similar tables that are adjusted periodically.

Payroll Period	One Withholding Exemption	Payroll Period	One Withholding Exemption
Daily or Misc.	$ 2.74	Monthly	$ 84.33
Weekly	19.23	Quarterly	250.00
Biweekly	38.46	Semiannually	500.00
Semimonthly	41.67	Annually	1,000.00

FIGURE 29-2

The employer may follow five simple steps to determine the amount to withhold for income taxes.

RULE FOR COMPUTING WITHHOLDING TAX

- Round off employee's earnings to the nearest dollar.
- Determine the exemptions claimed.
- Multiply the one withholding exemption in the table for the payroll period by the number claimed.
- Subtract this exemption amount from the rounded-off earnings.
- Multiply the difference by 20% or the prevailing withholding tax rate. Use the decimal equivalent (.20) to get the tax.

EXAMPLE: Determine the income withholding tax on a worker earning $348.41 weekly and claiming two dependents.

Step 1   Round off earned waged from $348.41 to $348.00

$19.23

Step 2   Multiply the one withholding exemption amount ($19.23) in figure 29-2 for a weekly payroll period by two exemptions.

×   2
$38.46

Step 3   Subtract the exemption amount ($38.46) from the rounded-off wage ($348.00).

$348.00
− 38.46
$309.54

Step 4   Multiply the difference ($310.00) by 20% (.20). The product ($62.00) is the income tax deduction.

$310.00
×   .20
$ 62.00   **Ans**

C. COMPUTING TAKE-HOME PAY

While only samples of Social Security and income taxes are covered in this unit, all other deductions are computed in a similar manner. The total is subtracted from the wages earned to determine the take-home pay.

RULE FOR COMPUTING TAKE-HOME PAY

- Determine the total number of hours worked in the payroll period.
- Multiply hours worked by the hourly rate and add any amount due from a bonus plan to get the wages due.
- Compute the Social Security, income tax, and any other taxes or deductions.
- Total any payment deductions for health, accident, or similar expenses taken directly out of wages.

• Add all deductions to be paid by the employee. Subtract the total of deductions from wages due. The difference is the take-home pay.

Review and Self-Test

ASSIGNMENT UNIT 29

A. Computing Payrolls and Wages (Practical Problems)

1. Determine from the time sheet the total hours worked and the amount due each worker. The overtime rate applies over 40 hours.

Worker	Hours Worked						Hourly Rate
	M	T	W	Th	F	S	
A	9	9	9	9	9	4	$18.60
B	8	8	8	9	9	4	14.30
C	8	8	8	8	8	4	11.60
D	8	0	8	8	8	4	8.30
E	9	9	9	9	9	6	4.75
F	8	8	8	8	8	0	4.35

B. Computing Payroll Taxes (Practical Problems)

1. Compute the deductions taken from wages earned for Social Security. Use the 7% rate. Overtime based on a 40-hour workweek is $1\frac{1}{2}$ times the hourly rate, up to 48 hours.

Worker	A	B	C	D	E
Hours Worked	44	42	48	40	32
Hourly Rate	$11.44	$8.28	$5.00	$4.60	$3.82

2. Figure the income tax deduction for each worker shown on the payroll sheet. Round off each wage to the nearest dollar. Use a weekly exemption allowance of $19.23 each and an income tax deduction of 20%

Worker	A	B	C	D	E
Amount Earned	$397.38	$277.92	$378.57	$221.78	$178.75
Exemption Claimed	3	1	4	3	2

C. Computing Take-Home Pay (Practical Problems)

1. Prepare a payroll sheet similar to the one shown. Use a standard workweek of 40 hours, overtime equal to time and a half, and double time for emergency Sunday work. Complete the information for all columns: total hours, weekly earnings, exemption allowance, tax A, and tax B on total earning and take-home pay.

Worker	Hours Worked							Regular Hours Worked	Overtime		Hourly Rate	Weekly Earning	Exemptions		Tax A 20%	Tax B 2%	Take-Home Pay
	M	T	W	Th	F	S	Su		Hours Worked	Hours Pay			at $18	Allowance			
A	8	8	8	8	8	6	6				$12.50		4				
B	9	9	9	9	9	4	0				8.70		3				
C	8	8	0	9	9	0	0				9.80		3				
D	9	9	9	9	9	8	4				5.24		4				
E	8	8	7	7	7	0	0				3.98		2				

2. Find the gross weekly salary for each of the five salespersons.

Salesperson	Salary and/or Commission Plan	Sales	Gross Salary
A	6% of all sales	$4,560	
B	5.75% of all sales	$5,325	
C	$8\frac{1}{4}$% of all sales	$2,698	
D	$310 salary, plus 2% of all sales	$2,342	
E	$218 salary, plus 3.5% of all sales over $1,500	$14,786	

Unit 30 Achievement Review on Finance

OBJECTIVES OF THE UNIT

This achievement review serves as an overall test for Section 6. The unit is designed to measure the student's/trainee's ability to

- *Solve everyday money problems that require one or more applications of basic mathematical processes.*
- *Calculate costs of manufactured products that involve material, labor, and overhead costs and discounts.*
- *Analyze data for preparing a payroll and applying wages, taxes, work schedules, and other considerations to compute take-home pay.*

A. MATHEMATICAL PROCESSES APPLIED TO MONEY (PRACTICAL PROBLEMS)

The average budget items and percents are given for a family of two, three, and four people.

Note. Figures following items are percentages of total monthly income.	Monthly Income					
	$1,000			$1,840		
	Exemptions					
Items	2	3	4	2	3	4
Food	30	35	38	24	28	32
Rent, Light, Fuel	25	25	26	22	23	23
Clothing	10	12	14	12	13	14
Church, Associations	2	2	2	3	3	3
Medical - Recreation	3	3	3	4	4	4
Auto - Transportation	7	7	7	8	8	8
Education	2	2	2	2	2	2
Savings - Insurance	9	6	4	10	7	5
Social Security - Taxes	12	8	4	15	12	9

1. Compute the amount spent monthly for each item for an income of $1,000 and a family of three.
2. Make the same calculations for an income of $1,840 and four exemptions.
3. Find the difference between the same items for the $1,000 and $1,840 incomes with three and four dependents, respectively.

B. APPLICATION OF MANUFACTURING COSTS AND DISCOUNTS (PRACTICAL PROBLEMS)

1. Use the information for articles A through E and determine the cost of materials, labor, and overhead.

	Quantity Produced	Total Mfg. Cost	Materials Cost (%)	Labor Cost (%)	Overhead (%)	Mark-up for List Price (%)	Discounts (%)
A	1,000	$2,400	30	65	5	20	10
B	750	3,250	25	68	7	25	6
C	10 gross	2,750	$16\frac{1}{2}$	$77\frac{1}{2}$	6	$37\frac{1}{2}$	15
D	15 doz	675	12.5	82.7	4.8	45	12 and 8
E	78 bbl	1,982	$8\frac{1}{2}$	$79\frac{1}{4}$	$12\frac{1}{4}$	$62\frac{1}{2}$	20, 10, 2

C. COMPUTING PAYROLLS, TAXES, AND TAKE-HOME PAY (PRACTICAL PROBLEMS)

1. Prepare a payroll form similar to the one illustrated. Compute the missing information in each column for each worker. Use a standard workweek of 40 hours, overtime equal to time and a half, and double time for all hours over 54. Tax B applies to weekly earnings.

Worker	Hours Worked						Regular Hours Worked	Overtime		Hourly Rate	Weekly Earning	Exemptions		Tax A 20%	Tax B 2%	Take-Home Pay
	M	T	W	Th	F	Su		Hours Worked	Hours Pay			at $20.00	Allowance			
A	10	10	10	10	10	10				$12.69						
B	10	10	9	9	9	9				7.93						
C	8	8	9	9	9	9				4.69						

2. Determine the actual cash each worker in problem C. 1. has left after additional taxes are paid out of the take-home pay. Assume the local tax of 3% and another state tax of $5\frac{1}{4}$% must be paid on everyday purchases.

GRAPHS AND STATISTICAL MEASUREMENTS

Unit 31 Development and Interpretation of Bar Graphs

OBJECTIVES OF THE UNIT

After satisfactorily studying this unit, the student/trainee will be able to
- *Translate technical information from tables for representation on bar graphs.*
- *Select appropriate scales to represent each value in practical problems in which X and Y axes are related to a common reference (origin) point.*
- *Prepare a bar graph from tabular data.*
- *Interpret bar graph values to meet specific conditions.*

Statistical data and other technical information are often represented in graphic form. In this way it is possible to compare one set of data with another. Equally important, values may be determined on the graph itself for various conditions. Graphs may be developed mechanically, by hand, or they may be generated by computer-aided design or other electronic equipment.

Types and Characteristics of Graphs

Many types of graphs are used in technical written materials, reports, and handbooks. The greatest number fall into one of four classifications: *picture graphs, bar graphs, circle graphs,* and *line graphs.*

Each type has certain advantages as well as disadvantages. The type of graph to use depends on the nature of the data to be presented and the skill of the person in portraying information graphically. The picture graph, sometimes called *pictogram*, is the easiest to read, but it is difficult to draw unless self-adhering, commercially available picture symbols are used. The steps required to produce and to read a bar graph are given in this unit. The development and interpretation of line graphs appear in Unit 32; the development and interpretation of circle graphs appear in Unit 33. Basic statistical concepts are covered in Unit 34.

Graph (coordinate) papers are obtainable with different spacings for varying conditions. These ruled sheets simplify the representation and interpretation of factual data. Graph papers often have scales printed on them. A 10 × 10 graph sheet indicates the number of equal spaces in a given area. In this instance, the 10 × 10 means 10 equal spaces vertically and 10 spaces horizontally.

Reference lines (sometimes called *base lines* or *axes*) of a graph intersect at a point called the *origin*. The horizontal lines on graph paper are generally associated with the *X axis*; the vertical lines, the *Y axis* values. Values on a graph are plotted according to a selected scale. The origin of the measurements may be zero (0) or any other appropriate starting value for those lines that are to be represented on the *X* and/or *Y* axes.

Wherever practical, graphs should be planned so that the units to be interpreted are read horizontally from left to right. Information presented in this manner is more easily readable than from a vertical position.

A. DEVELOPING BAR GRAPHS

The term bar graph merely signifies that solid lines or heavy bars of a definite length represent given quantities. Usually, graphs contain two *scales: a vertical (Y) scale* and a *horizontal (X) scale*. The scale indicates the value of each ruled line or lines. These specific values depend on the information to be presented.

RULE FOR DEVELOPING A BAR GRAPH

- Determine what information is to be presented and whether or not a bar graph is the best type to use.
- Range the data from smallest to largest or in some other logical sequence.
- Select a horizontal scale that makes it possible to represent the full range of data on the sheet.
- Select a vertical scale in the same manner.
- Determine the place on the sheet where both scales come together. Mark this point of origin as zero (0) or any other appropriate value.
- Write the vertical scale. Start at zero and add values at each major division on the graph paper.
  Note. The starting (base line) point does not necessarily have to be zero because in many instances only higher values are needed.
- Repeat the same process for marking the horizontal scale on the graph paper.
  Note. Plan the spacing on both vertical and horizontal scales so the graph is balanced.
- Plot the values on the horizontal or vertical scale from an original table or compilation.
- Draw the bars as solid lines to furnish the required data.
- Label the scales and give the chart a descriptive title.

EXAMPLE: Develop a bar graph to show the variation in production for a one-year period from figure 31-1.

Month	Jan.	Feb.	Mar.	Apr.	May	June	July	Aug.	Sept.	Oct.	Nov.	Dec.
Prod. Units	500	800	1200	1350	1450	1550	1300	900	850	800	750	1150

FIGURE 31-1

Step 1   Study the information in figure 31-1 and determine the high and low ranges in production units. These are 1550 and 500.
Note. The months are arranged in a logical sequence as they will appear on the graph.

Step 2   Select a graph sheet with spaces ruled to meet the job requirements.
Note. If no printed form is available, rule a sheet with light lines drawn to a predetermined scale.

Step 3   Determine whether to represent the months on the vertical or horizontal scale and the production units on the adjacent scale. In this case, represent the months on the horizontal scale.

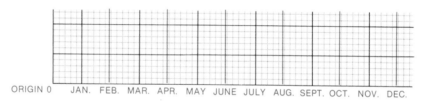

ORIGIN 0   JAN. FEB. MAR. APR. MAY JUNE JULY AUG. SEPT. OCT. NOV. DEC.

FIGURE 31-2

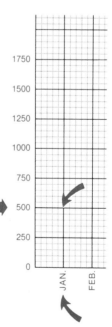

Step 4   Mark the major divisions or units on the vertical scale. Since the graph paper available is 10 × 10, each major vertical line is marked in intervals of 250 production units, starting at zero. Also, write what the scale represents.
Note. This scale is selected because, if a smaller one is used, the information will not fit the graph paper.

Step 5   Plot the height of the solid lines, starting with January. In this month, 500 units were produced. This value on the vertical scale is, as indicated, at the 500 point.

Step 6   Continue to plot the production units for the remaining months and recheck for accuracy.

Step 7   Label the graph so that the short descriptive title gives meaning to the facts.

FIGURE 31-3

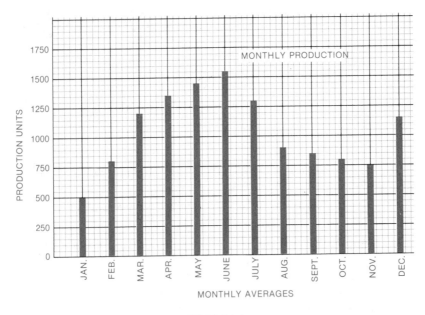

FIGURE 31-4

B. INTERPRETING BAR GRAPHS

The need for interpreting factual data from graphs already prepared makes the reading of graphs important. Regardless of the type of graph to be read, there are certain basic steps to be followed.

RULE FOR READING GRAPHS

- Study the problem to determine what information is given and what values must be determined from the graph.
- Read the title of the graph as a key to its organization and purpose.
- Determine the value which each major unit on the vertical scale represents.
- Read the horizontal scale for the value of each unit.
- Locate the given value or data on the horizontal scale.
- Follow an imaginary vertical line from this point to the end of the bar.
- Locate the length of the bar on the vertical scale.
- Continue to read other required values in the same manner.

Often the length of a bar does not fall on an even graduation. The extent to which a bar is above or below a graduated line may be estimated close enough for most practical purposes.

The same steps may be followed if the given values are on the vertical scale and the required value is represented somewhere on the horizontal scale.

EXAMPLE: Determine from the bar graph the number of pounds of brass plate used in manufacturing during the peak month and the lowest month.

FIGURE 31-5

Step 1	Determine which bar and month represent the peak.	**September**
Step 2	Read the value of each unit and subdivision on the vertical scale.	
	Note. The vertical scale reading = the unit reading (490 × 1000).	**490,000**
Step 3	Determine the shortest bar length.	**June**
Step 4	Read the value of this bar on the vertical scale (280) and multiply by 1000.	**280,000**
Step 5	State results in specific units of measure. The greatest production month was September when 490,000 pounds of brass plate were used. The lowest production month was June with 280,000 pounds of brass plate.	
Step 6	Continue to read the production for other months in the same manner.	

Review and Self-Test

ASSIGNMENT UNIT 31

A. Development of Bar Graphs (Practical Problems)

1. Make and label a bar graph to show how lumber is used according to the following average percents.

Fuel 15%   Building 50%   Paper Products 9%

Fires and Disease 11%   Miscellaneous 15%

2. Show the phenomenal motor car growth for periods A through J, according to the data given.

Period	Car Production	Period	Car Production
A	10,475,000	F	19,280,000
B	12,240,000	G	21,300,000
C	13,000,000	H	25,400,000
D	15,250,000	I	26,500,000
E	17,200,000	J	27,690,000

B. Interpretation of Bar Graphs (Practical Problems)

1. Study the horizontal bar graph.
   a. Select the greatest production year.
   b. Select the year with the smallest production.
   c. Determine the percent of increase in production between 1981 and 1984.

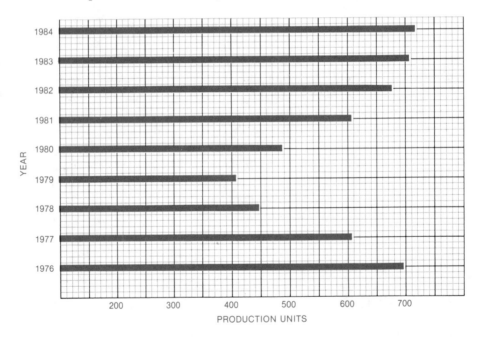

2. Study the bar graph of productivity.
   a. Select the three years of greatest productivity.
   b. Determine the average productivity for these three years, correct to two decimal places.
   c. Select the two years of lowest productivity.

d. Determine the average productivity for these two years.
e. Find the percent of increase in productivity between the average of the three highest years and the average of the three lowest years. (Round off answer to the nearest whole percent.)

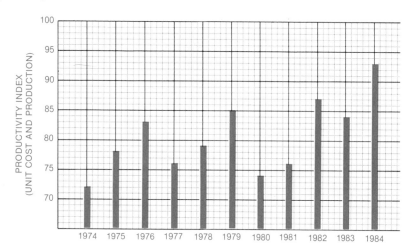

Unit 32 Development and Interpretation of Line Graphs

The *line graph* is the most widely used graph. It is easy to make, presents facts clearly, and is simple to interpret. Line graphs are of three general types: straight line, curved line, and broken line.

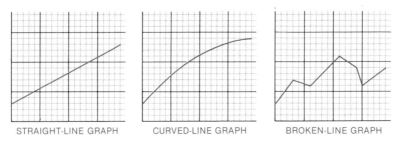

STRAIGHT-LINE GRAPH   CURVED-LINE GRAPH   BROKEN-LINE GRAPH

FIGURE 32-1

A. DEVELOPING LINE GRAPHS

The straight-line graph is used for related facts where there is some regularity in changes that take place. By contrast, the broken-line graph represents unrelated data where changes are irregular. The curved-line graph presents related information. In each instance, the graph is prepared from data that is either computed or available. One advantage of a graph is that once it is made, other values may be determined without additional computation.

RULE FOR DEVELOPING A LINE GRAPH

- Select appropriate horizontal and vertical scales so the facts may be presented clearly.
- Lay out the major graduations on graph paper for the vertical and horizontal scales and label each scale.
- Arrange the data in the same sequence as it will appear on the graph.
  Note. Sometimes these values must be computed from other given facts, using formulas.
- Plot the pairs of numbers on the horizontal and vertical scales. Place a dot or other identifying mark on the paper.
- Connect the points with either a straight edge or curve.
  Note. The line may be straight, broken, curved or a combination, depending on the relationship of data.

EXAMPLE: Make a line graph showing the variation in temperature from ground level to 8,000 feet.

Altitude in Feet	Ground Level	1,000	2,000	3,000	4,000	5,000	6,000	7,000	8,000
Air Temp. (°F)	85°	84°	79°	74°	64°	60°	47°	33°	10°

FIGURE 32-2

Step 1   Select suitable horizontal and vertical scales and label the graph.

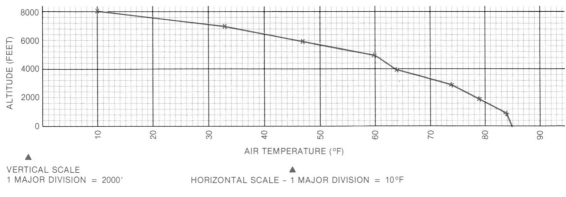

FIGURE 32-3

Step 2   Plot the points on the graph at which the pairs of numbers meet.

Step 3   Connect the points with a continuous line.

Step 4   Check the graph by taking altitudes of 2,500 feet and 6,500 feet, computing the air temperatures, and plotting these values on the graph.

Step 5   Check the original computations and plotting if the values being checked fall outside the graph line.

B. INTERPRETING LINE GRAPHS

The line graph furnishes information for comparisons of values without further computation. The results are, in most instances, accurate approximations, as it is impractical to represent data to too large a scale. The reading of a line graph is similar to that of a bar graph except that more data may be obtained.

RULE FOR READING A LINE GRAPH

- Determine what information is required.
- Locate the given value on either the horizontal or vertical scale.
- Visualize a horizontal or vertical line that passes through the given value and intersects the graph line.
- Determine the value of this point on the adjacent vertical or horizontal scale.
- Label the answer with an appropriate term.

EXAMPLE: Determine the air temperatures at altitudes of 7,500, 6,500, 5,500, 4,500, and 3,500 feet from the graph.

Step 1   Locate 7,500 feet on the vertical scale.

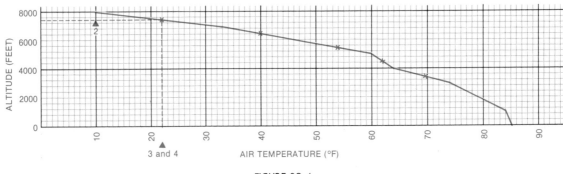

FIGURE 32-4

Step 2 Draw an imaginary horizontal line at 7500 feet until it intersects the graph line.

Step 3 Drop another imaginary vertical line from the intersecting point to the horizontal scale.

Step 4 Read the air temperature on the scale (22°F).

Step 5 Repeat these steps for the remaining altitudes.

Step 6 Label all answers and check.

C. DEVELOPING LINE GRAPHS WITH TWO OR MORE SETS OF DATA

A graph with two or more line graphs that are prepared from different sets of technical information is widely used to make comparisons and obtain other data. While common axes are used to represent particular factors and conditions, each series of values is plotted to develop a separate line graph.

RULE FOR DEVELOPING A GRAPH THAT COMBINES TWO OR MORE LINE GRAPHS

- Determine the maximum range of the data that are to be represented on the graph.
- Mark off the major divisions of the *X* and *Y* axes on the graph paper for the items and quantities to be represented.
- Start with one set of data. Plot the value of each *X* and *Y* axis item.
- Draw the appropriate straight-, curved-, or broken-line graph to connect each point as plotted.
- Label the first line graph.
- Continue to plot the second set of values.
- Connect the points with short dashes or other distinguishing type of line.
- Label the second line graph.
  Note. Additional line graphs may be drawn using the same procedure.
- Interpret values from any one or combination of line graphs, as required.
  Note. Follow the same steps as those that are used to read a single-line graph.

EXAMPLE: Construct a multiple line graph to show the variations in productivity for machine steel parts according to the following data for cutting fluids A and B.

Cutting Fluid A					
Surface Feet per Minute	20	30	40	50	60
Productivity (Parts per Hour)	150	155	160	165	166
Cutting Fluid B					
Surface Feet per Minute	20	30	40	50	60
Productivity (Parts per Hour)	150	165	175	183	186

FIGURE 32-5

Step 1   Determine the maximum ranges of production (150 to 186) and surface feet per minute (sfpm of 20 to 60).

Step 2   Lay out the horizontal scale to represent production values; the vertical scale, surface feet per minute. Label each scale.

Step 3   Plot productivity and cutting speed values for machining with cutting fluid A.
Note. Use a solid line to represent this production.

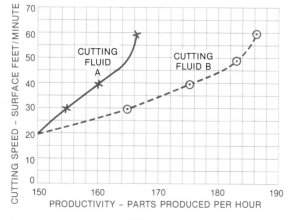

FIGURE 32-6

Step 4   Plot the values for machining with cutting fluid B.
Note. Use a dash line for easy identification and contrast.

Step 5   Make whatever comparisons are required by interpreting the differences in performance values between each line graph.

Review and Self-Test

ASSIGNMENT UNIT 32

A. Development of Line Graphs (Practical Problems)

1. Plot a line graph that shows how the specific gravity of a battery changes as the voltage is decreased.

Voltage	2.00	1.98	1.96	1.92	1.88	1.84	1.80	1.72	1.50
Specific Gravity	1.300	1.295	1.285	1.280	1.275	1.270	1.265	1.253	1.250

2. Show by a curved line graph the number of British thermal units (heat units) required to produce a temperature range of 10°F to 140°F.

Temperature (°F)	10°	50°	80°	110°	130°	140°
Btu	500	1000	1500	2000	3500	5000

B. Development and Interpretation of Line Graphs (Practical Problems)

1. Make a line graph of the surface speeds of grinding wheels. The diameters range from 8 inches to 16 inches inclusive. Increments in diameter are 1 inch. The constant speed is 2500 revolutions per minute ($\pi = 3.14$)

   a. Use the formula, Surface Speed $= \dfrac{\pi \times D \times RPM}{12}$

   b. Rounds off surface speed values to the nearest 50 feet per minute.

2. Locate on the line graph the surface speeds of wheels worn to diameters of $8\frac{1}{2}$, $9\frac{1}{2}$, $10\frac{1}{2}$, and $11\frac{1}{2}$ inches.

   Note. Give answer to closest 50 feet per minute.

3. Make a line graph that shows the relationship of the cross-sectional area of a square pipe to its length. Use the dimensions in the table to establish the cross-sectional area and length of side.

Lengths of Side (in inches)	4	5	6	7	8	9	10

4. Locate on the graph in problem B. 3 the cross-sectional area of square pipes $4\frac{1}{2}$, $6\frac{1}{2}$, $8\frac{1}{2}$, and $9\frac{1}{4}$ inches on a side. Check the area of each measurement by computation.

5. The straight-line graph shows the relationship between the diameter of a driving pulley and the surface speed of a belt (in feet per minute ['/min]). The driving pulley revolves at 300 RPM.

a. Determine the surface speed (in '/min) of the belt for the following drive pulley sizes (diameters): (1) $3\frac{1}{2}$, (2) 7, and (3) $10\frac{1}{2}$ inches.

b. Locate the driver pulley diameter sizes required to produce surface speeds of (1) 475 '/min, (2) 750 '/min, and (3) 875 '/min (approximate to the nearest $\frac{1}{4}$ inch diameter).

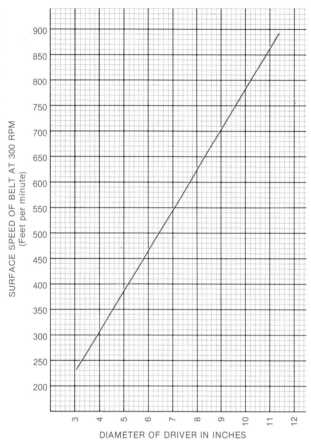

6. The curved (hyperbola) line graph represents the volume in cm^3 of certain materials. The volume is related to the density in grams per cubic centimeter (g/cm^3).

Estimate the density of materials A, B, C, and D.

	Material	Volume (cm^3)	Density (g/cm^3)
A	balsa	850	
B	cork	400	
C	maple	175	
	ice	110	

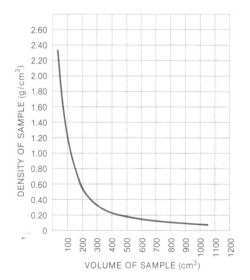

C. Interpreting Combination Multiple Line Graphs

1. The accompanying combination multiple line graph compares the production of electronic components from three different assembly systems. System A is semiautomatic. System B is numerically controlled using manual input. System C is computer-aided manufacturing.

   a. Identify the range of production hours involved in the study.
   b. Give the number of electronic components assembled in the first 42.5 hours in each system.
   c. State the productivity range of each system.
   d. Identify the system with the greatest variations in hourly production.
   e. Describe the effect on production among the three systems between 42.5 and 45.0 hours of operation and 45.0 to 47.5 hours.

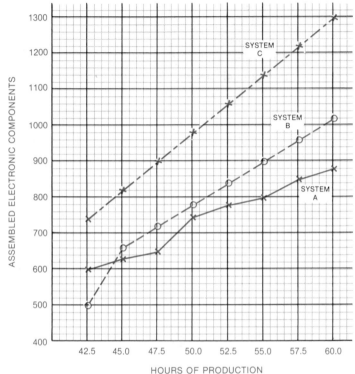

Unit 33 Development and Interpretation of Circle Graphs

OBJECTIVES OF THE UNIT

After satisfactorily studying this unit, the student/trainee will be able to
- *Translate technical information found on circle graphs to equivalent sector angles for portions of circles represented.*
- *Interpret data in order to develop circle graphs.*
- *Obtain required data from the interpretation of circle graphs.*

A circle graph, as the name implies, is a combination of a circle and the division of the circle into a given number of parts *(sectors)*. The circle graph (sometimes called a pie graph) is especially useful in showing how one part is related to another part and to the total.

A. DEVELOPING AND INTERPRETING CIRCLE GRAPHS

The circle graph differs from either the bar graph or the line graph in that it is used primarily for comparative purposes. It is impractical to determine other numerical values from the graph itself.

RULE FOR DEVELOPING CIRCLE GRAPHS

- Add together all of the items to be included on the graph. This sum is equal to 100%.
- Make a series of fractions of the data. The numerator represents one of the parts; the denominator, the total.
- Multiply the quotient of each fraction by 100 to get the percent equivalent each part is to the total.
- Multiply the quotient again, but this time by 360°, to get the equivalent angle of each part.
- Draw a circle large enough to divide easily into the required number of parts so that the graph may be read easily.
- Divide the circle into the number of degrees in each part *(sector)*.
- Label each part and the chart itself with a descriptive caption and the percent which that part represents.
- Determine the relationship of each part of the total. Then compare one part with another.

As an example, one of the best ways to make quarterly comparisons graphically is to use a circle graph.

EXAMPLE: Develop a circle graph to show quarterly production according to the information given in figure 33-1.

Quarter	First	Second	Third	Fourth
Production Units	12,500	25,000	31,250	18,750

FIGURE 33-1

Step 1 Total the production units.

12,500
25,000
31,250
18,750
87,500

Step 2 Determine the fractional part of the total each quarter represents.

First	Second	Third	Fourth
$\frac{12,500}{87,500} = \frac{1}{7}$	$\frac{25,000}{87,500} = \frac{2}{7}$	$\frac{31,250}{87,500} = \frac{5}{14}$	$\frac{18,750}{87,500} = \frac{3}{14}$

FIGURE 33-2

Step 3 Determine the percent of the total that each quarterly production represents.

First	Second	Third	Fourth
$\frac{1}{7} = 14\frac{1}{3}\%$	$\frac{2}{7} = 28\frac{2}{3}\%$	$\frac{5}{14} = 35\frac{5}{6}\%$	$\frac{3}{14} = 21\frac{1}{6}\%$

FIGURE 33-3

Step 4 Draw a circle and divide it into 14 parts. Then lay out each fractional part of the total.

FIGURE 33-4

Step 5   Label the chart with a descriptive title and the percent production each quarter.

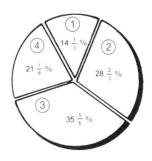

FIGURE 33-5

Step 6   Compare the data from the chart as may be required.

There are many ways of designating the parts of a circle graph. Another simple method is illustrated (*figure* 33-6). Colors or screens may be used to emphasize and to add interest. The technique depends on the purpose and the persons who are to translate the information.

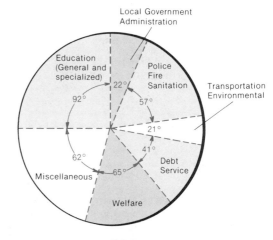

FIGURE 33-6

Review and Self-Test

ASSIGNMENT UNIT 33

A. Development of Circle Graphs (Practical Problems)

1. Prepare a circle graph to show the relationship between the number of men and women employed in a plant where there are 1,300 men and 300 women.

2. Make a circle graph to illustrate how each dollar is spent in a particular industry.

Item	Expenditure
Admin./Eng.	$100,000
Development	250,000
Tooling	180,000
Production	160,000
Sales and Service	90,000
Taxes	220,000

B. Interpretation of Circle Graphs (Practical Problems)

1. Study the circle graph.
   a. Determine total labor force in one industry of an area labor market.
   b. What percent of the total number employed is men? Women?
   c. How many men are in the 18–35 and 36–55 year age groups? How many women?

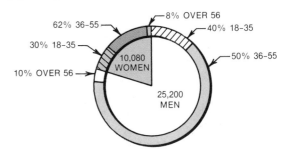

2. The following amounts were paid for raw materials: plastics, $7500; steel castings, $17,500; brass castings, $14,000; paints, $2500; and lumber, $8500.
   a. Prepare a circle graph to show the relationship of costs.
   b. Determine the percent of the total that was spent on each material.

Unit 34 Statistical Measurements

OBJECTIVES OF THE UNIT

After satisfactorily studying this unit, the student/trainee will be able to

- *Understand the importance of statistical measurements in the manufacture and control of interchangeable parts.*
- *Translate the importance of normal frequency distribution graphs in relation to quality control of measurements of manufactured parts and products.*
- *Interpret the value of single sampling, double sampling, and sequential sampling plans in statistical measurement.*
- *Solve practical problems of central tendency measurement involving computations for range, average and mean, median, and mode.*

The term *statistical measurement,* as used in this unit, relates to applications of basic mathematical processes in working with recorded data and a large number of measurements. These are widely used in quality control in the manufacture of interchangeable parts and components and in other everyday applications in agriculture, business, health, and marketing. This means that statistical measurements are applied to establish standards of acceptance of mass-produced parts, assembled units, and other products.

A. NORMAL FREQUENCY DISTRIBUTION OF DIMENSIONAL MEASUREMENTS

Size variations of a given number of manufactured parts, fabrics and other products may be established by directly measuring each unit. Graphs may then be prepared to project the number of cases and the actual sizes. For example, figure 34-1, a graph of dimensional size

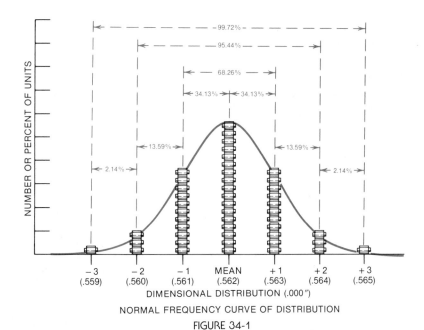

NORMAL FREQUENCY CURVE OF DISTRIBUTION

FIGURE 34-1

variations of 50 workpieces, shows the number of pieces that range in size from 0.565″ to 0.559″. The curve of the graph also presents the greatest number *(concentration)* of workpieces to be at the center line. In this case, the greatest number of parts is machined to what is called the *nominal (design) dimension* of 0.562″.

The curve of the graph is known as a *normal frequency distribution curve.* Verified over the years by a number of studies, the curve may be plotted mathematically. The parts are distributed so that 34.13% of a total production is within the first allowable tolerance zone of say, + 0.001″. An equal number of pieces fall within the − 0.001″ allowable tolerance zone. The total of 68.26% of the 50 parts on the graph is within ± 68.26%. Further, 95.44% of the parts are shown on the graph to be within the first and second allowable tolerance zones of ± 0.002″. In addition, 99.7% of the parts are within the maximum tolerance range of ± 0.003″. The cutoff point as to which parts meet the requirements of dimensional accuracy is determined by the parts or unit designer who specifies the maximum allowable tolerance that is acceptable.

B. SAMPLING PLANS AND STATISTICAL MEASUREMENT

Normal size variations are important to the designer, manufacturer, quality control inspector, and machine operator. Where parts are to be manufactured to extremely precise measurements, each part may be automatically inspected. Generally, a sampling plan, which is selected from technical tables, is used to establish the fixed number of parts to measure from the manufactured *lot (batch).* From the lot size sample, it is possible to determine the *average outgoing quality limit* (AOQL). The AOQL represents the percent of defective parts. If this percent is equal to or less than the allowable number of defective parts in the sampling plan (acceptable quality level, AQL), the lot is accepted. An excessive spread of rejects based on dimensional accuracy measurements indicates the need for more precise tooling, adjustments of machine conditions, or changes in other production practices.

Inspection Sampling Plans for Quantity Manufacturing
in Lot Sizes between 750 and 1250 Units (Parts)
(Based on an Acceptable Quality Level of .85 to 1.75)

Sampling Plans	Sample and Sequence	Sample Size	Combined Samples		
			Cumulative Size	Lot Acceptance Quantity	Lot Rejection Quantity
Single	Single	50	50	3	4
Double	First	34	34	1	5
	Second	68	102	4	5
Sequential	First	14	14	0*	2
	Second	14	28	0	3
	Third	14	42	1	4
	Fourth	14	56	3	5
	Fifth	14	70	3	5
	Sixth	14	84	3	5
	Seventh	14	98	4	5

*Two sample batches are required for acceptance.

FIGURE 34-2

Single Sampling Plan

There are three basic sampling plans: *single, double,* and *sequential.* The sampling plan table of lot sizes from 750 to 1250 parts (*figure* 34-2) shows that a random sample of 50 parts is required for a single sampling. If there are three or less parts in the 50 parts sampled that are defective, the whole lot (750 to 1200 parts) is accepted. If there are four or more rejects, the lot is rejected.

Double Sampling for Dimensional Accuracy

When the lot is rejected, each item in the sample is inspected. In the *double sampling plan,* a first sample of 34 parts is used. If one part in 34 is defective, the lot is accepted. But, if 2, 3, or 4 parts are rejected, a second sample of 68 additional parts is screened. The lot is accepted if the actual measurements of parts in the second sample produce up to four rejects; otherwise, the lot is rejected. The total number of parts *(cumulative size)* inspected in this double sampling example is 102.

Sequential Sampling of Measurements

In sequential sampling in this example *(figure* 34-2), there are seven different samples of 14 parts each for a total of 98 parts. If there is zero or one part defective in the first sample, a second sample must be used. Two or more rejects in the first sample calls for the lot to be rejected. Sampling continues as long as the parts fall within the lot acceptance and lot rejection numbers for each sample.

The product designer, engineer, or production department decides which sampling plan to use. The sampling plan depends upon the complexity of the part and the required degree of precision. Single sampling requires the inspection of the greatest number of units. This plan is used when manufacturing extremely accurate parts with high precision surface finish or other exacting specifications. Double and sequential sampling plans are practical with large lots where the part design has a comparatively wide range of acceptable tolerances.

C. CENTRAL TENDENCY TERMS AND APPLICATIONS

There are four common statistical terms used to describe representative variations in scores or measurements. The terms include *range, mean, mode,* and *median.*

Measurement Range

The simplest and most frequent way to report variations is to give the *range* between the highest and lowest measurement or score.

RULE FOR DETERMINING A MEASUREMENT RANGE

* Arrange the values in sequence from the lowest or smallest measurement or score to the highest or largest.
* Express the *range* (difference) from the highest to the lowest values.

EXAMPLE: Determine the range of sizes of the eight die cast part batches from the dimensions given in figure 34-3.

Batch Number	A	B	C	D	E	F	G	H
Number of Parts	30	26	24	18	16	2	2	2
Dimensional Measurements (inches)	1.125	1.126	1.124	1.127	1.123	1.128	1.122	1.120

FIGURE 34-3

Step 1 Arrange the measurements in sequence.	1.128″ 1.124″ 1.127″ 1.123″ 1.126″ ╱ 1.122″ 1.125″ 1.120″
Step 2 Determine the largest measurement.	(1.128″)
Step 3 Determine the smallest measurement.	(1.120″)
Step 4 State the range of the measurements from the largest to the smallest.	1.128″ to 1.120″ Ans

Determining the Average or Mean

The average is widely used in daily computations. The average is more technically known in statistical measurement as the *mean* (M).

RULE FOR DETERMINING THE AVERAGE OR MEAN

- Add all the scores or measurements.
- Divide the sum by the number of cases *(N)*.
- Label the measurement average in terms of the measurement unit.

EXAMPLE: Determine the average (mean) fuel consumption per hour from the data given in figure 34-4 for six diesel engines.

Diesel Engine	A	B	C	D	E	F
Fuel Consumption *(gal/hr)*	75.4	82.9	76.9	85.8	96.2	75.4

FIGURE 34-4

Step 1 Add the separate fuel consumption values.	75.4 85.8 82.9 ╱ 96.2 76.9 75.4 492.6
Step 2 Divide the sum (492.6) by the number of diesel engines (6).	6)492.6
Step 3 Express the mean or average in terms of the measurement unit.	82.1 82.1 gal/min Ans

Determining the Mode of a Measurement Series

The *mode* is another measurement of central tendency. Mode means the most frequently used score in a measurement series. Mode is the midpoint of the measurements. Mode is also referred to in statistics as the *class interval*. In the previous example, 75.4 gallons per hour represents the most frequent value even though there are only two cases. If there are a great number of values between a range, like 42 and 45, the mode is the value that is midway between the two values. In this instance, the mode is 43.5.

Determining the Median

The *median* is another measure of central tendency. The median is the midpoint in a distribution of measurements or other values that are arranged in the order of size. This means the distribution is bisected: 50% fall above the median; 50% fall below.

RULE FOR DETERMINING THE MEDIAN

- Arrange the scores or measurements (*figure* 34-4) in sequence according to size.
- Determine the midpoint in the series of scores.

$$50\% \text{ of Scores} \begin{cases} 75.4 \\ 75.4 \\ 76.9 \end{cases}$$

Midpoint - - - - - - - -

$$50\% \text{ of Scores} \begin{cases} 82.9 \\ 85.8 \\ 96.2 \end{cases}$$

- Add the value of the score immediately above and below the midpoint and divide the sum by 2.

$$\frac{76.9 + 82.9}{2} = 79.9$$

- Label the median answer in the required measurement units.

79.9 gal/hour Ans

Review and Self-Test

ASSIGNMENT UNIT 34

A. Normal Frequency Distribution of Dimensional Measurements (Practical Problems)

1. Refer to the graph, which shows the distribution of measurements for the production of 10,000 precision-ground ball bearings. The measurements and numbers are plotted and form a normal frequency distribution curve. First-quality bearings fall within the $+1$ to -1 distribution range. Second-quality bearings, having a higher dimensional tolerance, are identified between $+1$ and $+2$ and -1 and -2 ranges. Bearings with higher dimensional variations are rejected.

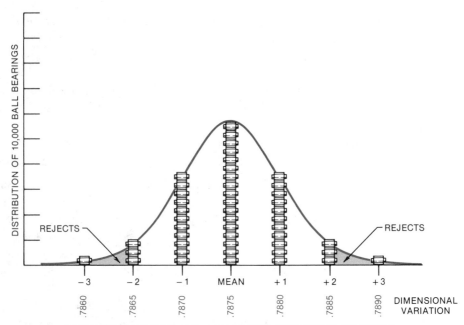

NORMAL FREQUENCY CURVE ACCORDING TO DIMENSIONAL VARIATIONS

a. (1) State the percent of first-quality bearings that fall into the + and − ranges. (2) Calculate the number of first-quality bearings in the + range of the distribution and the − range.

b. Calculate the total number of second-quality bearings that is acceptable.

c. Determine the total number of rejects.

B. Sampling Plans for Statistical Measurements (Practical Problems)

1. Use the data from the table of three sampling plans.

   a. Identify the size of single sample required for a 1500-parts lot within the acceptable quality level of the table.

   b. State the conditions under which the quantity is accepted or rejected.

   c. Identify the number of parts to be inspected in the first sample if a double sampling plan is used.

   d. Explain when a second sampling is required. Give the total number to inspect in the double sampling plan.

   e. Assume a sequential sampling plan is used. (1) Indicate the number of sequences and (2) the total number of the 1500 parts that are to be inspected.

Inspection Sampling Plans for Lot Sizes between 1,300 and 2,000
Parts with an acceptable quality level of 1.0 to 2.2

Sampling Plans	Sample and Sequence	Sample Size	Combined		
			Cumulative Size	Lot Acceptance Quantity	Lot Rejection Quantity
Single	Single	75	75	4	5
Double	First	42	42	2	5
	Second	84	126	5	6
Sequential	First	16	16	0*	2
	Second	16	32	0	3
	Third	16	48	1	4
	Fourth	16	64	3	5
	Fifth	16	80	3	5
	Sixth	16	96	3	5
	Seventh	16	112	4	5
	Eighth	16	128	4	5

*Acceptance is based on the inspection of two samples.

C. Determining Central Tendency Measurements (Practical Problems)

1. Refer to the table of results of physical laboratory tests of tensile strength of seven brass/bronze alloy specimens. The alloys are listed according to SAE specification numbers.
   a. Determine the range of tensile strength among the specimens.
   b. Compute (1) the average (mean) tensile strength, (2) the median tensile strength, and (3) the mode.

Brass/Bronze Alloys	A	B	C	D	E	F	G
SAE Specification Number	40	41	43	430	62	64	65
Tensile Strength (lb/sq in.)	26,000	20,000	65,000	90,000	30,000	25,000	20,000

2. Use the data given in the table for the weekly production of electronic devices for a six-week period.
   a. Give the (1) range of units produced each week and (2) the defects.
   b. Determine the average number of units produced each week.

c. Identify the production mode.

d. Calculate the median of the production number.

e. Find the number of defective parts for each weekly production run.

f. Determine the (1) mean, (2) median, and (3) mode percents for the defective electronic units produced over the six-week production period.

Weekly Production Units	1	2	3	4	5	6
	1,000	3,000	1,400	1,800	1,000	2,000
Number of Defects	2	9	5	6	3	6

Unit 35 Achievement Review on Graphs and Statistical Measurements

OBJECTIVES OF THE UNIT

This Achievement Review serves as an overall test for Section 7. The unit is designed to measure the student's/trainee's ability to

- *Understand the function of graphs and practical applications in industry.*
- *Prepare and read bar graphs from written and tabular data.*
- *Develop and interpret line graphs.*
- *Apply word statements and quantities in developing and reading circle graphs.*
- *Solve problems relating to central tendency and statistical quality control measurements.*

A. DEVELOPMENT AND INTERPRETATION OF BAR GRAPHS

1. Compute the Celsius melting points of metals A through E.

   Note. Find each Celsius reading by multiplying the Fahrenheit reading minus 32, by $\frac{5}{9}$; $\left(C = (F - 32) \times \frac{5}{9}\right)$.

	Metals	Fahrenheit	Celsius
A	Chromium	2,740°	
B	Cast Iron	2,300°	
C	Copper	1,940°	
D	Aluminum	1,200°	
E	Lead	620°	

2. Prepare a bar graph to show the melting point temperatures on both the Fahrenheit and Celsius scales for all metals in the table.

   Note. Use a solid bar for the Celsius readings and a dotted outline for Fahrenheit.

3. Determine from the graph what the differences are in Celsius readings between chromium and lead, cast iron and aluminum, and copper and lead.

B. DEVELOPMENT AND INTERPRETATION OF LINE GRAPHS

1. Statistics compiled on the exhaust temperatures of low-, medium-, and high-compression ratio engines, running at different RPM, are given in table form. Translate these facts into a line graph.

   Note. Use a solid line for medium compression, dotted line (– – –) for low compression, and an alternate solid-dash line (— –) for high compression.

Compression Ratio	Exhaust Temperature (°F)					
Low	780	870	950	1,030	1,100	1,160
Medium	825	920	1,000	1,080	1,150	1,220
High	880	970	1,060	1,130	1,210	1,275
RPM	1,000	1,500	2,000	2,500	3,000	3,500

2. What is the average difference in exhaust temperature between a medium- and a low-compression ratio engine as shown on the graph?

3. Locate on the graph the exhaust temperature of a high-compression ratio engine whose RPM is 1,750, 2,250, 2,750, 3,250, and 3,750.

4. Determine from the graph what the RPM of a low-compression engine is for these exhaust temperatures: a) 950°F; b) 1,050°F; c) 1150°F; d) 1250°F.

5. What is the difference in exhaust temperatures between a high- and a low-compression engine turning at 3250 RPM?

C. DEVELOPMENT AND INTERPRETATION OF CIRCLE GRAPHS

1. Draw a circle graph to compare the productive capacity of each country to manufacture a specific product.

Country	Productive Capacity
United States	175,000
France	90,000
Great Britain	125,000
Canada	100,000
All others	60,000

2. Give the percent which each country produces of the total.
3. Determine the percent of the total that is produced by the two largest producers (to the nearest two decimal places).
4. Show by a circle graph the percent of the total produced by the two largest, two smallest, and all other producers.

D. STATISTICAL MEASUREMENTS (PRACTICAL PROBLEMS)

1. a. State the purpose of using sampling plans in the production of automotive parts in quantities of 5,000 or more.
   b. Identify two differences between a single sampling plan and a sequential sampling plan in the manufacture of interchangeable parts.
2. a. Prepare a broken-line graph that shows the °F (and corresponding °C) tempering temperature range for the different plain carbon steel tools as recorded in the table.

Plain Carbon Steel Tools		Tempering Temperatures	
		°F	°C
A	Roughing mills	430	221
B	Counterbores	460	238
C	Knurls	485	251
D	Tube cutters	485	251
E	Taps	500	260
F	Threading dies	530	277
G	Pneumatic tools	580	302
H	Noncutting tools	640	338

   b. Compute the average (mean) tempering temperature in °F and °C.
   c. Determine (1) the median temperature and (2) the mode temperatures in °F and °C.

PART TWO

Fundamentals of SI Metric Measurement

METRICATION: SYSTEMS, INSTRUMENTS, AND MEASUREMENT CONVERSIONS

Unit 36 SI Metric Units of Measurement

OBJECTIVES OF THE UNIT

After satisfactorily studying this unit, the student/trainee will be able to

- *Understand why SI metrics is an evolving international system of measurement standards that is affected by high-technology developments.*
- *Translate the advantages of SI metrics over customary units of measure in mathematical computations.*
- *Apply the Scientific Notation System using prefixes, symbols, and power of ten values.*
- *Use established guidelines of writing problems involving SI metric units of measure.*
- *Solve practical problems in SI metrics relating to direct and computed linear measurements and computations of areas, volumes (including liquid quantities), mass (weight), and temperatures in Fahrenheit and Celsius degrees.*

This section has five units, including an end-of-section achievement review. SI metrics is introduced in the first of the units. Some advantages are given in contrast with customary inch and other British and American units of measurement. Applications are made of the Scientific Notation System, writing styles, and basic mathematics. These are related to problems involving linear, circular, area, volume and liquid, and temperature measurements.

Direct precision instruments and gages are applied in the Unit 37 to linear measurements. Principles of metric measurement with inside, outside, and depth micrometers are covered. Experience with vernier micrometers, height gages, and gage blocks is extended to cover the use of metric measuring instruments.

Unit 38 deals with the seven base units, common derived units, and supplementary units in SI metrics. Attention is directed to soft and hard conversion of measurements, tables of conversion values, and converting dimensions and other quantities between different measuring systems.

A. HISTORICAL PERSPECTIVE ON STANDARDS AND UNITS OF MEASURE

The initial system of metric measurement was developed by French scientists in 1799. The *meter* was chosen as the unit of length. The defined length of the meter in relation to the distance between the two geographic poles was accurate for measurements up through the middle 1980s. Interchangeable manufacture and later developments required more precise measurements. Over the years, slightly different standards and systems of measurement developed in other European countries. Currently, SI metrics is the system of measurement that is recognized worldwide. Measurements in the system are used alone or in connection with customary British/ American standards and units of measure.

The metric system is not new in the United States. It has been used for a number of years in the medical, pharmaceutical, and optometry industries, as well as in other sciences. Metric measurements have been adapted more recently to machine tools and to metal products manufacturing; the automotive, electronic, and aerospace industries; instrumentation and other business and industry applications.

The ''new'' SI metrics has evolved from French, British, German, and other European metric systems as a *compromise system*. Care must be exercised in interpreting the specific characteristics of SI metrics. When a new term or base unit is considered by the international standards-setting body and is included in its technical papers, this does not mean that the new term or unit of measure or a new standard is adopted and incorporated into the system. This action results after appropriate specialized groups study the implications and assess the value of any proposed change. After the item is resolved by committee action, it is brought up for adoption before the international body. Thereafter, its adoption by the member nations is a matter for each nation to decide.

For more than 125 years there has been an international body known as the *Conference Général des Poids et Mesures* (CGPM or the General Conference on Weights and Measures). This conference is the controlling body of the International Bureau of Weights and Measures. The Bureau is responsible for preserving metric standards and comparing the standards of different nations with the metric system. The bureau also conducts research and develops new standards of measurement. The member nations participate in research, assess findings and recommendations, and vote for international adoption by the *conference*. The matter of acceptance and implementation by the individual member nations rests with responsible groups in each nation. The National Bureau of Standards is the United States representative to the CGPM.

The original metric system of the eighteenth century contained the meter as the keystone measurement. All other elements in the European metric system were derived from the meter, a *unit of length*. The *kilogram* was later added as a *unit of mass (weight)*. With these two base units, other units were ''derived'' for the measurements of length, area, volume, capacity, and mass.

Impact of Technology on Standards for Measurement

Standards of measurement are made in response to growing needs. Some result from accelerated developments in industry, agriculture, health, and science and other major economic growth areas. Other standards are brought about by increased consumer needs. Technological progress usually creates a need for new base units and other derived units of measure.

For example, new developments and demands within the electrical and electromagnetic fields required the adoption of the *ampere* (A) as a base unit of measure. The ampere was accepted in 1950 by the tenth CGPM. Four years later, the *candela* (cd) was recognized as the unit of light intensity and the *degree kelvin* (°K) as the base unit of temperature. These base units, together with the *second(s)* were redefined and revised at the twelfth and thirteenth (1964 and 1967) CGPM. In 1964 the liter (ℓ) was adopted as the synonym for cubic decimeter (dm^3). At the same time, the use of the name liter was discouraged for precision measurements.

The *mole* (mol) is the seventh base unit in SI metrics. It was added during the fourteenth CGPM in 1971. Other descriptive measurement terms for derived SI units, like *pascal* (Pa), *siemens* (S), and *newton* (N) were also accepted. It is important to note the period over which SI metrics has been evolving. A number of years elapses between recognition of need, research, recommendations, acceptance, and adoption by CGPM. Added to this time is the interval required within each nation to determine its own system of measurement.

B. ADVANTAGES OF SI METRICS

SI metrics is recognized as an international system of accurately measuring and quantitatively defining all measurable objects. "SI" was universally adopted as the abbreviation of the International System of Units *(Système International d'Unités)* at the 1960 CGPM. SI is also referred to as "SI metrics," the term used in this book. Terms and certain units are spelled according to accepted American standards. The following are some of the most important advantages of SI metrics.

- Seven base units are used in the *measurement of every known physical quantity*. These seven include the *meter* (length), *kilogram* (mass), *second* (time), *kelvin* (temperature), *ampere* (electric current), *candela* (light intensity), and the *mole* (substance of a system). In addition, there are two supplementary units, *radian* and *steradian,* and a number of derived units.
- A well-defined set of symbols and abbreviations is established, which is adequate to define all conditions and phenomena.
- Mathematically, a set of prefixes, multiples, and submultiples is used to simplify computations involving large values or number of digits.
- SI is a "coherent system." The product or quotient of any two unit quantities in the system is a unit of the resulting quantity. For example, in a coherent system, multiplying unit length by unit length produces unit area.
- SI base units are precisely defined and are reproducible in laboratories in each country. The one exception is the kilogram (mass) standard. This is still preserved in the International Bureau of Weights and Measures.
- The SI system may be related by powers of ten to other units that are not a part of the system.
- SI provides for a single international standards system that affects the interchangeability of parts, processes, components, and systems.

C. SCIENTIFIC NOTATION SYSTEM: SI PREFIXES, SYMBOLS, AND POWER OF TEN VALUES

The term *SI* or *SI metric* is used in this textbook as an abbreviation of *le Système International d'Unités*. The evolving system was named by the *Conference Général des Poids et Mesures* (CGPM) in 1960. Quantities are measured and computed in SI metrics according to the *scientific notation system*. This system has many advantages. Mathematical processes involving many multiple-digit values are simplified. Errors in calculations are reduced. Quantities are easier to read and write.

The quantity 10 is used as the foundation of the scientific notation system. The writing of mathematical quantities and mathematical computations are simplified by using what are called *power of ten multiple* (greater than 1) *and submultiple* (smaller than 1) *values*. A number of the powers are further identified by a *prefix* or a *symbol* for the prefix. A prefix identifies a specific value or quantity associated with a particular unit of measurement. For instance, *centi* in centimeter means one-hundredth part of a meter. *Centi* is called the prefix.

(A) Power of Ten Multiple Values			
Prefix	**Symbol**	**Value as Power of Ten**	**Multiplication Factor**
deka-	da	10	10
hecto-	h	10^2	100
kilo-	k	10^3	1 000
mega-	M	10^6	1 000 000
giga-	G	10^9	1 000 000 000
tera-	T	10^{12}	1 000 000 000 000
(B) Power of Ten Submultiple Values			
deci-	d	10^{-1}	0.1
centi-	c	10^{-2}	0.01
milli-	m	10^{-3}	0.001
micro-	μ	10^{-6}	0.000 001
nano-	n	10^{-9}	0.000 000 001
pico-	p	10^{-12}	0.000 000 000 001
femto-	f	10^{-15}	0.000 000 000 000 001
atto-	a	10^{-18}	0.000 000 000 000 000 001

FIGURE 36-1 Selected SI prefixes, symbols, and power-of-ten multiple and submultiple values

Selected prefixes, symbols, and power of ten values are given in figure 36-1. Section (A) shows decimal *multiple values* of SI units. Note the simplified way of writing one million (1,000,000) as (10^6). The 10^6 is the same as the value obtained by multiplying 1 by 10 for 6 times. The designation for 1,000,000 is *mega*. *Mega* (M) is called a prefix.

Every measurement is stated in terms of (1) a *quantity* and (2) a *unit of measure*. An answer of 10.2×10^6 states a quantity. It must be more descriptive, however, and also relate to a specific unit of measure. If the problem deals with waveforms that are measured by cycles (or frequency), the unit of measure is *hertz* (Hz). Thus, the answer is stated as 10.2 MHz (megahertz). This quantity quickly defines 10,200,000 hertz.

Section (B) in figure 36-1 shows decimal *submultiple* values of SI units. Note that there is a different set of prefixes and symbols required. In either instance, a prefix and a symbol are followed by a particular unit of measure. The many mathematical processes and practical applications that are based on the scientific notation system follow.

D. GUIDELINES FOR COMMUNICATING IN SI METRICS

As a coherent system, SI metrics requires a uniform set of rules for communication. *Numerical* and *literal* (letter) *quantities* are identified with specific SI units of measure. The use of SI metric units follows prescribed guidelines.

- Specific measurements (quantities) are given in terms of base units, derived units, supplementary units, and combinations of these.
- Digits and decimals, in excess of a required degree of precise measurement, are eliminated.
- Multiple and submultiple power of ten quantities are used in computations.
- Prefixes express the "order of magnitude" of a quantity. For example, 200 kg and 24.26 MV define precisely a 200 kilogram mass and a 24.26 megavolt circuit, respectively.
- A prefix is considered to be combined to the symbols to which it is attached.
  Example: $1 \text{ cm}^3 = (10^{-2} \text{ m})^3 = 10^{-6} \text{ m}^3$
- Prefix values of 1 000 are preferred over lesser or greater values. For example, milli- (m, 10^{-3}), micro- (μ, 10^{-6}), and pico- (p, 10^{-12}) are used in expressing submultiple values. The most common and preferred power of ten multiples are the kilo- (k, 10^3), mega- (M, 10^6), and giga- (G, 10^9).
- Units of measure that are derived from proper names are capitalized; for example, volts (V), siemens (S), and henry (H).
- Numerical prefixes of mega-, giga-, and tera- are capitalized: (M), (G), and (T), respectively.
- SI metric symbols of units of measure are written in singular form. For example, 125 megavolts is written as 125 MV; 12.7 kilometers as 12.7 km.
- Numerical quantities are written in groups of three digits. Commas are omitted. A quantity like 96243.8275 pascals of pressure is written 96 243.827 5 Pa.

E. APPLYING METRIC UNITS OF LINEAR MEASURE

The standard units of linear measure in the metric system are given in figure 36-2. Unit values appear with appropriate symbols that simplify the writing of each unit.

	Metric System		
	Common Linear Units	Symbol	Value of Unit in Terms of the Meter
	1 meter	m	Standard Unit of Length
Submultiple Values	1 decimeter $= \frac{1}{10}$	dm	0.1 meter
	1 centimeter $= \frac{1}{100}$	cm	0.01 meter
	1 millimeter $= \frac{1}{1000}$	mm	0.001 meter
Multiple Values	1 dekameter $= 10$	dam	10. meters
	1 hectometer $= 100$	hm	100. meters
	1 kilometer $= 1\ 000$	km	1 000. meters

FIGURE 36-2 Values of metric units of linear measure

Where dimensions in the metric system are expressed in more than one kind of unit in a system of measure, the different kinds may be combined to simplify the mathematical processes. For example, a dimension 3 decimeters 4 centimeters 6 millimeters long (3 dm 4 cm 6 mm) may be expressed as 3.46 decimeters. This value is based on the fact that 10 cm = 1 dm and 100 mm = 1 dm. The 4 cm is equivalent to $\frac{4}{10}$ and the 6 mm to $\frac{6}{100}$ of a decimeter. The dimension then is equal to $3 + \frac{4}{10} + \frac{6}{100}$ decimeters. Expressed as a decimal, this value is 3.46 dm.

Dimensions in the metric system that are given as units of linear, surface, or volume measure may be simplified in the same way.

Direct Linear Measurements in the Metric System

Metric measurements may be computed or taken directly. Measuring tools and precision instruments are similar to those used for customary units. The measurements are different, however. Whole and fractional part values in the customary system are related to the inch as the base unit. Metric measurements are in terms of the meter as the base unit.

Comparatively rough measurements may be taken directly with meter sticks and rulers graduated in centimeters and fractional parts. More accurate measurements are made with other line-graduated measuring tools. Metric triangular engineering and flat rules, drafting machine rules, and metal rules are examples. Some of these tools are graduated for measurements in one millimeter to $\frac{1}{2}$ millimeter (mm) range. Measurements to accuracies of 0.02 mm, 0.002 mm, and finer are covered in Unit 37 in relation to vernier measuring tools and gage blocks.

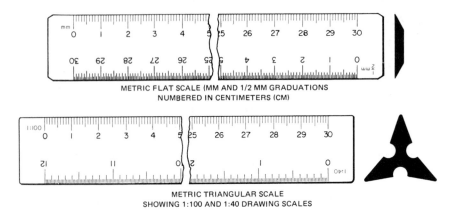

METRIC FLAT SCALE (MM AND 1/2 MM GRADUATIONS
NUMBERED IN CENTIMETERS (CM)

METRIC TRIANGULAR SCALE
SHOWING 1:100 AND 1:40 DRAWING SCALES

FIGURE 36-3

F. COMPUTING SQUARE MEASURE (AREAS) IN SI METRICS

The principles of surface measure that are used with the British customary units are applied in the same manner with the metric units of measure. The names of the units and the values of each differ in both systems, however.

The area of a surface in the metric system is the number of square metric units that it contains. The three metric units most commonly used for small areas are the square meter, square centimeter, and square decimeter.

One square meter represents the area of a square figure that is one meter long and one meter high. The linear meter, in turn, is equal to 10 decimeters or 100 centimeters. By substituting these values for the meter, the area of a square 10 decimeters on a side is 100 square decimeters. The area of a square 100 centimeters on a side is 10,000 square centimeters. Thus, the value of the square meter, which is the standard unit in terms of decimeters and centimeters, is 100 square decimeters or the equivalent, 10,000 square centimeters.

The three common metric units of square measure and the value of each unit in the metric system are given in figure 36-5.

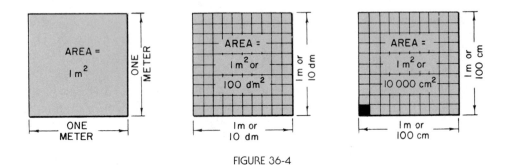

FIGURE 36-4

Metric Unit of Surface Measure	Value of Unit in Metric System
1 sq meter (m^2)	Standard Unit of Measure
1 sq decimeter (dm^2)	0.01 square meter (sq m)
1 sq centimeter (cm^2)	0.000 1 sq m
1 sq millimeter (mm^2)	0.000 001 sq m

FIGURE 36-5 Values of metric units of square measure

The principles of surface measure that apply to the area measurement of squares, rectangles, parallelograms, trapezoids, triangles, circles, sectors of a circle, and cylinders in British units are the same for the metric system. The only difference is in the name of the unit and its size. The same rules for determining the area of a surface apply for both the British and metric systems.

G. DETERMINING VOLUME MEASURE IN SI METRICS

The volume of a body is the measurement of its cubical contents. These are expressed in cubic units of the same kind as the linear units. When the linear dimensions are given as metric units, the volume is the number of cubic metric units that the body contains. The common metric

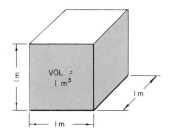 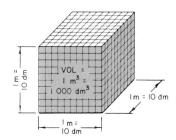

FIGURE 36-6

units are the meter, decimeter, centimeter, and millimeter. The volume is computed as so many cubic meters (m^3), cubic decimeters (dm^3), cubic centimeters (cm^3), or cubic millimeters (mm^3).

Since 1 meter = 10 decimeters, the volume of a cube 10 decimeters on a side is 1 000 cubic decimeters (dm^3), which is equivalent to 1 cubic meter (m^3).

One cubic decimeter is the volume of a cube one decimeter long, one decimeter high, and one decimeter deep. One cubic decimeter is equivalent also to the volume of a cube that measures 10 centimeters on a side (10 cm = 1 dm) or 1 000 cubic centimeters.

One cubic centimeter is the volume of a cube one centimeter long, one centimeter high, and one centimeter deep. Since the centimeter is equal to 10 millimeters, one cubic centimeter is equal to the volume of a cube 10 millimeters on a side, or 1 000 cubic millimeters. Values in the metric system are given in figure 36-7 for the cubic meter, cubic decimeter, and cubic centimeter.

Metric Unit of Volume Measure	Value of Unit in Metric System
1 cu meter (m³)	1 000 cu decimeters (dm³)
1 cu decimeter (dm³)	1 000 cu centimeters (cm³)
1 cu centimeter (cm³)	1 000 cu millimeters (mm³)

FIGURE 36-7 Values of common metric units of volume measure

The measurement of the cubical contents of squares, rectangular solids, and cylinders is the same for both the British and metric systems of volume measure. The volume of a solid is always computed in the same manner regardless of the system in which the linear dimensions are expressed.

H. PRINCIPLES OF VOLUME MEASURE APPLIED TO LIQUID MEASURE (METRIC SYSTEM)

The standard unit of liquid measure in the metric system is the liter (ℓ). This unit, as defined by law, is equivalent to a volume of 1 000 cubic centimeters or 1 cubic decimeter.

	Value of the Liter
Metric Unit of Liquid Measure	Value of Unit in Metric System
1 liter (ℓ)	Standard Unit of Liquid Measure
	1 000 cm³
	1 dm³

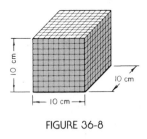

FIGURE 36-8

The same methods of computing the liquid capacity of a container are used regardless of whether the volume is expressed in British or metric units. Known values in one system may be changed readily to a desired unit in the other system.

I. DETERMINING METRIC UNITS OF MASS (WEIGHT) MEASUREMENT

Mass in physical science terms means the measurement of the earth's force in attracting a material. Mass and weight are often used interchangeably. The *gram* (g) is the base unit of mass measurement in SI metrics. Technically, one gram represents the mass (weight) of one cubic centimeter of water at 4° Celsius temperature.

Prefixes are used to identify larger and smaller quantities than one gram. The prefixes *deka, hecto,* and *kilo,* and the *metric ton* are used for larger units. Smaller weight units are identified by the prefixes *deci, centi,* and *milli.* The value of each unit of mass measurement in relation to the gram is recorded in the table.

1 000 kilograms	=	1 metric ton
1 000 grams	=	1 kilogram (kg)
100 grams	=	1 hectogram (hg)
10 grams	=	1 dekagram (dag)
1 gram (g)	=	Base Unit of Measure
0.1 gram	=	1 decigram (dg)
0.01 gram	=	1 centigram (cg)
0.001 gram	=	1 milligram (mg)

Smaller Units ↓   Larger Units ↑

FIGURE 36-9 Common units of metric mass (weight) measure

RULE FOR CHANGING METRIC UNITS OF MASS (WEIGHT) MEASURE

* Determine whether the given unit of measure is to be changed to a larger or a smaller unit.
* Multiply by 10 (or add a zero digit while moving the decimal point one place to the right) to change from one SI metric unit of mass measure to the next smaller unit.
  Note. When changing to successively smaller units of measure, add additional zeros to the right according to the quantity relationship.
* Divide by 10 to change from one unit of mass measure to the next larger unit or by a multiple of ten depending on the required larger unit.
* Express the changed value in terms of the required unit of measure.

EXAMPLE: *Case 1.* Change 1.24 grams to equivalent centigram units.

Step 1	Determine the relationship between a gram and a centigram.	**1 centigram (cg) = 0.01 gram (g)**
Step 2	Change the larger gram unit to the smaller centigram unit by multiplying by 100.	**100 cg = 1 g** **1.24 × 100 = 124**
Step 3	Express the answer in the required measurement unit.	**124 grams (g)** Ans

EXAMPLE: *Case 2.* Change 4 746 kilograms to the equivalent metric tons.

Step 1	Determine the number of kilograms in a metric ton.	**1 000 kg = 1 metric ton**
Step 2	Divide by 1 000 by moving the decimal point three places to the left.	**4 746**
Step 3	Label the answer.	**4.746 metric tons** Ans

J. DETERMINING CELSIUS AND FAHRENHEIT TEMPERATURE MEASUREMENTS

The metric Celsius and the Fahrenheit temperature systems of measuring temperature changes are the two most commonly used systems. The unit of temperature measurement in both systems is called *degree* and is denoted by the symbol (°). However, the value of 1° on the Fahrenheit scale differs from 1° on the Celsius scale.

Temperature scales are calibrated in relation to the boiling and freezing points of water at a standard pressure. On the Fahrenheit scale, 32° is used to indicate the temperature at which water freezes; 212° indicates the boiling point. The corresponding points on a Celsius scale are 0° for freezing and 100° for boiling. By comparison, the freezing points are 0°C and 32°F and the boiling points are 100°C and 212°F. This means there are 100 one-degree graduations on the Celsius scale and 180 one-degree graduations on the Fahrenheit scale, beginning at 32°F. Celsius degree temperatures that are below the equivalent 32°F are always negative. For example, a temperature like 27°F is equal to -5°C. Formulas requiring simple mathematical processes are used to convert temperature values from one scale to the other.

RULE FOR CONVERTING TEMPERATURE MEASUREMENTS BETWEEN SYSTEMS

- Use the formula $°C = (°F - 32) \times \frac{5}{9}$ to change °F to °C.

- Use the formula $°F = \left(°C \times \frac{9}{5}\right) + 32$ to change °C to °F.

- Substitute given temperature values in the appropriate formula.

- Perform the mathematical processes. Label the answer in the required temperature system.

EXAMPLE: *Case 1.* Determine the equivalent temperature in degrees Celsius for subzero heat treating steel gages at -103°F.

Step 1	Write the formula for changing °F to °C.	$°C = (°F - 32) \times \frac{5}{9}$
Step 2	Substitute °F values.	$°C = (-103 - 32) \times \frac{5}{9}$
Step 3	Perform the required mathematical processes.	$= (-135) \times \frac{5}{9}$
Step 4	Express the answer in the required temperature measurement.	$= 75°C$ Ans

EXAMPLE: *Case 2.* Compute the equivalent °F temperature required to anneal an SAE 1040 steel that requires heating to 665°C.

Step 1	Use the formula.	$°F = \left(°C \times \frac{9}{5}\right) + 32$
Step 2	Substitute the given °C temperature value of 665.	$°F = \left(665 \times \frac{9}{5}\right) + 32$
Step 3	Perform the mathematical processes.	$= (1197) + 32$
Step 4	Label the answer in °F.	$= 1229°F$ Ans

Review and Self-Test

ASSIGNMENT UNIT 36

A. Applying the Scientific Notation System and SI Metric Units of Measurement (General Applications)

1. Match each term in column I with its corresponding value in column II.

Column I	Column II
Term	*Value*

1. Multiple power of ten   a. 10 (power of -1 or less)   f. 10 (power of -1 or more)
2. Submultiple power of ten   b. 0.000 1   g. 100
3. M (mega)   c. 0.000 000 1   h. 10^3
4. k (kilo)   d. 10^{-1} i. 10^6
5. deci (d)   e. 10^{-6} j. 1 billion
6. micro (μ)

2. Take each of the four computed quantities in the table.
   a. Write each one as an SI metric quantity.
   b. Round off each value to the indicated degree of accuracy. State this new quantity.

Computed Quantity		Written as an SI Metric Quantity	Required Accuracy (Decimals)	Rounded-off SI Metric Quantity
A	7624.2796		2	
B	1729685.499		0	
C	300002.9768392		4	
D	1753.0501		1	

B. Direct Linear Measurements in the Metric System

1. Read and record each metric measurement (A) through (J) as indicated. The first rule has graduations in millimeters. On the second rule the numbered graduations are in centimeters. Each division represents 1 millimeter. State dimensions (F) through (J) in terms of centimeter values.

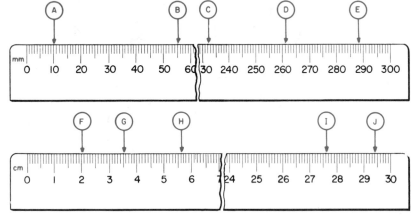

2. Measure the length of lines (a) through (i) to the degree of accuracy indicated. Measurements are to be metric.

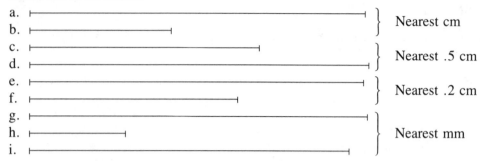

a. } Nearest cm
b.

c. } Nearest .5 cm
d.

e. } Nearest .2 cm
f.

g.
h. } Nearest mm
i.

3. Measure and record lengths Ⓐ, Ⓑ, Ⓒ, and Ⓓ. Dimensions are in mm. Add the distance to the first center line to the sum of Ⓐ, Ⓑ, Ⓒ, and Ⓓ. Check this overall measurement with dimension Ⓔ.

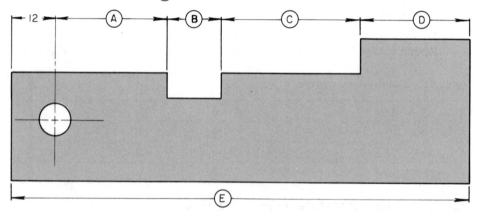

4. Measure the outside diameters of bars A, B, C, D, and E with a caliper. Transfer this measurement to a rule. Record the diameter to the nearest 0.1 centimeter.

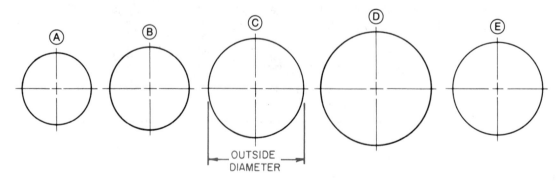

5. Use a metric architect's scale of 1:100.
   a. Measure the lengths of sections Ⓐ, Ⓑ, Ⓒ, and Ⓓ. Record each measurement to the nearest centimeter (cm).
   b. Give the overall length Ⓔ of the structure. Then, add each dimension and check the sum against the overall length.

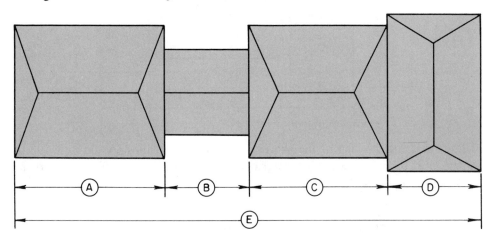

C. Indirect Metric Measurements

1. Compute dimensions Ⓐ, Ⓑ, Ⓒ, Ⓓ, Ⓔ, and Ⓕ to the nearest 0.01 mm. State each dimension in terms of centimeters.

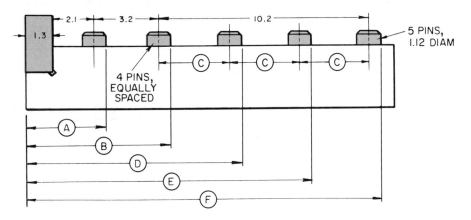

2. Calculate dimensions Ⓐ through Ⓙ to which the shaft must be rough turned. Allow 1.6 mm on all diameters and 0.8 mm on all faces for finish machining.

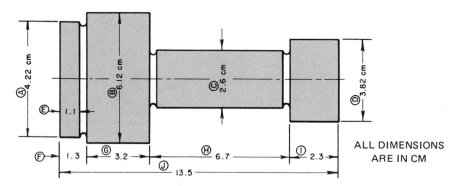

ALL DIMENSIONS ARE IN CM

3. Use the same drawing of the shaft. Determine dimensions Ⓐ through Ⓙ to which the part must be finish turned before grinding. Allow 0.2 mm on all diameters and 0.16 mm on all faces for grinding.

D. Application of Square Measure in the Metric System

1. Determine the area of square Ⓐ in square centimeters and of rectangle Ⓑ in square decimeters.

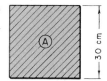

2. Determine the area of circles A and B from the given diameters and the area of C and D from the given radii.

	A	B	C	D
Diameter	4 dm	7.5 cm	—	—
Radius	—	—	2 m	22.4 cm

3. Determine the area of the sector of circle Ⓐ and the shaded area of Ⓑ.

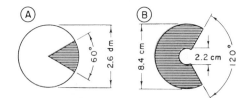

4. Compute the cross-sectional area of the link. Express the result in square inches correct to two decimal places.

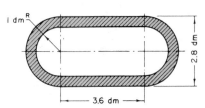

5. A form is to be built for a circular pool. The pool is 4.34 meters in diameter by 1.5 meters high. Find (a) the lateral surface area and (b) the number of boards required to construct the form. The boards are 7.62 cm wide. Round off the surface area to two decimal places and the boards to the next board.

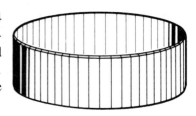

E. Application of Volume Measure in SI Metrics

1. Determine the volume of cube A in cubic decimeters and rectangular solid B in cubic meters.

	A	B
Length	6.2 dm	1.2 m
Depth	6.2 dm	18 dm
Height	6.2 dm	9.4 dm

2. Find the volume of the rectangular parts (A) and (B).

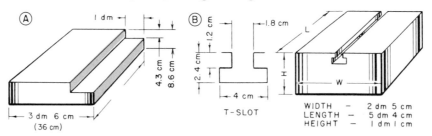

3. Determine the volume of cylinders A, B, and C. Express the volume correct to two decimal places in the unit of measure indicated in each case.

			Volume in
A	Diam	8.2 cm	cm^3
	Length	24.6 cm	
B	Radius	2.4 dm	dm^3
	Length	3.2 dm	
C	Diam	6 cm 4 mm	cm^3
	Length	4 dm 4 cm 8 mm	

4. Find the volume of the bronze bushing in cubic decimeters.

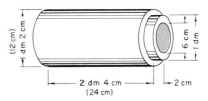

F. Application of Liquid Measure in the Metric System

1. Find the liquid capacity of the rectangular coolant tank to the nearest half liter.

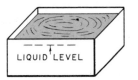

INSIDE DIMENSIONS
LENGTH — 7.4 dm
WIDTH — 3.2 dm
LIQUID HEIGHT–2.6 dm

LIQUID LEVEL

2. Determine the liquid capacity of the outer shell of the quenching tank to the nearest gallon.

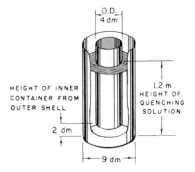

O.D.
4 dm

HEIGHT OF INNER CONTAINER FROM OUTER SHELL

2 dm

1.2 m
HEIGHT OF QUENCHING SOLUTION

9 dm

3. Compute the liquid capacity of containers A, B, C, and D. State each capacity in the unit of measure indicated in the table, rounded to two decimal places.

Container	Diameter (d) or Radius (r)	Height (h)	Required Unit of Measure
A	(d) 25.4 cm	38.1 cm	cm³
B	(d) 15.24 cm	11.43 cm	liters
C	(r) 64.77 cm	76.84 cm	qt
D	(d) 1.3 m	2.36 m	gal

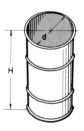

G. SI Metric Units of Mass (Weight) Measurement (General Applications)

1. a. Identify two metric mass (weight) measurement units that are (a) larger than a gram and (b) two smaller units than a gram.

	Larger Units		Smaller Units	
a. Designation				
b. Unit Symbol				
c. Value in Relation to 1 Gram				

   b. Give the measurement symbol for each larger and smaller unit.
   c. State the value of each larger and each smaller unit in relation to one gram.

2. Change each of the following weight measurements (a through f) to the unit of measurement specified in each case.
   a. 4 978 milligrams to grams
   b. 396.2 centigrams to grams
   c. 52.5 grams to decigrams
   d. 22.5 kilograms to grams
   e. 6.8 metric tons to kilograms
   f. 96 hectograms to kilograms

H. Converting Celsius and Fahrenheit Temperature Measurements (Practical Problems)

1. Convert the °F at which the normalizing and hardening heat treating processes are carried on to equivalent degree Celsius temperatures.

SAE Steel Number 1080	Heat Treating Processes in °F			
	Normalizing	Annealing	Hardening	Tempering
	1,550		1,450	
	Equivalent Heat Treating Temperatures in °C			
		760		232.2

2. Compute the equivalent °F temperatures to those Celsius degree heat treating temperatures given in the table for annealing and tempering the SAE 1080 steel part.

Unit 37 Precision Measurement: Metric Measuring Instruments

OBJECTIVES OF THE UNIT

After satisfactorily completing this unit, the student/trainee will be able to
- *Understand how the linear movement of a spindle or vernier beam (in relation to a fixed anvil or base) is transferred to become a precision measurement.*
- *Read and set metric inside, outside, and depth micrometers to an accuracy of ±0.01 mm.*
- *Read and set metric vernier micrometers to accuracies of 0.002 mm and 0.001 mm.*
- *Read and set metric vernier calipers and height gages to within 0.01 mm.*
- *Use the vernier bevel protractor to read angular dimensions within tolerances of ±5 minutes for applications in any system of measurement.*
- *Establish metric gage block combinations for setting instruments and other gages within accuracies of ±0.000 02 mm.*

Design features, principles used in graduating instruments, and applications of metric micrometers, verniers, and gage blocks are similar to those employed in the customary inch system. The main difference is that metric instruments are graduated in relation to the millimeter as the standard unit of measure.

The most commonly used linear metric measuring instruments, besides the steel rule and metal measuring tape, include the standard (0.01 mm) and vernier (0.002 mm) micrometers. The graduated standard micrometer head (consisting of a barrel and thimble) is adapted to inside, outside, depth, and other special micrometers. The metric calibrated vernier caliper beam and scales are also used on height and depth gages, gear-tooth calipers, and other precision calipers.

This unit deals with metric graduations on measuring instruments and practical applications to linear measurements in the metric system. The vernier principle is applied to the bevel protractor for precision angle measurements. Previous experiences with inch-standard gage blocks are extended to include metric gage block combinations.

A. GRADUATIONS: READING AND SETTING THE STANDARD (0.01 mm) METRIC MICROMETER

The standard metric micrometer is graduated to read to an accuracy of 0.01 mm (equivalent to 0.0004″). The addition of a vernier scale on the barrel makes it possible to improve the accuracy to within 0.002 mm (0.000 08″). The principles of reading measuring rules, micrometers, vernier instruments, and gage blocks are the same, regardless of the system. The standard metric micrometer has line graduations along the barrel at intervals of 0.5 millimeter. Enlarged graduations are shown on the line drawing (*figure* 37-1). The graduations indicate the distance the thimble and spindle move each complete revolution $\left(\frac{1}{2} \text{ mm or } 0.5 \text{ mm}\right)$. Each graduation above the index line represents a measurement of 1 mm. Each graduation below the index line represents 0.5 mm.

The thimble (*figure* 37-2) has 50 graduations cut into and around the beveled edge. Each graduation represents $\frac{1}{50}$ of 0.5 mm or 0.01 mm. The linear measurement is read directly by combining the barrel and sleeve readings. The line drawing (*figure* 37-3) shows a reading of 5.5 mm on the barrel and another 41 (mm) graduations of the sleeve. The micrometer reading is 5.5 mm plus 41 mm, totaling 5.91 mm.

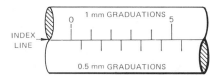

FIGURE 37-1 Graduations on barrel (enlarged)

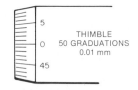

FIGURE 37-2 Graduations on micrometer thimble (enlarged)

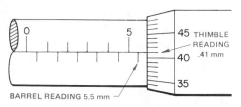

FIGURE 37-3 Combined barrel and thimble reading (enlarged)

RULE FOR READING AND SETTING A STANDARD (0.01 mm) METRIC MICROMETER

- Read the last visible 1 mm graduation above the index line on the barrel.
- Add 0.5 mm for any visible graduation below the index line.
- Read and add the 0.01 mm graduation line on the thimble at the index line of the barrel.

  Note. The sum of the readings represents the linear measurement correct to 0.01 mm.

EXAMPLE: *Case 1.* Read the standard metric micrometer head setting as illustrated.

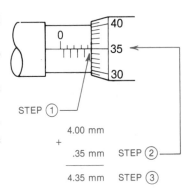

    Step 1   Record the number of whole millimeters on the top barrel scale.

    Step 2   Add the reading of the division on the thimble as the 0.01 mm value.

           Note. Since no half-mm value shows on the scale below the index line, the top reading (mm) and thimble division represent the actual measurement.

    Step 3   Express the sum of the readings as the required 0.01 mm metric micrometer reading.

          4.00 mm
+   .35 mm STEP ②
       4.35 mm STEP ③

EXAMPLE: *Case 2.* Set a standard metric micrometer head to a measurement of 7.92 mm.

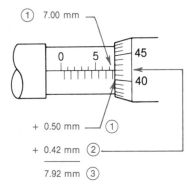

    Step 1   Turn the thimble to the 7 mm graduation (above the index line) and slightly beyond the next 0.5 mm graduation (below the index line).

    Step 2   Continue to turn the thimble until the 42 graduation (0.42 mm) cuts the index line.

    Step 3   Read the barrel and thimble settings for the 7.92 mm measurement.

① 7.00 mm
+ 0.50 mm ①
+ 0.42 mm ②
   7.92 mm ③

Estimating Readings of 0.005 mm on a Standard Metric Micrometer

It is possible to estimate to a fractional part of one division on the sleeve. A judgment is made about the additional distance between two graduations on the thimble at the index line. The illustration (*figure 37-4*) shows a reading of 5.915 mm. The additional precision of 0.005 mm is the estimated quantity.

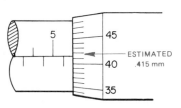

FIGURE 37-4 Reading of 5.915 (enlarged)

B. READING AND SETTING METRIC VERNIER (0.002 mm and 0.001 mm) MICROMETERS

As stated before, the vernier principle of measurement applies equally to instruments graduated in customary inch (0.0001″) and SI metric millimeter units of measure. A metric vernier micrometer is read in the same manner as the standard micrometer. The three-place decimal millimeter reading is read on the vernier scale. The reading is taken at the graduation on the vernier scale on the barrel that coincides with one of the graduations on the thimble.

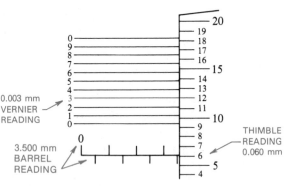

The representative line drawing (*figure 37-5*) shows a 0.001 mm metric vernier micrometer reading of 3.563 mm. The barrel reading is 3.5 mm (3.0 mm + 0.5 mm). The sleeve reading is 0.06 mm. The vernier reading is 3 (0.003 mm) because the 3 line coincides with a graduation on the thimble. The final reading of 3.563 mm consists of the 3.0 mm + 0.5 mm + 0.06 mm + 0.003 mm.

FIGURE 37-5 Reading of 3.563 mm on a 0.001 mm metric vernier micrometer

<u>E</u>XAMPLE: Read the metric vernier micrometer setting as illustrated.

Step 1 Read the last visible whole millimeter graduation on the barrel.

Step 2 Determine whether a half millimeter graduation appears on the lower scale on the barrel. Add 0.50 mm if the graduation is visible.

Step 3 Read the dimension on the thimble at the index line.

Step 4 Determine which vernier line coincides with a graduation on the thimble. Read the vernier graduation value.

Step 5 Combine the separate readings of 6.000 mm + 0.500 mm + 0.330 mm + 0.008 mm

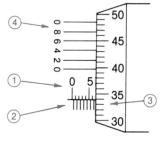

= **6.838 mm** Ans

Direct Reading Metric Vernier Micrometers

Micrometers are designed for direct reading. Such a micrometer, showing sections of the barrel and thimble, is illustrated in figure 37-6. Each graduation shown on the barrel represents readings of 1.0 mm. Note that the barrel reading shows 5+ millimeters. A mechanism in the thimble moves numerals that display the additional fractional parts of a millimeter in the measurement. The numeral in the left digit of the thimble represents 0.10 mm units. The right digit shows 0.01 mm units. Further, each 0.01 mm graduation is divided into five parts for readings of 0.002 mm.

The direct reading of the last exposed numeral on the barrel and the numerals and fractional part read on the thimble is the linear measurement of the part. In this example, the total reading is 5.374 mm. This three-place millimeter value carries the same degree of precision measurement as a vernier micrometer graduated in customary units of 0.000 1″.

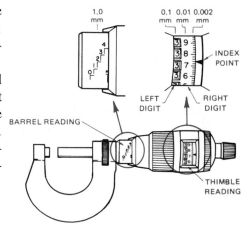

FIGURE 37-6 Direct reading micrometer

C. READING AND SETTING INSIDE METRIC MICROMETERS

The inside micrometer consists of a head (barrel and thimble), precision extension rods, and a rod locking screw. Inside micrometers are used to take internal measurements between a reference surface and a measured surface as shown in figure 37-7.

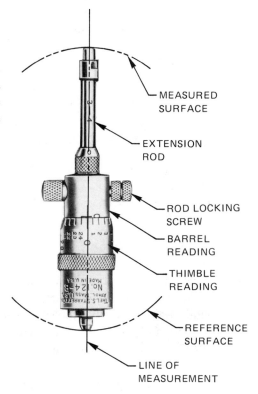

FIGURE 37-7 Setup for taking an internal measurement (*Courtesy of the L. S. Starrett Company*)

RULE FOR MEASURING WITH AND READING CUSTOMARY-INCH AND METRIC-GRADUATED INSIDE MICROMETERS

- Select the appropriate extension rod. Test the inside micrometer for accuracy.
- Place one leg of the micrometer against the reference point.
- Adjust the thimble. Turn it slowly while swinging a slight arc to bring the leg of the rod to the measured point. Continue until a slight contact is made.
- Remove the micrometer and read the measurement.
  Note. The inside measurement includes the graduation readings on the barrel + thimble + the length of the extension rod.

EXAMPLE: Determine the inside diameter reading of the measurement shown in figure 37-7.

Step 1 Take the reading on the barrel. ① – – – – – – – – – – **0.125**

Step 2 Add the reading on the thimble. ② – – – – – – – – – – **0.020**

Step 3 Add the size of the extension rod. The sum of these readings represents the inside micrometer measurement. ③ – – – – – – – – – **3.000**

④ – – – – – – – – – **3.145″** Ans

Note. The same procedure is followed for inside measurements using metric graduated instruments.

D. READING MEASUREMENTS ON A METRIC-DEPTH MICROMETER

The graduations on the micrometer head portion of a depth micrometer are in a *reverse order* from the conventional micrometer. In other words, the zero reading is at the topmost position of the barrel. The usual one inch of 25.0 mm depth measurement range may be increased by changing to an appropriate size interchangeable measuring (extension) rod.

RULE FOR MEASURING WITH A DEPTH MICROMETER AND READING MEASUREMENTS

- Select the length measuring rod to use according to the depth of the measurement.
- Position the depth micrometer. Slowly turn the thimble to lower the measuring rod until it almost touches the surface to be measured. Then use the ratchet mechanism to bottom the rod on the measured surface.
- Read the numbered graduation and any other exposed graduation value on the barrel.
- Add the reading on the thimble and the length of the measuring rod.
- State the depth as the total of the readings and the rod length.

E XAMPLE: Determine the depth as represented by the readings shown in figure 37-8.

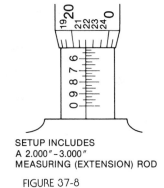

SETUP INCLUDES
A 2.000″–3.000″
MEASURING (EXTENSION) ROD

FIGURE 37-8

Step 1 Read the value of the last graduation shown on the barrel.

0.525″ ①

Step 2 Read the value of the thimble graduation at the index line.

0.022″ ②

Step 3 Add the length of the measuring (extension) rod.

2.000″ ③

Step 4 State the measurement in the required unit of measure.

.525 + .022 + 2.000 =
2.547″ Depth
Measurement Ans

E. MEASURING WITH METRIC VERNIER CALIPERS

Vernier calipers with either metric or customary inch units of measure are read in the same manner. In the case of metric vernier calipers, each graduation represents 1 mm. Each graduation on the vernier scale equals 0.02 mm. One of the vernier scales is positioned for outside (external) measurements; the other scale is positioned for inside (internal) measurements. Regardless of the system, dimensional readings are taken by combining the beam and vernier reading on either the inside or outside scale, as required.

RULE FOR READING VERNIER CALIPER MEASUREMENTS

- Note which graduation on the appropriate vernier scale coincides with a graduation on the beam.
- Read the beam graduation at the zero index point.
- Add the vernier scale reading. The sum equals the required vernier caliper measurement.

EXAMPLE: Read both the external and internal metric measurements as represented by the line drawing.

Step 1 Read the external millimeter measurement (80 mm) at the zero (0) vernier scale graduation.

Step 2 Determine which graduation on the beam scale coincides with a graduation on the vernier scale.

Step 3 Add the vernier scale reading (0.08 mm). The sum equals the required outside measurement (80.08 mm).

Step 4 Repeat the steps 1 through 3 to obtain an internal metric measurement. In this instance, if the measurement were an internal one, the reading would be 88.08 mm.

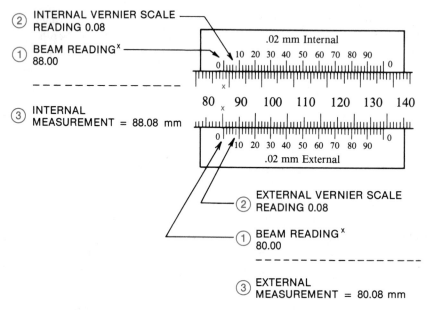

EXTERNAL/INTERNAL VERNIER CALIPER MEASUREMENTS

Metric and Customary Inch Combination Vernier Calipers

Another functionally designed vernier caliper includes graduations in millimeters on one edge of the beam. The other edge is graduated in customary inch units of measurement of 0.050″. The vernier plate attachments permit measurements to be taken on the metric scale to an accuracy of 0.02 mm and to 0.001″ on the inch standard vernier (*figure* 37-9).

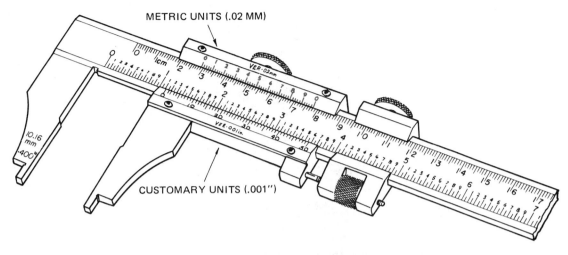

FIGURE 37-9 Combination outside/inside metric/customary inch vernier caliper

F. READING ANGLES WITH A VERNIER BEVEL PROTRACTOR

The vernier principle is also applied to measuring angles with a vernier bevel protractor (*figure* 37-10). This instrument permits the measurement of angles to a precision of 5 minutes or $\frac{1}{12}$ of one degree.

The vernier bevel protractor consists of a body, a turret, an adjustable blade, and a clamp. Degree graduations appear near the outer rim of the body. The vernier scale is attached to a movable, rotating turret. The measuring blade is positioned for measuring an angle and is locked to the turret. The angle formed is measured by the combination of the outer scale and vernier scale readings. There are 24 divisions on the vernier scale shown in figure 37-11. The divisions are numbered 60 to 0 to 60 in increments of 5 minutes.

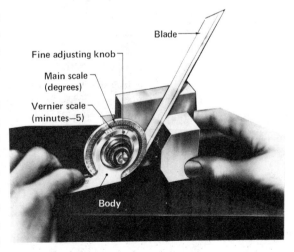

FIGURE 37-10 Application of vernier bevel protractor in measuring angles to 5' (*Courtesy of the L. S. Starrett Company*)

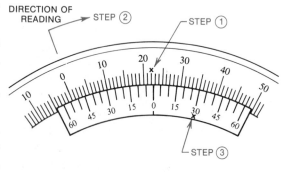

FIGURE 37-11 Graduations on the vernier bevel protractor

RULE FOR READING VERNIER BEVEL PROTRACTOR MEASUREMENTS

- Read the number of degrees on the main scale at the zero index line.
- Follow in the same direction and read the minute graduation on the vernier scale which coincides with a graduation on the main scale.
- Combine the main scale degree reading and the vertical scale (5′) reading. This value represents the angle being measured in degrees and minutes.

E XAMPLE: Read the angle for the vernier bevel protractor setting as illustrated.

Step 1 Read the whole number of degrees on the main scale at the 0 index line.

Step 2 Note the direction from which the reading is made.

Step 3 Continue in the same direction and determine which graduation on the vernier scale coincides with a graduation on the main scale.

Step 4 Combine the degree and minute values to establish the measured angle.

Reading = 22° 30″ Ans

G. DETERMINING METRIC GAGE BLOCK COMBINATIONS

Metric gage blocks are commercially available in sets that cover the same range of measurements as the customary inch gage blocks. Metric gage blocks are produced to dimensional accuracies of ± 0.000 05 mm (0.05 microns, 0.05 μm) for size, flatness, and parallelism. Generally, the gage blocks in a set permit measurements in increments of 0.001 mm to 25.0

mm. Figure 37-12 gives the size increment for each series of blocks in a set and the range of sizes in each series.

Increment for Each Series in a Set	Range of Lengths for Each Increment Series
0.001 mm	1.001 mm to 1.009 mm
0.01 mm	1.01 mm to 1.49 mm
0.5 mm	1.0 mm to 24.5 mm
1.0 mm	1.0 mm to 9.0 mm
10.0 mm	10.0 mm to 100.0 mm
25.0 mm	25.0 mm to 100.0 mm

FIGURE 37-12

Metric gage blocks are selected, combined, and wrung together following the same steps as for inch standard precision gage blocks. In general practice, *wear blocks* are placed on each end of a gage block combination. Also, the largest possible sizes and the least number of blocks are combined.

RULE FOR ESTABLISHING A PRECISION MEASUREMENT WITH METRIC GAGE BLOCKS

- Follow the same procedure in selecting the gage block combination as for inch standard gage blocks.
- Select the first gage block from the series that has the same number of decimal places as the last right decimal digit.
- Select one or more gage blocks from the series having the same number of decimal places as the next decimal digit.
- Proceed to select gage blocks from blocks in the 0.5, 1.0, 10.0, or 25.0 mm series, as may be required.
- Add the combination of blocks. Check this measurement against the required measurement.
- Clean and ''wring'' the gage blocks together.

EXAMPLE: Select the metric gage block combination for a measurement of 120.887 mm. *Note.* Use 2 mm wear blocks as the end blocks. Refer to a table of metric gage block thicknesses in a set.

Step 1   Subtract the two wear blocks from the required measurement.
Step 2   Select a gage block in the 0.001 series to eliminate the last right decimal digit.
Step 3   Proceed to eliminate each remaining decimal digit.

Step 4   Add the blocks in the combination. Then, subtract the sum from 116.887 mm to find the remaining blocks.

Step 5   Eliminate the two right digits (__ 1 3) by selecting two blocks in the 1 mm series that total 13 mm.

Step 6   Eliminate the last digit (1 __ __) by selecting the 100 mm block in the 25.0 mm series.

Step 7   Recheck the combination of blocks against the required measurement.

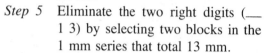

```
2.00      (1)
1.007     (2)
1.08
1.8       (3)
6.0
7.0       (5)
100.0     (6)
2.0       (1)
_____
120.887 mm (7)
```

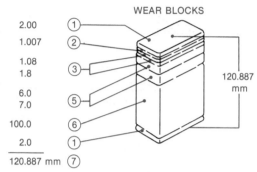

Review and Self-Test

ASSIGNMENT UNIT 37

A. Reading Standard Metric Micrometer Measurements

1. Read the linear measurements displayed on the standards metric micrometer settings at Ⓐ, Ⓑ, and Ⓒ. Estimate the third decimal-place measurement for settings Ⓓ and Ⓔ. (*Note.* The barrel graduations in the illustrations are enlarged for easier reading.)

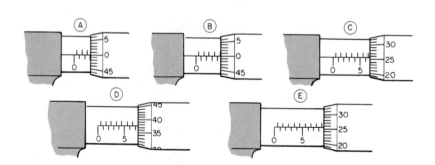

B. Reading and Setting Metric Vernier Micrometers

1. Read the metric vernier micrometer settings as represented by line drawings Ⓐ, Ⓑ, and Ⓒ to 0.01 mm; Ⓓ and Ⓔ to 0.002 mm; and Ⓕ to 0.001 mm.

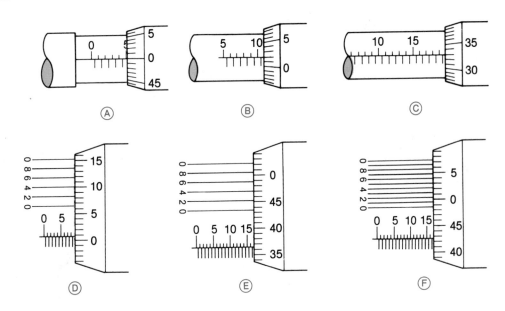

2. Indicate the required barrel, thimble, and vernier scale settings of a 0.002 mm outside micrometer for measurements A, B, and C.

	Required Measurement (mm)	Barrel Graduations		Thimble Setting (0.01 mm)	Vernier Scale Setting (0.002 mm)
		1 mm	0.5 mm		
A	7.50				
B	12.77				
C	125.926				

C. Reading Depth Micrometer Measurements

1. Read the 0.001″ depth micrometer measurements as illustrated at (A) and (B).

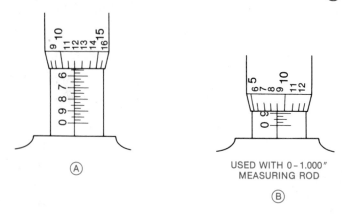

USED WITH 0–1.000″
MEASURING ROD

(B)

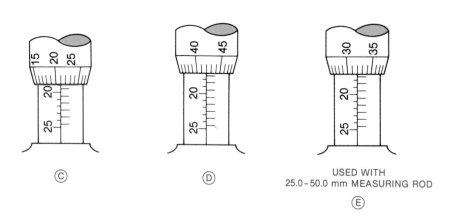

USED WITH
25.0–50.0 mm MEASURING ROD

(E)

2. Read the SI metric depth micrometer measurements at (C), (D), and (E).

D. Reading Metric Vernier Caliper Measurements

1. Compute the overall linear measurements (A), (B), and (C) from the information provided on the drawing and in the table of hole sizes.

Hole Diameters	
Hole #	Nominal Hole Diameter (mm)
A	25.4
B	37.12
C	28.58

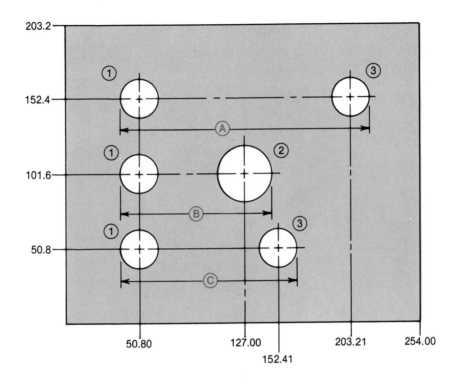

2. Give the beam (main scale) and inside vernier scale settings for each overall measurement.

Hole #	Overall Measurement (mm)	Vernier Caliper Scale Settings	
		Main Scale (mm)	Vernier Scale (0.02 mm)
A			
B			
C			

E. Reading Metric Vernier Height Gage Measurements

1. Read the 0.02 mm vernier height gage settings Ⓐ, Ⓑ, and Ⓒ. Each vernier scale graduation is equal to 0.02 mm.

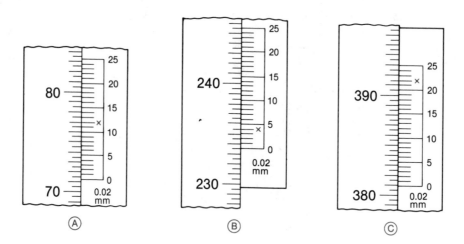

F. Reading Angular Measurements with the Vernier Bevel Protractor

1. Read the angular settings as displayed on the main beam and vernier scales for bevel protractor measurements Ⓐ, Ⓑ, and Ⓒ.

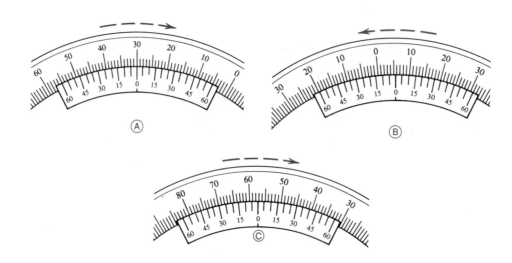

G. Determining Metric Gage Block Combinations

1. Establish the series and determine the sizes of metric gage blocks to combine to produce measurements A and B.

	Required Linear Dimension (mm)	Series of Gage Blocks (mm)	Size(s) of Gage Blocks in Each Series (mm)
A	58.555		
B	155.863		

Unit 38 Base, Supplementary, and Derived SI Metric Units

OBJECTIVES OF THE UNIT

After satisfactorily studying this unit, the student/trainee will be able to
- *Understand how each of the seven base units in SI metrics is defined and used for length, mass, time, electric current, temperature, luminous intensity, and amount of substance measurements.*
- *Apply supplementary and derived SI metric units to solve general, practical industrial, and physical science problems.*
- *Work with nonsignificant zeros and dual dimensioning.*

Units 36 and 37 relate to problem solving using base decimal ten multiple and submultiple values. Quantities, problems, and solutions are stated according to guidelines for communicating in SI metrics. Symbols are used in the system. The symbols provide a convenient, consistent, simple method of designating a specific unit or combination of units. Prefixes further simplify the unit designations, mathematical processes, and solutions.

Each SI metric unit may be converted to a customary unit or other conventional metric unit of measure. Tables are readily available that give

- The *quantity to be measured* (for example, density, *D*)
- The *unit of measure* (kilograms per cubic meter, kg/m^3)
- The *formula* (*D* = kg/m^3)
- The *conversion units* that may be in SI metrics and conventional metric or customary units (for example, g/cm^3 or lb (mass)/ft^3
- The *conversion factor and mathematical processes,* for example,

$$kg/m^3 = lb \ (mass)/in.^3 \cdot (2.768 \times 10^4)$$

In Unit 38, attention is directed to seven base units, two supplementary units, and a series of derived units in the SI metric system. The combination of base, supplementary, and derived units in the full system is adequate to measure all known physical quantities.

A. SEVEN BASE UNITS OF MEASURE IN SI METRICS

The base units of measure cover (1) length, (2) mass, (3) time, (4) electric current, (5) temperature, (6) luminous intensity, and (7) the amount of a substance. Each of these base units is identified and, in some cases, defined technically.

1. Unit of Length

Figure 38-1 shows how the meter, as a measure of length, is standardized. In October 1983 the *Conference Général des Poids et Measures* (GCPM) adopted a new definition for the meter as the unit of length in the International System of Measurement (ISO). The *new meter* is defined as the distance traveled by light in a vacuum in 1/299,792,458 of a second. This combination of scientific activity, transferred to the measuring bar, represents the precise measurement of the meter.

The *meter* (m) is the standard unit of length in the SI metric and all other conventional metric measurement systems. One major difference among the standards for the early meter standard, the SI metric standard until late 1983, and the present one is in the laboratory method of producing a standard meter measuring bar by using time as the most accurate measurement to define length. The last meter standard was defined in terms of a wavelength of orange-red light emitted from a krypton$_{86}$ lamp.

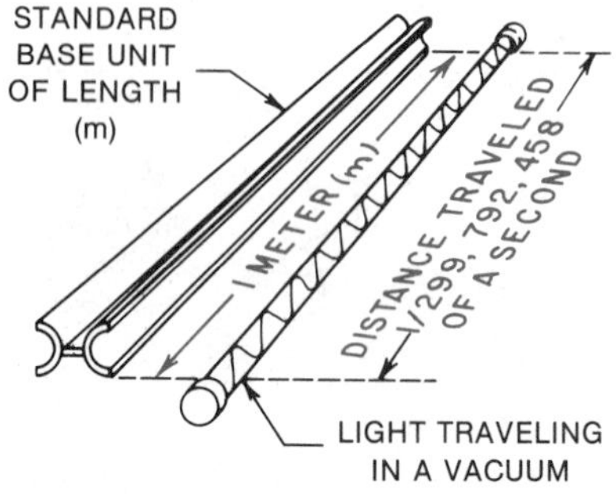

The early meter represented a fixed value in relation to the earth's circumference. Initially, standard meter measur-

FIGURE 38-1

ing bars were compared for precision and accuracy against a standard bar held by the International Bureau and the predecessor academy. The duplicate measuring bars were called *prototypes*. The prototypes became the standards for each participating nation. The advantage of the new meter is that it may be reproduced anywhere in the world in a laboratory.

Measurements must be taken that range all the way from minute particles finer than a billionth part of a meter to earth-outer space dimensions that represent multidigit quantities. A series of more practical units of measure, derived from the meter as the base unit, are described later.

2. Unit of Mass

The *kilogram* still represents a base *unit of mass*. The kilogram (kg) is approximately the equivalent of 2.2 pounds. Other recommended multiples of the SI kilogram unit are the *megagram (Mg), gram (g), milligram (mg),* and *microgram (µg)*. The *metric ton (t)* is another unit that is often used (1 t = 10^3kg). Additional units that relate to mass are discussed later.

3. Unit of Time

The *second* remains as the unit of time. This unit is defined scientifically in terms of periods of radiation of a cesium atom and two levels of activity. The activity is overly simplified in *figure 38-2*.

The duration or time interval of a second is still 1/86 400 part of the mean solar day. The designation for the second is (s). Recommended decimal multiples and submultiples of the SI unit include the *kilosecond* (ks), *millisecond* (ms), *microsecond* (µs), and the *nanosecond* (ns). Other units that may be used are the customary ones of *minute, hour,* and *day*.

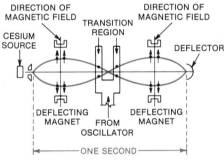

FIGURE 38-2

4. Unit of Electric Current

The *ampere* is the unit of intensity of electric current. As a common name, the ampere is designated by the capital (A). Other recommended values of this SI unit include the *kiloampere* (kA), *milliampere* (mA), *microampere* (µA), *nanoampere* (nA) and *picoampere* (pA). Note that most of these examples express submultiple decimal values of an ampere.

The illustration (*figure 38-3*) shows two straight parallel conductors of negligible cross section and infinite length. These are placed one meter apart in a vacuum. The *ampere* is that constant, which, if maintained in the two conductors, produces a force equal to 2×10^{-7} newton/meter of length.

5. Unit of Temperature

The *kelvin* (K) is the SI unit of thermo-dynamic temperature. A kelvin temperature is expressed without the degree (°) symbol. The boiling point of water on the kelvin scale is 373.15 K degrees (373.15 K). Water freezes at 273.15 K.

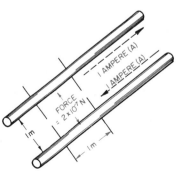

FIGURE 38-3

The more commonly used temperature measurement unit is the derived *Celsius* degree. Temperatures are readily convertible between the kelvin and Celsius and the kelvin and Fahrenheit readings by transposing values in these two formulas.

$$t_K = t_C + 273.15 \qquad t_K = \frac{(t_F + 459.67)}{1.8}$$

6. Unit of Luminous Intensity

The *candela* is the SI unit of luminous intensity. The candela was adopted at the 13th CGPM in 1967. By scientific definition the candela represents

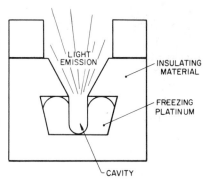

FIGURE 38-4

- the luminous intensity
- of a 1/600 000 square meter surface of a black body
- at the temperature of freezing platinum under a pressure of
- 101 325 newtons per square meter.

7. Units for the Amount of Substance of a System

The *mole* (mol) is the most recent SI base unit. It was introduced to physical chemistry and molecular physics in relation to mass, internal energy, and heat capacity. The mole represents a unit of measure of the quantity or the *amount of substance* in such an elementary unit as a molecule, atom, ion, or electron.

B. SUPPLEMENTARY UNITS OF ANGULAR MEASURE

There are two types of angles in SI metrics for which supplementary units of measure have been developed. The *radian* unit (rad) is used for measurements of *plane angles* (*figure* 38-5). The *steradian* (sr) unit provides for the measurement of *solid angles*.

In simple mathematical terms, 1 rad $= 180°/\pi$, or $57.295\ 78°$. The SI recommended units for radian are *milliradian* (mrad) and *microradian* (μrad). Other customary units that may be used include *degree* (°), *minute* ('), *second* ("), and *grad* (^{g}).

By contrast, the steradian encloses an area of the spherical surface as shown in figure 38-6. The enclosed area is equal to that of a square. The sides of the square are equal in length to the radius.

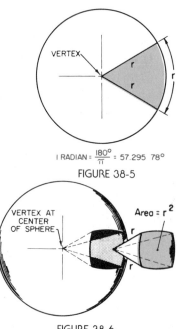

FIGURE 38-5

FIGURE 38-6

C. DERIVED UNITS OF MEASURE

There are many instances when base or supplementary units of measure are not suitable and *derived* units are required. These are derived from base or supplementary units. In general, the preferred units of measure in SI metrics are *multiples of three* (^{3}): *kilo* (k, 10^3), *mega* (M, 10^6), *giga* (G, 10^9) or *milli* (m, 10^{-3}), *micro* (μ, 10^{-6}), and *pico* (p, 10^{-12}).

1. Derived Length, Area, and Volume Units

The decimal multiple and submultiples of SI units that are used for linear measurements include the *kilometer* (km), *millimeter* (mm), *micrometer* (μm) and *nanometer* (nm). The *decimeter* (dm) and *centimeter* (cm) are also accepted.

The derived units of area and volume are also related to the meter as the base unit. In area measure, the *square kilometer* (km^2) and *square millimeter* (mm^2) are recommended multiples. The dm^2 and cm^2 are accepted. For larger measurements the *are* and *hectare* are other units. One are (a) $= 10^2 m^2$ (100 square meters); 1 hectare (ha) $= 10^4 m^2$ (10 000 square meters).

In volume measure, the cubic millimeter (mm^3) is a recommended multiple of the base unit of m^3. The cubic decimeter (dm^3) and (cm^3) are accepted units of volume measure. The *liter* (ℓ) has been declared an acceptable SI unit. One liter equals one cubic decimeter (1 $\ell =$ 1 dm^3). In terms of the meter, 1 $\ell = 10^{-3}$ m^3 or 0.001 m^3. The *centiliter, milliliter,* and *hectoliter* are other accepted units of volume measure.

$$1 \text{ centiliter (c}\ell) = 10^{-5} \text{ m}^3 \text{ (or 0.000 01 m}^3)$$
$$1 \text{ milliliter (m}\ell) = 10^{-6} \text{ m}^3 \text{ (or 0.000 001 m}^3)$$

2. Units Derived from the Unit of Mass

Units derived from the unit of mass deal with force, work, power, pressure and other physical science measurements. The units are derived from the *kilogram* (kg) as the base unit of mass. Recommended multiples of the kg are the *megagram* (Mg), *gram* (g), *milligram* (mg), and *microgram* (μg). Other widely used common units, which are derived from the kilogram as the unit of mass, are discussed.

FORCE.   Force is measured in *newtons* (N). The SI unit of force is N · m. A force of one newton applied to a mass of one kilogram produces an acceleration of one meter per second. Expressed as a formula

$$N = kg \cdot m/s^2$$

WORK, ENERGY, QUANTITY OF HEAT.   Work is measured in *joules* (J). One joule of work is produced when a force of one newton is applied through a distance of one meter (J = N · m). Work is usually expressed in *gigajoules* (GJ), *megajoules* (MJ), *kilojoules* (kJ), and *millijoules* (mJ). Another practical unit of measure is the *kilowatt-hour* (kW·h) of work.

$$1 \text{ kW·h} = 3.6 \times 10^6 \text{ J or 3.6 MJ}$$

POWER. The *watt* (W) is the SI unit of power. It represents one joule of work completed in one second.

$$W = J/s \text{ or } N \cdot m/s$$

PRESSURE. The *pascal* (Pa) is the SI unit for measuring pressure. A pascal is equal to one newton per meter squared.

$$Pa = N/m^2$$

3. Units Derived from the Unit of Time

Three commonly needed units of measure that are derived from the second relate to frequency (f), velocity (v), and acceleration (a).

FREQUENCY. Frequency denotes the number of cycles per second. The *hertz* (hz) is the SI unit of measure of frequency.

$$1 \text{ Hz} = 1 \text{ s}^{-1}$$

VELOCITY. Velocity relates to distance and time. It is measured in terms of *meters per second* (v = m/s). Velocity is also expressed in the derived unit of *kilometer per hour*.

$$1 \text{ km/h} = \frac{1}{3.6} \text{ m/s}$$

ACCELERATION. Acceleration (a) denotes a rate of change in velocity. Acceleration is equal to distance (meters) divided by time in seconds squared.

$$a = m/s^2$$

4. Units Derived from the Base Electrical Unit

Six common electrical measurements relate to *potential, resistance, capacitance, quantity, inductance,* and *magnetic flux*. These characteristics are measured by units that are derived from the *ampere* (A) as the base unit. Figure 38-7 indicates the derived unit in each case and the formula.

Base Electrical Unit: Ampere (A)		
Electrical Measurement	Derived Unit of Measurement	Formula
Potential	volt (V)	$V = W/A$
Resistance	ohm (Ω)	$\Omega = V/A$
Capacitance	farad (F)	$F = A \cdot (s/V)$
Quantity	coulomb (C)	$C = A \cdot s$
Inductance	henry (H)	$H = Wb/A$
Flux	weber (Wb)	$Wb = V \cdot s$

FIGURE 38-7

5. Units Derived from the Base Unit of Luminous Intensity

The *candela* (cd) is the base unit of luminous intensity. There are two major derived units from the candela: *illumination* lux (lx) and the lumen (lm, luminous flux). The *watt* (W), as a common unit of electrical power, equals 17 lumens (1 W = 17 lm).

6. Units Derived from the Base Unit of Temperature

The *degree Celsius* is the derived unit of temperature. It is used in general everyday applications. Unlike the kelvin (K) SI unit, a temperature on the Celsius scale is followed by the °C symbol. The degree Celsius replaced the centigrade designation. The freezing and boiling points of water on the Celsius scale are 0°C and 100°C, respectively. The mathematical relationships between computing Celsius and customary Fahrenheit degrees remain the same.

Many problems still require the conversion of temperature units between customary and other metric systems. The same simple formulas that were used between the former Centigrade and Fahrenheit scales apply to Celsius/Fahrenheit conversions.

$$t_C = (t_F - 32) \cdot \tfrac{5}{9} \qquad\qquad t_F = \left(t_C \times \tfrac{9}{5} + 32\right)$$

$$\text{or} \qquad\qquad\qquad \text{or}$$

$$t_C = (t_F - 32) \div 1.8 \qquad\qquad t_F = 1.8 t_C + 32$$

7. Units Derived from the Base Unit of Time for Ordinary Time Measurement

The *day* (d), *hour* (h), *minute* (min) and *week*, *month* and *year* are commonly used units of time. Each is derived from the *second* as the base unit. The value of each of these derived units and its use in mathematical computations follows the customary or the conventional metric system measurement in the problem.

8. Units Derived from the Basic Molar Unit

The *mole* (mol) is the most recent base unit in SI metrics. Other derived mole units that measure quantity are related to the nature of each quantity. For example, the kg/mol is a derived unit for measuring *molar mass;* j/mol for *molar internal energy;* and J/(mol K) or J/(mol °C) for *molar heat capacity.*

D. NONSIGNIFICANT ZEROS IN MEASUREMENT

The term *nonsignificant zeros* refers to the practice of adding zeros to decimal digits. Nonsignificant zeros appear primarily on drawings where a tolerance is specified. One or more nonsignificant zeros may be added to the decimal value of a dimension or a tolerance. This means that the dimension and the tolerance have the same number of digits. For instance, in a dimension like 5.3125″ ± 0.0010″ the last right decimal digit is a nonsignificant zero.

It has no value since the tolerance is not changed by its addition. The last decimal digit is important in stating precisely what the tolerance is.

Nonsignificant zeros are not added in SI metrics. It is common practice to see values or quantities in tables or on drawings with a different number of decimal digits in a dimension and the tolerance. For instance, an object may be dimensioned 272.38 mm ± 0.025 mm. The three-place decimal tolerance of 0.025 mm tells that the part can be accepted if machined between 272.38 + 0.025 mm or 272.405 mm as the upper limit and 272.38 − 0.025 mm or 272.355 mm as the lower limit.

E. SI METRIC AND CUSTOMARY UNIT DUAL DIMENSIONING

Dual dimensioning implies that a product or part is dimensioned in equivalent SI metric and customary units of measure. This practice applies particularly to industry drawings where the worker must interpret and use the dimensions to produce a part precisely using either system of measurement.

Special dimensioning guidelines are followed. Two sets of dimensions, SI metric and customary units, are used. Generally, a dimension that appears first and that is separated from the dimension that follows by a slash or bracket or a dimension that appears on a drawing above another dimension is considered the *controlling dimension*. This dimension represents the system of measurement in which the product is designed. Dimensions that are given after a slash symbol or within brackets are usually noncontrolling dimensions. A note is often used on a drawing to indicate the controlling dimensioning system or the position of the dimensions.

Each technique of positioning a controlling dimension or indicating the system of dimensioning is illustrated in figure 38-8. Placement of dimensions one above another is shown at Ⓐ; the use of slash and bracket symbols at Ⓑ; and the identification of measurement systems as a note at Ⓒ.

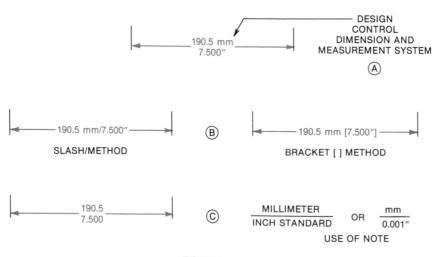

FIGURE 38-8

Review and Self-Test

ASSIGNMENT UNIT 38

A. Base, Supplementary, and Derived Units of Measure (Practical Problems)

1. Identify four different kinds of information that are generally found in SI metric tables.
2. a. Give two applications of derived units of measure for each base unit.
   b. Identify the measurement symbol that is used with each application of a derived unit.

Base Units				Derived Unit Applications			
Measure	Unit	Symbol	Application 1	Measurement Symbol	Application 2	Measurement Symbol
1	Length	meter	(m)				
2	Mass	kilogram	(kg)				
3	Electric Current	ampere	(A)				
4	Luminous Intensity	candela	(cd)				
5	Time	second	(s)				
6	Temperature	degree kelvin	(K)				
7	Substance	mole	(mol)				

3. a. Name the two SI supplementary units of angular measure.
   b. Identify the symbol for each unit of measure.
   c. Explain briefly what each supplementary unit measures.
   d. Give the mathematical value of either supplementary unit.

B. Nonsignificant Zeros (Practical Problems)

1. Identify one purpose for using nonsignificant zeros in customary unit dimensioning.
2. State one difference in dimensioning practices followed in SI metrics as contrasted with customary unit dimensioning in relation to the use of nonsignificant zeros.
3. Give an example of dimensioning and tolerancing in SI metric and another example for customary unit practices in using nonsignificant zeros.

C. SI Metric and Customary Unit Dual Dimensioning (Practical Problems)

1. State what function is served by the controlling dimension on a drawing.
2. Give the controlling dimension and identify the measurement system in examples Ⓐ, Ⓑ, and Ⓒ.

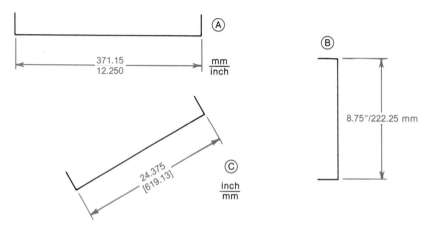

Unit 39 Conversion: Factors and Processes (SI and Customary Units)

OBJECTIVES OF THE UNIT

After satisfactorily studying this unit, the student/trainee will be able to
- *Interpret measurements obtained by soft or hard conversion.*
- *Use conversion tables and reciprocals to obtain measurements and values in either SI metrics or customary units of measure or in both systems.*
- *Solve practical problems requiring the conversion of linear, square, mass, and other volume measurements and quantities.*

Today, it is common practice to use dimensions and other measurements in either the SI metric or customary system or in both systems at one time. The measurement values are usually computed, although, in some applications, it is possible to use direct-reading conversion scales and rules.

A. SOFT AND HARD CONVERSION OF SI METRIC AND CUSTOMARY UNITS OF MEASURE

Equivalent measurements, which are computed directly without any changes, are known as *soft conversion*. While the conversion process involves simple mathematics, in most manufacturing processes there are few instances where each standard dimensional size in one system is commercially available in the other system. Further, there is the question of dimensional accuracy. For example, the exact equivalent metric linear measurement to a dimensional reading of 3.125″ on a standard inch micrometer is 79.375 mm. The decimal 0.375 mm identifies a higher degree of accuracy.

In this case, the three decimal-place reading is equivalent to an accuracy of 0.000 04″. This precision may be far beyond production requirements. Equal precision in both systems requires that one less decimal digit be used with an SI metric measurement. Thus, the 79.375 mm becomes 79.38 mm. For everyday conversion applications, it is understood that soft conversion is required unless stated otherwise.

Hard Conversion

Hard conversion means that a product is designed to conform to measurements in one system, regardless of comparable sizes in the other system. A quart in British measure is equal to 0.9463 liters in metric measure. One millimeter in linear measure equals 0.03937″. A table of drill sizes shows that the closest metric size drill to $\frac{15''}{32}$ (0.4688″) is 12 mm, which is the equivalent of 0.4724″. These values provide examples of problems in production resulting from soft conversion.

Hard conversion requires complete design changes. As a result, products designed to conform to standards in one system are not interchangeable in the second system. Preferred sizes are used in hard conversion, such as 25 mm in place of 25.4 mm; 4 liters, not 3.785, and 1 kg, not 453.6 g. It is apparent that ISO standards and those adopted by the American National Standards Institute (ANSI), the Canadian Standards Association (CSA), and special bodies like the American Gear Manufacturer's Association (AGMA) and others are not identical.

B. CONVERSION FACTORS AND PROCESSES IN METRICATION

It is possible to remember some of the commonly used values and formulas for computing specific quantities. However, it is a more accurate practice to use tables containing both formulas and numerical values. These tables are identified as "tables of conversion factors."

Different tables are available depending on the degree of precision required. In engineering and scientific problems, tables are used that have a great number of least significant digits (decimal places). Tables ranging from six- to eight-place decimal values are adequate for general applications. The final result is then rounded off, depending on the required accuracy.

A smaller unit of measure is changed to a larger unit in the metric system by moving the decimal point to the left. For example, in changing 1 700 millimeters to meters, the value is stated as 1.7 m. In reverse, zeros are added and the decimal point is moved to the right (12 MHz = 12 000 000 Hz).

When a conversion table is used, a conversion factor like 1 mm = 0.039 37″ may be changed to 1 cm = 0.393 7″, or 1 km = 39 370″. The prefix indicates the relationship of the numerical value to the unit of measure.

Reciprocal Processes in Conversion

While a table may give one set of conversion factors, other units of measure may be computed. The reverse process is performed by using reciprocals.

Mathematical processes involving reciprocals are covered in later units. For the present, "using the reciprocal" means dividing (1) by a numerical value. For example, to convert feet to meters the conversion factor is 0.304 8. Thus, 10 ft = 10 · (0.304 8) or 3.048 m. The answer in meters may be converted to its feet equivalent by using the reciprocal of the multiplier (0.304 8). In this instance, the reciprocal is 1/0.304 8. Thus, 3.048/0.304 8 = 10 ft.

Tables are available for conversion of common quantities from SI or conventional metrics to customary units and the reverse. Other tables are detailed for specialized application. Part of Conversion Table 5 that appears in the Appendix is reported in figure 39-1. Selected units of measure are grouped according to use by categories (for example, acceleration, area, and density). Conversion factors are indicated for units of measure in the three common measurement systems.

	Conversion of Customary or Metric Units to SI Metrics			Conversion of SI Metrics to Customary and Conventional Metric Units		
Category	From Customary or Conventional Metric Unit	To SI Metric	Factor (A) (Multiply by)	From SI Metric	To Customary or Conventional Metric Unit	Factor Multiply by the reciprocal of the multiplier (Factor A) which is used in conversion to the SI Metric unit
Acceleration	ft./s^2	m/s^2	0.304 8	m/s^2	ft./s^2	
	in./s^2		$2.540\ 0 \times 10^{-2}$*		in./s^2	
Area	ft.^{2}	m^2	$9.290\ 3 \times 10^{-2}$	m^2	ft.^{2}	
	in.^{2}		$6.451\ 6 \times 10^{-4}$*		in.^{2}	
Density	g/cm^3	kg/m^3	$1.000\ 0 \times 10^3$*	kg/m^3	g/cm^3	
	lb. (mass)/ft.^{3}		16.018 5		lb. (mass)/ft.^{3}	
	lb. (mass)/in.^{3}		$2.768\ 0 \times 10^4$		lb. (mass)/in.^{3}	

*Exact Values

FIGURE 39-1 Partial table of conversion factors for SI and conventional metric and customary units of measurement

The abstracted contents of Table 5 from the Appendix summarize the information presented to this point. Symbols, powers of ten decimal multiple and submultiple quantities, positive and negative values, and simple mathematical processes are illustrated in figure 39-1. The balance of this unit now deals with soft conversions of commonly used measurement units in practical problems involving linear, surface, volume, and liquid measures.

C. CHANGING UNITS OF MEASURE FROM ONE SYSTEM TO ANOTHER

No uniform relationship exists between the units of measure in the British (customary) and metric systems. In changing values from one system to the other it is necessary to know what the equivalent of one unit is in terms of the other system. The desired unit may then be computed by either multiplying or dividing.

Units that are commonly used to convert a measurement from the British (customary) to the metric system are given in figure 39-2.

British Units	Equivalent in Metric Units (Approximate)
One inch (in.)	2.54 centimeters (cm) or 25.4 millimeters (mm)
One foot (ft)	0.304 8 meter (m)
One yard (yd)	0.914 4 meter (m)

FIGURE 39-2 Comparison of British and metric units of linear measure

RULE FOR CHANGING CUSTOMARY UNITS OF MEASURE TO METRIC UNITS (INCHES, FEET, OR YARDS TO METERS, CENTIMETERS, DECIMETERS, OR MILLIMETERS)

- Determine the equivalent value of one metric unit in the customary system.
- Divide the given customary unit by the metric equivalent of one unit.
- Express the result in the required metric unit.
  Note. The same result is obtained if the number of metric units in one customary unit is determined and the given number is multiplied by the metric units. This second method is also illustrated.

EXAMPLE: Change 9″ to centimeters (cm).

Method 1

Step 1   Determine the value of one centimeter in terms of inches. $1\ cm = 0.393\ 7″$

Step 2   Divide the given unit (9) by this number (0.3937). $\frac{9}{0.393\ 7} = 22.86$

Step 3   Express the result (22.86) in terms of the required unit of metric measure (cm). $9″ = 22.86\ cm$ Ans

Method 2

Step 1   Determine the number of metric units (cm) in one given unit (1″). $1″ = 2.54\ cm$

Step 2   Multiply the given unit (9) by this number (2.54). $9 \times 2.54 = 22.86$

Step 3   Express the product 22.86 in terms of the required unit (cm). $9″ = 22.86\ cm$ Ans

The mathematical processes required to change units of measure from one system to the other are the same. This condition applies regardless of whether the unit is expressed in terms of square or volume measure.

RULE FOR CHANGING METRIC UNITS OF MEASURE TO CUSTOMARY UNITS (MILLIMETERS, DECIMETERS, CENTIMETERS, OR METERS TO INCHES, FEET, OR YARDS)

- Determine the number of equivalent metric units in one of the required units.
- Divide the given metric unit by this number.
- Express the result in terms of the required customary unit.

EXAMPLE: Change 228.6 millimeters (mm) to inches.

Step 1   Determine the number of metric units (mm) in one required unit (1"). $1'' = 25.4$ mm

Step 2   Divide the given metric unit (228.6) by (25.4) mm. $\frac{228.6}{25.4} = 9$

Step 3   Express the result (9) in terms of the required customary unit (inches). 228.6 mm $= 9''$ Ans

D. CONVERTING SQUARE (SURFACE) MEASURE VALUES

The same principles of surface measure for determining customary units are applied in the same manner to SI and conventional metric units of measure. The names and value of the units differ in each system, however. The area of a surface in the metric system is the number of square metric units that it contains. The three metric units most commonly used for small areas are the square meter, square centimeter, and square decimeter. Examples are given in the table of comparative values for common metric and customary units of surface (area) measure.

Metric Unit	Equivalent Customary Unit	Customary Unit	Equivalent Customary Unit
1 m^2	1550 in.^{2}	1 in.^{2}	6.452 cm^2
1 dm^2	15.50 m^2		0.0645 dm^2
1 cm^2	0.155 m^2		0.0929 m^2
1 mm^2	0.001 55 m^2	1 ft^2	9.29 dm^2
			929.03 cm^2
		1 yd^2	0.836 m^2

FIGURE 39-3 Comparative values of common metric and customary units of surface measure

RULE FOR CHANGING METRIC UNITS OF SURFACE MEASURE TO CUSTOMARY UNITS (SQ m, SQ dm, SQ cm TO SQ IN., SQ FT, OR SQ YD)

- Determine the number of metric units of surface measure in one of the required customary units.
- Divide the given metric units by this number.
- Express the quotient in terms of the required customary unit.

EXAMPLE: The area of a sheet of brass is 11.148 square meters. Express this value in units of surface measure (sq. ft).

Step 1 Determine the number of metric units of surface measure (m^2) in one of the required customary units (sq ft).

$$0.092\ 9\ m^2 = 1\ sq\ ft$$

Step 2 Divide the given number of metric units (11.148) by the number of square meters in 1 sq ft (0.092 9).

$$\frac{11.148}{0.092\ 9} = 120$$

Step 3 Express the quotient (120) in terms of the required customary unit (sq ft).

$$11.148\ m^2 = 120\ sq\ ft\ (120\ ft^2)\quad \text{Ans}$$

E. CONVERTING VOLUME AND LIQUID MEASUREMENTS

Figures 39-4 and 39-5 provide values for converting common units of volume and liquid measure in SI metric and customary unit systems.

Metric Unit	Equivalent Customary Unit Value (Approximate)	Customary Unit	Equivalent Metric Unit Value (Approximate)
1 cu meter (1 m^3)	35.314 cu ft (ft^3)	1 cu in. (in^3)	16.387 cm^3
1 cu dm (1 dm^3)	61.023 cu in. ($in.^3$)	1 cu ft (ft^3)	0.028 3 m^3
1 cu cm (1 cm^3)	0.061 cu in. ($in.^3$)	1 cu yd (yd^3)	0.764 6 m^3

FIGURE 39-4 Values of common units of volume measure

RULE FOR CHANGING METRIC UNITS OF VOLUME MEASURE TO BRITISH (CUSTOMARY) UNITS (m^3, cm^3, dm^3 TO CU IN., CU FT, OR CU YD)

- Determine the number of British (customary) units of volume measure in one given metric unit.
- Multiply the given metric unit by this number.

* Express the product in terms of the required British (customary) unit of volume measure.

EXAMPLE: Change 17.3 cm^3 to cu in.

Step 1   Determine the number of British units (cu in.) in one given metric unit (cm^3).

$$1\ cm^3 = 0.061\ cu\ in.$$

Step 2   Multiply given metric unit (17.3 cm^3) by this value .061.

$$17.3 \times 0.061 = 1.055$$

Step 3   Express the product in terms of the required British units.

$$17.3\ cm^3 = 1.055\ cu\ in.\ (1.055\ in.^3)\quad Ans$$

British units of volume measure may be changed to metric units by multiplying the given unit by the number of metric units in one British unit. For example, to change 3.2 cubic feet to cubic meters, simply multiply by (0.028 3), the metric unit equivalent of one cubic foot. The product (0.090 56) is in terms of metric units (m^3). Thus, 3.2 ft^3 = 0.090 6 m^3.

Metric Unit	Equivalent Customary Unit Value (Approximate)	Customary Unit	Equivalent Metric Unit Value (Approximate)
1 liter (ℓ)	1.057 qt	1 qt	0.946 ℓ
	61.023 in.^{3}	1 gal	3.7085 ℓ

FIGURE 39-5 Values of common units of liquid measure

RULE FOR CHANGING TO METRIC AND BRITISH UNITS OF LIQUID MEASURE

* Determine the equivalent value of one British or metric unit from a comparison table.
* Multiply the given value by the equivalent value in the desired unit.
* Express the product in the desired unit.

EXAMPLE: How many liters are there in 3 quarts?

Step 1   Determine the equivalent value in liters of one quart.

$$1\ qt = 0.946\ liters$$

Step 2   Multiply the (3) quarts by (0.946).

$$3 \times 0.946 = 2.838$$

Step 3   Express the product as liters.

$$3\ qt = 2.838\ liters\quad Ans$$

Review and Self-Test

ASSIGNMENT UNIT 39

A. Soft and Hard Conversions: SI and Customary Measurements

1. Use soft conversion to change linear dimensions A, B, and C to equivalent SI metric mm values. Round off each converted value to three decimal places.

Part	Linear Measurement	Decimal Value (.001")	Equivalent mm Value (.001 mm)	Equivalent (.01 mm) Measurement to (.001") Precision
A	$7\frac{7}{8}''$			
B	$4\frac{13}{68}''$			
C	$\frac{31''}{32}$			

2. Round off each metric dimension to two decimal places to establish a similar degree of accuracy to each three-place customary decimal measurement value.
3. Describe briefly the effect of hard conversion on the interchangeability of manufactured parts.
4. Determine the hard conversion SI dimensions that a designer might use for parts A, B, and C in problem A. 1.
5. Change each given linear dimension to the system and unit of measurement as specified.
   a. 200 cm to m   c. 5.14 in. to cm   e. 15.3 m to in.   g. 50.8 mm to in.
   b. 575 cm to dm   d. 6.5 yd to m   f. 12.7 cm to in.   h. 70 dm to ft
6. Convert surface measurements (a through h) to equivalent measurements as specified.
   a. 3,100 sq in. to m²   e. 9 sq ft to m²

   b. 62 sq in. to dm²   f. $3\frac{1}{2}$ sq yd to m²

   c. 127.875 sq in. to dm²   g. 3.2 dm² to in.²

   d. $3\frac{1}{2}$ sq in. to cm²   h. 725.6 cm² to in.²

7. Convert volume or liquid measurement (a through h) to equivalent measurements as specified.
   a. .75 dm to m³   c. 4.5 m³ to yd³   e. 17 cℓ to liters (ℓ)   g. 3.171 qt to ℓ
   b. 10 m³ to ft³   d. 3.6 ft³ to m³   f. 122.046 in.³ to ℓ   h. 31.5 gal to ℓ
8. Establish the weight of each machine (A through E) in its metric equivalent. Round off each value to one decimal place. Give the metric weight of the five machines.

2,713 lb

612 lb

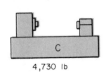
4,730 lb

1,819 lb

12,674 lb

B. Applications of Conversion Tables to Physical Science Problems

1. Identify three specific items of information that are contained in conversion tables of standards of different units of measure.

2. Provide the following technical information for three derived units of measure for the three categories of physical science measurements: mass (A), electric current (B), and time (C).
   a. Indicate the nature or area of the measurement.
   b. Give the symbol of the selected unit of measure.
   c. State the formula for the derived unit of measure.
      Note. The answer to A.1 under mass provides an example.

	Base Units of Measure									
	A. Mass (kilogram, kg)			B. Electric Current (ampere, A)			C. Time (second, s)			
	Derived Units of Measure									
	Nature of Measurement (a)	Symbol (b)	Formula (c)	Nature of Measurement (a)	Symbol (b)	Formula (c)	Nature of Measurement (a)	Symbol (b)	Formula (c)	
1	force	N	$N = kg \cdot m/s^2$							
2										
3										

3. Give the reciprocal factor to use to convert the three required unit values (A, B, C, and D) to the given value. The table provides the initial conversion information to change from the given to the required unit value.

	Measurement Values		Conversion Factor and Process	Required Reciprocal
	Given Unit Value	Required Unit Value		
A	in.	mm	25.4 (x)	
B	kg/s	lb · mass/min	0.007 560 (÷)	
C	psi	pascal (Pa)	6 894.757 (x)	
D	grains (g)	kilograms (kg)	15 432.9 (÷)	

4. Use the table of conversion factors for SI metrics in the Appendix.
   a. Record the factors for converting the given A, B, C, and D quantities to the required SI units.
   b. Compute the required unit values.

	Category	SI Metric Units		Conversion Factor	Converted Value
		Given	Required		
A	Density	10^{12} g/cm^3	kg/m^3		
B	Energy	75 kW·h	J		
C	Pressure	10^2g (force)/cm^2	N/m^2		
D	Volume	$7\ell \cdot 10^6$	m^3		

5. Compute (a) the equivalent SI metric pressures of systems A and B. (b) Determine the total force exerted by each piston to the nearest two decimal places.

$$\text{Pa} = \text{psi}\ (6\ 894.757) \qquad \text{kg(force)/m}^2 = \text{Pa}\ (9.806\ 650)$$

Piston System	System Pressure	Equivalent SI Metric Pressure (a)	Piston Area	Total Piston Force (b)
A	10 psi	Pa	10 in.^{2}	
B	64 Pa	kg (force)/m^2	12.2 m^2	

Unit 40 Achievement Review on SI Metrics, Measurement, and Conversion

OBJECTIVES OF THE UNIT

This achievement review serves as an overall test for Section 8. The unit is designed to measure the student's/trainee's ability to

- *Understand how base and derived units of measure are used and the procedures of adoption into the SI metric system.*
- *Identify both advantages and disadvantages of SI metric units of measure as compared with customary units.*
- *Use conversion factors and processes to translate given quantities in customary units to SI metric values.*
- *Read direct metric linear measurements with line-graduated steel rules and drafting room scales.*
- *Read measurements on outside, inside, and depth micrometers, vernier calipers, and height gages.*
- *Interpret angle readings on the vernier bevel protractor.*

A. HISTORICAL PERSPECTIVE ON STANDARDS AND UNITS OF MEASURE

1. State three reasons why international units of measure and standards change slowly but continually.

B. ADVANTAGES OF SI METRICS

1. a. List five of the seven base units in SI metrics.
   b. Identify the physical property that each unit measures.

SI Base Unit					
Physical Property					

2. Give three advantages of SI metrics over other systems of weights and measure.

C. SCIENTIFIC NOTATION SYSTEM

1. Translate each of the four given values to its equivalent SI value in the required unit.

	Given Quantity and Unit	Required Unit	Quantity SI and Abbreviation
A	1 250 grams	kilograms (kg)	
B	2 500 000 volts	megavolts (MV)	
C	1.75 meters	millimeters (mm)	
D	9 835 800 000 hertz	gigahertz (GHz)	

2. Cite three rules for writing SI metric units of measure.

D. DIRECT LINEAR MEASUREMENTS IN THE METRIC SYSTEM

1. Measure the diameter of each of the six bored holes (A through F) with an inside caliper. Transfer each measurement to a rule. Record each diameter to the nearest millimeter.

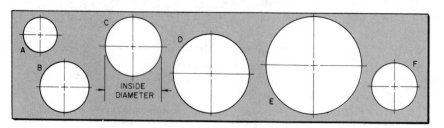

2. Read measurements Ⓐ through Ⓕ on the 1:50 metric drafting scale. State each measurement in terms of meter and fractional decimal values.

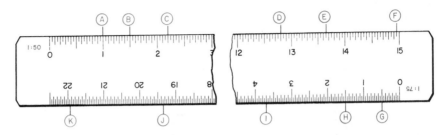

3. Read measurements Ⓖ through Ⓚ on the bottom 1:75 metric drafting scale. Give each reading in terms of its meter/centimeter value.

4. Determine the linear dimensions on the vernier micrometer setting Ⓐ, Ⓑ, and Ⓒ. (*Note.* The 0.1 and 0.05 mm sleeve graduations in the illustration are enlarged for easier reading.)

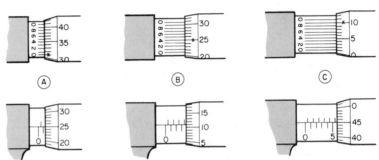

5. Read the three metric vernier caliper settings as indicated at Ⓐ, Ⓑ, and Ⓒ.

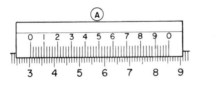

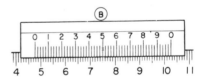

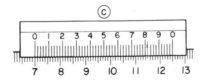

6. Read the metric vernier height gage measurements displayed at Ⓐ and Ⓑ.

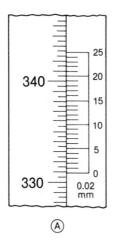

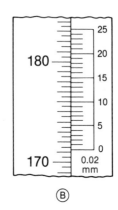

7. Determine the metric depth gage measurements (0.02 mm) shown at Ⓐ and Ⓑ and the inside micrometer measurements using the measuring rods as indicated at Ⓒ and Ⓓ.

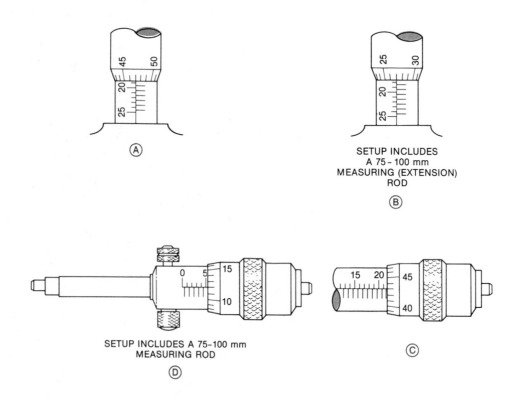

E. READING ANGLES WITH THE VERNIER BEVEL PROTRACTOR

1. Determine the vernier bevel protractor readings displayed at Ⓐ and Ⓑ.

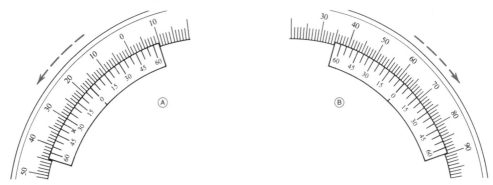

F. COMPUTED MEASUREMENTS IN THE METRIC SYSTEM

1. Determine the weight of 10 cast iron blocks to the nearest pound.

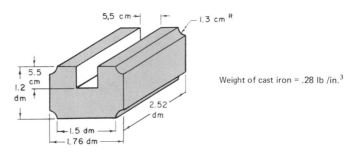

Weight of cast iron = .28 lb /in.^{3}

2. Compute the cost of the bronze castings, correct to two decimal places.

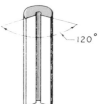

Outside Diameter	1.2 dm
Inside Diameter	2.54 cm
Height	3.6 dm
Number Required	12
Weight of Brass	.26 lb/in.^{3}
Cost	$3.48/lb

3. a. Secure a set of metric gage blocks or a manufacturer's table to establish the series and sizes of gage blocks in the set.
   b. Select a gage block combination to use in setting up for a measurement of 70.824 mm. (*Note.* Use a 1.5 mm wear block at each end.)
   c. List each gage block in the position (order) in which a craftsperson would set up (wring) the combination.
4. Compute the nominal weight in kilograms of each quantity of thermoplastic pipe according to the sizes, weights, and quantities given.

5. Determine how many metric tons of thermoplastic piping is required, correct to two decimal places.

	Outside Diameter (cm)	Wall Thickness (cm)	Nominal Weight (kg/m)	Standard Length (m)	Quantity Required	Nominal Weight (kg)
A	10.2	0.638	2.2	16	36	
B	15.2	0.754	3.8	24	12	
C	30.5	1.09	10.7	20	15	

6. Determine the weight in kilograms of the 15 bronze bushings described in the drawing. Round off the answer to the nearest whole kilogram value.

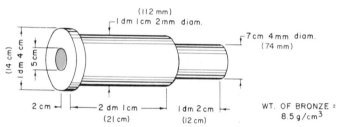

G. BASE AND DERIVED UNITS OF MEASURE IN SI METRICS

1. Make two brief statements about the need for derived units of measure in SI metrics.
2. a. Name any one of the units of measure in SI metrics.
   b. Give the base unit for the group or category.
   c. Name four recommended decimal multiple and submultiple units and give the symbol for each unit.

H. CONVERSION FACTORS AND PROCESSES IN METRICATION

1. Convert each of the temperature measurements (A, B, C, and D) to its equivalent value in each of the two other temperature measurement units.

	SI Metric Temperature		Customary Temperature
	Celsius (°C)	Kelvin	Fahrenheit (°F)
A	− 10		
B	40		
C		1 753.15	
D	3.43×10^3		

2. a. Compute the equivalent Fahrenheit or Celsius degree temperatures that are missing in the table for soft-solder alloys A and B.

Soft Solder Alloy	Melting Range			
	Completely Solid		Completely Liquid	
	°F	°C	°F	°C
A	361		460	
B		270		312.22

    b. Give the temperature span (range) between the completely solid and completely liquid states of the two alloys in (1) degrees Celsius and (2) Fahrenheit degrees.

3. Secure a table of conversion factors.
    a. Record the conversion factor and mathematical process for values A, B, C, and D.
    b. Convert the given measurement to the required measurement, correct to two decimal places.
    c. Give the reciprocal and mathematical process to use in converting the computed measurement back to the given measurement.
    d. Recompute the original given value to the nearest whole number.

	Measurements		Conversion Factor and Process	Computed Measurement	Reciprocal and Mathematical Processes	Recomputed Given Value
	Given	Required				
A	465 mm	in.				
B	150 kg	lb (mass) avoirdupois				
C	125 Btu/s	kW				
D	240 km/h	m/s				

4. Compute the required volume (mass) in systems A, B, C, and D. Use the two formulas as indicated.
    a. State the conversion factor in each case.
    b. Compute and record each volume in the unit indicated in the table, correct to two decimal places.

	Volume (Mass)		Conversion Factor and Process
A	4 313.80 kg/m^3	lb (mass)/gal (US)	
B	kg/m^3	22.6 lb (mass)/gal (US)	
C	kg/m^3	32 lb (mass)/ft^3	
D	7 220 kg/m^3	lb (mass)/ft^3	

$$\text{lb (mass)/gal (US)} = \frac{\text{kg/m}^3}{119.8264}$$
$$\text{kg/m}^3 = \text{lb (mass)/ft}^3 \times 16.0185$$

PART · THREE

Fundamentals of Electronic Calculators

CALCULATORS: BASIC AND ADVANCED MATHEMATICAL PROCESSES

Unit 41 Four-Function Calculators: Basic Mathematical Processes

OBJECTIVES OF THE UNIT

After satisfactorily completing the unit, the student/trainee will be able to
- *State the advantages of electronic calculators over conventional methods of solving mathematics problems.*
- *Understand algebraic (logic) entry, algebraic rules of order, and important components and general operation of basic electronic calculators.*
- *Round off values of significant and nonsignificant figures based on a knowledge of precision and accuracy meanings.*
- *Solve general and practical problems involving the basic arithmetical processes of addition, subtraction, multiplication, and division and other problems requiring chain processes.*

A *calculator* is a computing machine that performs either basic arithmetical or advanced mathematical processes. All calculators may be used to compute the four basic processes of addition, subtraction, multiplication, and division. There are, however, larger and more complicated calculators. These models are identified as engineering and scientific calculators, advanced business computers, or electronic slide rules.

A. CHARACTERISTICS OF ELECTRONIC CALCULATORS AND TERMINOLOGY

Calculators that are small and pocket-sized are classified as "hand-held" or "pocket calculators." Larger calculators that are operated from a desk are called "desk-top" models. Some calculators are battery-operated; others depend on electric power, or a combination. Since calculators consist of modern, compact integrated electronic circuits that perform the mathematical operations, they are also called *electronic calculators*. Modern electronic calculators have many advantages over the older forms of calculators, slide rules, and adding and other business machines. These advantages include

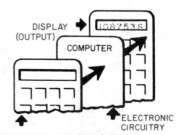

FIGURE 41-1

- Portability: lightweight, small size, comparative inexpensiveness
- Higher calculating speed and greater accuracy
- Performance in carrying out calculations and processes that are not possible by other machines.

Every calculator has a *keyboard*. The keyboard consists of a number of keys. There are ten keys for the basic numerals: 0, 1, 2, 3, 4, 5, 6, 7, 8, and 9. In addition, there is a decimal point and another set of keys for instructions relating to mathematical processes. The numerical keys, when pressed, enter the values to be used in the computation. The mathematical keys indicate the processes ($+$, $-$, $\times$, etc.) that are to be performed.

The mechanical operations of the keys provide the *input*. The computer component of the calculator performs the mathematical processes as directed by the input. The answer appears as an easy-to-read numerical value in the *display*.

FIGURE 41-2 Four basic mathematical function calculator *(Reprinted by permission of Sharp Electronics)*

The four basic arithmetical processes *(functions)* are addition, subtraction, multiplication, and division. These processes are required in more than 90% of all calculations. The electronic calculator that performs these processes is identified as a *four-function* type. This four-function calculator can also be used to perform higher mathematical processes. This is accomplished by breaking down the problem into a form that requires the four basic processes. The four-function calculator usually includes a percent (%) key to perform this particular arithmetical process.

Programmable Calculators

Programmable (scientific/engineering) calculators contain a *memory* to store numbers and *instructions* for performing a sequence of operations. The sequence in which the instructions are to be executed on the numerical values forms a *program* for solving the problem.

The program is stored in the calculator. Problems are entered and all mathematical processes are performed electronically according to the specified sequence. The program may be repeated any number of times using different numerical values. At the completion of the problem the correct answer is displayed. There is no further operator input into the problem.

Programmable calculators perform other sophisticated processes such as *decision making* and *branching*. Under specified conditions quantities are analyzed. Certain numerical values are then automatically processed by branching through a different program. Programmable calculators have the capacity to execute a program that is stored in a memory of the calculator. A depth study is made in Unit 43 of scientific calculator operations, functions, and advanced applications.

FIGURE 41-3 Handheld engineering and scientific calculator *(Reprinted by permission of Sharp Electronics)*

Algebraic Entry

The sequence in which the numerals of a problem are entered on the calculator is referred to as the *algebraic (logic) entry*. This means that a problem is solved according to standard *algebraic rules of order*. Thus, the calculator uses the same logical sequence in which an individual solves a problem. The keys on an algebraic calculator are identified as $+$, $-$, $\times$, $\div$, and $=$.

Numerical values and mathematical processes are entered on the calculator in the sequence in which they occur in the problem. Basic rules of order must be followed. For example, all of the multiplication and division processes in the problem must be performed first. Then, the addition and subtraction calculations are completed as required.

The advantages of this system are simplicity and consistency. Its disadvantages lie in the fact that the individual must arrange the sequence of processes, write down and remember intermediate results, and apply these values later in the solution of the problem.

Some of the more advanced calculators use another form of algebraic entry. Engineering calculators are able to perform a series of *hierarchy* operations. Internal storage registers eliminate intermediate storage as well as the need to remember the rules of order. Information is entered as an algebraic entry in the same sequence in which the problem is to be solved. The correct answer appears when the equals ($=$) key is depressed.

Arithmetic Entry

Computations for addition and subtraction are carried on differently on the arithmetic calculator. The arithmetic calculator is distinguished by its $\boxed{+ =}$ key and $\boxed{- =}$ key. The $\boxed{+ =}$ key is pressed after all positive numbers. The $\boxed{- =}$ key is pressed after all negative numbers or if the numbers are to be subtracted. For instance, if 27 is to be subtracted from 85, the key sequence on the arithmetic calculator is $85 + = 27 - = 58$.

The procedures for multiplication and division are the same on both the algebraic and the arithmetic calculators.

True-Credit Balance

A calculator is said to have a *true-credit balance* when the display indicates whether the resulting numerical value is positive or negative. Usually, a negative result or difference on a true-credit balance calculator is displayed with a minus sign.

Registers

A *register* is a memory circuit. It is used to remember multidigit numbers. Four-function calculators have at least three *operating registers*. These registers are used to enter the numbers of a problem, display the numbers, and carry out the arithmetical processes. The most significiant register is known as the *accumulator* or *display register*. The numbers that the operator enters via the keyboard are stored in the accumulator register. The numbers appear on the display register.

When the arithmetical process is entered on the keyboard, followed by the next value in the problem, the first

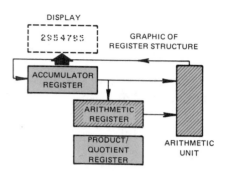

FIGURE 41-4

number is transferred from the accumulator register into the *arithmetic register*. When the equal key is pressed, the numbers (which are stored in the accumulator register and the arithmetic register) are transmitted to the *arithmetic unit*. Here the values are added or subtracted. The addition or subtraction result is stored in the accumulator. It is also displayed.

Since multiplication and division processes are shortened methods of adding and subtracting, a third register known as the *product/quotient register* is used to give the calculator additional memory storage.

Memory: Processing Store/Recall Data

All calculators have registers (memories) that store values to be used in a problem. The registers are fixed because they are permanently tied into the calculator circuitry. The registers are adequate for everyday applications.

Other calculators have greater capacity with long multidigit numbers that normally require complex calculations. Auxiliary memory is provided in such calculators to permit the storage and later recall of intermediate calculation results.

Separate keys are used to control the auxiliary memory features. A $\boxed{\text{CM}}$ or $\boxed{\text{MC}}$ key clears all numbers from the memory register, returning to zero. Pressing the $\boxed{\text{RM}}$ or $\boxed{\text{MR}}$ key recalls and transfers the value in memory to the accumulator register. The numerical value is displayed and is ready to be used in some computation without clearing or disturbing the memory accumulation.

The sum or difference of the memory content and the content of the accumulator register may be computed by pressing the appropriate $\boxed{\text{M}+}$ or $\boxed{\text{M}-}$ key. The sum or difference calculation is then contained in the memory. A number is stored in memory by first clearing the memory with the $\boxed{\text{CM}}$ key. The $\boxed{\text{M}+}$ key is pressed to transfer the accumulator contents to the memory. The $\boxed{\text{M}-}$ key subtracts numbers in display from memory. On some calculators, when memory is in use, a decimal place appears in the furthest left (''left-most'') digit space. Additional information on memory processes is presented in Unit 43 in relation to scientific calculators.

Clear, Clear-Entry, and Clear Display Key Functions

These keys are used to clear or erase the results of previous calculations and to reset the calculator. The clear $\boxed{C}$ key is pressed on many calculators to reset the contents of the accumulator display register and the arithmetic register to zero. The circuitry of the calculator is then set to start a new calculation.

The clear-entry $\boxed{CE}$ or clear-display $\boxed{CD}$ key is used to correct number-entry errors without disturbing any previous part of the calculating sequence. The $\boxed{CE}$ key is pressed to erase an error and to enter a correct new value. On some models the clear-error and total-erase processes are performed by a key marked $\boxed{C/CE}$. An entry mistake is corrected by pressing the $\boxed{C/CE}$ key once and entering the correct value. This action does not disturb any previously recorded data or calculations. The $\boxed{C/CE}$ key is pressed twice to erase all previous data and to clear the machine completely.

Calculator Display

The numerical readout on a calculator is found in a lighted *display*. The display serves three main purposes.

- Answers to problems are presented as a series of numerical digits in a readout display.
- Intermediate numbers used in a calculation are entered on the keyboard and presented on the display as a visual check.
- Information is fed back to a calculator operator through the display.

The number of digits in a calculator display varies. Six-digit displays are found in the less expensive models. Twenty or more digits are used in scientific calculators where greater precision is required. A practical common electronic calculator has an eight-digit display. This calculator has the capacity to display any number from 00 000 000 to 99 999 999. Fractional values are displayed by using a decimal point preceding the appropriate digit. Calculators are available in a variety of lighted color-digit displays. The displays operate on extremely small quantities of power.

Some accounting, scientific, and other business applications require a printed record. Calculators are designed with impact ribbon, heat-sensitive mechanical printers, ink injectors, and other mechanisms to meet such requirements and to produce a record tape.

B. SIGNIFICANT FIGURES, PRECISION, AND ACCURACY

Precision refers to exactness in defining a quantity. Precision is determined by the smallest increment that can be distinguished in the change of a quantity. When a higher degree of precision is required, there must be greater *resolution* and number of *significant figures* in a problem result.

A linear dimension like 3.01″ may be measured directly on a steel rule to a precision of roughly one one-hundredth part of one inch. If a precision of .001″ or 0.0001″ or 0.0254 mm or 0.00254 mm is required, measuring instruments with capability to measure to these degrees of precision are needed. In either instance, the number of significant digits in a required dimension is determined by the required precision. In these two examples, the steel rule may not be used to resolve any finer, more precise differences in length. Its limits of precision are reached at ±.01″. The micrometer, by comparison, has capacity to increase the precision from 10 (0.001″ or 0.0254 mm) to 100 (0.0001″ or 0.00254 mm) times. Precision is determined by the number of significant figures in a quantity, particularly those relating to decimal values.

The number of significant digits in whole numbers is determined by counting the number of digits with values from 0 through 9 that appear in the display to the left of the decimal point. For whole numbers and decimal values, all of the digits are counted.

At this point a statement should be made about accuracy as contrasted with precision. *Accuracy* is associated with degree of closeness to precision. While a micrometer may be used to measure to a precision of 0.001″ or 0.0001″ (0.2 mm

Significant Digits (Figures)	Examples	
1	0.000 005	
2	98.	or 0.098
3	268.	or 0.000 268
4	3.192	or 0.003 192
5	29 323	or 29.323
6	29.3230	or 0.293230

FIGURE 41-5

or 0.02 mm), the operator's handling of the instrument may cause an inaccurate measurement to be taken. Accuracy is the relationship between a computed measurement and any variance with the true required measurement. Accuracy is usually stated as a percent of the actual (or calculated) value to the true value. If a measurement is computed to be 7.65 meters and its true length is known to be 7.77 meters, the calculated value is $7.65/7.77$ of 100% or 98.5% accurate.

C. BASIC MATHEMATICAL PROCESSES ON A FOUR-FUNCTION CALCULATOR

All calculators may be used for the four basic arithmetical processes of addition, subtraction, multiplication, and division. To do this, an individual must know the basic arithmetic principles and their application. The calculator simplifies all of the basic processes, ensures accurate results, and saves time.

Addition

The first of the processes to be discussed is that of addition. Two common methods of carrying on addition are described.

Method 1   *Simulated Example.* Add quantities $A_1 + A_2 + A_3 + A_4$
   Step 1   Turn the calculator on.
   Step 2   Press the numbered keys to enter the numerical value of A_1. Read the visual display to check that the correct number is recorded.
   Step 3   Press the $\boxed{+}$ key.

Step 4 Press the numbered keys to enter the second numerical value (A_2). Read the visual display for accuracy.

Step 5 Press the $\boxed{=}$ key. This value is the sum of ($A_1 + A_2$).

Step 6 Press the $\boxed{+}$ key.

Step 7 Press the numbered keys to enter the value of A_3. Read the visual display to check the entry.

Step 8 Press the $\boxed{=}$ key. This is the sum of ($A_1 + A_2$) + A_3. The display shows this sum.

Step 9 Press the $\boxed{+}$ key.

Step 10 Enter the value of A_4. Check this value on the display.

Step 11 Press the $\boxed{=}$ key. The display total is the answer, representing the sum of $A_1 + A_2 + A_3 + A_4$.

Note. If whole numbers and decimal values are involved, the decimal key is used to enter the value. The *floating-decimal point* within the calculator automatically indicates the position of the decimal point in the answer.

Method 2 A number of key strokes may be eliminated by a second method.

Example. Add 1 763 (A_1) + 23 825 (A_2) + 197 (A_3) + 192 368 (A_4)

Step 1 Turn the calculator on.

Step 2 Press the numerical keys to enter the value of A_1 of 1 763. Check this number on the display.

Step 3 Press the $\boxed{+}$ key.

Step 4 Press the numerical keys to enter the value of A_2 of 23 825. Check the display.

Step 5 Press the $\boxed{+}$ key.

Step 6 Enter the value of A_3 of 197. Again, check this entry on the display.

Step 7 Press the $\boxed{+}$ key.

Step 8 Enter the value of A_4 of 192 368. Check the display.

Step 9 Press the equals key. Read the answer $\boxed{218\ 153}$ on the display. In this example, **1 763 + 23 825 + 197 + 192 368 = 218 153.**

Note. All problems should be checked by repeating the steps and comparing the answers.

Subtraction

The same two methods may be used to solve problems involving subtraction. Instead of pressing the $\boxed{+}$ key, the $\boxed{-}$ key is used.

Correcting an Error

If a wrong numerical value is entered, two simple steps are required to correct the error.

- Press the $\boxed{C}$ (or $\boxed{CE}$ key on some models) key once. This removes the last entry.
- Enter the correct value on the keyboard. Check this value on the display. Then proceed with the remaining part of the problem.

EXAMPLE: Add 79 and 67.

	The Display Reads
Step 1 Enter the 79.	79
Step 2 Press the $\boxed{+}$ key. Check the display.	79
Step 3 Assume you enter 65 by mistake.	65
Step 4 Press the $\boxed{C}$ key.	0
Step 5 Enter the 67. Check the display.	67
Step 6 Press the $\boxed{=}$ key. The display quantity is the answer.	146 **Ans**

FIGURE 41-6

Multiplication and Division

In addition to regular multiplication and division problems, a number of calculators have what is called a *constant feature*. Some calculators have a constant $\boxed{K}$ panel switch to enter a required constant with the keyboard. This means that if a constant number is used in a series of multiplication or division problems, the constant is entered once. For example, in the accompanying series of multiplications

$$12 \times 12 \qquad 12 \times 16 \qquad 12 \times 47$$

12 is the constant. Figure 41-7 shows how each value is computed.

Step	Press/Enter	Display Reads	
1	CC (Press C key twice)	0	
2	12 (and the × key)	12	
3	=	144	(**Ans** to 12 × 12)
4	16	16	
5	=	192	(**Ans** to 12 × 16)
6	47	47	
7	=	564	(**Ans** to 12 × 47)

FIGURE 41-7

A similar series of steps is followed in division. For example, if -10 is to be divided three times by 3.2,

$$-10 \div 3.2 \div 3.2 \div 3.2$$

the 3.2 would be a *constant*. If an answer to three significant digits is required, the -0.3051757 would be rounded off to -0.305.

Step	Press/Enter	Display Reads
1	CC	0
2	− 10	− 10
3	÷	− 10
4	3.2	3.2
5	=	− 3.125
6	=	− 0.9765625
7	=	− 0.3051757

FIGURE 41-8

Combination or Chain Arithmetical Processes

Many practical problems require a series of mathematical processes. Such series are known as *chain calculations*. They may involve addition, subtraction, multiplication, and division in varying combinations. In chain processes one of the goals is to solve the problem without having to store or write down intermediate values. Another goal is to keep the number of key strokes to a minimum.

There are rules of order to follow in establishing a sequence for carrying out chain calculations.

- Multiplication and division calculations are completed before addition and subtraction. For example, to solve the problem

$$7 + 9 \times 5 - 8 \div 4,$$

carry out the multiplication ($9 \times 5 = 45$) and division ($8 \div 4 = 2$) processes first. Then add $7 + 45 - 2 = 50$ **Ans**

Note. If the rules of order were not followed, an incorrect answer is obtained. For example,

$$7 + 9 = 16; 16 \times 5 = 80; 80 - 8 = 72; 72 \div 4 = 18.$$

- Portions of the problem are grouped before any mathematical processes are started. In the example, the values may be grouped and enclosed in parentheses.

The correct order or sequence for more complicated problems may be clarified by using brackets and parentheses. The operations in the parentheses are performed first. Those within the brackets are carried out next. The division process (which refers to all items within the brackets) is last.

The keys, processes, and display readouts for this problem are shown in figure 41-9.

$$[27 \times (19 + 17) + 68] \div 56$$
$$[27 \times \quad (36) \quad + 68] \div 56$$
$$[\quad \quad 972 \quad \quad + 68] \div 56$$
$$1040 \quad \quad \quad \div 56 = \boxed{18.571428} \quad \textbf{Ans}$$

Step	Press	Display
1	CC	0
2	19	19
3	+	19
4	17	17
6	×	36
6	27	27
7	+	972
8	68	68
9	÷	1040
10	56	56
11	=	18.571428

FIGURE 41-9

Review and Self-Test

ASSIGNMENT UNIT 41

A. Characteristics of Electronic Calculators

1. Indicate the one process in column II that correctly relates to each machine or function (column I)

Column I	Column II
Calculator or Function	*Process*
(1) Four-function calculator	a. Contains memory and instructions capability
(2) Display	b. Problem solving in the same logical sequence a person usually follows
(3) Programmable calculator	c. Performs addition, subtraction, multiplication, and division processes only
(4) Program	d. Total-erase process
(5) Algebraic entry	e. Visible colored entry or problem solution
	f. Registers that store problem values
	g. Corrects a number entry error
	h. Sequence for executing instructions entered in a computer

2. State the functions performed by the (a) $\boxed{\text{CM}}$, (b) $\boxed{\text{RM}}$, and (c) $\boxed{\text{M}+}$ memory keys.
3. Indicate (a) how a calculator may be reset to zero and (b) a method of correcting an entry error.
4. (a) Describe what en eight-digit display means. (b) Give a numerical example of the capacity of this model calculator.

B. Significant Figures, Precision, and Accuracy

1. Give four examples of whole quantities and four of mixed number quantities. The quantities must contain the required significant digits given in the table for A, B, C, and D.

2. Define (a) precision and (b) accuracy.

	Required Significant Digits	Quantity Examples	
		Whole Number Values	Mixed Number Values
A	1		
B	3		
C	5		
D	7		

C. Basic Arithmetical Processes: Addition, Subtraction, Multiplication, Division

1. Add the linear measurements given in the table with a calculator. List (a) the sequence of steps, (b) the key(s) and values to be entered, and (c) the display readings. Mark the answer with the appropriate unit of measurement.

Problem (Add)	
29 375.	Kilometers (km)
7 645.2	km
987.96	km
1 969.7	km

Step (a)	Keyboard Entry (b)	Display Reads (c)

2. Add the four quantities given in the table with a calculator. Indicate (a) the sequence of steps, (b) keys to be pressed or value to be entered, and (c) the display reading.

Between the second and third quantity introduce an error. Include the steps to show the wrong entry, the display, and how the error is corrected.

		797.75 RPM
Problem		8 756.90 RPM
(Add)	+95 682.	RPM
		439.6 RPM

Step (a)	Press (Keyboard) (b)	Display Reads (c)

3. Set up a problem with a quantity that is to be divided three times by a constant value. State the problem, the keyboard inputs, and the display readout for each step.

Problem

Step	Keyboard Input	Display Readout

4. Solve the problem, using the rules of order. Round off the decimal value to two significant places.

$$27 + 19.2 \times 31.3 \div 79$$

Unit 42 Four-Function Calculators: Advanced Mathematical Processes

OBJECTIVES OF THE UNIT

After satisfactorily completing the unit, the student/trainee will be able to
- *Understand how a calculator truncates values greater than the capacity of the instrument as underflow/overflow conditions.*
- *Use the calculator to convert fractions and decimals.*
- *Add and subtract numbers that have a greater number of digits than the capacity of the calculator.*
- *Deal with algorithms in solving calculator problems.*
- *Raise numerical values to higher powers and calculate reciprocals.*
- *Calculate square, cube, and higher root problems using a four-function calculator and use rounded-off values of π.*

Persons who regularly deal with financial, business, engineering, scientific, slide rule, or other advanced mathematical processes usually perform them on more expensive, special calculators. The electronic circuitry is more complex in these special calculators and there is additional built-in capacity and accuracy. Also, additional keys are included in order to carry on many combinations of mathematical functions.

It is possible, however, to perform higher mathematical processes involving algebraic, trigonometric, or geometric calculations using the four-function calculator. Step-by-step procedures are described in this unit for performing advanced mathematical processes.

Rounding Off Values

Regardless of whether a value is rounded off by conventional mathematical computing processes or on a calculator, error results. For most practical purposes, dimensions and values that are rounded off to two, three, and four decimal places are well within the usual range required. Parts produced within dimensional tolerances of 0.001″ and 0.0001″ (0.2 mm and 0.02 mm) are considered to be practical to measure and may be reproduced at reasonable cost. Movable units, machined to such tolerance limits, will mesh and work properly.

The degree of required precision should be determined first, before the solution of a mathematical problem. The precision will then establish the number of significant figures in the answer and at what point the computed value may be rounded off.

The calculator also *truncates* values. For example, the display of the decimal value for ⅓ is *truncated*. This means the display value on an eight-digit calculator reads .33333333, rounded off by eliminating the least significant digits. While this degree of accuracy in a final result is acceptable in most everyday problems, added care must be taken in making a long series of calculations. When values are rounded off after each calculation, the *cumulative* error may result in an inaccurate measurement.

A. OVERFLOW AND UNDERFLOW CALCULATOR CONDITIONS

When the numerical capacity of a calculator is exceeded, a condition of *overflow* exists. Overflow occurs when a quantity greater than the number of digits in the calculator is fed into it or when the result of a calculation exceeds the calculator capacity. This overflow condition is indicated in the display by the appearance of the letter **E** or **C** in the furthest left position of the display.

The overflow may be cleared by dividing by 10. This is repeated enough times to bring the decimal point into the display. Calculations are then continued. The result will be multiplied by 10^N. N represents the number of times the overflow was divided by 10.

Underflow denotes a condition where, again, the capacity of the calculator is exceeded. In this instance, the least significant digits may be lost. For example, in an eight-digit calculator, the display would register zero if the keys were pushed to record a value like **.000 000 003 256.**

B. CONVERSIONS OF FRACTIONS AND DECIMALS

Calculators are used with whole numbers or decimal parts. Sometimes, it is necessary to convert a fractional quantity to its equivalent decimal value. This process is easily accomplished by dividing the numerator by the denominator. Where the floating decimal is incorporated into the calculator, the decimal is automatically located. For instance, to convert the fraction $\frac{7}{64}$ to the equivalent decimal value, divide the numerator 7 by 64 = 0.109375.

To convert a decimal quantity to its fractional value, determine the number of decimal digits in the fractional part. A quantity like 1.104 indicates that there are 104 thousandths in the fractional part. So, the 1.104 quantity, expressed as a mixed number, $= 1\frac{104}{1\,000}$. The fraction is further reduced by dividing by the greatest common divisor (8). Expressed in its lowest terms, the $1\frac{104}{1\,000} = 1\frac{13}{125}$.

C. DOUBLE-PRECISION CALCULATIONS

Addition of Numbers with Digits Exceeding Calculator Capacity

Occasionally, it is necessary to add numbers that have more significant digits than the capacity of the calculator. Two numbers, like

$$140\ 569\ 074\ 279. \quad \text{and}$$
$$\underline{9\ 435\ 829\ 308\ 695.}$$

can be added on an eight-digit calculator by following these steps.

Step 1   Align the decimal points.

Step 2   Split each number into two parts: most significant digits and least significant digits. Keep each set of digits properly aligned. The number of least significant digits should be one less than the number of digits in the calculator display.

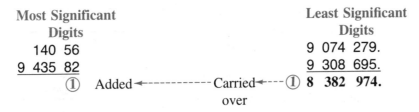

Most Significant Digits		Least Significant Digits
140 56		9 074 279.
9 435 82		9 308 695.
①	Added ←----------- Carried ←---- ① 8 382 974.	
	over	

Step 3   Add the least significant digit values. Record the answer. Note that there are eight digits in this answer. The eighth place digit value ① is carried over and added to the most significant digit value in the furthest right digit column.

$$
\begin{array}{r}
140\ 56 \\
+\,9\ 435\ 82 \\
+\quad\quad ① \quad \text{(carryover)}\\
\hline
9\ 576\ 39
\end{array}
$$

Step 4   Combine the sums of the two parts. The sum of the two combined display readings of **9576398382974** may be more easily read and accurately used when it is written as **9,576,398,382,974.** In SI metrics, the sum is given in this form: **9 576 398 382 974.**

Subtraction of Numbers with Digits Exceeding Calculator Capacity

Large quantities may be subtracted in a similar manner. The decimal points are aligned. The values are placed in two columns (1) the most significant part and (2) the least significant part. There must be one less number of digits than the capacity of the calculator in the least significant part. The values (1) and (2) are then subtracted.

It should be noted that if the top number is less than the lower number, one (1) must be *borrowed* the same as in simple arithmetic. When (1) is borrowed in the last (left) digit position of the least significant part, the right (and first) digit in the most significant part is reduced by (1). After the two parts are subtracted, they are combined to obtain the answer.

<u>E XAMPLE</u>:

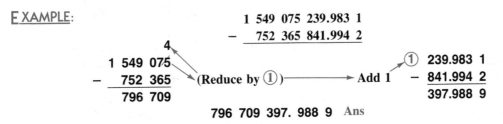

$$
\begin{array}{r}
1\ 549\ 075\ 239.983\ 1 \\
-\quad 752\ 365\ 841.994\ 2 \\
\end{array}
$$

4

$$
\begin{array}{r}
1\ 549\ 075 \\
-\quad 752\ 365 \\
\hline
796\ 709
\end{array}
$$
(Reduce by ①) ────→ Add 1

$$
\begin{array}{r}
① \ 239.983\ 1 \\
-\quad 841.994\ 2 \\
\hline
397.988\ 9
\end{array}
$$

796 709 397. 988 9   Ans

D. ALGORITHMIC PROCESSES APPLIED TO FOUR-FUNCTION CALCULATORS

An *algorithm* is a step-by-step routine used in solving a problem. It is this same routine that is used internally in calculators. The electronic circuitry within a calculator performs three basic processes to solve or implement any mathematical problem. The circuitry can perform only addition, subtraction, and shifting. The term *shifting* means moving a number one or more digits to the right or left in relation to another number. As stated many times, multiplication and division processes require successive addition and subtraction, respectively.

In summary, an algorithm is a set of mathematical rules, numerical procedures, or logical decisions. It provides variable input data to a calculator. Once an algorithm is mastered, advanced problems that would involve a single key on a sophisticated calculator may be easily performed on the four-function calculator.

The *reciprocal* interchanges a numerator and denominator. This interchanging is a valuable technique in chain processes. Since division is involved in finding a reciprocal, the value (1) is divided by a designated value. For instance, the reciprocal of 16 is 1 divided by 16 = 0.0625. Another procedure for finding the reciprocal of 16 is illustrated by the five steps shown in figure 42-1.

When used in a chain process, the algorithm (*figure* 42-2) applies.

Example: Find the reciprocal of (275.392 75 ÷ 17.687).

Step	Keyboard Entry	Display Readout
1	CC	0
2	16	16
3	÷	16
4	=	1
5	=	0.0625

FIGURE 42-1

Step	Keyboard Entry	Display Readout
1	CC	0
2	275.392 75	275.392 75
3	÷	275.392 75
4	17.687	17.687
5	=	15.570 348
6	÷	15.570 348
7	=	1
8	=	0.064 224 6

FIGURE 42-2

Raising Numerical Values to Higher Powers

One of the simplest problems of raising a number to a higher power is squaring the number. This process requires the use of the exponent (^{2}) with the number. Squaring problems are computed on the calculator the same as in ordinary multiplication. The square of 6.7 is 6.7 × 6.7 = 44.89.

There are two common algorithms that may be used when a number (N) must be raised to a higher power like the (nth) power. The first sequence is practical when the value of the power is smaller than ten. The number to be raised to a given power is entered on the keyboard. Then the multiplication $\boxed{\times}$ key is pressed. This step is followed by pressing the equals $\boxed{=}$ key one less time than the required power. The final value represents N^n, the quantity (N) raised to the required (^{n}) power.

If there is a chain calculation involving the raising of a quantity to a power, the problem should be restated in the sequence in which the combined mathematical processes are performed. The example eliminates going back to parts of a problem.

EXAMPLE:

$$\frac{(17.96 + 8.53)\, 6.2^3}{(7.23 \times 3.45)}$$

Restated as $(17.96 + 8.53) \times 6.2 \times 6.2 \times 6.2 \div (7.23 \times 3.45)$, the problem is keyed as a chain process into a calculator.

Raising a Number to a Very High Power

Raising a quantity to a very high power by using the first method is cumbersome and subject to possible error of a lost count. A simple algorithm to follow requires the breaking down of the high power into smaller factors. Assuming a quantity N is to be raised to the 180th power $(N)^{180}$, the 180 is broken down into smaller factors, like $3 \times 3 \times 4 \times 5$.

EXAMPLE: Raise N (value 1.3) to the 45th power.
$(1.3)^{45} = [(1.3^3)^3]^5$.

- Enter N (1.3).
- Press $\boxed{\times}$ key.
- Press $\boxed{=}$ key one less time than the value of first power.
- Press $\boxed{\times}$ key.
- Press $\boxed{=}$ key one less time than the value of the second power.
- Press $\boxed{\times}$ key.
- Press $\boxed{=}$ key one less time than the value of the third power.
- Read the display.

Step	Keyboard Entry	Display Readout
1	CC	0
2	1.3	1.3
3	$\times$	1.3
4, 5	=, =	1.69 2.197
6	$\times$	2.197
7, 8	=, =	4.826 809 10.604 499
9	$\times$	10.604 499
10	=	112.455 539
11	=	1 192.533
12	=	12 646.215
13	=	134 106.77

FIGURE 42-3

The thirteen steps shown in the table may be reduced if the calculator has a key for squaring $\boxed{x^2}$. After the required power has been factored, each factor may be further broken down. For example, $N^4 = N^2 \times N^2$; $N^3 = N^2 \times N$; $N^8 = N^2 \times N^2 \times N^2$. The required quantity is factored by using the $\boxed{x^2}$ key in combination with the $\times$ key to produce the equivalent exponential factor value.

E. CALCULATING SQUARE ROOT ON A FOUR-FUNCTION CALCULATOR

Square root problems are easily solved on calculators that have a square root function key. The required quantity for which the square root is needed is entered on the keyboard. The $\boxed{\sqrt{}}$ key is pressed. The display readout is the answer.

The process is more complicated on the four-function calculator because a trial-and-error procedure and a formula are used. An estimate is made of the square root of the required quantity. Suppose the square root of 97 is required. An estimate of 10 is made. Substitute the first estimate (10) for E and 97 for N in the formula.

$$\sqrt{R} = [(N \div E) + E] \div 2$$
$$= [(97 \div 10) + 10] \div 2$$
$$= 9.85$$

$\sqrt{R}$ = square root value
N = number for which the square root is required
E = estimated square root of N

- The approximate root (9.85) is now squared. The result 97.022 5 is too inaccurate. So, the process is repeated. This time the second estimated new root of 9.85 is substituted.

$$\sqrt{R} = [(N \div E) + E] \div 2$$
$$= [(97 \div 9.85) + 9.85] \div 2$$
$$= 9.848\ 857\ 5$$

- Again, squaring the resulting third estimated root of 9.848 857 5 gives a product of 96.999 994. With this degree of precision, the third estimate is accepted as the answer. Normally, a four-place decimal value is sufficiently accurate.
- If the last estimate does not produce a product within the limits of accuracy required in the problem, the estimating procedure is repeated one or more times. When still higher degrees of accuracy are needed, a calculator with a greater number of digits or built-in square root capacity should be used.
- The percent of error may be calculated by using a formula and substituting values.

$$\frac{(N) - (R)^2}{(N)} \times 100 = \text{Percent Error}$$

For example, if the second estimated root of 9.85 is squared and the product (97.022 5) is substituted for R^2

$$\frac{97 - 97.022\ 5}{97} \times 100 = -0.023\ 19\% \text{ Error}$$

F. CALCULATING CUBE ROOT AND HIGHER ROOTS ON A FOUR-FUNCTION CALCULATOR

Another algorithm is needed to obtain the cube and higher roots. While a similar procedure is followed as for determining square root, a different formula is used.

$$R^x = [(N \div A^{n-1}) + (n - 1)A] \div n$$

R^x = cube root or higher root
N = quantity for which root is to be computed
n = order of the root
A = approximate value of the nth root

As an example of the steps and entries that are made on a calculator, assume the cube root of 500 is required. The root value should be within .001% of accuracy. The estimated value (A) of the nth (^{3}) root is 8.

Formula		Step	Keyboard Entry	Display Readout
Item	Quantity			
		1	CC	0
[N	500	2	500	500
÷		3	÷	500
(A^{n-1})	(8^{3-1})	4	64	7.8125
+		5	+	7.8125
($n-1)A$]	$(3-1)8$	6	16	23.8125
÷		7	÷	23.8125
n	(3)	8	3	3
=	R^x	9	=	7.9375

FIGURE 42-4

The cube of $7.937\ 5 = 500.093\ 5$. If the error of .093 5 is too great, the process may be repeated. In this case, the second estimated root value should be slightly smaller. For closer precision, an estimated value smaller than the first estimate of 7.9375 (like 7.937) may be used $(7.937)^3 = 499.999$. This value is accurate to within 0.0002%.

While a cube root example was used to show the steps involved with a four-function calculator, higher root values may be computed using the same formula and processes.

G. ROUNDED PI (π) VALUES

Throughout the text, pi (π) has been used in many different types of practical mathematical problems. Pi is an *irrational number* whose value is 3.141 592 653 589 793. . . . It is obvious that most calculations are not this precise. So, the exact value is rounded off to a required number of decimal places. Values of 3.1416, 3.142, and the fractional value $\frac{22}{7}$ are widely used for most everyday problems. The highest degree of accuracy possible using a four-function, eight-digit calculator is to round the value of π to 3.141 592 7.

Review and Self-Test

ASSIGNMENT UNIT 42

A. Overflow, Underflow, and Rounded-Off (Truncated) Values

1. Define (a) overflow and (b) underflow. Give a numerical example in each case.
2. State how rounding off errors in high precision calculations result from truncated display values.
3. Subtract the three numerical quantities, using an eight-digit calculator. Round off the metric measurement to five significant digits.

8 957.625   mm
−2 868.732 9   mm
−4 927.843 42 mm

B. Conversions of Fractions and Decimals

1. Compute the missing fractional, decimal, or metric (mm) equivalent values from the information provided for measurements A through F. Round off decimal values to four places.

Kind of Measurement	Equivalent Measurements					
	A	B	C	D	E	F
Fractional (")	$\frac{1}{16}$	$\frac{1}{64}$				
Decimal (")			0.125	0.3125		
Metric (mm)					37.7	108.346

C. Double-Precision Calculations

1. Add the following quantities with a calculator having fewer digits than any of the three quantities. (a) Arrange the quantities into two columns of most significant digits and least significant digits. Then proceed to find the sum. (b) State the answer in megahertz (MHz) to four decimal places (least significant digits).

765 372 987.37   Hz
8 849 235.967 89 Hz
26 795 869.758 7   Hz

D. Algorithmic Processes Applied to a Four-Function Calculator

1. Describe briefly the importance of an algorithm in using a calculator.
2. a. Calculate the reciprocal of the four quantities given. List the display readouts.
   b. State the answer, rounding each decimal value to four significant digits.

		Reciprocal Value	
	Quantity	Display Readout (a)	Rounded Answer (b)
A	18		
B	$\frac{1}{7.4}$		
C	$\frac{176.33}{16\ 206.4}$		
D	$\frac{422.17 \times 16.7}{14\ 692.9}$		

3. a. Raise each quantity (A through H) to the required power. Record the display readout.
   b. Indicate each measurement by rounding off the decimal display values to two significant digits, unless otherwise indicated.

	Quantity	Power Value	
		Display Readout (a)	Measurement (b)
A	8^6		
B	7.2^3		
C	$8.4^2 \times 10.6^2$		
D	$10.04^3 \times 16.2^2$		
E	2.2^{12}		
F	0.46^{15}		
G	π^4 Use $\pi = 3.1416$		
H	α^6 Use $\alpha = 2.7183$		

E. Calculating Square and Higher Roots on a Four-Function Calculator

1. Compute the square root and the higher roots for quantities A, B, C, and D. Note the required degree of precision.
   a. Give the display readout of each final root value.
   b. Round each root value to four decimal places.

	Quantity	Root Value		
		Required Precision	Display Readout (a)	Required Dimension (b)
A	$\sqrt{86}$	±.001		
B	$\sqrt{\dfrac{(6.2 \times 9)}{1.74}}$	±.01		
C	$\sqrt[3]{(5.07 \times 4.02)^2}$	±.02		
D	$\sqrt[6]{716.88}$	±.005		

Unit 43 Scientific Calculators: Operation, Functions, and Programming

OBJECTIVES OF THE UNIT

After satisfactorily completing this unit the student/trainee will be able to
- *Interpret the range of a scientific calculator with respect to memory usage, basic mathematical functions, statistical functions, and advanced algebraic, trigonometric, and geometric functions*
- *Identify each key and key sequence function and read display register data.*
- *Establish the calculator mode; clear a scientific calculator; isolate numerical expressions; and enter data.*
- *Evaluate and solve practical problems involving combinations of basic mathematical processes.*
- *Program a scientific calculator for statistical (central tendency) measurements of mean, standard deviation, and variance functions.*

Particular attention was given in preceding units to the use of a simple four-function calculator. The applications related principally to basic arithmetical functions of addition, subtraction, multiplication, division, and combinations of these processes.

A. INTRODUCTION TO THE SCIENTIFIC/ENGINEERING CALCULATOR

More advanced problems encountered by technicians, engineers, and scientists, who formerly used slide rules, are now solved with sophisticated and powerful electronic hand calculators. While each calculator incorporates different design features, similar mathematical processes are performed.

A representative *scientific calculator,* designed for data entry by *algebraic notation,* is used as a model in this unit. According to the *Algebraic Operating System* (AOS)™, each number and function is programmed into the calculator. Thus, each entry follows the same mathematical sequence in which each function and quantity is written. Intermediate results are obtained by pressing the *equals key* $\boxed{=}$

Calculators in this family or level have the ability to accept entries in *scientific notation* (or *powers of ten*). The significant number of digits in internal calculator capacity is eleven. However, eight digits are displayed. The last number displayed for most functions is generally rounded off to ± 1 for the eighth-digit value.

FIGURE   43-1 Representative model of a scientific/engineering electronic calculator *(Courtesy of Texas Instruments, Inc.)*

First and Second (Alternate) Function Keys

Some calculator keyboards have keys that provide a *second (alternate) function.* The first function is printed on each key. The second function appears above a key. The drawing (*figure* 43-2) shows one row of second function keys on a calculator as designed by one manufacturer. The first function keys

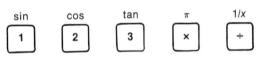

FIGURE 43-2 Sample row of keys with identification of first and second (alternate) functions

$\boxed{1}$, $\boxed{2}$, and $\boxed{3}$ are numeric; the following $\boxed{\times}$ and $\boxed{\div}$, arithmetic. The alternate (second) functions of sin, cos, and tan relate to trigonometry. The $\boxed{\pi}$ key is used for circle and angle measurement and the $\boxed{1/x}$ key for reciprocal functions. First and second function keys may be positioned horizontally or vertically, depending on the design of the calculator.

Identification and Functions of Keys

Each key on the representative model as illustrated in figure 43-2 is identified by name, function, and processing practices. Similar information with procedures follow in later units for those key and key sequences that apply to algebra, trigonometry, and geometry. This unit builds upon the skills developed with a four-function calculator.

B. BASIC CALCULATOR OPERATIONS

A *power-on condition* is provided in the model scientific calculator by pressing the $\boxed{\text{ON/C}}$ key. A $\boxed{0}$ *display register* indicates that the calculator is *clear* of *all pending operations and statistical registers.* Some calculators are designed so that *user memory* is not cleared when power is removed from the calculator by the $\boxed{\text{OFF}}$ key.

Mode Indicator

Scientific/engineering calculators use a *mode indicator* in the display. During short periods of a calculation, the abbreviation DEG (degree), RAD (radian), and GRAD (grad) may appear. These letters are *angular mode indicators*. Similarly, STAT in the register indicates the computer is in the *statistical mode*. The statistical mode indicator may be turned off, the STAT register cleared, and the calculator set for normal calculations by pressing the [2nd] [CSR] *clear statistical register key sequence*.

Data Entry

- The [·] *decimal point key* permits a fractional part of a number to be entered. Each digit value in the whole number is entered. The decimal point then *floats* with the additional numeric values added at each digit position. On this model (*figure* 43-1), a decimal with a maximum of eight digits may be entered.
- *Numeric Keys* [0] through [9], like the earlier four-function calculator, enter values of 0 through 9 as appropriate digits.
- The [+/−] *change sign key* is pressed after a number is entered in a calculator to change the sign of the number on display.
- The constant value of (π) pi (rounded to 3.1415927 on the display) is entered by pressing the [2nd] [π] *pi key sequences*. The *internal calculator value of pi* in the model is correct to eleven digits.
- The [K] *constant key* is used to store a number and its associated operations for repetitive calculations.
- Values up to eleven digits may be entered from the keyboard. A value like 5232.7684713 is entered as the sum of two numbers (5232 plus .7684713).

Enter (Value)	Press (Function)	Display
5232	[+]	5232
.7684713	[=]	5232.7685

Clearing the Calculator

The [ON/C] *clear entry/clear key* is pressed before any function or operation key to remove an incorrect entry from the display. The display, constant, and all pending operations are cleared when the [ON/C] key is pressed after an operation key or a function key or if the [ON/C] key is pressed twice. It should be noted that the [ON/C] key does not affect the user memory or statistical registers.

An incorrect number entry may be cleared by pressing the [ON/C] key before any non-number key. This action does not affect any calculations that are in progress.

C. ISOLATING NUMERICAL EXPRESSIONS

Parentheses serve the important function of grouping *(isolating)* particular mathematical values and functions. The closing of parentheses () indicates that all necessary information is complete. Each group is isolated so that the numerical values and required mathematical processes may be evaluated and performed on the calculator just as they are written.

EXAMPLE: Evaluate $6 \times (7 + 4) \div (7 - 2)$.

Numerical Entry	Key or Key Sequence	Display	Notes
6	$\times$ $($	6	The 6 $\times$ is stored
7	$+$	7	(7 + is stored
4	$)$	11	(7 + 4) is stored
	$\div$	66	6 $\times$ (11) is evaluated and the operation is performed
	$($	7	
7	$-$	5	(7 − is stored
2	$)$		(7 − 2) is evaluated and then divided into
	$=$	13.2	6 $\times$ (7 + 4)

D. MEMORY FUNCTIONS AND USAGE

Scientific calculators are designed with *memory storage*. Data and operations may be stored in the *memory register* even when the calculator is turned off. Operations in progress are not affected by the use of memory. The advantage of memory is that complicated computations and high powers of numbers and multiple calculations may be performed once, stored, and recalled any number of times.

Memory Store

The STO *memory store key* is used to store a displayed value in the memory. When the key is pressed, the new entry replaces any previous value or quantity in memory.

Memory Recall

Contents of the memory register are recalled into display by pressing the RCL *memory recall key*. The content of the memory register is not affected.

EXAMPLE: Store and recall the value 46.375.

Numerical Entry	Key or Key Sequences	Display
46.375	STO	46.375
	OFF ON/C	0
	RCL	46.375

Adding a Display Value (Sum) to Memory

The SUM *sum to memory key* permits the display value to be added to the content in memory register. The number on display or any calculation in progress is not affected by this key. When pressed, the SUM key accumulates the quantities from a series of independent calculations.

EXAMPLE: Store the sum to memory of the calculations:

$$(16.36 \times 4.2) + (12.974 + 7.286) + (3.14156 - 2.1347 + 6.42)$$

Numerical Entry	Key or Key Sequences	Display	Sum to Memory	
16.36	×	16.36	0	
4.2	= STO	68.712	68.712	
	+ ((			
12.974	+	12.974	68.712	
7.286	) SUM	20.26	88.972	
	+ ((			
3.14156	−	3.14156	88.972	
2.1347	+	1.00686	88.972	
6.42	) SUM	7.42686	96.39886	
	RCL	96.39886	96.39886	Ans

Memory Exchange

The content of the memory is exchanged for the display value by pressing the EXC *exchange key*. This key is used to store numbers or a result and to recall such values for comparison calculations or further computations.

E. SQUARING NUMBERS

The $\boxed{x^2}$ *square key* is used to calculate the square of the number entered into the display.

E XAMPLE: Square the value 27.435.

Numerical Entry	Key to Press	Display	
27.435	$\boxed{x^2}$	752.67922	**Ans**

F. STATISTICAL FUNCTIONS

Data related to quality control in manufacturing, cost analyses, sales and marketing, and other business activities are usually evaluated by statistical methods. Such statistics show central tendency. Such common terms as mode, median, standard deviation, and curves of distribution were described and applied earlier. Values obtained were computed "long hand" using basic mathematical processes.

The scientific electronic calculator is designed for effectively evaluating data and calculating *mean, standard deviation,* and *variance* for a total population or a sample of the population.

The *second function keys* for statistical functions appear on the keyboard as $\boxed{2nd}$ $\boxed{\bar{x}}$ for the *mean key;* $\boxed{2nd}$ $\boxed{\sigma n}$ for the *population standard deviation key sequence;* and $\boxed{2nd}$ $\boxed{\sigma n\text{-}1}$ for the *sample standard deviation key sequence.* The *variance key sequences* appear on the keyboard as the *population variation key sequence,* which includes three keys: $\boxed{2nd}$, $\boxed{\sigma n}$, $\boxed{x^2}$; the *sample variation key sequence* involves pressing the $\boxed{2nd}$, $\boxed{\sigma n\text{-}1}$, $\boxed{x^2}$ keys. The keys and sequential data are internally programmed according to the appropriate mathematical processes and data-gathering formulas.

Statistical Data Entry and Removal

The term *population* relates to a large quantity of values or sets of items. A *sample* or *sampling* represents a smaller portion selected from the population. The calculator is set in the *statistical mode* by the $\boxed{\Sigma+}$ *sum plus key* after the first entry is made. The mode is identified as STAT on the display.

Resetting the Calculator for Regular Functions

The statistical registers, the STAT indicator, and the calculator may be reset for regular calculations and functions by pressing the $\boxed{2nd}$ $\boxed{CSR}$ keys.

The $\boxed{2nd}$ $\boxed{\Sigma-1}$ *sum minus key sequence* is used to remove any unnecessary data points. After removal, the X register displays the current number of data points.

G. PROGRAMMING FOR STATISTICAL MEASUREMENTS OF CENTRAL TENDENCY

Calculating Mean, Standard Deviation, and Variance

Once the calculator is clear and set in the statistical mode, data points are entered by pressing the $\boxed{\Sigma +}$ *sum plus key*. If an error is entered, the data point entry is removed by pressing the $\boxed{\text{2nd}}$ $\boxed{\Sigma -}$ keys. This action produces an $\boxed{\text{ERROR}}$ display and causes the statistical registers to be cleared. After clearing and all new data are entered, the *mean value* is computed and displayed by pressing the $\boxed{\text{2nd}}$ $\boxed{\bar{x}}$ keys. Similarly, the $\boxed{\text{2nd}}$ $\boxed{\sigma n}$ keys are pressed for *standard deviation* and the $\boxed{x^2}$ *square key*, to obtain the *variance*.

It is important to remember that the $\boxed{\text{2nd}}$ $\boxed{\text{CSR}}$ keys are pressed before entering data points for arithmetical calculations.

EXAMPLE: Quality control inspection of a total population run of ground bearing races produces quantities of parts that meet dimensional and form requirements (*figure* 43-3). Calculate the mean, standard deviation, and variance.

Category	A	B	C	D	E
Accepted Quantities	990	950	894	943	996

FIGURE 43-3

Numerical Entry	Key Sequence (Programming)	Display	Notes and Functions
	$\boxed{\text{ON/C}}$		Applies power Clears calculator
	$\boxed{\text{2nd}}$ $\boxed{\text{CSR}}$	0	Clears the statistical register Clears the STAT indicator Resets calculator for manual operation
990	$\boxed{\Sigma +}$	1 STAT	Enters first data point
950	$\boxed{\Sigma +}$	2 STAT	Enter, add, and store second data point in statistical memory
894	$\boxed{\Sigma +}$	3 STAT	Enter, add, and store third data point
997	$\boxed{\Sigma +}$	4 STAT	Enter, add, and store fourth data point *Note.* Incorrect entry
997	$\boxed{\Sigma -}$	3 STAT	Remove fourth entry

Numerical Entry	Key Sequence (Programming)	Display	Notes and Functions
943	$\boxed{\Sigma+}$	4 STAT	Enter, add, and store correct fourth data point
996	$\boxed{\Sigma+}$	5 STAT	Enter, add, and store fifth data point
	$\boxed{2\text{nd}}\ \boxed{\bar{x}}$	954.5 STAT	Mean*(average number of parts that satisfactorily meet standards) **Ans**
	$\boxed{2\text{nd}}\ \boxed{\sigma\text{n}}$	36.86516 STAT	Standard deviation **Ans**
	$\boxed{x^2}$	1359.040 STAT	Variance **Ans**

*Values are rounded to the whole number in the display

H. KEYS FOR PROGRAMMING ALGEBRAIC, TRIGONOMETRIC, AND GEOMETRIC FUNCTIONS

Figure 43-4 identifies the remaining keys on the model. The keys in figure 43-4 are grouped according to functions in each branch of mathematics and do not appear in the position occupied on the calculator.

Algebraic Functions		Trigonometric and Geometric Functions			
Key	**Applications**	**Key**	**Applications**	**Key**	**Applications**
$\boxed{1/x}$	Reciprocals	$\boxed{\log}$	Common logarithm	$\boxed{\sin}$	Trigonometric functions
$\boxed{x!}$	Factorial	$\boxed{\ln x}$	Natural log	$\boxed{\cos}$	
$\boxed{\sqrt{x}}$	Square root	$\boxed{e^x}$	Natural antilog	$\boxed{\tan}$	
$\boxed{Y^x}$	Powers of numbers	$\boxed{\text{DRG}}$	Degree/Radian/ Grade mode		
$\boxed{\sqrt[x]{Y}}$	Roots of numbers				
$\boxed{\text{EE}}$	Enter exponent				
$\boxed{\text{INV}}$	Inverse				
$\boxed{\text{K}}$	Constant				
$\boxed{\%}$	Percentage				

FIGURE 43-4

Review and Self-Test

ASSIGNMENT UNIT 43

A. Scientific Calculator Operation

1. State two important differences between a simple four-function calculator and a scientific calculator.
2. Identify (a) two mode indicators and (b) the display readout(s) for each indicator.
3. a. Tell how to remove an incorrect entry.
   b. Tell how to clear the display and any constant and/or pending operation on a scientific calculator.
4. Identify the key sequence to the following operations:
   a. Clearing the statistical registers, turning off the STAT display indicator, and setting the calculator for normal operation.
   b. Entering the value of (π) pi.

B. Isolating Numerical Expressions: Chain Calculations

1. List the scientific calculator process by (a) numerical entry, (b) key(s), (c) display, and (d) notes for solving the following problem.

$$3.1416 \times (8.2 + 5.73) \div (9.84 - 3.652).$$

2. Solve problems (a) and (b). Round off the final value to two decimal places.
   (a) $(4.264 + 2.40 - 0.96) \times (13.568 \div 2.6)$
   (b) $(79.8 - 13.47 \times 0.22) \div (12.9 + 6.7 - 3.142)$
3. Compute the weight of metal required to stamp and form parts A, B, and C to the nearest pound.
   Note. The weight per piece is equal to the mass (volume) of each part multiplied by the weight factor given in the table. The scrap allowance is added to the length of each part stamped.

Parts	Quantity Required	Surface Area		Thickness	Metal	Weight Factor	Scrap Allowance
		Length	Height				
A	10,000	2.00"	2.00"	0.03196"	Brass	0.306	0.25"
B	5,275	3.40"	1.50"	0.05082"	Steel	0.284	0.35"
C	20,625	5.687"	3.25"	0.02535"	Copper	0.322	0.032"

4. Calculate the change in overall length (in inches) of current carrying conductors A and B for the (a) lowest and (b) highest temperatures recorded in the table. Round off the final answers to three decimal places.

Conductor	Metal	Length of Conductor at 25°C	Coefficient of Linear Expansion (inches/foot)	Temperature Range	
				Low (a)	High (b)
A	Pure copper	1,000 ft	0.000 009 2	− 40°C	50°C
B	Copper alloy	1,000 ft	0.000 009 9	− 40°C	50°C

C. Processing Reciprocals and Pi Values

1. Give the processes involved in determining the reciprocal of a physical measurement like 25.4 millimeters. Indicate the (a) numerical entry, (b) name and symbol of the key, and (c) the display.

2. Calculate the pitch of the following national standard screw threads, correct to three decimal places.

   (a) $\frac{1}{2}$-13 UNC   (b) 1″-12 UN   (c) 2-56 UNF

   Note. The pitch is equal to $\dfrac{1}{\text{No. threads/inch}}$.

   The number of threads per inch is 13, 12, and 56, respectively.

3. List each (a) entry, (b) name and sequence of keys, and (c) display in calculating the mass (volume) of the cored bronze casting.

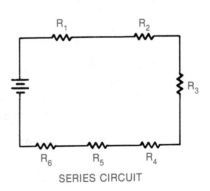

$$\text{Core diameter} = 17.78 \text{ cm diam}$$
$$\text{Outside diameter} = 27.94 \text{ cm diam}$$
$$\text{Length} = 20.32 \text{ cm}$$

Note. Mass (volume) = $(\pi) \times (\text{radius})^2 \times \text{length} - (\pi) \times (\text{radius})^2 \times \text{length}$

D. Calculating Statistical Measurements

1. Find the total resistance (R_T), mean, standard deviation, and variance of series circuits A and B. Round off each answer to three decimal places.

   Note. $R_T = \Sigma$ of $R_1 + R_2 + R_3 + R_4 + R_5 + R_6$

Series Circuit	Resistance in Ohms					
	R_1	R_2	R_3	R_4	R_5	R_6
A	5.00	12.50	6.72	12.97	32.63	17.26
B	41.70	26.0	38.0	46.42	18.40	9.20

SERIES CIRCUIT

2. a. Read and record the revolutions per minute (RPM) as displayed on each of the five tachometer test instruments.
   b. Calculate the (1) mean of the high speed turbine engine speeds, (2) the standard deviation, and (3) the variance.

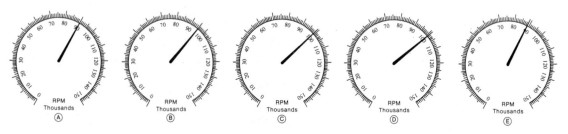

Unit 44 Achievement Review on Basic and Scientific Calculators

OBJECTIVES OF THE UNIT

This achievement review serves as an overall test for Section 9. The unit is designed to measure the student's/trainee's ability to
- *Describe the characteristics, advantages, and selected features of an electronic calculator.*
- *Deal with applications of significant figures with relation to precision and accuracy.*
- *Work with a four-function calculator to solve typical arithmetic and chain process problems.*
- *Apply algorithms with a four-function calculator to solve practical problems involving the substitution of values in formulas and chain processes.*
- *Isolate values for separate mathematical processes.*
- *Identify keys and functions of a scientific calculator.*
- *Calculate statistical measurements using a scientific calculator.*

A. CHARACTERISTICS OF ELECTRONIC CALCULATORS

1. State four advantages of modern electronic calculators over older calculating machines and devices.
2. a. Describe the purpose served by the registers of a calculator.
   b. State the function of three operating registers in a four-process calculator.
3. Cite two purposes that are served by the calculator display.

B. SIGNIFICANT FIGURES, PRECISION, AND ACCURACY

1. Tell and illustrate how the number of digits in a least significant decimal value indicates the precision of a quantity.
2. State a problem with three mixed numbers having an overall total of 12 digits and requiring addition and subtraction processes. Show how the three quantities may be grouped for an eight-digit calculator for double-precision calculations.

C. THE FOUR-FUNCTION CALCULATOR AND BASIC ARITHMETICAL PROCESSES

1. Solve the problem. (a) Number the steps and indicate the (b) keyboard entries and (c) display readings. Round the answer to six significant digits and mark the answer with the appropriate unit of measurement.

Problem	Step (a)	Keyboard Entry (b)	Display Readout (c)
865.7 volts			
+1 720. volts			
− 440.80 volts			
− 720.095 volts			

2. Add and subtract the quantities given in the problem. Use an eight-digit calculator.
   a. Show how the quantities are arranged in two columns.
   b. Label and give the answer corrected to eight significant digits.

$$
\begin{array}{lll}
 & 4\,3\,5\,8\,7\,2.5 & \text{cycles} \\
+ & 8\,9\,4\,2\,5.7\,6 & '' \\
+ & 4\,9\,8\,7.5\,3\,7 & '' \\
- & 2\,0\,9.6\,7\,9\,6 & '' \\
- & 3\,2\,6\,8.7\,8\,2\,5 & ''
\end{array}
$$

3. Multiply the quantity of 208 volts by the constant value 2.67 for three times. Use a hand calculator and give the answer correct to three decimal places.

D. ALGORITHMS USED WITH THE FOUR-FUNCTION CALCULATOR

1. Determine the ignition pulse (distributor point) frequency (ipf) for the two- and four-stroke cycle engine specifications given in the table.

Use the formula, $\text{ipf} = \dfrac{(n) \times (\text{RPM})}{(X)}$

n = number of cylinders

X = 120 for 4-stroke cycle engines

    = 60 for 2-stroke cycle engines

Note. Round off the pulses per second (ips) to one decimal place.

	Engine			Specifications
	Cylinders (n)	Crankshaft (RPM)	Stroke Cycle	Distributor Point Frequency (ipf)
A	8	2 000	4	
B	12	1 625	4	
C	6	1 375	2	
D	4	1 840	2	

2. The rise time (t_r) and the cutoff frequency (f_2) of a circuit may be computed by the following formulas.

$$t_r = \sqrt{(c^2) - (a^2 + b^2)} \qquad f_2 = \frac{0.35}{t_r}$$

Use a calculator to compute the (t_r) and (f_2) values for the A and B circuit conditions given in the table. Round off megahertz values to two decimal places.

	Rise Time (in nanoseconds)				Upper Cutoff
Circuit	Oscilloscope (a)	Square Wave (b)	Pulse (c)	Circuit (t_r)	Frequency (f_2) in mega-hertz (MHz)
A	40	20	110		
B	32	18	90		

E. OPERATION OF A SCIENTIFIC CALCULATOR

1. (a) Name and (b) state the function of each key or key sequence as represented by items 1 through 8.

	Key or Key Sequence	Name (a)	Function (b)		Key or Key Sequence	Name (a)	Function (b)
1	$1/x$			5	2nd CSR		
2	$\Sigma+$			6	$+/-$		
3	SUM			7	2nd π		
4	(( ))			8	x^2		

F. CHAIN CALCULATIONS: ISOLATING VALUES FOR SEPARATE PROCESSES

1. Solve the two problems (a and b) by isolating the bracketed values for separate mathematical processes. Round off the final value to three decimal places.

 a. $(\pi) \cdot (6.50)^2 - (\pi) \cdot (4.75)^2 \times (12.25 + 2.50 - 0.92)$

 b. $(72.4 + 26.37 - 14.67) \div (\pi \times 17.2) \times (14)^2 - (16.4 - 10.26 - 14.38)$

G. CALCULATING STATISTICAL MEASUREMENTS

1. a. Read the 0.01 mm metric dial indicator readings for five sample parts.

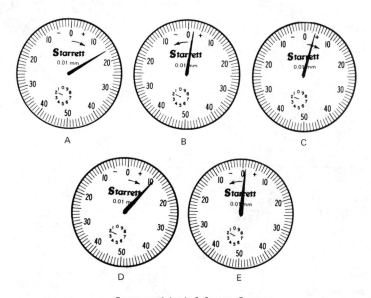

Courtesy of the L. S. Starrett Company

 b. Calculate the (1) mean measurement, (2) standard deviation, and (3) variance.

PART FOUR

Fundamentals of Applied Algebra (with Calculator Applications)

SYMBOLS, TERMS, AND SIGNED NUMBERS

Unit 45 The Concepts of Symbols and Terms

OBJECTIVES OF THE UNIT

After satisfactorily studying this unit, the student/trainee will be able to

- *Interpret and apply letter symbols, grouping symbols, mathematical symbols, and constant value symbols.*
- *Express quantities in literal terms consisting of numerical and literal factors.*
- *Apply literal terms in dimensioning shop and laboratory drawings.*
- *Understand the use of the signs:* $=$, $\neq$, $>$, $<$, $\geq$, *and* $\leq$ *in relation to different quantities.*
- *Use variables to express numbers, symbols, and other values.*

Algebra is a branch of mathematics that extends the applications of the four arithmetical processes to complex mathematical problems. Algebra involves the use of symbols, terms, and signed numbers and permits the user to solve problems that cannot be solved by arithmetic alone.

The several sections in Part Four of the text provide technical information, rules, examples, and practical applications that relate to equations, ratio and proportion relationships, scientific notation, and exponents, radicals, and formulas and complex equations.

A. MEANING AND IMPORTANCE OF SYMBOLS

Symbols are used as a shorthand and simplified way of saying what operations are to be performed. Symbols identify the quantities and units of measurement that are involved in the various steps of a problem. Typical examples of the everyday use of symbols are found in money problems (\$), (¢), and (%) and the arithmetical signs of ($+$), ($-$), ($\times$), and ($\div$).

309

Also, symbols are another form of worldwide communication. By their use it is possible to develop mathematical formulas. Symbols have the same interpretation regardless of the part of the world in which they are applied.

Symbols may be used with ease and accuracy if, at first, certain basic concepts are developed. The concepts relate to (1) the grouping symbols, (2) letters and subnumbers used as symbols, (3) signs and symbols that denote mathematical processes, (4) methods of handling symbols, (5) techniques of expressing quantities, and (6) symbols for constant values. These concepts and the application of symbols in solving practical problems of addition, subtraction, multiplication, and division are reviewed in this unit.

B. SYMBOLS USED FOR GROUPING QUANTITIES

Many problems are simplified by using what are called *grouping symbols*. The term refers to symbols that are used to group quantities together for ease in reading and to simplify mathematical processes. There are four common grouping symbols. While they serve the same function, each symbol is also used to separate or differentiate one quantity from another.

The first of these grouping symbols, and the one most widely applied, is the *parenthesis*. Some problems contain a number of parentheses ( ) and additional mathematical operations are to be performed with the quantities enclosed in the parentheses. A second grouping symbol, called the *bracket* [ ], is used in such instances. Still more complex mathematical practices involve the quantities in parentheses ( ) and brackets [ ]. A third grouping symbol, known as *braces* { }, is used with parentheses and brackets. The *bar* ——— (which is a straight line) is usually placed over or under a quantity to indicate that a required mathematical operation must be performed for the whole group or quantity.

Applying Grouping Symbols

E XAMPLES:

Case 1. Parentheses ( )

$$(4C) + (5D) + (6C - 4D) =$$

Case 2 Bar ——— and Parentheses ( )

$$\frac{(3A)}{2} + \frac{3.14\,(D) + 3.14\,(3D)}{4} - (6A) =$$

Case 3 Bar ———, Parentheses ( ), and Brackets [ ]

$$5X + \frac{(7X + 4X)}{2} + \left[\frac{(8C)}{4} - (C) + \frac{12X + 10C}{2} + \frac{1.158C + 4X}{4}\right] =$$

Case 4 Bar ———, Parentheses ( ), Brackets [ ], and Braces { }

$$\frac{16}{1.157\,(4D) + (2D)} + (4X) + \{[8X + (3X + 2D) - (4X + D)] + 2X\} =$$

RULE FOR SOLVING PROBLEMS USING GROUPING SYMBOLS ( ), [ ], { } ───────

* First perform the mathematical operations indicated within the *parentheses* ( ) and under or over a bar ──.
* Follow with the operations required for the quantities in the *brackets* [ ].
* Complete the operations indicated for the *braces* { }. Combine like terms.
  Note. The mathematical processes may be simplified by following this practice of solving the quantities in the parentheses, brackets, and braces, in that order.

EXAMPLE: Remove the signs of grouping and reduce the algebraic statement to lowest terms.
$5 + \{3X - [2 - (5 + 5X)]\}$

> *Step 1*   Remove the parentheses ( ).
> *Note.* The negative (subtraction) sign preceding the ( ) indicates that all mathematical signs in the ( ) are changed.

$5 + \{3X - [2 - 5 - 5X]\}$

> *Step 2*   Remove the brackets [ ]. Change all signs within the brackets as indicated by the negative sign.

$5 + \{3X - 2 + 5 + 5X\}$

> *Step 3*   Remove the braces { }. Since all values are to be added, the signs remain unchanged.

$5 + 3X - 2 + 5 + 5X$

> *Step 4*   Combine like terms.
> *Note.* The result expresses the algebraic statement in its lowest terms.

$$①$$
$$5 + 3X - 2 + 5 + 5X$$
$$②$$

① $5 - 2 + 5 = 10 - 2 = 8$
② $3X + 5X = 8X$
$8X + 8$ Ans

C. LETTERS AND SUBNUMBERS USED AS SYMBOLS

Close examination of problems in mathematics and shop formulas reveals that the formula or problem in many instances is a combination of numbers and symbols. These, in a convenient form, express the relationship of one quantity to another. The symbols that group the quantities may appear in any one of, or a combination of, the four forms: parentheses ( ), brackets [ ], braces { }, or bar ──.

A problem may be simplified further by using *letters*. In the case of a formula, the numerical values of letters or other symbols are substituted in the final solution of a problem. These letters are sometimes an abbreviation of a term or word. In other instances, just a single letter is used. The letter may become a *standard* when it always denotes a definite dimension. The dimension is established and accepted by national and international standards-setting organizations.

Notations or letters that appear in mathematical expressions or formulas have a fixed meaning. For example, in dealing with surface measurement, the area of a rectangle is determined by multiplying the linear dimension of the base by the height. This arithmetical process can be expressed in a simple and compact form in the formula $(A = L \times H)$.

In this case, the (A) is the letter symbol for area, (L) for length, and (H) for height. The letters are selected to represent definite values and convey the same meaning as the written value.

Capitals and Lowercase Letters as Symbols

Capital letters and *lowercase letters* are used as symbols to designate different quantities. For example, the capital letter (D) may denote the diameter of a bar of steel and (d), a lowercase letter, that of another. Care must be taken when using both capital and small or lowercase letters as symbols to make sure that the correct value of the letter is used in computations. Usually, it is easier to read lowercase letters when combined with numbers. Capital letters are most often used as constants or for specific terms in formulas, for example $I = E \times R$.

Signs Showing Relationship of Order

A simplified way of showing the relationship between two values is to use such signs as $=, \neq, >, <, \geq$, and $\leq$. While it is known that $=$ means "is equal to," the $\neq$ sign between two numbers, letters, values, or quantities that are grouped indicates "is not equal to." For example, $4AB \neq 5AB$. This is read "$4AB$ is not equal to $5AB$." The $\neq$ expresses the relationship between the two values.

An open arrowhead symbol like $(>)$ indicates a difference between two quantities. An arrowhead directed to the right $(>)$ shows that a quantity is "greater than." A left pointing arrowhead $(<)$ means "less than." An expression like $6A > 3B$ is an algebraic shorthand way of saying that the quantity $6A$ is greater than $3B$. The addition of a bar under an arrowhead indicates that the value of a number or symbol on the left of the $\geq$ is larger than or equal to the value on the right side. Similarly, $\leq$ indicates that the value on the left is smaller than or equal to the value on the right side.

Variables and Subscripts Used as Symbols

A *variable* is a letter (a lowercase letter, a capital letter, a letter with a subscript, such as b_1 (read as "b sub-one"); a frame ($\square$, $\square$, $\bigcirc$, and $\triangle$); or a blank (which holds a place open for a number). Sometimes the first letter of a key item is used as the variable. A variable may represent any number or value and, under specific conditions, a very definite number or value.

For example, in physical science, subscripts are widely used in different laws. In the case of Boyle's law, the variable letters P and V are directly related to values for pressure (P) and volume (V). In electric circuits, the total resistance is equal to the sum of the separate resistances. These are usually identified as R_1, R_2, R_3, etc. The value of each resistance (variable R) is substituted in a formula or an algebraic expression when solving a problem. $R_t = R_1 + R_2 + R_3 +$ etc.

Boyle's Law

$$P_1 \times V_1 = P_2 \times V_2$$

$P_1 =$ original pressure of a gas
$P_2 =$ pressure of a gas under a second set of conditions
$V_1 =$ original volume of a gas
$V_2 =$ volume of a gas under a second set of conditions

Subnumbers Used as Symbols

In many formulas and problems the same letter may be used to denote a number of values. When this is the case, a *subnumber*, called a subscript, follows the letter or symbol for purposes of identification. These subnumbers are placed to the right of and slightly below the letter. The symbol (A_1) would be read "A sub-one." This quantity is differentiated from the next which may be (A_2) "A sub-two" or (A_3) "A sub-three" by using the subscript.

D. SYMBOLS DENOTING MATHEMATICAL PROCESSES

In common use are the four arithmetical signs. The signs are symbols that denote the mathematical operations required. The plus sign ($+$) indicates addition; the minus sign ($-$), subtraction; the times sign ($\times$), multiplication; and the division sign ($\div$), division.

In addition to these signs, other symbols are used to denote multiplication. For instance, if the quantity (A) is to be multiplied by the quantity (B), the problem may be written five different ways to express the same process (*figure* 45-1).

In case 1, the multiplication process is denoted by the sign ($\times$). The symbol ($\cdot$) in the second case indicates that A is to be multiplied by B. In case 3, it is understood that A and B are to be multiplied together. Finally, parentheses are used to indicate multiplication. The methods shown in cases 4 and 5 are used when combinations of quantities must be multiplied.

Case	Method of Expressing Multiplication Process
1	$A \times B$
2	$A \cdot B$
3	AB
4	$(A)(B)$
5	$A(B)$ or $(B)A$

FIGURE 45-1

While the multiplication process may be expressed in five different ways, the quantities indicated in each case have the same value. In other words

$$A \times B = A \cdot B = AB = (A)(B) = A(B) \text{ or } (B)\,A$$

The operations involving the quantities within parentheses are carried out before the final value of a parentheses is multiplied by another value in a second parenthesis.

E. METHODS OF HANDLING SYMBOLS AS NUMBERS

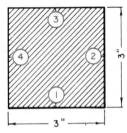

FIGURE 45-2

The letters or numbers that are used to express a fixed mathematical relationship are sometimes given or stated, as in the case of formulas. Otherwise, arbitrary symbols are selected to simplify the writing and solution of problems.

In figure 45-2, each side is three inches long. The distance around this square (which is called the *perimeter*) may be determined a number of different ways.

The perimeter may be found by adding the four sides (method 1). In method 2, the number of sides (4) is multiplied by the length of each side (3″). The multiplication process is preferred because it is easier and more accurate.

Method 1: Addition

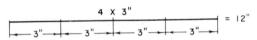

Method 2: Multiplication

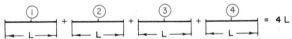

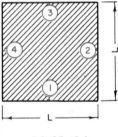

FIGURE 45-3

In the second square (*figure* 45-3), the length of each side is given as dimension (L). The perimeter of this square is also equal to the sum of the four sides. However, instead of using a number to indicate the dimension of a side, the letter (L) is substituted. The perimeter, which may be found either by multiplication or addition, is equal to (4L).

Method 1: Addition

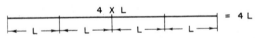

Method 2: Multiplication

Symbols may be used as readily as numbers. In fact, where the numbers in a problem are large and unwieldy, it is far simpler to perform the mathematical operations by using symbols. The value of each symbol is substituted in one of the last steps.

The perimeter of a rectangle is equal to the sum of the four sides. Since both pairs of sides are equal, the perimeter, in terms of the base whose length is (L) and height is (H), is equal to 2(L) + 2(H).

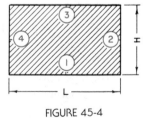

FIGURE 45-4

Method 1: Addition

Method 2: Multiplication

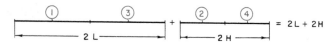

The relationship of the sides to the perimeter may be expressed as a formula where (P) denotes the perimeter of the rectangle: $P = 2L + 2H$. When simplified, $P = 2(L + H)$.

Methods of Expressing Quantities

A quantity may be expressed either in terms of numbers or a combination of numbers and symbols. Where only numbers are used, the quantity is said to be expressed in *numerical terms*. Where letters are used in combination with numbers, like $(8A)$ and $(4B)$, the expression is called a *literal term*. The parts of a literal term, namely the *numerical factor* and the *literal factor,* are shown in the illustration figure 45-5.

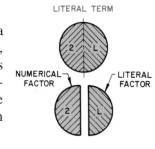

FIGURE 45-5

In the formula for the perimeter of a rectangle (*figure 45-6*), the literal terms are $(2L)$ and $(2H)$.

The numerical value of a literal factor is substituted in the final solution of a problem.

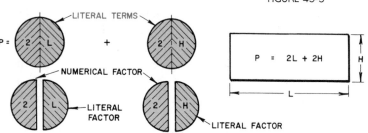

FIGURE 45-6

EXAMPLE: Determine the value of the literal term $(4A)$ when dimension (A) is equal to (4).

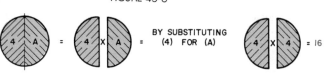

FIGURE 45-7

The terms used in expressing quantities may be identified easily. Remember that the numerical factor refers to a number value and the literal factor to a letter or other symbol. Symbols may be applied to the mathematical solution of triangles, circles, or any other shaped part, or to unknown quantities. In fact, symbols may be used in most problems where simplicity of expression and ease of solution are required.

The word *expression* mathematically refers to the abbreviated form (method) of stating a problem in numerical and literal terms. For example, the total area of the two rectangular parts is found by adding the area of each rectangle.

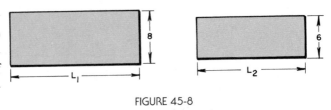

FIGURE 45-8

Since the area of a rectangle is equal to the length of the base multiplied by the altitude, this relationship may be expressed as

$$A = 8L_1 + 6L_2$$

where (A) refers to the total area and L_1 and L_2 refer to the respective lengths of each rectangle. The value $A = 8L_1 + 6L_2$ is called an *expression*.

G. SYMBOLS USED FOR CONSTANT VALUES

Where a value has been established scientifically and a specific symbol is used universally to denote a fixed mathematical relationship, the value is known as a *constant*. A constant represents a number or other value that is definite and never changes.

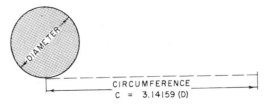

FIGURE 45-9

For example, to find the circumference of a circle, the diameter is multiplied by the constant 3.14159. This value is numerically correct to four decimal places. This relationship is sometimes expressed in the formula $C = \text{Pi}\ (D)$.

The symbol (C) denotes the circumference; (D) the diameter; and the English letters (Pi) are the symbol for 3.14159. In most formulas, the Greek letter for (Pi), which is (π), is used. Since 3.14159 is a fixed quantity that has been proved and is an established fact, it is called a constant value for (Pi).

When symbols indicate constant values in problems, the mathematical processes are simplified. The symbols may be added, subtracted, multiplied, or divided with greater ease than the numerical values. In most instances, it is comparatively simple in the final steps of solving a problem to substitute the numerical value of each literal factor.

Review and Self-Test

ASSIGNMENT UNIT 45

A. Grouping Quantities and Expressing Problems in Literal Terms

Express problems 1, 2, and 3 in numerical and literal factors. Use appropriate grouping symbols.

1. Required: Area (*A*) of the rectangular casting.

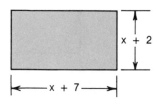

2. The circuit diagram shows three resistors: R_1, R_2, and R_3.
   The current each resistor draws is expressed in terms of
   (*y*) amperes; voltage, in (*x*) terms.
   Power (watts-W) = Current (A) × Voltage (V)
   Required: a. Power (W) to operate R_1, R_2, and R_3.
   b. Total power (W_3) for the three resistors.

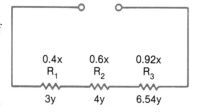

3. Refer to the literal dimensions used on
   the drawing of a Holder Plate.
   Required: Center distances for Ⓐ, Ⓑ,
   and Ⓒ.

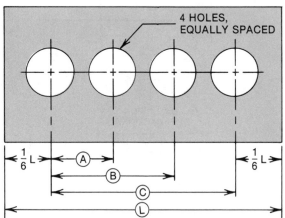

B. Applications of Symbols and Terms in Dimensioning and Measurement

1. Determine the dimensions of each shaded portion in figures (a through h). Note that the
   objects are divided equally into the number of parts indicated in each instance.

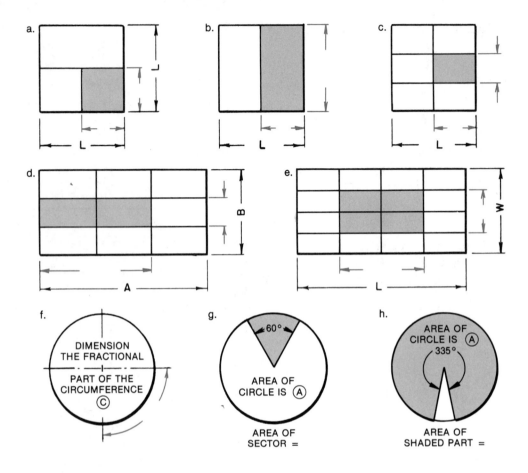

2. Dimension the copper stamping. Use the symbol (*L*) to denote length and (*W*) its width. The inner square is (3*X*) inches on the side. The rectangle is (5*X*) inches long and (2*X*) inches wide.

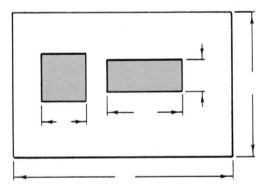

3. A step block that has four equal steps is (H cm) high, (W cm) wide, and (L cm) long. Dimension the sketch according to these known sizes.

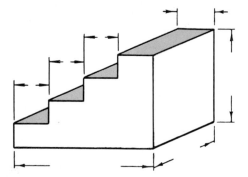

4. Dimension the square, rectangle, and triangles (a through d) according to the specifications given in each instance.

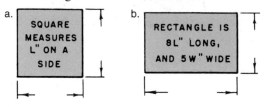

a. SQUARE MEASURES L" ON A SIDE

b. RECTANGLE IS 8L" LONG, AND 5W" WIDE

c. INDICATE THE LENGTH OF THE LONGEST SIDE OF THE TRIANGLE WHEN IT IS 2.5 TIMES THE SHORTEST SIDE

d. SIDES ARE 3A, 4A, AND 5A UNITS LONG

5. Dimensions on shop drawings are often given in terms of letters or symbols. The numerical value of each dimension is stated in table form. Where this practice is followed, one drawing is used for many different sizes of the same-shaped part. Indicate on the drawing the letter symbol given in the table for each dimension.

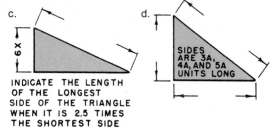

Linear Dimension on Drawing		Letter Symbol
Overall length		L
Overall height		H_1
Width (depth)		W
Radius at corners		R
Length of step	①	A
	②	B
	③	C
	④	D
	⑤	E
	⑥	F
Elongated slot	width (depth)	G
	length	H

Unit 46 The Concept of Signed Numbers

OBJECTIVES OF THE UNIT

After satisfactorily completing this unit, the student/trainee will able to
* *Relate signed numbers to appropriate values on line-graduated measuring instruments.*
* *Transfer the concept of signed numbers to everyday business, industry, and consumer applications.*
* *Translate problems of signed numbers and decimal and fractional values to a number line or other graduated measurement scale.*
* *Write absolute numbers.*

A. THE MEANING AND IMPORTANCE OF SIGNED NUMBERS

Positive (+) and *negative* (−) *numbers* serve two prime functions in algebra. These numbers (1) indicate direction from a fixed reference point (usually zero) and (2) identify the mathematical operation. Positive and negative numbers are called *signed* or *directed numbers*. The algebraic (+) and (−) signs differ from those in arithmetic that are used to indicate the *sign of operation*.

For example, temperature scales measure degrees in relation to a fixed boiling point of water (100°C or 212°F) and the freezing point (0°C or 32°F). Temperatures are generally recorded as above (+) or below (−) the freezing point. A −10°C temperature indicates the *direction* (below zero) and the *quantity* (10°C). The −10°C is a *negative* or *signed number*. It is understood that unless the sign is (−), the number is a *positive number*.

The intent of this unit is to help the student develop a concept of signed numbers, number lines, and make applications to graduated lines on measuring instruments and absolute values. Unit 47 deals with applications of signed numbers requiring the four basic mathematical processes. Signed numbers are also basic to later numerical control applications in positioning cutting tools and programming manufacturing processes.

B. THE USE OF SIGNED NUMBERS ON MEASURING TOOLS AND INSTRUMENTS

A dial indicator is a typical example of an instrument that uses signed number values. These values are indicated by the movement of a pointer across the graduated dial face. Referring to the illustration in figure 46-1, note that the pointer may be actuated to move clockwise to record a specific plus (+) measurement from a zero reference point. If the pointer moves counterclockwise ⟲, the measurement is a negative value. The dial face on this instrument is grad-

FIGURE 46-1

uated in increments of 0.001". The drawing shows the pointer movement to be in a clockwise (+) direction from zero to 0.015". The signed number in this case is +0.015" (or, because it is positive, 0.015").

Readings on many other electronic/electrical instruments and visual digital registers on machines express + and − number values.

C. THE NUMBER LINE (INSTRUMENT GRADUATED LINE)

Signed Numbers as Major (Whole) Graduations

Numbers can be placed on a horizontal or vertical straight line or on an arc or circle. A zero, or other suitable reference point, may be located in the middle or at another location. In general, numbers (values) are represented by graduations on a line. Such a line is called a *number line*.

By convention, number values to the right of the zero (index or reference) on the number line are positive (+). All numbers extending to the left of zero are negative (−). The illustration (*figure* 46-2) shows each numbered graduation to be equal and to represent one unit on the number line. Clockwise movements on dial indicators are recorded as positive numbers; counterclockwise, as negative (−) numbers.

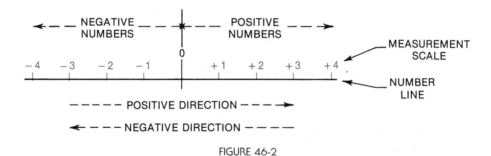

FIGURE 46-2

Consideration must be given to *direction* when using positive and negative numbers. A signed number may be moved in a positive direction as seen on the drawing (*figure* 46-3), starting from the zero reference point or from any other position along the number line. The same practice is followed in moving a signed number in a negative direction.

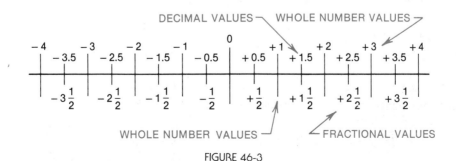

FIGURE 46-3

Decimal and Fractional Signed Numbers

Decimal and fractional values can also be positive or negative because they represent parts of whole numbers. Whole value graduations on a number line are often divided to represent fractional and decimal parts. Actually, a number line (graduated beam or instrument dial) can have one or a combination of several scales, as illustrated. Normally, just one (+) and one (−) sign appear on an instrument scale.

RULE FOR DETERMINING SIGNED NUMBER VALUES ON A NUMBER LINE

- **Starting at Zero Reference Point**
  - Identify the value on the number line that corresponds to the positive or negative number in the problem.

    Note. Positive number values are read to the right of the zero reference point; negative numbers, to the left. If the number line is an arc or circle, positive number values are read in a clockwise direction; negative numbers, counterclockwise.

- **Starting a + or − Value from a Reference Point Other Than Zero**
  - Locate the graduation on the number line that corresponds to the starting point of the measurement.
  - Count off the value from the starting point of the next quantity according to the indicated direction of the signed number.

EXAMPLE: *Case 1.* Indicate two readings on the number line (graduated dial) of an ohmmeter. The readings are 3.5 ohms and −2.5 ohms.

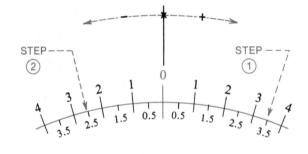

Step 1   Read the positive number value of 3.5 ohms by moving clockwise from the zero reference point.

Step 2   Read the negative value of −2.5 ohms counterclockwise from the zero reference point.

Case 2. Locate two successive measurements on the pressure measurement gage. The pressure measurements are ① 75 newtons per square meter (75 N/m²) and ② a decrease in pressure of 50 N/m².

Step 1   Read the positive number value of 75 N/m² on the number line.

Step 2   Subtract 50 N/m² for the negative pressure starting at the 75 graduation. Position ② represents the negative value from position ①.

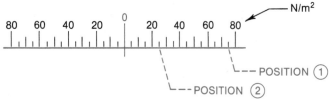

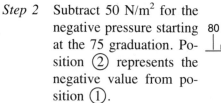

D. ABSOLUTE VALUES AND SYMBOL

Each value or number *(integer)* on the number line has two functions. Position on the line indicates quantity and direction. Figure 46-4 shows that each integer has a corresponding or paired opposite value. For instance, $(+1)$ is paired with (-1); (-2) with $(+2)$, etc. In other words, each value is the same distance from the zero reference point but in the opposite direction.

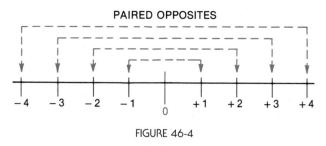

FIGURE 46-4

The numerical value without the sign is known as the *absolute value*. The absolute value is the distance between the number and zero on the number line. A symbol (| |) is used with a number to express an absolute value. The number (or other value) is placed between the two short vertical bars. For example, $|x|$ is read as the absolute value of x; $+6 = |6|$; $-4 = |4|$; $0 = |0|$. In determining the absolute value of a number, the sign is not regarded. Applications of positive and negative numbers are covered in detail in the Unit 47.

Review and Self-Test

ASSIGNMENT UNIT 46

A. Signed Numbers: Meaning and Importance

1. State two functions that are served by the use of signed numbers in algebra.
2. Mark the positions on the graduated scale to match the signed number millivolt (mV) values recorded in the table.

    Note. Use the same circled number as the one in the table to identify each reading.

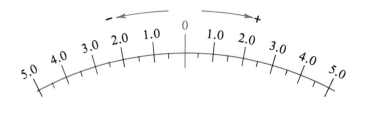

Measurement Identification	Meter Reading (mV)
①	2.5
②	5.0
③	−3.5
④	0
⑤	−1½
⑥	−3½

B. Quantity and Direction of Signed Numbers

1. Give the size (quantity) and direction represented by each of the signed numbers (a through h) on a number scale.

   a. −6 mm
   b. +27″
   c. 1500 volts (V)
   d. −0.006 seconds(s)
   e. +27.5 N/m²
   f. 0°
   g. −2.75 km
   h. $57\frac{5''}{8}$

2. Determine which of the statements (a through g) are true.

   a. +3 > −4
   b. 0 < −5
   c. +2 < +3
   d. −12 > +10
   e. −6 < −3
   f. −7 < −7
   g. +6 > +1

3. Write each of the values (a through d) as a signed number.

   a. Negative 14
   b. Positive 37
   c. Positive $\frac{1}{3}$
   d. Negative eight thousandths

C. Representing Absolute Values

1. Express each of the signed numbers (a through g) as an absolute value.

   a. 3
   b. −15
   c. 439.75
   d. −0.5869
   e. 0
   f. $\frac{1}{2}$
   g. $12\frac{9}{16}$

D. Practical Applications of the Number Line: Interpreting Values

1. The letters that appear above the graduations on the number line of an ammeter indicate a series of test readings.

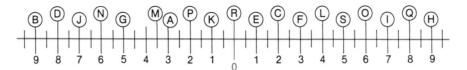

a. Give the ampere values that correspond to the following test readings.

(1) Ⓐ   (4) Ⓡ

(2) Ⓜ   (5) Ⓚ

(3) Ⓘ   (6) Ⓢ

b. Identify the letter from the number line that corresponds to the following ampere readings.

(1) −3   (5) +4

(2) +1   (6) −7

(3) −6   (7) +8

(4) −9   (8) −2

c. Determine which reading represents the larger positive or negative number value.

(1) Reading Ⓝ or Ⓠ   (3) Reading Ⓡ or Ⓖ

(2) Reading Ⓢ or Ⓓ   (4) Reading Ⓐ or Ⓙ

E. Translating Problem Statements into Number Lines and Signed Numbers

Use a signed number in answer to problems 1 through 7. Relate each signed number to a number line or other graduated measuring device. The appropriate measurement unit is to be given with each signed number.

1. A reading of +18 miles per hour (mph) indicates a tail wind. Represent a head wind of 23 mph.

2. A loss of $3\frac{1}{2}$ points on a stock is indicated by $-3\frac{1}{2}$ points. Indicate a gain of $2\frac{3}{4}$ points.

3. The positive camber on front automobile wheels at factory setting is 1/2°. After 15,000 miles of use, there is a negative camber of 1/4°. State the negative camber as a signed number.

4. A motor rated at 60 horsepower (hp) is actually delivering 72 hp. State the amount of increased power delivery as a signed number.

5. A deposit of $125 represents a positive number. Express a withdrawal of $75.

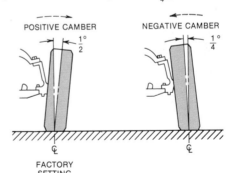

6. Express a temperature change from five degrees below zero to eight degrees above zero.

7. A metal part expands in length 0.000 000 9″ per degree rise in Fahrenheit temperature. Determine how much the part contracts when the temperature drops 100°F. Express the amount of contraction as a signed number.

Unit 47 Application of Positive and Negative Signed Numbers

OBJECTIVES OF THE UNIT
After satisfactorily completing the unit, the student/trainee will be able to
- *Add signed numbers with positive or negative values in the combination.*
- *Solve problems of addition and subtraction of signed numbers involving grouping symbols.*
- *Apply rules for multiplying or dividing positive and negative signed numbers.*
- *Perform any combination of the four basic mathematical processes using positive and negative quantities.*

A. SIGNS OF QUALITY AND OF OPERATION

The four mathematical signs of $(+)$, $(-)$, $(\times)$, and $(\div)$ denote the same processes regardless of whether used in simple arithmetical problems or in computations requiring the use of higher mathematics. The $(+)$ and $(-)$ signs are referred to here as *signs of operation*. Signs of operation are used to indicate the operation that is to be performed with specific quantities.

In performing mathematical operations with equations and formulas, the result is often a minus quantity. The minus quantity is referred to as a *negative quantity*. A negative quantity is indicated by a minus $(-)$ sign; a *positive quantity,* by a plus $(+)$ sign. These signs are now used to indicate the quality of a number, that is, whether a number is a *positive* or a *negative quantity*. Such signs are called *signs of quality*.

EXAMPLE: Add the following
$$(+10) + (+15) + (+6) = +31$$

In the $(+31)$, the $(+)$ is the sign of a positive quantity or a sign of *quality*. The plus $(+)$ signs between the parentheses are signs of *operation*.

Signs of Quality

$$(+10) + (+15) + (+6)$$

Signs of Operation

B. ADDITION OF POSITIVE AND NEGATIVE QUANTITIES

In addition, either numerical or literal quantities are combined in a unit. If all the numbers added are positive and the sign of operation is $(+)$, the answer is a positive number.

The words *positive number* refer to all numbers greater than zero. All numbers less than zero are *negative numbers*. Positive numbers are preceded by a $(+)$ sign; negative numbers, a $(-)$ sign. A quantity like (-10) is a negative number as indicated by the minus $(-)$ sign. When no sign appears before a number it is assumed to be positive.

When it is necessary to find the sum of two or more quantities with different signs, subtract the smaller value from the larger. Give the result the same sign as the larger unit.

There are four possible combinations in addition.

- Two positive quantities.
- Two quantities with unlike signs where the positive term is larger than the negative one.
- Two quantities with unlike signs where the negative quantity is larger than the positive.
- Two negative quantities.

The possible combinations are illustrated as four cases (*figure* 47-2).

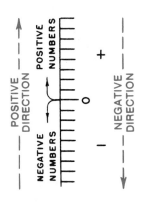

FIGURE 47-1

Adding Like Signs		Adding Unlike Signs	
Case 1	Case 2	Case 3	Case 4
+4	−4	+4	−4
+1	−1	−1	+1
+5	−5	+3	−3

FIGURE 47-2

RULE FOR ADDING SIGNED NUMBERS WITH LIKE SIGNS

- Add each signed number.
- Place (prefix) the common (+) sign before this sum.

EXAMPLE: *Case 1.* Add (+4) + (+5).

Step 1   Find the arithmetical sum.   (4 + 5) = 9

Step 2   Place (prefix) the common sign (+) before this   (+4) + (+5) = +9   Ans
sum.

EXAMPLE: *Case 2.* Add (−4) + (−5).

Step 1   Find the arithmetical sum.   (4 and 5) = 9

Step 2   Place (prefix) the common sign (−) before this sum.   (−4) + (−5) = −9   Ans

RULE FOR ADDING TWO SIGNED NUMBERS WITH UNLIKE SIGNS

- Subtract the smaller signed number from the larger.
- Place the sign of the larger signed number before the sum.

EXAMPLE: *Case 3*. Add $(-9) + (+5)$.

Step 1   Subtract the smaller signed number $(+5)$ from the larger signed number (-9).   **$(9 - 5) = 4$**

Step 2   Place the sign of the larger signed number $(-)$ before the sum.   **$(-9) + (+5) = -4$**   Ans

EXAMPLE: *Case 4*. Add $(+8) + (-4)$.

Step 1   Subtract the smaller signed number (-4) from the larger $(+8)$.   **$(8 - 4) = 4$**

Step 2   Place the sign of the larger signed number $(+)$ before the sum.   **$(+8) + (-4) = +4$**   Ans

RULE FOR ADDING MORE THAN TWO ABSOLUTE NUMBERS WITH UNLIKE SIGNS

- Add all the positive absolute numbers.
- Add all the negative absolute numbers.
- Subtract the smaller quantity from the larger.
- Place the sign of the larger before the final sum.

EXAMPLE: Add $(+6) + (+5) + (-4) + (-3)$.

Step 1   Add all the positive absolute numbers.   **(6) and (5) = (11)**

Step 2   Add all the negative absolute numbers.   **(4) and (3) = (7)**

Step 3   Subtract the smaller quantity (7) from the larger (11).   **$(+11) + (-7) = 4$**

Step 4   Place the sign of the larger $(+)$ before the final sum.   **$+4$**   Ans

C. SUBTRACTION OF POSITIVE AND NEGATIVE QUANTITIES

The subtraction of any quantity or expression is indicated by the minus $(-)$ sign. However, the subtraction of positive and negative signed numbers is the inverse of addition. This fact is illustrated in the six cases in figure 47-3 that show possible combinations of two signed numbers.

Case 1		Case 2		Case 3		Case 4		Case 5		Case 6	
Subtract	Add	Subtract	Add	Subtract	Add	Subtract	Add	Subtract	Add	Subtract	Add
$(+4)$	$+4$	$(+4)$	$+4$	(-4)	-4	(-4)	-4	(0)	0	(0)	0
$-(+1)$	-1	$-(-1)$	$+1$	$-(-1)$	$+1$	$-(+1)$	-1	$-(-4)$	$+4$	$-(+4)$	-4
$(+3)$	$+3$	$(+5)$	$+5$	(-3)	-3	(-5)	-5	$(+4)$	$+4$	(-4)	-4
Ans $+3$		Ans $+5$		Ans -3		Ans -5		Ans $+4$		Ans -4	

FIGURE 47-3

RULE FOR SUBTRACTING SIGNED NUMBERS

- Reverse (change) the sign of the quantity to be subtracted (subtrahend).
- Proceed as in addition.
- Prefix the answer with the common sign for like numbers.
  Note. Use the sign of the larger number when subtracting positive and negative signed numbers.

EXAMPLE: Subtract $(-8) - (+4) - (-3)$.

Step 1	Change the signs of the quantities to be subtracted (subtrahends).	$(+4)$ becomes (-4) $(-3$ becomes $(+3)$
Step 2	Change the signs of operation to plus $(+)$.	$(-8) + (-4) + (+3)$
Step 3	Proceed as with addition. Add all negative quantities.	$(-8) + (-4) = -12$
Step 4	Subtract the smaller quantity (3) from the larger (12).	$(-12) + (+3) = -9$
Step 5	Place the sign of the larger quantity $(-)$ before the final sum.	-9 Ans

D. ADDITION AND SUBTRACTION INVOLVING GROUPING SYMBOLS

When the plus sign precedes parentheses or other grouping symbol, all the terms inside the parentheses are to be added to the preceding terms. The parentheses may be removed without changing the signs of any of the terms if it is preceded by a $(+)$ sign.

EXAMPLE: Add $8 + (5 - 3) + (6 - 2)$.

Step 1	Remove all parentheses preceded by a plus sign. The quantity $+ (5 - 3)$ is the same as $5 - 3$; $+ (6 - 2) = 6 - 2$.	$(8 + 5 + 6) = 19$
Step 2	Add all the positive numbers.	$(8 + 5 + 6) = 19$
Step 3	Add all negative numbers.	$(-3) + (-2) = -5$
Step 4	Subtract and place the sign of the larger before the final sum.	$19 - 5 = 14$ Ans

When a minus $(-)$ sign precedes parentheses or other grouping symbol, it indicates that all terms within the parentheses are to be subtracted from preceding terms. The signs of all the terms in the parentheses are changed and the same procedure is followed as in addition.

EXAMPLE: Perform the operations indicated: $6 - (5 - 2) - (3 - 6)$.

Step 1	Change the signs of quality in each parentheses preceded by a $(-)$ sign.	$-(5 - 2)$ becomes $(-5 + 2)$
Step 2	Remove parentheses and proceed as in addition.	$-(3 - 6)$ becomes $(-3 + 6)$ $6 - 5 + 2 - 3 + 6 = +6$ Ans

E. MULTIPLICATION OF POSITIVE AND NEGATIVE QUANTITIES

When two numbers with *like signs* are multiplied, the product is positive (+) as illustrated in Cases 1 and 2 in figure 47-4. The product of two numbers with *unlike signs* is negative (−) (Cases 3 and 4).

Multiplication: Like Signs		Multiplication: Unlike Signs	
Case 1	Case 2	Case 3	Case 4
$+4$ $\times(+3)$ $+12$	-4 $\times(-3)$ $+12$	-4 $\times(+3)$ -12	$+4$ $\times(-3)$ -12

FIGURE 47-4

RULE FOR MULTIPLYING POSITIVE AND NEGATIVE SIGNED NUMBERS

- Multiply the absolute value of each number.
- Prefix the sign of the answer.
  Note.
  - The product is positive (+) when any number of positive signed numbers are multiplied.
  - The product is positive (+) when an even number of negative signed numbers are multiplied.
  - The product is negative (−) when an odd number of negative signed numbers are multiplied.

EXAMPLE: Multiply the following signed numbers.

$(+4) \times (+3) \times (-5) \times (-1.5)$

$|4| \times |3| \times |5| \times |1.5| = 90$

Step 1 Multiply each absolute value.
Step 2 Prefix the answer with a sign based on an even number of negative signed numbers.

$+90$ **Ans**

EXAMPLE: Multiply $(+6) \cdot (+5) \cdot (-4)$.

Step 1 Multiply each absolute value.

$|6| \cdot |5| \cdot |4| = 120$

Step 2 Prefix the answer with a sign based on an odd number of negative signed values.

-120 **Ans**

F. DIVISION OF POSITIVE AND NEGATIVE QUANTITIES

The rules governing the use of signs for multiplication apply also in division because of the relationship between these two processes. When dividing quantities with *like signs*, the quotient is plus (+); for quantities with *unlike signs*, minus (−).

RULE FOR DIVIDING TWO SIGNED NUMBERS WITH LIKE SIGNS

- Divide the absolute value of each signed number.
- Place (prefix) the plus (+) sign before the quotient.

RULE FOR DIVIDING TWO SIGNED NUMBERS WITH UNLIKE SIGNS

- Divide the absolute value of each signed number.
- Place (prefix) the minus (−) sign before the quotient.

These rules are applied in Cases 1, 2, 3, and 4. The quotient zero divided by any number is zero as shown in Case 5.

Case 1	Case 2	Case 3	Case 4	Case 5
$\dfrac{+12}{+4} = +3$	$\dfrac{-12}{-4} = +3$	$\dfrac{-12}{+4} = -3$	$\dfrac{+12}{-4} = -3$	$\dfrac{0}{+12} = 0$

FIGURE 47-5

Review and Self-Test

ASSIGNMENT UNIT 47

A. Addition of Positive and Negative Signed Numbers and Quantities

1. Add the following combinations of positive and negative signed numbers for the numerical and literal terms as indicated.

a. $\begin{array}{r} +5 \\ +4 \\ \hline \end{array}$
b. $\begin{array}{r} +8 \\ -3 \\ \hline \end{array}$
c. $\begin{array}{r} -6 \\ +4 \\ \hline \end{array}$
d. $\begin{array}{r} -10 \\ -\ 4 \\ \hline \end{array}$
e. $\begin{array}{r} +6X \\ +7X \\ \hline \end{array}$
f. $\begin{array}{r} +9Y \\ -2Y \\ \hline \end{array}$

g. $\begin{array}{r} -6AB \\ +3AB \\ \hline \end{array}$
h. $\begin{array}{r} -\ 4XYZ \\ -12XYZ \\ \hline \end{array}$
i. $\begin{array}{r} +6\frac{1}{2}L \\ +9\frac{1}{4}L \\ \hline \end{array}$
j. $\begin{array}{r} +6.25A \\ -3.75A \\ \hline \end{array}$
k. $\begin{array}{r} -5\frac{1}{8}L \\ -8\frac{1}{4}L \\ \hline \end{array}$
l. $\begin{array}{r} -3.063B \\ -8.875B \\ \hline \end{array}$

m. $\begin{array}{r} +5\frac{1}{2} \\ +6\frac{1}{2} \\ +9 \\ -3 \\ \hline \end{array}$
n. $\begin{array}{r} 17 \\ +6\frac{1}{4} \\ -3\frac{1}{4} \\ -4\frac{1}{2} \\ \hline \end{array}$
o. $\begin{array}{r} -12\frac{7}{8} \\ -5\frac{3}{8} \\ -4\frac{1}{4} \\ +\ 6\frac{1}{2} \\ \hline \end{array}$
p. $\begin{array}{r} +9.625A \\ -7.125A \\ -2.250A \\ +3.375A \\ \hline \end{array}$

2. The gains and losses in temperature are recorded in degrees Celsius. Give the temperature readings at each of the following four points. Each temperature reading starts at −17°C.

   a. Temperature drop −10°C   c. Drop −17°C
   b. Rise +33°C   d. Rise +21°C

3. A merchant showed the following six monthly profits (+) and losses (−). Determine the total profit or loss for this six-month period.

Month	Profit (+) or Loss (−)
1	+$4,275
2	+2,017
3	−200
4	−639
5	+2,794
6	+5,226

4. Instrument readings for heat treating an SAE carbon steel are recorded in the chart for four time periods.

	Period 1	Period 2	Period 3	Period 4
Temperature Changes (Degrees Fahrenheit)	Rise from 82° to 1570°	Drop of 72°	Drop of 39°	Rise of 120°

   a. Determine (1) the temperature change during Period 1 and (2) the final temperature during Period 4.
   b. Indicate by a signed number (°F) how much the final temperature must be raised or lowered to reach the hardening temperature of 1625°F.

B. Subtraction of Positive and Negative Quantities

1. Subtract each of the combinations a through h.

   a. +6   c. −7   e. +5   g. −35
      +4   +4   +7   +16

   b. +8   d. −5   f. +9   h. −17
      −5   −7   −12   −33

2. Subtract each of the following.
   a. $12 - (+7)$ d. $(5) + (6) + (3)$ from $(22) - (-9)$

   b. $(-16) - (+6)$ e. $(-6 + 9)$ from $(24 - 7)$

   c. $(-32) - (-10)$ f. $(5 - 3)$ from $(8 + 2)$
3. Solve problems a through d.
   a. $8 - (+5 + 2)$ c. $4 - (-2 - 3)$

   b. $3 + 7 - (-6 + 2)$ d. $-3 - (-12 - 6)$

C. Multiplication of Positive and Negative Quantities

1. Multiply the numerical values a through d.

   a. $\begin{array}{r} 4 \\ \times 3 \\ \hline \end{array}$ b. $\begin{array}{r} (+2.2) \\ \times(\ -5) \\ \hline \end{array}$ c. $\begin{array}{r} (-6.4) \\ \times(+3.4) \\ \hline \end{array}$ d. $\begin{array}{r} (\ -8) \\ \times(-6\frac{1}{4}) \\ \hline \end{array}$

2. Multiply each of the combinations a through i.
   a. $(+8)(+9)$ d. $(-5)(-8)$ g. $(-3)(+6)(-4)$

   b. $(+7)(-8)$ e. $(+6)(+4)(+3)$ h. $(-17)(-2)(-3)$

   c. $(-6)(+3)$ f. $(+15)(-10)(+2)$ i. $(+12)(-11)(-5)(+8)$
3. Perform the operations indicated at a, b, c, and d.
   a. $(5) + (6)(-4)(-3)$ c. $-(+75) + (80) + (-3)(12)(-4)$

   b. $(-35) - (-4) + 3(4)(-5)$ d. $(-3)(+2) - [3(-5)(-2) + 6]$
4. The electrical circuit as illustrated contains the lamps, refrigerator units, and heating coils listed in the table. The total current (amperes-A) in a circuit equals the sum of the individual currents. The system is designed to carry a total load of 100 amperes.

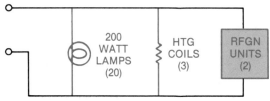

Item	Quantity (Units)	Load for Each Unit of (1) in Amperes
200 watt lamps	20	0.824
Refrigeration	2	17.82
Heating coils	3	8.42

a. Calculate the total current required to operate all of the electrical lamps and equipment.

b. Use a signed number to indicate any additional ampere capacity that is left of the amount the system is overloaded.

D. Division of Positive and Negative Quantities

1. Divide.

   a. $\dfrac{14}{7}$ b. $\dfrac{28}{-2}$ c. $\dfrac{-16}{4}$ d. $\dfrac{-75}{-15}$

2. Divide.

   a. $(-72) \div 9$ c. $(-36) \div (-2.4)$
   b. $64 \div (-8)$ d. $(-11.9) \div (-1.7)$

3. Perform the operations indicated.

   a. $[(5) + (6)] \div (.3)$ c. $[(+16) - (5)] \div (-.08)$
   b. $[(-8) + (7)] \div (-5.6)$ d. $[(7)(-4) + (17)] - (-69) \div (-1.6)$

E. Addition, Subtraction, Multiplication, and Division of Positive and Negative Quantities

1. Perform the operations indicated for the three positive and negative quantities given in the table. Round off each answer to one decimal place.

a.	$[(17)(11) - (14.2)(15) \div (.6)(.04)] \times (-4.3)$
b.	$[(-25.6)(.4) + (-17.338)(20) \times (14.5)(-4)] \div (-.6)(4.2)$
c.	$[(12)(-4.6)(3.5) - (15)(13.7)(-.09) \times (19)(-4.3)(5.2)] \div (.02)(-35)(10.1)$

Unit 48 Graphic Solutions to Signed Numbers

OBJECTIVES OF THE UNIT

After satisfactorily studying this unit, the student/trainee will be able to

- *Apply vectors to represent the direction and magnitude of signed numbers on a number line.*
- *Design a number line with appropriate scales.*
- *Visualize and graphically solve general and concrete algebraic problems requiring the addition, subtraction, multiplication, and division of signed numbers.*

The graphic method of carrying on the four basic mathematical processes with signed numbers provides another technique of problem solving. This method involves the representation of positive and negative numbers or other quantities on a number line and the use of *vectors*.

A. APPLICATION OF VECTORS IN GRAPHICALLY SOLVING SIGNED NUMBER PROBLEMS

A *vector* is a line segment that represents *direction* (positive or negative) and *magnitude* (size or quantity). The absolute value of a signed number provides the length of the vector. The sign indicates the direction of a vector.

A vector (line) extends from a starting, reference point (sometimes called a *point of origin*) on a number line in the direction shown by an arrowhead to the end point of a measurement. Successive vector measurements are identified as vector ①, vector ②, etc. To repeat, vector measurements may be positive or negative. Vector measurements may originate at the zero or any other reference point on a number line. The absolute value of a signed number indicates the length of the vector.

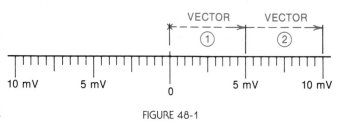

FIGURE 48-1

B. GRAPHIC SOLUTIONS TO ADDING SIGNED NUMBERS

RULE FOR ADDING AND SUBTRACTING SIGNED NUMBERS GRAPHICALLY

- Draw a number line with integers (values) plotted according to an appropriate scale.
- Apply the rules for the addition or subtraction of signed numbers as required.
- Group all positive and negative signed numbers into subgroups.
- Start at the zero or other specified reference point. Draw the first vector line equal in units to the first *addend* of the (+) subgroup.
  Note. Addend refers to the number or quantity to be added to another number or quantity.
- Continue to draw additional addends starting at the end of the previous addend for the (+) subgroup entries.
- Draw vectors for each negative (−) subgroup quantity starting at the furthest positive point.
- Read the answer at the *coordination point*. This represents the last vector entry on the number line.
  Note. The coordination point corresponds with the location or position of the last vector arrow.

EXAMPLE: Add the signed numbers of +2 and +3 graphically.

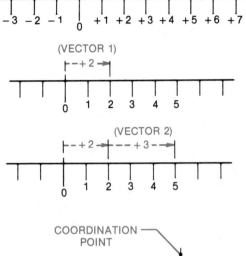

    Step 1   Draw a number line with an appropriate scale.

    Step 2   Group all positive signed numbers.

    Step 3   Start at zero. Draw the first vector for the first signed number (addend).

    Step 4   Continue drawing an additional vector for each addend.

              Note. Since there is no negative signed number subgroup, no negative value vector is drawn from the last addend.

    Step 5   Read the answer on the number line at the graduation that corresponds to the coordination point.

C. GRAPHIC SOLUTIONS TO SUBTRACTING SIGNED NUMBERS

Refer to the preceding combined rule for the steps required to solve problems involving the subtraction of signed numbers.

EXAMPLE: *Case 1.* Use the rule for graphically subtracting signed numbers and add +4.5 R, −6.5 R, +4.0 R, and −3.0 R.

    Step 1   Draw a number line. Identify the quantities represented on the graduated scale.

    Step 2   Arrange the signed numbers into positive and negative subgroups.

    Step 3   Start at zero. Draw the first vector for (+4.5 R).

    Step 4   Continue to draw vectors for each positive number addend.

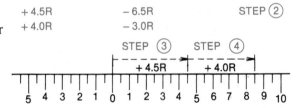

Step 5   Draw vectors for each negative
number addend (−6.5 R and
−3.0 R).

Step 6   Read the answer at the coordination point.

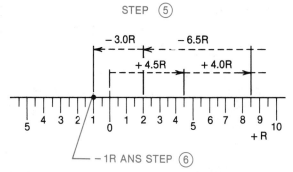

E XAMPLE: *Case 2.* A building has a 100 ampere (A) current supply. The circuit includes a
drying unit that draws 30 amperes (A), a small motor, 6 A, and a cooling unit, 11 A. Draw a
number line and solve the problem graphically, using signed numbers.

Method A

Step 1   Draw a number line.
Identify the scale and
quantities represented in
the scale.

Step 2
3 Plot each vector as an
4 addend.
5

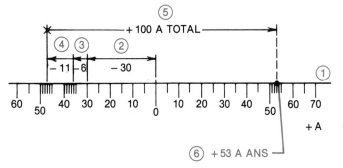

Step 6   Read the balance of amperes in the circuit on the
number line.

Method B

Step 1   Identify 100 A as the
point of origin on the
number line.

Step 2 Plot each vector addend
3 for each negative signed
4 number quantity.

Step 5   Read the answer on the
graduated scale at the coordination point.

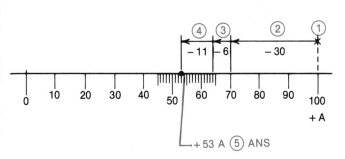

D. GRAPHIC SOLUTIONS OF PROBLEMS INVOLVING THE MULTIPLICATION AND/OR DIVISION OF SIGNED NUMBERS

Problems requiring the multiplication and/or division of signed numbers may also be solved graphically. The practice is to first multiply or divide absolute values, as the case may be. The sign of the product or quotient is then determined. The result represents the magnitude; the sign, the direction. Vectors are next stepped off on the number line the same as for adding or subtracting signed numbers.

RULE FOR GRAPHICALLY SOLVING THE MULTIPLYING OR DIVIDING OF SIGNED NUMBERS
_____

* Multiply or divide quantities that are contained within grouping symbols, as required.
* Determine the magnitude and direction of each product or quotient and the grouping value.
  Note. This value represents a vector.
* Mark off the first vector from the reference point on the number line.
* Establish the magnitude and direction of each successive grouping in the problem.
* Step off each successive vector.

_____

EXAMPLE: Solve graphically:

$(6 + 2) + (+2 \times 3) - (4 \div 2) - [(-4 \times 2) + (36 \div 3)]$

Step 1 Perform the required mathematical processes within the parentheses and braces.

$$8 + 6 - 2 - [-8 + 12]$$

VECTORS

Step 2 Lay out a number line to accommodate the range of quantities.

Step 3 Plot vectors Ⓐ, Ⓑ, Ⓒ, and Ⓓ on the number line.

Step 4 Read the answer on the number line at the termination point of the last vector.

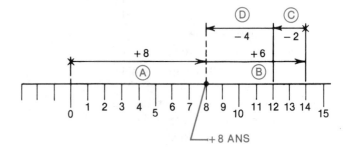

Review and Self-Test

ASSIGNMENT UNIT 48

A. Graphical Solutions to Problems Involving the Addition of Signed Numbers

1. a. Draw a number line to accommodate positive and negative signed numbers that range from +6.5 to −3.5 in increments of 0.5.
   b. Plot vectors on the number line to graphically (algebraically) add each of the following signed number values.
      (1) −1 mm and +3 mm
      (2) −3.5 mm and +2.5 mm
      (3) −1.5 mm and −4.0 mm
2. The RPM of a motor starting at rest accelerates at the RPM indicated in the table for each of five one-second intervals.
   a. Determine the vectors that represent the increased speed for each interval.
   b. Draw a number line and plot the RPM at the end of each second.
      Note. Since all the signed numbers are positive, all RPM graduations on the number line are positive.
   c. Determine graphically the RPM at maximum speed at the end of the six seconds.

Starting RPM	Acceleration Each Second					
	1 Sec	2 Sec	3 Sec	4 Sec	5 Sec	6 Sec
0	750	1,500	2,250	1,500	750	0

3. a. Mark off a second scale of values above or below the RPM graduations in problem 2. The surface speed in feet per minute for a 4″ diameter part is 785 sfpm for each 750 RPM increment.
   b. Convert each RPM on the number line to a corresponding sfpm value.
   c. Read the sfpm for each of the five one-second intervals.

B. Graphic Solution of Problems Involving the Subtracting of Signed Numbers

1. Use the number line to graphically subtract signed numbers (a) through (e).
   a. −4 and −6   d. −0.25 V and −2.00 V
   b. −6.5 and −4.5   e. +10.75″ and −3.625″
   c. 3.5 and −2.5
2. a. Lay out a number line with two graduated scales that represent only positive signed number values. One scale is to display the diameters in mils (0.001″) of eight different copper conductor wires. The second scale is to represent the corresponding resistance values in ohms per 1,000 feet. The diameters and resistances of the wires are identified in the table.

Copper wire	1	2	3	4	5	6	7	8
Diameter (mils)	18	27	36	54	72	108	144	216
Resistance (Ohms/1,000″)	32.96	16.48	8.24	4.12	2.06	1.03	.52	.26

b. Express as signed numbers the drop or increase in resistance across 1,000 feet of conductor wire for the following comparisons:

(1) ① and ②   (4) ⑥ and ④
(2) ① and ⑤   (5) ⑦ and ①
(3) ③ and ⑧

C. Graphic Solutions of Problems Involving the Multiplying and Dividing of Signed Numbers

1. Use the information provided in the table for problems A, B, and C.
   a. Perform the mathematical processes specified for each set of grouped values.
   b. Set up an appropriate number line and identify the units of measurement.
   c. Plot each vector to graphically determine each answer.
   d. State each answer in the appropriate unit of measurement.

Problem	Unit of Measurement	Signed Number Groupings, Quantities, and Processes
A	Liters	$(10 + 2) + (6.5 \times 4) - (2.8 \div 0.7) - (-3 + 7)$
B	Milliamperes (mA)	$(-2.5 + 4.5) - (1.0 \times 3.5) + (-6.6 \div 2.2) + (7.4 - 9.9)$
C	Force (newton-N)	$(10.5 - 7.5) + (-20 \times .5) + [-(77 \div 14) - (47 - 4.5 - 68.5)]$

Unit 49 Application of Symbols in Addition of Terms

OBJECTIVES OF THE UNIT

After satisfactorily completing this unit, the student/trainee will be able to
- *Interpret the meaning and use of like and unlike algebraic terms.*
- *Apply symbols in solving general problems in addition in which like and unlike terms are used.*
- *Perform the basic algebraic operation of addition in industrial problems.*

In any mathematical expression, the parts that are separated by any of the signs like $(+)$ and $(-)$ are called *terms*. A *term* in algebra is that part of an expression that is not separated by a $(+)$ or $(-)$. When the various terms of an expression have the same literal factor, they are called *like terms*. In the expression $8A + 9A + 4A$, all of the terms have the same literal factor (A). These terms are like terms and may be added.

A. ADDING LIKE TERMS

In any problem of addition involving the use of symbols, the like terms are added the same as any other numerical values. However, attention must be given to expressing the sum in terms of the literal factor. In the above case, when the three terms are added, the sum of $(21A)$ has the same literal factor (A) as each of the terms.

RULE FOR ADDING LIKE TERMS

- Group all like terms in the expression.
- Add the numerical factors as in signed numbers.
- Express the sum in terms of the literal factor.

EXAMPLE: Add $5S + 3S + 4S + 6S$.

Step 1   Group all like terms in the expression.
Step 2   Add the numerical factors $(5 + 3 + 4 + 6)$.
Step 3   Express the sum (18) in terms of the literal factor (S).

$$5S + 3S + 4S + 6S = 18S \quad \textbf{Ans}$$

B. ADDING UNLIKE TERMS

The words *unlike terms* are used when the literal factors are different or unlike. Unlike terms cannot be added or subtracted. For example, in the rectangle illustrated (*figure* 49-1), the base is $8L$ units long and $5W$ units wide. The perimeter, which is equal to the sum of all the sides, may be expressed as

$$8L + 5W + 8L + 5W$$

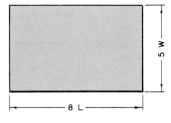

FIGURE 49-1

In this expression, the first and third terms ($8L$ and $8L$) and the second and fourth ($5W$ and $5W$) are like terms and can be added: $(8L + 8L = 16L)$; $(5W + 5W = 10W)$.

The $16L$ and $10W$ are unlike terms and cannot be added because the value of the literal factor is different in each instance. The perimeter, which is equal to $10L + 16W$, is expressed in as simple terms as possible.

RULE FOR ADDING EXPRESSIONS WITH UNLIKE TERMS

* Group all like terms.
* Add grouped like terms as in signed numbers.
* Express the sum in terms of the literal factors.

EXAMPLE: Add $-5x$, $1.5y$, $3.5x$, $-3.2y$.

Step 1 Group all like terms.	$(-5x + 3.5x)$, $(1.5y + [-3.2y])$
Step 2 Add grouped like terms.	$-5x$ $\qquad$ $+1.5y$
Step 3 Express the sums in terms of the literal factor of each term.	$+3.5x$ $\qquad$ $+(-3.2y$
	$-1.5x$, $\qquad$ $-1.7y$
	$-1.5x$ $\qquad$ $-1.7y$ Ans

Review and Self-Test

ASSIGNMENT UNIT 49

A. General Problems in the Use of Symbols in Addition of Like and Unlike Terms

1. Add the like terms in columns A through F.

A	B	C	D	E	F
$9x$	$3\frac{1}{2}L$	$7.2W$	$13.63h$	$19.25a$	$16.5ab$
$16x$	$14\frac{3}{4}L$	$19.4W$	$17.0\ h$	$13.88a$	$17.7ab$

2. Add the following like terms.
   a. $X + X + 4X =$
   b. $9H + 7H + H =$
   c. $1\frac{1}{2}B + 9\frac{1}{4}B + 6B =$
   d. $3C + 4.7C + 5.2C =$
   e. $(5.5X) + (92.6X) + (4.7X) + .5X =$
   f. $\left(7\frac{1}{2}X\right) + \left(8\frac{1}{4}X + 5\frac{1}{8}X\right) =$
   g. $(3.2A) + [8.5A + (19.2A + 6.4A)] + 2.8A =$
   h. $\left[2\frac{1}{2}B + \left(7\frac{1}{4}B + 3\frac{3}{4}B\right)\right] + 16\frac{1}{2}B =$

3. Simplify the following expressions by grouping all like terms and then find the sum for problems a through e.
   a. $6l + 7l + 5w$
   b. $7ab + (8b - 15ab - 3b)$
   c. $14\frac{1}{4}cd + 12\frac{1}{2}cd + \left(-9\frac{1}{6}d\right)$
   d. $19.25xy + (-17.45y) + 12.75xy - 32.5z$
   e. $[-9lw + 3lh + 14lw]$, $[10\frac{1}{4}lw + 9\frac{3}{8}lh + -12\frac{3}{4}l] + 6\frac{1}{4}lw$

B. Industrial Applications Using Symbols in Addition

1. Determine dimensions (B) through (F).

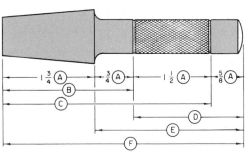

2. Find the length of the hardened plugs (A) and (B), the length of the body (C), and the overall length of the Plug Gage (D).

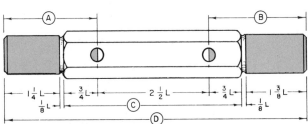

3. Find the literal values of dimensions (A) through (G).

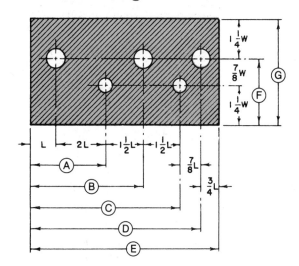

4. Determine the literal dimensions (A) through (F) for the Plate.

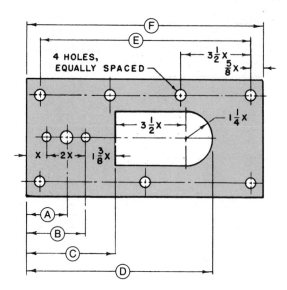

5. The total resistance in a series electrical circuit is equal to the sum of all the separate resistances.

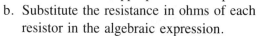

   a. Express the total resistance (R_T) for the seven different resistances (as shown in the series circuit diagram) in algebraic terms.
   Note. Group like resistors (X, Y, or Z). Identify each resistor in each group by the letter R and the appropriate subscript.
   b. Substitute the resistance in ohms of each resistor in the algebraic expression.
   c. Determine the following group and total resistance values.
      (1) Resistor group X (R_X)
      (2) Resistor group Y (R_Y)
      (3) Resistor group Z (R_Z)
      (4) Resistor groups (X + Y) and (Y + Z)
      (5) R_T for the series circuit.
6. Use a calculator to add the separate resistances and to calculate the mean resistances for the following combinations of resistors in problem 6. Round off the values in ohms correct to three decimal places.

   (a) R_X group   (d) R_X and R_Y groups
   (b) R_Y group   (e) R_Y and R_Z groups
   (c) R_Z group   (f) R_T

Unit 50 Application of Symbols in Subtraction of Terms

OBJECTIVES OF THE UNIT

After satisfactorily completing the unit, the student/trainee will be able to
- *Lay out problems using symbols and expressing quantities algebraically.*
- *Apply rules for subtracting like and unlike terms and solve general problems.*
- *Solve practical industrial problems involving symbols and the adding and/or subtracting of like or unlike terms by algebraic or calculator methods.*

A. APPLICATION OF SYMBOLS IN SUBTRACTING LIKE TERMS

The two mathematical operations that workers use daily are addition and subtraction. Most answers can be easily obtained by simple arithmetic. There are numerous occasions, however, when it is necessary to subtract quantities that contain symbols. Here, as in addition, only like terms may be subtracted.

For example, an 8″ length of wood can be subtracted from a 10′ length as both numerical factors are in linear units of measure. By contrast, 8 kg of wood may not be subtracted from a 10 m length because the units of measure are unlike.

RULE FOR SUBTRACTING LIKE TERMS

- Locate the like terms in the expression.
- Subtract the numerical factors, as in signed numbers.
- Express the difference in terms of the literal factor.

EXAMPLE: Subtract $\left(17A - 3\frac{1}{2}A - 5A - 4\frac{1}{2}A\right)$.

Step 1 Locate all like terms in the expression. $\left(17A, 3\frac{1}{2}A, 5A, 4\frac{1}{2}A\right)$

Step 2 Subtract the numerical factors (all numbers are signed positive numbers). $\left(17 - 3\frac{1}{2} - 5 - 4\frac{1}{2}\right) = 4$

Step 3 Express the difference (4) in terms of the literal factor (A). $\left(17a - 3\frac{1}{2}A - 5A - 4\frac{1}{2}A\right) = 4A$ **Ans**

When the terms in the same expression are connected by both plus and minus signs, the plus terms are collected separately. The minus terms are next collected. Then the difference between the plus and minus quantities is determined. The practice of collecting terms that involve the same mathematical operation simplifies the actual work involved in arriving at the answer.

Here again, the mathematical processes indicated in the parentheses are performed first, those in the brackets next, and finally, those processes indicated in the braces. Once these operations are completed, all like terms may be combined by adding or subtracting.

B. APPLICATION OF SYMBOLS IN SUBTRACTING LIKE TERMS AND COMBINING WITH UNLIKE TERMS

Problems in subtraction often include expressions that involve like and unlike terms. In such instances, like terms are subtracted algebraically. The unlike terms are then combined with the difference for the like terms to form the answer.

RULE FOR ADDING OR SUBTRACTING ALGEBRAIC EXPRESSIONS WITH LIKE AND/OR UNLIKE TERMS

- Locate and group all like terms and all unlike terms.
- Perform the required mathematical processes with the grouped like items.
- Combine the sum or difference with the remaining unlike terms to form the answer.

EXAMPLE: Determine the value of the following expression that involves addition and subtraction: $(6D - 4D + 1.5D + 8B) + (5B - 1.25B - 2.75B + 6D) =$

Step 1	Locate all like terms in each set of parentheses.	$(6D, 4D, 1.5D)$
		$(5B, 1.25B, 2.75B)$
Step 2	Add or subtract the numerical factors as indicated by the signs in the parentheses.	$(6D - 4D + 1.5D) = 3.5D$
		$(5B - 1.25B - 2.75B) = 1B$
Step 3	Combine any remaining like terms in the original parentheses $(8B)$ and $(6D)$.	$(8B + 1B) = 9B$
		$(6D + 3.5D) = 9.5D$

The value of the expression
$$(6D - 4D + 1.5D + 8B) + (5B - 1.25B - 2.75B + 6D) = 9B + 9.5D \quad \text{Ans}$$

Occasionally the literal factors contain more than one letter, like $6AB$ and $9XY$. In such cases, where the terms in the expression are identical, they are still referred to as like terms and may be added or subtracted.

EXAMPLE: Subtract $(14AB - 7AB + 8BC - 4.2BC - 2.4AB)$

Step 1	Locate all like terms.	$(14AB - 7AB - 2.4AB)$
Step 2	Subtract the numerical factors in each case.	and $(8BC - 4.2BC)$
		$(14 - 7 - 2.4) = 4.6$
		and $(8 - 4.2) = 3.8$
Step 3	Express the differences in terms of the literal factor.	$(14AB - 7AB - 2.4AB) = 4.6AB$
		$(8BC - 4.2BC) = 3.8BC$

The value of the expression
$$(14AB - 7AB + 8BC - 4.2BC - 2.4AB) = 4.6AB + 3.8BC \quad \text{Ans}$$

Review and Self-Test

ASSIGNMENT UNIT 50

A. Applications of Symbols in Subtracting Like Terms

1. Simplify the quantities A through F.

A	B	C	D	E	F
$25X$ $-12X$	$19\frac{3}{4}W$ $-10\frac{1}{2}W$	$16\frac{3}{8}L$ $-7\frac{9}{16}L$	$20.6N$ $-14.0N$	$23.38YZ$ $-17.86YZ$	$144.2XYZ$ $-92.8XYZ$

2. Simplify the following expressions.
   a. $35X - 16X - 7X =$
   b. $75A - 19\frac{1}{2}A - 12\frac{1}{2}A =$
   c. $144\frac{1}{2}B - 64\frac{1}{4}B - 17\frac{7}{8}B =$
   d. $8Y - 4.6Y - 1.62Y =$
   e. $(92.4A) - (16.8A) - (12.6A) =$
   f. $\left(15\frac{1}{2}C\right) - \left(9\frac{3}{4}C + 2\frac{1}{4}C + 1\frac{1}{4}C\right) =$
   g. $[(8.8D + 6.6D) - (1.4D + 6D)] - 3.9D =$
   h. $76.75BC - [2.50BC + (7.25BC + 3.38BC)] =$

3. Find the numerical value of A through F for each given value by subtracting (8.6) in each case.

	Given Value				
A	B	C	D	E	F
12	9.8	12.2	$10\frac{5}{8}$	$16\frac{1}{4}$	17.312

4. Determine the value of A through F by subtracting $\left(2\frac{1}{2}\right)$ from each numerical value.

A	B	C	D	E	F
12	$20\frac{3}{4}$	$11\frac{3}{8}$	4.5	9.38	12.125

B. Subtracting Like Terms and Combining with Unlike Terms

1. Simplify algebraic expression a through f by combining (where required) the unlike terms with the difference found by adding or subtracting like terms.
   a. $17P + 9L - 6P$
   b. $16ab - (9ab + 12bc - 2ab)$
   c. $14\frac{1}{2}X - 10\frac{1}{4}X + (3X + 7Y + X)$
   d. $75.38lw - 20.25wd - 20.63lw + 43.12wd$
   e. $(16.25HD) - (17.50CD) + (12.62HD + 7.88HD)$
   f. $\left[9\frac{1}{8}l + \left(22\frac{3}{4}wd - 6\frac{3}{8}l - 8\frac{1}{2}wd - 2\frac{1}{8}wd\right)\right] - 1\frac{1}{16}l$

C. Practical Applications of Symbols in Adding and/or Subtracting Like Terms

1. Determine the value of dimensions A through G in terms of the literal dimension L.

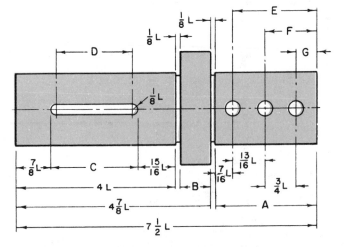

2. Determine the literal dimensions A through G for the drill plate.

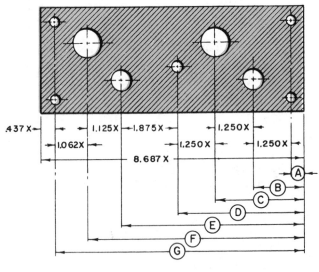

3. Six basic deductions from an employee's gross wage (G_W) are recorded in the accompanying chart.
   a. Express the employee's net wage (N_W) as an algebraic expression.
   b. Determine the net wage in relation to the gross wage.

Payroll Deductions from Employee's Gross Wage					
Federal Income Tax	Social Security Deduction	State Income Tax	Disability Insurance	Retirement Plan	Club Plan
0.15	0.075	0.035	0.045	0.06	0.01

4. The graph reports the rejects in the production of electronic components. The base unit = Y. The quality control batches are identified as Q_1, Q_2, etc.

   a. Express the decreases (D) in rejects in algebraic terms between the following quality control batches.

      (1) Q_1 and Q_2
      (2) Q_3 and Q_4
      (3) Q_5 and Q_6
      (4) Q_1 and Q_6

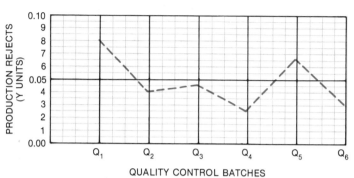

   b. Use a calculator to compute the means in relation to the rejects for batches Q_1 through Q_6 for

      (1) Decreases in rejects   (2) Increases in rejects

      Round off each mean value correct to three decimal places.

Unit 51 Application of Symbols in Multiplication of Terms

OBJECTIVES OF THE UNIT

After satisfactorily completing this unit, the student/trainee will be able to
- *Understand the process of multiplying quantities that are represented by symbols containing literal terms.*
- *Apply the multiplication process to general problems involving like and/or unlike terms.*
- *Solve practical multiplication problems where dimensions or quantities are represented by symbols, literal terms, and numerical terms.*

The multiplication of quantities involving the use of symbols differs somewhat from either addition or subtraction. It is possible in multiplication to multiply all terms in an expression regardless of whether they are like or unlike.

To perform this mathematical operation, the numerical factors and expressions are multiplied together the same as any whole or mixed numbers. The literal factors are next multiplied. The answer is expressed in terms of the numerical product followed by the product of the literal factors.

For example, if the quantities A, B, and C are multiplied, the product is ABC. In the case of a triangle, its area is equal to the product of one-half the base multiplied by the altitude (*figure* 51-1).

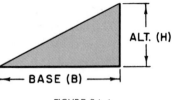

ALT. (H)

BASE (B)

FIGURE 51-1

If the letter (B) is used to denote the length of the base and (H) its altitude, then the area $A = B/2 \times H$ or $BH/2$.

In the rectangle illustrated (*figure* 51-2), if the length of the base is $5L$ and the height $4H$, the rectangle contains five units of (L) length and 4 units of (H) height. The area of the rectangle is equal to $5(L) \times 4(H) = 20LH$. This area is identical with that composed of 20 of the smaller shaded rectangles. Since the area of one of the shaded rectangles is equal to its length $(L) \times$ height (H), the total area is equal to $20LH$.

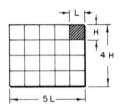

FIGURE 51-2

There are times when the signs in an expression indicate that certain of the quantities must be added, subtracted, and multiplied. When such is the case, like quantities that must be added are added. Next, those quantities to be subtracted are subtracted. Finally, the multiplying of like and unlike quantities is performed. This method of doing each of the three operations separately in the sequence indicated simplifies the solution of a problem.

RULE FOR MULTIPLYING LIKE OR UNLIKE TERMS

* For all numbers included in parentheses, add all like terms in the expression where indicated by $(+)$ signs.
* Subtract all like terms where indicated by $(-)$ signs.
* Multiply either like or unlike quantities where indicated by $(\times)$ signs.
* Multiply numerical factors.
* Multiply literal factors.
* Combine numerical and literal factors.
* Express product in literal terms.

EXAMPLE: *Case 1.* Multiply (6A) by 7B.

Step 1   Multiply *numerical factors* 6 and 7. $6 \times 7 = 42$
Step 2   Multiply the literal factors A and B. $A + B = AB$
Step 3   Combine numerical and literal factors 42 and AB. $(6A) \cdot (7B) = 42AB$ **Ans**

EXAMPLE: *Case 2.* Solve $(2A + 4A - 3A - A)(2.2B)(4C)$.

Step 1   Add all like terms as indicated. $2A + 4A = 6A$
Step 2   Subtract all like terms, as indicated, from this $6A - 3A - A = 2A$
           sum.
Step 3   Multiply the result $(2A)$ by the literal term $(2.2B)$.

Multiply numerical factors.	$(2) \cdot (2.2) = 4.4$
Multiply the literal factors.	$(A) \cdot (B) = AB$
Combine the numerical and literal factors.	$(2A) \cdot (2.2B) = 4.4AB$

Step 4 Repeat these processes with the remaining
literal term (4C).

Multiply numerical factors.	$(4.4) \cdot (4) = 17.6$
Multiply the literal factors.	$(AB) \cdot (C) = ABC$
Combine and express in literal terms.	$(4.4AB) \cdot (4C) = 17.6ABC$

$(2A + 4A - 3A - A) \cdot (2.2B) \cdot (4C) = 17.6ABC$ **Ans**

Review and Self-Test

ASSIGNMENT UNIT 51

A. Application of Symbols in Multiplying Like or Unlike Terms (General Applications)

1. Multiply each term given in problems (A) through (F).

A	B	C	D	E	F
$10X$	$9.2A$	$6.8L$	$5.25AB$	$8\frac{1}{2}X$	$6\frac{1}{4}AD$
$\times\ 5Y$	$\times\ 5B$	$\times 8.4M$	$\times 7.5C$	$\times 12Y$	$\times 3\frac{1}{2}C$

2. Perform the operations indicated in problems a through e and simplify.
   a. $(6p)(7) + 5(p)$ c. $[(9l)(8w)(6h)] + (12lwh)$
   b. $(12x)(5y) + (17xy)$ d. $16\frac{1}{2}lh + \left[5\frac{1}{4}w - \left(3\frac{1}{2}h\right)\left(4\frac{1}{2}l\right)\right]$
   e. $32.5xy + [(8.25x - 3.13x - 6.38z)(2.5y)]$

3. Compute the value of (A) through (F) by substituting the values for N given in the table in the formula $12(N + 6.5)$.

	A	B	C	D	E	F
Value of N	$7\frac{1}{2}$	5	$4\frac{1}{4}$	6.5	7.75	5.063

4. Determine the area of the shaded portion of each object (A) through (E). Note the number of equal parts in each object. Round off answers, where possible, to two decimal places.

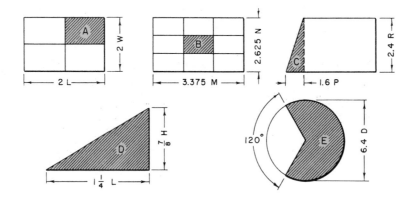

B. Practical Applications of Symbols in Multiplication of Terms

1. Compute the volume of parts Ⓐ through Ⓕ accurate to two decimal places.

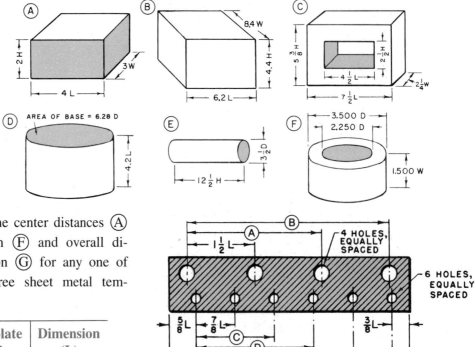

2. Find the center distances Ⓐ through Ⓕ and overall dimension Ⓖ for any one of the three sheet metal templates.

Template Number	Dimension (L)
1	10 cm
2	15 cm
3	20 cm

3. The percent by which the weights of five copper conductors vary from a standard are recorded in the table. The weight of the standard conductor (S) is 0.314 pound.

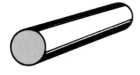

Standard Volume (S)	A	B		D	E
Percent Variation in Volume	+10	+15	−6.0	−7.4	+8.25
Quantity Required	10	5	10	6	10

  a. Express algebraically how the following values may be determined.
     (1) The volume (V) and
     (2) The weight (W) for the specified quantity *of each conductor*.
     (3) The total volume (V_T) and
     (4) The total weight (W_T) for all conductors.
     (5) The mean volume (V_m) and
     (6) The mean weight (W_m) for all conductors.
  b. Substitute values from the table. Compute either by the conventional method or by using a calculator to change each algebraic expression to its equivalent numerical value.
    Note. Round off each value (1) through (4) to three decimal places; (5) and (6) to two decimal places.

Unit 52 Application of Symbols in Division of Terms

OBJECTIVES OF THE UNIT
After satisfactorily completing this unit, the student/trainee will be able to
* *Understand the process of dividing numerical factors in a numerator and denominator.*
* *Cancel literal factors that appear in both numerator and denominator.*
* *Solve general and practical problems involving symbols and the division process.*

While the three other fundamental processes of addition, subtraction, and multiplication are used more extensively than division, there are many times when both the like and unlike terms of an expression must be divided. The division process is comparatively simple if the basic steps described in this unit are followed. The numerical factors of like or unlike terms are divided first. These are followed by the division of any literal factor that appears in both the numerator and denominator.

RULE FOR DIVIDING LIKE TERMS

- Place the numerator above the denominator.
- Divide the *numerical factors* in numerator and denominator.
- Cancel the *literal factors* that are identical for both numerator and denominator.
- Express the result as a numerical factor where the literal factors cancel each other.

EXAMPLE: Divide (24XY) by (8XY).

Step 1 Place the numerator (24XY) above the denominator (8XY).

$$\frac{\overset{3}{\cancel{24}}XY}{\cancel{8}XY}$$
$$_1$$

Step 2 Divide the numerical factors (24) and (8).

Step 3 Divide through by (X) in both numerator and denominator to get (1). Then divide through with Y.

$$\frac{\overset{3\cdot1\cdot1}{24\,\cancel{X}\cancel{Y}}}{8\,\cancel{X}\cancel{Y}}$$
$$_{1\cdot1\cdot1}$$

Step 4 Express the result as a numerical factor (3) because the literal factors cancel each other.

3   Ans

RULE FOR DIVIDING UNLIKE TERMS

- Form a fraction.
- Divide the numerical factors in both numerator and denominator.
- Cancel any literal factor that appears in both the numerator and denominator.
- Express the result in terms of the numerical quotient followed by all literal factors that do not cancel out.

EXAMPLE: Divide 12ST by 4S.

Step 1 Form a fraction.

$$\frac{\overset{3}{\cancel{12}}ST}{\cancel{4}S}$$
$$_1$$

Step 2 Divide the numerical factors.

Step 3 Cancel any literal factor which appears in both numerator and denominator.

Note. In this case the symbol (S) is common to numerator and denominator.

$$\frac{\overset{3}{\cancel{12}}\cancel{S}T}{\cancel{4}\cancel{S}} =$$
$$_1$$

Step 4 Express the combined results in terms of the numerical quotient (3) followed by the literal factor (T).

$$\frac{\overset{3}{\cancel{12}}\cancel{S}T}{\cancel{4}\cancel{S}} = 3T \quad \text{Ans}$$
$$_1$$

Occasionally, both the multiplication and division signs appear in the same expression. When this happens, the multiplication process must be completed before division.

E XAMPLE: Divide $(6L \times 4H \times 6D)$ by $(12D)$.

Step 1   Form a fraction and indicate the required mathematical processes.

$$\frac{6L \times 4H \times 6D}{12D}$$

Step 2   Multiply all literal terms in the numerator.

$$\frac{144LHD}{12D}$$

Step 3   Divide the numerical factors in both numerator and denominator.

$$\frac{\overset{12}{\cancel{144}}LHD}{\cancel{12}D}$$

Step 4   Divide the literal factors common to both the numerator and denominator (D in this case).

$$\frac{\overset{12}{\cancel{144}}LH\cancel{D}}{\cancel{12}\cancel{D}}$$

Step 5   Express the combined result in terms of the numerical quotient (12) followed by all literal factors that did not cancel out (LH).

$$12LH \quad \text{Ans}$$

$$\frac{6L \times 4H \times 6D}{12D} = \frac{\overset{12}{\cancel{144}}LH\cancel{D}}{\underset{1}{\cancel{12}\cancel{D}}} = 12LH \quad \text{Ans}$$

Review and Self-Test

ASSIGNMENT UNIT 52

A. Symbols in Division (General Applications)

1. Find the result in problems a through i.

   a. $LH \div H =$
   b. $12A \div 6 =$
   c. $16\frac{1}{2}B \div 4 =$
   d. $9.62X \div 6 =$
   e. $16C \div 4C =$
   f. $12.75AB \div 4.25B =$
   g. $6\frac{3}{8}BC \div 1\frac{1}{16}BC =$
   h. $.76lhw \div .019 w =$
   i. $(.3x + .6xy) \div (.02x) =$

2. Compute the value of Ⓐ through Ⓔ according to the formula $D = C \div 3.14$.

	A	B	C	D	E
Value of C	31.40	25.12	20.41	$10\frac{1}{2}$	$30\frac{3}{8}$

B. Using Symbols in Division (Practical Problems)

1. Determine dimension Ⓜ.

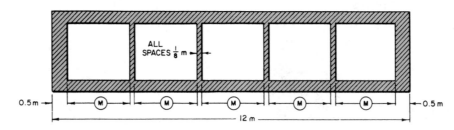

2. A gallon of paint covers 35.2 square meters of surface area.
   a. (1) Compute the surface area of each room and (2) the total area for both rooms A and B.
   b. Determine the total amount of paint required for the job.
   c. Convert the gallons required to equivalent full liters.

Room	Wall/Ceiling Areas (m)	Number of Surfaces	Surface Area (A)(m²)
A	30.5 × 3.7	2	
	24.5 × 3.7	2	
	30.5 × 24.5	1	
B	27.4 × 2.9	2	
	18.3 × 2.9	2	
	27.4 × 18.2	1	
		Total	

3. Determine dimensions Ⓐ, Ⓑ, Ⓒ, Ⓓ, and Ⓔ for each of five sections when L = 254.0, 381.0, 317.5, 479.4, and 622.3 millimeters, respectively.

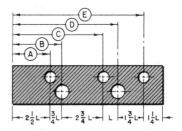

4. Compute the center distances Ⓒ between the holes on Links A through E.

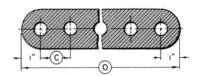

Link	Overall Distance (O)	No. Equally Spaced Holes
A	14	6
B	32	20
C	$24\frac{11}{16}$	11
D	16.4	8
E	20.576	12

5. Determine dimensions Ⓐ, Ⓑ, Ⓒ, Ⓓ, and Ⓔ for the four different Test Bars indicated in the table.

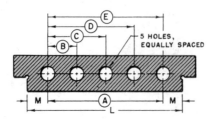

Test Bar	Value of	
	L	M
1	4.500	.750
2	6.500	1.000
3	8.750	1.250
4	10.875	1.750

6. Compute dimensions (A) through (E) for Drill Templates (1) and (2).

Drill Template	Value of (X)
1	1.4
2	2.25

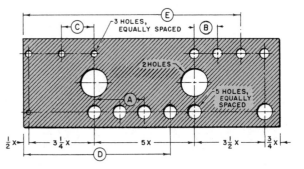

7. The total resistance $(R)_T$ for three resistors (R_1, R_2, and R_3) connected in parallel equals the product of the three resistors divided by the sum: $(R_1R_2 + R_1R_3 + R_2R_3)$.
   a. Express the total resistance (R_T) algebraically.
   b. Substitute the ohm (Ω) value for each resistor as given in the table.
   c. Use a calculator to compute the total resistance in *ohms* for parallel circuits (A) and (B). Round off the R_T values to two decimal places.

Circuit	Resistances		
	R_1	R_2	R_3
A	20 Ω	80 Ω	120 Ω
B	4200 Ω	2.3 kΩ	0.5 kΩ

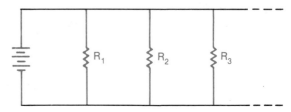

Unit 53 Achievement Review on the Use of Symbols, Terms, and Signed Numbers

OBJECTIVES OF THE UNIT

This achievement review serves as an overall test for Section 10. The unit is designed to measure the student's/trainee's ability to
- *Use symbols to represent dimensions and other quantities on drawings and for other business and industry problems.*
- *Visualize how signed numbers and number lines are used to establish measurements on gages, instruments, and drawings.*
- *Use graphic methods to solve shop, laboratory, and consumer problems that involve signed numbers and values.*
- *Solve general and practical problems using symbols with the four basic mathematical processes.*
- *Use a four-function or scientific calculator as an optional method of calculating specific quantities for problems involving signed numbers and symbols.*

A. THE CONCEPT OF SYMBOLS

1. Determine dimensions Ⓐ through Ⓔ in the sketch of the train of discs.
2. Place the letter symbols in the table in one set of circles on the sketch of the Trip Plate. Place the numerical value of each symbol in the second set of circles.

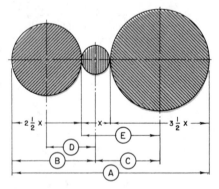

Linear Dimensions on Drawing	Letter Symbol	Numerical Value
Overall Length	L	4.6
Overall Height	H	2.8
Width of Plate	W	.5
Radius at Corners	r	.6
Included Angle	△	45°
Length to Angle	A	2.5
Elongated Slot-Width	B	.6
Elongated Slot-Length	C	1.0

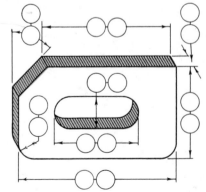

B. SIGNED NUMBERS AND NUMBER LINES: GRADUATED MEASUREMENTS (PRACTICAL PROBLEMS)

1. Read the sine wave graph and record each of the following required voltages as signed numbers.
   a. The peak positive voltage
   b. The peak negative voltage
   c. The peak to peak range of voltage
   d. The voltage at the midpoint of the cycle.

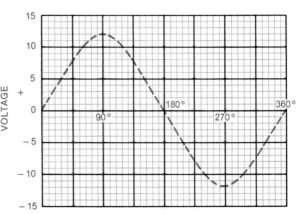

2. Write each signed number in problem 1 as an absolute value.

3. Determine the total measurement as recorded on dial indicators Ⓐ, Ⓑ, Ⓒ, and Ⓓ. Express each measurement as a signed number.

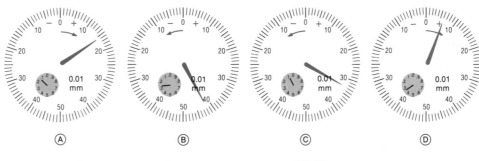

4. The constant force exerted by a spring on a piston in a hydraulic system is 4.75 kilograms (kg). A variable auxiliary force of 3.75 kg is also exerted against the piston. The reaction (opposite) force is 7.00 kg.
   a. Express each force as a signed number.
   b. Lay out a simple number line (force line) and graduate it in kg.
   c. Mark off the line the ① constant force, ② the value of the constant and auxiliary forces, and ③ the reaction force.

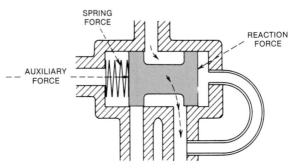

C. GRAPHIC SOLUTIONS TO SIGNED NUMBER PROBLEMS

1. Production costs fluctuate monthly over a six-month period as recorded in the chart. The starting index point is 101.5.

Month	①	②	③	④	⑤	⑥
Starting Index (101.5 points)	102.5	104.0	101.0	99.5	103.0	104.5

a. Set up a number line with a range appropriate to reflect ($+$) and ($-$) costs in relation to the starting index point.

b. Plot each monthly cost on the number line.

c. Make the following analyses of increases ($+$) and decreases ($-$) in manufacturing costs.

   (1) Final manufacturing cost

   (2) Changes between the starting index and each of the six production months.

   (3) Maximum overall cost increase and decrease

   (4) Signed number values indicating increases or decreases between the months of: ① and ③, ① and ⑥, ① and ④, ② and ④, ⑥ and ④

2. a. Plot the data in the table on a graduated number line. The unit of measurement is to be identified in each case.

   Note. Solve each problem graphically by first performing the processes within each grouping symbol and then by using vectors and signed numbers.

  b. Read the final signed number quantity in terms of the required unit of measurement.

Problem	Unit of Measurement	Signed Number Quantity, Grouping, and Processes
A	Velocity (ft/sec)	$(30 \times 4) + \left(32 \times 2\frac{1}{2}\right) + \left[\left(32 \div 8\right) - \left(32 \div \frac{1}{2}\right)\right]$
B	Power (kW)	$(37.5 + 12.5 \times 2) + (480 - 435 \div 4.5) -$ $[(176 \div 2.2) - (12 \times 2.5)] - (-30 \div 0.5)$

D. APPLICATION OF SYMBOLS IN ADDITION

1. Determine the literal dimensions Ⓐ through Ⓖ for the Parallel Bar.

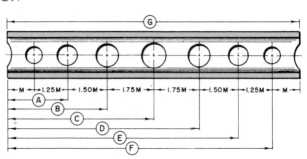

2. Find the literal dimensions (A) through (H) for the End Plate.

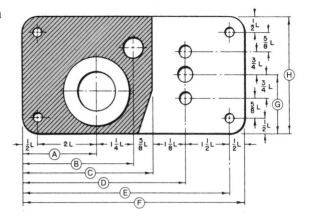

E. APPLICATION OF SYMBOLS IN SUBTRACTION

1. Determine the literal values of dimensions (A) through (E).

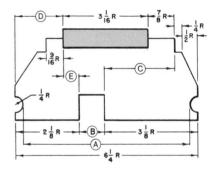

2. a. Express dimensions (A), (B), (C), and (D) in literal terms that relate to the overall length of the house.

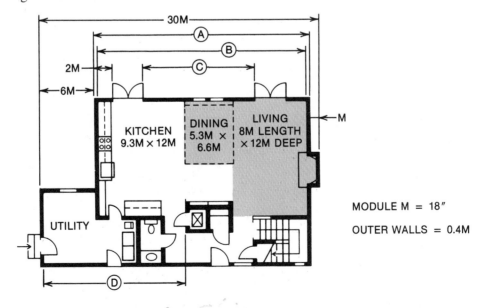

MODULE M = 18"

OUTER WALLS = 0.4M

b. Use 18″ as the (M) factor. Compute dimensions Ⓐ, Ⓑ, Ⓒ, and Ⓓ. Express each answer in feet and to the nearest whole inch measurement. Use of a hand calculator is optional.

3. Compute literal dimensions Ⓐ through Ⓔ for the Gage Plate.

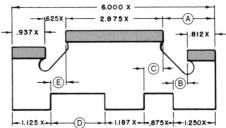

F. APPLICATION OF SYMBOLS IN MULTIPLICATION

1. Use the values of A_1 and A_2 in the formula $2\frac{1}{2}(A_1 - A_2) + 2$ to determine values for dimensions Ⓐ through Ⓓ.

Dimension	A_1	A_2
A	20	10
B	$8\frac{1}{2}$	$4\frac{1}{2}$
C	$6\frac{3}{8}$	$3\frac{7}{8}$
D	10.50	6.25

2. By using letters as symbols, only one drawing is required for different sizes of the same template. Determine dimensions Ⓐ through Ⓞ, which are needed for machining each of the five templates.

Table of Values		
Template No.	T	U
1	1	.66
2	1.5	1.0
3	2	1.32
4	2.5	1.66
5	3	1.98

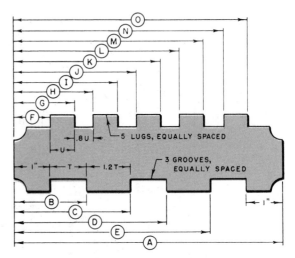

G. APPLICATION OF SYMBOLS IN DIVISION

1. Determine the dimensions (A) through (K) to which the three different sizes gages must be machined.

   Note. Express all dimensions as decimals, correct to three places (maximum).

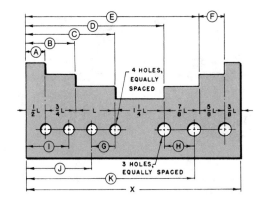

	Gage No.		
	1	2	3
X =	107.5 mm	40.96 cm	19.146 cm

11

EQUATIONS

Unit 54 A Concept of Equations

OBJECTIVES OF THE UNIT

After satisfactorily studying this unit, the student/trainee will be able to
* *Interpret the parts of equations and the functions served in algebra and higher mathematics.*
* *Balance and check an equation.*
* *Collect terms from the members of an equation.*
* *Remove common factors to simplify or reduce an expression to its lowest terms.*
* *Transpose terms and solve literal equations and practical problems.*

Equations, like symbols, are used in higher mathematics to solve problems that otherwise would be difficult and complicated. Symbols are widely applied in equations to represent unknown quantities that must be computed. Symbols also represent given known quantities that would be difficult to use in a problem except in the final steps in the solution.

In this section, the term *equations* is defined. The common principles that apply in the daily use of equations are also stated. The concept regarding the term *equations* and the principles underlying their use are applied in problems involving addition, subtraction, multiplication, and division. The use of equations is essential to a study of basic algebra. In this unit, equations are treated as a method of applying higher mathematics to the solution of everyday problems.

A. DEFINING EQUATIONS

An equation is a statement expressed in mathematical terms. An equation indicates that the quantities or expressions on both sides of an equal ($=$) sign are equal. An equation is said to be an equality of two quantities. An expression like $(C = \pi D)$ is a simple equation. The (C) is the symbol for circumference; $(D),$ for the diameter of the circle; and $(\pi),$ the Greek letter symbol, for the constant $(3.1416).$ The equation $C = \pi D$ indicates that the circumference (C)

is equal to the product of (**π**) times the diameter (**D**). The symbols (**C**) and (**D**) are used in this case to represent unknown quantities.

An equation is, therefore, a combination of numerical quantities and symbols that when added, subtracted, multiplied, or divided (as indicated by the signs in an expression), is equal to another stated quantity. An equation usually asks a question:

- What number added to (7) will equal (21)? The letter (*X*) is commonly used to represent the unknown quantity. This problem can be stated as a simple equation that may be solved by inspection.

$$X + 7 = 21$$
$$\textcircled{14} + 7 = 21$$

Therefore, $$X = 14 \quad \text{Ans}$$

Where subtraction, multiplication, and division processes are indicated, a question is also asked. For example:

- What number subtracted from 10 equals 4?
- What number multiplied by 6 equals 30?
- What number divided by 5, a given number (or any symbol), equals 10, a stated quantity?

B. PARTS OF AN EQUATION

In all equations, the expressions that appear on either side of the equality sign (=) are called *members*. Usually, the term *first member* is used to indicate the quantity to the left of the sign, and *second member,* the expression on the right of the equality sign.

The first member in figure 54-1 is (5*A* + 4); the second member is (24).

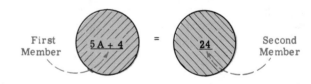

First Member · 5 A + 4 · = · 24 · Second Member

FIGURE 54-1

C. BALANCING AN EQUATION

In every equation both members must be equal and the equation is said to *balance*. To keep an equation in balance, the equal members must be increased or decreased, multiplied or divided by equal amounts.

After an equation has been solved, the value obtained for an unknown quantity is substituted in the equation. If the equation is *balanced,* as indicated by the equality of both members, the solution is correct.

This process of checking is essential at all times to prove the accuracy of dimensions and quantities. It is comparatively simple to substitute computed values in the original equation to determine the accuracy of the computation.

RULE FOR CHECKING AN EQUATION

- Substitute the computed value of any letter or symbol in the original equation.
- Perform each operation as indicated.
  Note. The equation is balanced when the values on both sides of the equation are equal.

EXAMPLE: *Case 1.* Check the answer $B = 4$ in the equation $6B = 24$.

Step 1   Substitute the value of B in the original equation.
$(B = 4)$ $6(4) = 24$

Step 2   Multiply. $6 \times 4 = 24$ Ans

Step 3   Check the answer. Since both sides of the equation are $24 = 24$ Check
equal, the answer $B = 4$ is correct.

EXAMPLE: *Case 2.* Solve the equation $8Y + 7Y = 45$ for the variable Y and check.

Step 1   Perform the operations indicated for the first mem- $8Y + 7Y = 15Y$
ber (left side) of the equation. $15Y = 45$

Step 2   Divide the value of both members by 15, the coef- $\dfrac{15Y}{15} = \dfrac{45}{15}$
ficient of the variable Y.

Step 3   Check by substituting the value of Y in the original $Y = 3$ Ans
equation. $(Y = 3)$ $8(3) + 7(3) = 45$ Check

Step 4   Perform the mathematical operations as indicated.
Multiply. Then add. $24 + 21 = 45$

Step 5   Check the final value on both sides of the equation. $45 = 45$ Check
Note. If these values are not equal, start at the be-
ginning in the original problem and check each successive step.

D. COLLECTING TERMS

The expression *collecting terms* refers to the process of combining letters or numbers within brackets, parentheses, or other grouping symbols. The terms are collected or combined as indicated by the signs of operation either to determine a final result or to simplify additional mathematical operations.

RULE FOR COLLECTING TERMS TO DETERMINE FINAL RESULT

- Combine letters, numbers, or other symbols in the first member; then do the same for the second member.

- Add, subtract, multiply, or divide as indicated in each case.
  Note. Follow the practice of starting with the quantities in parentheses first, brackets next, and, finally, those in braces.
- Write the resulting simplified equation.
- Perform the operations as indicated.
- Simplify the result by reducing to lowest terms when needed.

EXAMPLE: Solve the equation $2R = 7 + (4 + 3) + 6$ for the variable R.

Step 1 Collect terms in the right member of the equation by adding the numerical terms. $7 + (4 + 3) + 6 = 20$

Step 2 Restate the simplified equation to simplify the next step. $2R = 7 + (4 + 3) + 6$

Step 3 Solve this equation $(2R = 20)$ for (R) by inspection.
$2R = 20$
$R = 10$ Ans

Collecting Terms to Simplify Operations

EXAMPLE: Solve the equation $(4X + 7X) - 3X = 5 + (18 - 3) + (14 - 2)$ for the variable X.

Step 1 Perform the mathematical operations indicated within the grouping symbols of the right (second) member of the equation.
$(18 - 3) = 15$
$(14 - 2) = 12$

Step 2 Collect terms in the right member of the equation. $5 + 15 + 12 = 32$

Step 3 Perform the mathematical operations indicated in the parentheses of the left member. $(4X + 7X) = 11X$

Step 4 Collect terms in left member. $11X - 3X = 8X$

Step 5 Restate equation in simplified terms. $8X = 32$

Step 6 Solve for the unknown by dividing each side of the equal (=) sign by the coefficient of the variable. $X = 4$ Ans

Check

The accuracy of the computed value of X may be checked by substituting 4 for X in the original equation.

Step 1 Start with the original equation. $(4X + 7X) - 3X = 5 + (18 - 3) + (14 - 2)$

Step 2 Substitute 4 for X. $(4 \times 4 + 7 \times 4) - 3 \times 4 = 5 + (18 - 3) + (14 - 2)$

Step 3 Perform the mathematical operations as indicated.
$(16 + 28) - 12 = 5 + (15) + (12)$
$44 - 12 = 5 + 15 + 12$

Check both sides of equation to be sure that they balance.
$32 = 32$ Check

E. REMOVING COMMON FACTORS

Formulas or other expressions of equality may often be reduced to simplest terms without changing their value. Where a factor is common to an expression, it may be removed. For example, the perimeter of a rectangle is equal to the sum of all four sides. Expressed as a formula, $P = 2L + 2H$. This formula may also be written $P = 2(L + H)$. The (2) is common to both literal terms, and the way in which it is written does not change its value: $2L + 2H = 2(L + H)$.

FIGURE 54-2

This practice may be applied also when only letter symbols or combinations of letters and numbers are used in an expression. The common factor may be a single numeral, letter, or symbol, or any combination of one or more of these.

RULE FOR REMOVING FACTORS

* Examine the letters, numbers, and symbols on the left side of the equation.
* Determine which number or letter (if any) may be removed as a common factor.
* Place the common factor outside parentheses.
* Remove the common factor from each term.
* Perform the same operations for the right side of the equation.

EXAMPLE: *Case 1.* Simplify the expression $AB - AC + AD$.

Step 1 Determine which factor is common to AB, AC, and AD.

Step 2 Place the common factor (A) outside parentheses. $A($ $)$

Step 3 Factor each term by removing the common factor and $A(B - C + D)$ Ans
writing the remainder in the parentheses. The new expression
$A(B - C + D)$ is equal to $AB - AC + AD$.

EXAMPLE: *Case 2.* Simplify the expression $(3XY - 6Y)$.

Step 1 Determine the factor that is common to both terms ($3Y$).

Step 2 Place common factor ($3Y$) outside the parentheses. $3Y($ $)$

Step 3 Factor each term. $3XY = 3Y(X)$

$6Y = 3Y(2)$

Step 4 Combine terms and keep the original signs. $3Y(X) - 3Y(2) =$

$3Y(X - 2)$ Ans

The new expression $3Y(X - 2)$ is equal to $3XY - 6Y$.

F. SOLVING LITERAL EQUATIONS

A *literal equation* is one in which some or all of the quantities are represented by letters instead of numbers. Generally, beginning letters of the alphabet (like *A, B, and C*) are used in equations to represent *known quantities*. Letters toward the end of the alphabet (*X, Y, and Z*) represent *unknown quantities*. The same procedures are used to solve a literal equation as those used to solve other equations. *Literal factors* are handled the same as though they are numbers in a regular equation.

RULE FOR SOLVING LITERAL EQUATIONS

* Start with the mathematical or other processes required of the quantities in the innermost parentheses within an equation.
* Combine like terms.
* Perform the required mathematical processes and solve for the unknown quantity by adding, subtracting, multiplying, and/or dividing.
  Note. Make the unknown quantity positive by (a) multiplying each side of the equation by -1 or by interchanging the left and right members of the equation.
* Mark the answer with the appropriate literal terms or specified quantity.
* Check the solution by substituting the computed values in the original equation.

EXAMPLE: Solve the equation $9y - 12k = 3y + 9k - y$.
Note: y is the variable; k is the constant.

Step 1 Transpose like unknowns to each side of the equality sign.
  Note. In problems where a number of literal factors are contained in parentheses, the mathematical processes for these quantities are performed before the quantities are transposed.

$$9y - 3y + y = 9k + 12k$$

Step 2 Combine like terms.

$$7y = 21k$$

Step 3 Divide both sides by any common factor.

$$\frac{\cancel{7}y}{\cancel{7}} = \frac{\overset{3}{\cancel{21}}k}{\cancel{7}}$$

Step 4 Identify the answer.

$$y = 3k \quad \text{Ans}$$

Step 5 Check the solution.

$$9y - 12k = 3y + 9k - y$$
$$9(3k) - 12k = 3(3k) + 9k - 3k$$
$$27k - 12k = 9k + 9k - 3k$$
$$15k = 15k \quad \text{Check}$$

G. TRANSPOSING TERMS

Equations may also be solved by transposing terms. By this method, all known terms are moved to one side of the equation (usually the right) and all unknown terms to the other side.

The sign of each transposed quantity is changed to the opposite whenever a quantity moves across the equal sign. Hence, $+$ becomes $-$; and $\times$ becomes $\div$. Transposing is a simple form of adding, subtracting, multiplying, or dividing each side of the equation by like amounts.

RULE FOR SOLVING EQUATIONS BY TRANSPOSING AND CHECKING

- Determine which terms should be moved from one side of an equation to the other.
- Take one term at a time, of those to be transposed, and move it to the opposite side.
- Change the sign of each term as it is moved from one side of the equal sign to the other. Continue to transpose by collecting the known and unknown terms.
- Divide the equation by the numerical value or letter (known as the coefficient) preceding the unknown quantity.
- Check the equality of the equation by substituting the computed value of the unknown in the original equation.

EXAMPLE: Solve the equation $5X + 4 = -2X + 25$ for the variable X.

Step 1 Transpose known quantity (4) from left to right in the original equation.

Step 2 Transpose unknown quantity from right to left.

$$5X + \boxed{4} = -2X + 25 \quad \boxed{-4}$$

Change sign $+$

$$5X \boxed{+2X} = \boxed{-2X} + 25 \quad -4$$

Change sign

$$5X \boxed{+2X} = 25 \quad \boxed{-4}$$

Equation with known and unknown terms transposed.

Step 3 Collect all terms.

$$7X = 21$$

Step 4 Divide the known term (21) by the coefficient of the unknown (7).

$$X = \frac{21}{7}$$
$$X = 3 \quad \text{Ans}$$

Check

Step 1 Substitute the computed value 3 for the unknown X in the original equation.

Step 2 Collect terms on both sides of the equation by performing the operations as indicated.

Step 3 Check final values on both sides of the equation to see that they are equal.

$$5X + 4 = -2X + 25$$
(Original Equation)
$$5(3) + 4 = -2(3) + 25$$
$$15 + 4 = -6 + 25$$
$$19 = 19 \quad \text{Check}$$

Review and Self-Test

ASSIGNMENT UNIT 54

A. Developing a Concept of Equations

1. Develop an equation to show the amount of metal machined from the gear casting as illustrated. Use the letter (X) to denote the amount of metal removed during machining.

 WEIGHT WHEN CAST
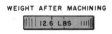 WEIGHT AFTER MACHINING

2. Express the circumferences (C_1) and (C_2) as equations in terms of diameters (D_1) and (D_2), respectively.

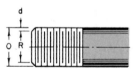

3. Develop an equation for finding the center distance (C) for the discs in the illustration.

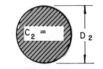

4. Indicate by an equation how the root diameter of the thread may be computed.

O = OUTSIDE DIAMETER
R = ROOT DIAMETER
d = SINGLE DEPTH OF THREAD

5. Show by an equation how the pitch of a thread may be computed. The pitch is equal to the unit (1) divided by the number of threads per inch.

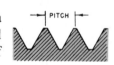

P = PITCH
N = NUMBER OF THREADS PER INCH

6. Write an equation in terms of the inside diameter (I) and wall thickness (W) for computing the outside diameter (O) of the tube illustrated.

7. Determine the equations for computing dimensions Ⓐ, Ⓑ, Ⓒ, Ⓓ, and Ⓔ given on the drawing. Express each equation in terms of the letter symbol indicated in the table. *Note.* Simplify each equation by collecting terms and removing any common factor.

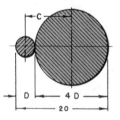

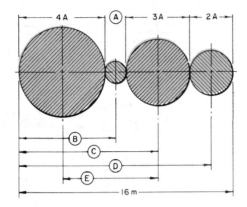

Required Dimen.	Terms
A	16 m
B	A
C	B and A
D	C and A
E	16 and A

B. Collecting Terms and Removing Common Factors

1. Simplify the expressions (a) through (j).

   a. $2X + Y$

   b. $4n - 2ln$

   c. $3XY + 6LM$

   d. $\pi D_1 + \pi D_2$

   e. $2\pi R_2 - 2\pi R_3$

   f. $3\frac{1}{7}D - 3\frac{1}{7}D$

   g. $1.28P_4 - .64P_5$

   h. $1\frac{1}{2}l + 3n$

   i. $6A + 18B - 30C$

   j. $1.57(ab) + 1.57(bc) - 3.14(bd)$

2. Solve equations (a) through (f) by inspection. Check results.

   a. $4A = 24$

   b. $21 = 3x$

   c. $22 + 6 = 7B$

   d. $5\frac{1}{2}y = 34 + 32$

   e. $18C - 13\frac{1}{2}C = 126 - 58\frac{1}{2}$

   f. $7m - 3m = 44 + 16$

UNIT 55 Solving Equations by Addition and Subtraction

OBJECTIVES OF THE UNIT

After satisfactorily studying this unit, the student/trainee will be able to

- *Simplify the solution of equations by applying the additive or subtraction axiom.*
- *Form equations from data supplied on drawings or in tables.*
- *Solve shop and laboratory problems in algebra that require the addition and/or subtraction of literal and numerical factors in equations.*
- *Check the solutions to problems by substituting a computed value in the original equation.*

A. SOLVING EQUATIONS BY ADDITION

The addition of the same term or quantity to both sides of an equation does not change the value of the equation. This statement, known as the *additive axiom,* is a self-evident truth. The solution of equations is often simplified by adding the same quantity to both members.

RULE FOR SOLVING AN EQUATION BY ADDITION

- Add the same quantity to both sides.
- Solve for the unknown.
- Check the answer by substituting the computed value for the unknown quantity in the original equation.

EXAMPLE: *Case 1.* Solve the equation $(A) - (4) = 15$.

Step 1   Add (4) to both members of the original equation.

$$A - 4 = 15$$
$$+\ 4 = +4$$

Step 2   Add both sides of the equation.

$$A \quad = 19 \quad \text{Ans}$$

Step 3   Check by substituting the numerical value (19) for (A) in the original equation.

$$A - 4 = 15$$
$$19 - 4 = 15$$
$$15 = 15 \quad \text{Check}$$

EXAMPLE: *Case 2.* Solve the equation $(2\frac{1}{2}) = (B) - (7)$.

Step 1   Add (7) to both members of the original equation.

$$2\frac{1}{2} = B - 7$$
$$+7 \qquad +\ 7$$

Step 2   Determine the sum $(9\frac{1}{2} = B)$.

$$9\frac{1}{2} = B \quad \text{Ans}$$

$$2\frac{1}{2} = B - 7$$

Step 3   Check. Substitute the numerical value $(9\frac{1}{2})$ for (B) in the original equation.

$$2\frac{1}{2} = 9\frac{1}{2} - 7$$
$$2\frac{1}{2} = 2\frac{1}{2} \quad \text{Check}$$

B. SOLVING EQUATIONS BY SUBTRACTION

As in the case of addition, the *subtraction axiom* simplifies the solutions of equations. The axiom states that the value of an equation remains unchanged when the same term or numerical quantity is subtracted from both sides of an equation.

RULE FOR SOLVING AN EQUATION BY SUBTRACTION

- Subtract the same number or quantity from both sides.

* Solve for the unknown.
* Check the computed value by substituting it in the original equation.

E XAMPLE: *Case 1.* Solve the equation $X + 9 = 13$ for the variable X.

Step 1   Subtract (9) from both sides of the equation.
Note. The difference (4) indicates the value of (X).

$$\begin{array}{r} X + 9 = 13 \\ -9 \quad\; -9 \\ \hline X = 4 \quad \text{Ans} \end{array}$$

Step 2   Check by substituting the numerical value (4) for (X) in the original equation.

$$\begin{array}{r} X + 9 = 13 \\ (4) + 9 = 13 \\ 13 = 13 \quad \text{Check} \end{array}$$

E XAMPLE: *Case 2.* Solve the equation $22.50 = M + 6.25$ for the variable M.

Step 1   Subtract (6.25) from both sides of the equation.
Note. The difference (16.25) indicates the value of (M).

$$\begin{array}{r} 22.50 = M + 6.25 \\ -6.25 \qquad\quad - 6.25 \\ \hline 16.25 = M \quad \text{Ans} \end{array}$$

Step 2   Check. Substitute the numerical value (16.25) for (M) in the original equation.

$$\begin{array}{r} 22.50 = M + 6.25 \\ 22.50 = 16.25 + 6.25 \\ 22.50 = 22.50 \quad \text{Check} \end{array}$$

Review and Self-Test

ASSIGNMENT UNIT 55

A. Solving Equations by Addition

Note. Each dimension computed for each problem must be checked for accuracy.

1. Determine dimensions Ⓐ and Ⓑ for the block illustrated.

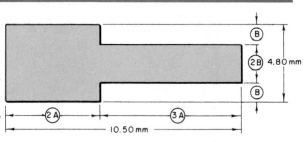

2. Find the dimensions Ⓐ, Ⓑ, and Ⓒ for the shaft.

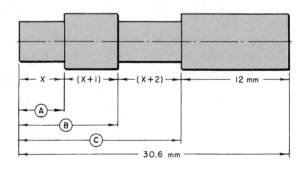

3. Determine the outside diameter (mm) of each size bushing given in the table. Round off each value to two decimals.

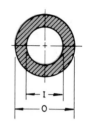

	Inside Diameter (*I*)	Single Wall Thickness (*T*)
A	31.75 mm	6.35 mm
B	15.875 mm	4.7625 mm
C	14.2875 mm	3.175 mm
D	26.9875 mm	4.7625 mm
E	27.7876 mm	3.9624 mm

4. Determine dimensions Ⓐ, Ⓑ, Ⓒ, and Ⓓ for the special roll.

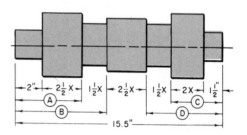

5. Solve for dimension Ⓛ for each value of M and H given in the table of dimensions for the plug.

Dimension	A	B	C	D	E	F
M	8	10	$6\frac{1}{4}$	2.500	$1\frac{3}{4}$	1.250
H	2	3	$2\frac{1}{4}$	.500	$\frac{1}{4}$	.375

6. Find the overall dimension Ⓞ for each size given in the table of dimensions for the special gage.

Dimension			Dimension		
	(P)	(N)		(P)	(N)
A	1	3	D	$1\frac{3}{4}$	$4\frac{1}{4}$
B	2	$4\frac{1}{2}$	E	.800	2.600
C	$1\frac{1}{2}$	$3\frac{1}{2}$	F	.625	2.375

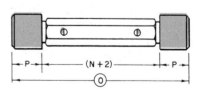

7. Compute the total resistance (R_t) and total voltage (V_t) of the four dry cells connected in series.

$$V_t = e_1 + e_2 + e_3 + e_4$$
$$R_t = r_1 + r_2 + r_3 + r_4$$

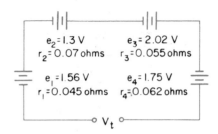

B. Solving Equations by Subtraction

1. Find the center distance $\textcircled{L}$ for each set of dimensions given in the lever table. Check all computed center distances.

	Overall Length (O)	Radius (R_1)	Radius (R_2)
A	12"	2	$1\frac{1}{2}$
B	11"	$1\frac{1}{2}$	$1\frac{3}{8}$
C	10"	$1\frac{1}{4}$	$1\frac{1}{16}$
D	$9\frac{1}{2}''$	$1\frac{1}{8}$	$\frac{15}{16}$
E	8.25"	1.125	.875
F	7.625"	.875	.687

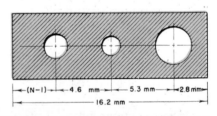

L = Center Distance
O = Overall Length
R_1 and R_2 = Radii of Ends

2. Convert each computed center distance in problem B.1 to its equivalent metric dimensions. Round off each dimension to an accuracy of two decimal places.

3. Determine the value of $\textcircled{N}$ for the drill plate. Check the answer.

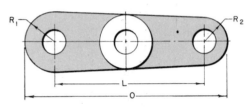

4. Solve for dimension $\textcircled{E}$ by using equations. Check the computed value of $\textcircled{E}$. All dimensions are in millimeters (mm).

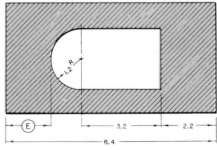

5. Solve for the internal resistance (r) or external resistance (R_ℓ) of circuits A, B, and C.

Circuit	Total Circuit Resistance (R_t) (ohms)	External Resistance (R_ℓ) (ohms)	Internal Resistance (r) (ohms)
A	2.25	1.7	
B	3.262		0.787
C	6.073		1.294

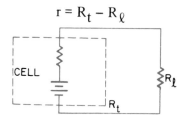

$$r = R_t - R_\ell$$

6. Find wall thickness T and $2T$ for each size of tubing specified in the table.

	Diameters	
	Outside (OD)	Inside (ID)
A	2	$1\frac{3}{4}$
B	4	$3\frac{1}{2}$
C	$1\frac{1}{2}$	$1\frac{1}{8}$
D	5.000	4.500
E	3.750	3.375
F	1.063	1.000

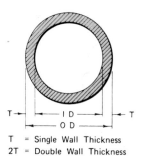

T = Single Wall Thickness
2T = Double Wall Thickness

7. Determine the minimum taper of machined parts A through E. The minimum taper (T) is equal to the difference between the large diameter (D_1) and the small diameter (D_2) at the ends of the taper, minus the tolerance (t).

	Large Diameter (D_1)	Small Diameter (D_2)	Tolerance (t)
A	1.000"	.875"	.005"
B	$8\frac{1}{2}"$	$7\frac{5}{16}"$	$\frac{1}{64}"$
C	4.4x	3.2x	.03x
D	4.25A	3.125A	.19A
E	2.4 cm	.9 cm	.07 cm

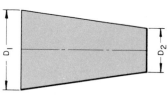

$$T = (D_1 - D_2) - t$$

Unit 56 Analyzing, Expressing, and Solving Equations by Multiplication and Division

OBJECTIVES OF THE UNIT

After satisfactorily completing this unit, the student/trainee will be able to

- *Apply the multiplication axiom to simplify the solving of equations.*
- *Solve practical shop and laboratory equations by multiplication and cross multiplying to remove fractions.*
- *Apply the division axiom and solve equations that involve division processes.*
- *Analyze, express, and solve practical equation problems by any combination of the four basic processes.*

Solving Equations by Multiplication

The *multiplication axiom* is used to simplify the solution of equations. The axiom states that the value of an equation remains unchanged when both sides are multiplied by the same quantity.

A. MULTIPLYING BOTH MEMBERS BY THE SAME QUANTITY

RULE FOR SOLVING AN EQUATION INVOLVING A FRACTION BY MULTIPLICATION

- Combine like quantities on each side of the equation.
- Use the multiplication axiom. Select a quantity that will cancel out one of the factors.
- Multiply both sides of the equation by the same quantity.
- Cancel like factors where possible from both sides of the equation.
- Solve for the unknown value(s).
- Check by substituting the computed answer for the unknown in the original equation.

EXAMPLE: Solve the equation $A/5 = 10$ for the variable A.

Step 1 Multiply both sides of the equation by 5.

Step 2 Cancel like factors in the numerator and denominator. The computed value of A is 50.

Step 3 Check by substituting 50 in the original equation.

$$\frac{A}{5} \cdot (5) = 10 \cdot (5)$$

$$\frac{A}{5} \cdot (5) = 10 \cdot (5)$$

$$A = 50 \quad \text{Ans}$$

$$\frac{50}{5} = 10$$

$$10 = 10 \quad \text{Check}$$

B. CROSS MULTIPLYING

Where the right and left members of an equation are fractions, the denominators may be removed by *cross multiplying*. The term implies that the numerator of each member of the equation is multiplied by the denominator of the other.

RULE FOR REMOVING FRACTIONS BY CROSS MULTIPLYING

- Multiply the numerator of the right member by the denominator of the left.
- Multiply the numerator of the left member by the denominator of the right.
- Collect terms as may be required.
- Check accuracy of result by substituting the computed value in the original equation.

EXAMPLE: Solve and check the equation $\frac{4}{6+B} = \frac{2}{7}$.

Step 1   Multiply the numerator of the right number (2) by the denominator of the left member $(6 + B)$.

$$\frac{4}{6+B} = \frac{2(6+B)}{7} = \frac{12+2B}{7}$$

Step 2   Multiply the left numerator (4) by the right denominator (7).

$$4(7) = \frac{12+2B}{7}$$

Step 3   Move known term (12) to the left side of the equation and change sign to $(-)$.

$$28 - 12 = 2B$$
$$16 = 2B$$

Step 4   Divide known term (16) by the coefficient (2) of the unknown term.

$$\frac{16}{2} = B; \; B = 8 \quad \textbf{Ans}$$

Step 5   Check by substituting the computed value of B in the original equation.

$$\frac{4}{6+B} = \frac{2}{7} \; original \; equation$$
$$\frac{4}{6+8} = \frac{2}{7} \; substituting \; (8) \; for \; B$$
$$\frac{4}{14} = \frac{2}{7}; \frac{2}{7} = \frac{2}{7} \quad \textbf{Check}$$

C. SOLVING EQUATIONS BY DIVISION

According to the *division axiom,* when both sides of an equation are divided by the same number, symbol, or other quantity, the value remains unchanged.

RULE FOR SOLVING AN EQUATION BY DIVISION

- Divide both members of the equation by the same quantity.
- Cancel like factors in each numerator and denominator.
- Check the answer in the original equation.

EXAMPLE: Solve the equation $5A = 20$ for the variable A.

Step 1   Divide both members of the equation by 5.

Step 2   Cancel like factors in each numerator and denominator.

Step 3   Check by substituting the computed value of A in the original equation.

$$5A = 20 \qquad \frac{5A}{5} = \frac{20}{5}$$

$$\frac{\cancel{5}A}{\cancel{5}} = \frac{\overset{4}{\cancel{20}}}{\underset{1}{\cancel{5}}} \qquad A = 4 \quad \textbf{Ans}$$

$$5A = 20 \text{ \textit{original equation}}$$
$$5(4) = 20 \text{ \textit{substituting (4) for A}}$$
$$20 = 20 \quad \text{Check}$$

D. ANALYZING, EXPRESSING, AND SOLVING PRACTICAL PROBLEMS WITH EQUATIONS

A definite sequence for analyzing practical problems should be followed because the solution may require more than one method of computation. This practice simplifies the solution of a problem and reduces the possibility of error.

RULE FOR ANALYZING, EXPRESSING, AND SOLVING PRACTICAL PROBLEMS INVOLVING EQUATIONS

- Read the entire problem through.
- Represent unknown quantities either by standard symbols assigned or select symbols that may be easily associated with unknown dimensions.
- Determine whether a formula may be used to express the required relationship or if an equation is needed to show an equality between two expressions.
- Solve the equation by any one or combination of the four fundamental processes.
- Check all results by substituting the computed values for the unknowns in the original problem.

EXAMPLE: Determine the number of revolutions per minute that a 12-inch grinding wheel will revolve for an allowable surface speed of 6280 feet per minute.

Step 1   Select symbols to use.
- (S) surface speed of wheel
- (D) diameter of wheel
- (N) RPM of grinder spindle

Step 2   Select or derive a formula to determine the surface speed.

Step 3   Substitute known values in the formula. $S = 6280;\ \pi = 3.14;\ D = 12$.

Step 4   Cancel like factors in numerator and denominator of the right member.

Step 5   Divide both members by (3.14).

$$S = \frac{\pi \times D \times N}{12}$$

$$6280 = \frac{3.14(12)(N)}{12}$$

$$6280 = \frac{3.14(\cancel{12})(N)}{\cancel{12}}$$

$$\frac{\overset{2000}{\cancel{6280}}}{\cancel{3.14}} = \frac{\cancel{3.14}(N)}{\cancel{3.14}}$$

Step 6 Check. Substitute the computed value of N and other known values in the original formula.

$(N) = 2000$ **Ans**

$S = \dfrac{\pi \times D \times N}{12}$ *original formula*

$6280 = \dfrac{3.14\ (12)\ (2000)}{12}$ *substituting values*

Step 7 Cancel like factors in numerator and denominator.

$6280 = \dfrac{3.14\ (12)(2000)}{12}$

Step 8 Multiply the remainder and compare the results. The equation is balanced when both sides are equal.

$6280 = 6280$ **Check**

Review and Self-Test

ASSIGNMENT UNIT 56

A. Multiplying Both Sides of an Equation

1. Solve equations a through h by multiplication. Check each answer.

   a. $\dfrac{X}{4} = 12$

   b. $\dfrac{M}{2\frac{1}{2}} = 10$

   c. $\dfrac{U}{5} = 2.2$

   d. $\dfrac{1}{6}n = 5$

   e. $\dfrac{c}{3.14} = 2$

   f. $.62 = \dfrac{a}{8.2}$

   g. $3.1416 = \dfrac{c}{2.25}$

   h. $\dfrac{B}{.5} = .063$

B. Cross Multiplying

1. Find the value of the unknown in equations (a) through (e) by cross multiplying. Check the value of each computed value.

   a. $\dfrac{X}{2} = \dfrac{6}{4}$

   b. $\dfrac{2}{4Y} = \dfrac{12}{72}$

   c. $\dfrac{10}{8.4} = \dfrac{4A}{60.48}$

   d. $\dfrac{4B + 6}{3} = \dfrac{4}{5}$

   e. $\dfrac{1 - 4(L - 3L)}{200} = \dfrac{.6}{1}$

C. Equations Used in Practical Problems

1. A circuit with a total line resistance of 0.54 ohms (R) draws 72 amperes (I) of current. Compute the power (P) of the circuit in watts.

$$P = (I) \cdot (I) \cdot (R)$$
$$P = I^2(R)$$

2. Find the wattage for appliances A, B, and C. The amperes and volts in each circuit are given in the table.

$P = E \times I$

$P = $ watts (W)
$E = $ volts (V)
$I = $ amperes (A)

Appliance	Amperes (I,A)	Volts (E,V)	Watts (P,W)
A	1.136	110	
B	0.68	220	
C	2.174	230	

3. Determine the number of threads per inch for screws A through F from the single depths indicated in the table.

Note. Compute to the nearest whole number of threads.

	Single Depth (D)		
A	.0649	D	.0361
B	.1299	E	.0101
C	.0203	F	.0232

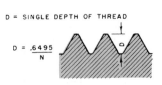

D = SINGLE DEPTH OF THREAD

$$D = \frac{.6495}{N}$$

4. Determine the cutting speed (*CS*) in feet per minute for machining parts A, B, C, D, E, and F. Use the given equation and the 3.14 value of (π) in solving for (*CS*).

$$CS = \frac{\pi \times D \times RPM}{100} = \text{m/min}$$

CS = Cutting Speed in ft./min.
D = Diameter of Work
RPM = Revolutions per minute for work

	Diameter of Work (D)	Revolutions per Minute (RPM)		Diameter of Work (D)	Revolutions per Minute (RPM)
A	20 cm	24	D	38. mm	260
B	8.2 cm	96	E	19.1 mm	1200
C	2.86 cm	514	F	158.75 mm	44

D. Solving Equations by Division (Practical Problems)

1. Solve for the unknown value in each problem (a) through (f). Check each answer.
   a. $4X = 8$
   b. $5Y = 32$
   c. $21 = 4Z$
   d. $1\frac{1}{2}N = 30$
   e. $4\frac{1}{8}L = 33.66$
   f. $7.6875 = .375A$

2. The perimeter of a square die block is 40.88″. Find the length of each side. Check the result.

PERIMETER
40.88″

L

3. Determine the outside diameters of Gear Blanks A, B, C, and D, correct to three decimal places. *Note.* Use the equation $C = \pi D$ to solve for *D*. Check each computed value.

	Circumference (C)
A	31.416
B	57.124
C	4.7124
D	5.652

OUTSIDE DIAMETER

4. The area of a rectangular blank 10.6 cm high is 261.82 sq cm. Determine the length of the blank and check the result.

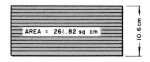

5. Convert the five Celsius degree temperatures to equivalent Fahrenheit degrees.

$$C = \frac{(F - 32)5}{9}$$

	Oven Temperatures (°C)	Equivalent Temperature (°F)
A	176.67°	
B	204°	
C	429.33°	
	Motor Temperature Rise (°C)	
D	72°	
E	97.30°	

6. One method used to calculate a child's dosage of certain medication is based on weight.

Child	Weight (lb)	Adult Dose	Child Dose
A	50	12.6 mg	
B	85	27.0 mg	
C	97	1.3 cm³ (cc)	

$$\frac{\text{Weight of child in lb}}{150} \times \text{adult dose} = \text{child's dose}$$

Give the dosages of three different medications for A, B, and C.

7. Determine the unknown quantity in each of the following equations (A through E). Check computed value.

	Given Equation	Given Values	Determine
A	$G = AB$	$G = 3, A = 6$	B
B	$L = \frac{2n}{N}$	$L = 40, N = 25$	n
C	$4X = \frac{A - b}{3P}$	$X = 1.25, A = 4.75, b = 2.13$	P
D	$1\frac{1}{2}M = \frac{Fd}{r}$	$M = 4.6, d = 1\frac{1}{4}, r = \frac{3}{4}$	F
E	$2.2C = \frac{HM}{25}$	$H = 2200, M = 1.8$	C

8. What is the efficiency of engines A, B, and C, correct to two decimal places?

Efficiency $= 100 \cdot \dfrac{\text{Useful output } (O)}{\text{Input } (I)}$

$$E = 100 \cdot \dfrac{O}{I}$$

Engine	Useful Output (O)	Input (I)
	Horsepower	
A	20	60
B	$1\frac{1}{2}$	$2\frac{1}{2}$
C	.162	2.12

9. Find the brake horsepower (B) of engines A, B, and C, correct to the nearest $\frac{1}{4}$ horsepower. Check each answer.

$$B = \dfrac{2(\frac{22}{7})(L)(W)(R)}{33,000}$$

L = Length of arm ($''$)
W = Weight (lb)
R = Revolutions per minute

Engine	(L) inches	(W) pounds	(R) RPM
A	14	15	550
B	$17\frac{1}{2}$	$30\frac{1}{2}$	750
C	$36\frac{3}{4}$	$75\frac{1}{4}$	66

Unit 57 Achievement Review on Equations

OBJECTIVES OF THE UNIT

This achievement review serves as an overall test for Section 11. The unit is designed to measure the student's/trainee's ability to

- *Interpret the functions of equations, balance equations, collect and transpose terms, remove common factors, and cross multiply.*
- *Solve practical shop, laboratory, business, and industry problems involving equations and applications of the additive, subtraction, multiplication, and division axioms.*
- *Follow definite procedures for analyzing, expressing, and solving practical problems with equations.*

Note. The achievement review problems may be solved in the conventional manner or by using a four-function or scientific calculator. Each calculated value should be checked.

A. APPLICATION OF POSITIVE AND NEGATIVE QUANTITIES

1. Add the following combinations.

   a. $3A + 2B$
      $\underline{4A + 6B}$

   c. $4L + 6R$
      $\underline{-2L - 8R}$

   e. $-5X - 3Y$
      $+4X + 2Y$
      $\underline{+3X + 8Y}$

   g. $-1.2xy - 3.22$
      $-2.2xy - 8.02$
      $\underline{-2.6xy - 9.62}$

   b. $6a - 4b$
      $\underline{-2a + 7b}$

   d. $-5AB + 4CD$
      $\underline{+3AB - 6CD}$

   f. $5AB - 3C$
      $4AB - C$
      $\underline{-2AB + 8C}$

   h. $-4.25yz + 3.50pq$
      $-2.75yz - 1.75pq$
      $\underline{+3.25yz - .50pq}$

2. Perform the operations indicated at A and B.

A	$17.2A - [3A + (5A - 2B) - (-3A - 4B)]$
B	$-16\frac{1}{2}L - [-5\frac{1}{2} - (8 + 3L) - (-15 + 14\frac{1}{2}L)]$

3. Two liter vessels contain 75% and 35% solutions, respectively, of the same nutrient. Determine the volume in milliliters (mℓ) of each solution that is required to produce one liter of a 55% solution. Use the equation

   $(X) + (l - X) = $ required solution

   $X = $ volume of 75% solution
   $(\ell - X) = $ volume of 35% solution

B. SOLVING EQUATIONS BY ADDITION

1. Determine the numerical value of ⓧ and each dimension. Check the accuracy of the computed value.

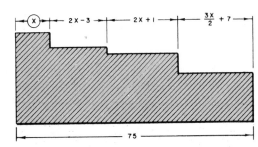

2. Solve for dimension ⓛ and check.

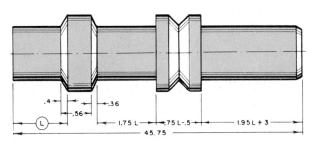

3. Determine the decimal value of dimensions O, T_1, and T_2, correct to three places. Then determine the value of each unknown dimension on the large drawing of the perforating and blanking die and the small drawing of the perforated ring.

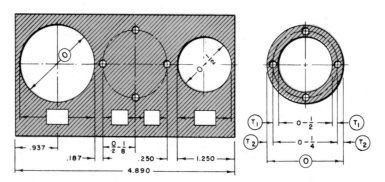

C. SOLVING EQUATIONS BY SUBTRACTION

1. Find dimensions X and Y for the bearing illustrated in both customary and metric units. Check each dimension.

   Note. Dual dimensions: $\dfrac{\text{customary (inches)}}{\text{metric (millimeters)}}$.

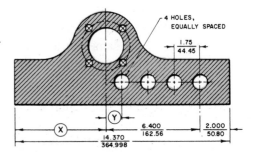

2. Solve for dimension Y in both customary and metric units. Use the numerical value $\dfrac{1.26''}{32.0 \text{ mm}}$ for dimension A. Give each answer correct to two decimal places. Check.

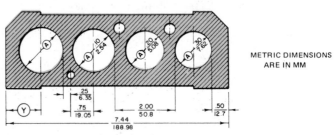

METRIC DIMENSIONS
ARE IN MM

D. SOLVING EQUATIONS BY MULTIPLICATION

1. Solve for the unknown term by cross multiplying.

   a. $\dfrac{1.6}{2.1} = \dfrac{14.4}{6.2B}$

   b. $\dfrac{8.2}{5X + 4} = \dfrac{2.2}{16.1}$

   c. $\dfrac{6}{22} = \dfrac{15}{8(r + 1)}$

   d. $\dfrac{6 - 3(5 - 2P)}{6} = \dfrac{7(4P - 20)}{32}$

2. Determine the revolutions per minute of a milling machine spindle to produce the cutting speeds indicated for operations A, B, C, and D.

   Note. State answers in terms of the nearest whole number. Use 3.14 for the value of (π).

Expressed as an equation, $\text{RPM} = \dfrac{12\,CS}{\pi D}$

RPM is the number of revolutions per minute
CS is the cutting speed in feet per minute
D is the diameter of the milling cutter

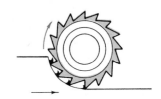

Milling Operations	A	B	C	D
Diameter of Milling Cutter (D)	4	6	$3\frac{1}{2}$	4.75
Cutting Speed in Ft/Min (CS)	100	35	800	1500

E. SOLVING EQUATIONS BY DIVISION

1. Substitute the given value in each equation and solve for the unknown. Check results.

	Given Equation	Given Value
A	$\dfrac{L_1}{T_1} = \dfrac{L_2}{T_2}$	$L_1 = 1\frac{5}{8},\ L_2 = 3\frac{1}{4},\ T_2 = 280$
B	$Xa = \dfrac{1 - (3\,Xa - 7)}{3\pi t_1 z}$	$Xa = 72,\ \pi = \frac{22}{7},\ z = .5$
C	$5C = \dfrac{12(C - A)}{B_1 - B_2(2A - 3B_1 + 4)}$	$C = 4.2,\ B_1 = 3.6,\ B_2 = 2.8$

2. Determine dimensions (A), (X), (R_1), and (r_2) for the spacing gage. Check the accuracy of each computed dimension.

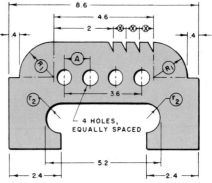

3. The total resistance (R_t) of the two line wires in the diagram is 0.415 ohms (R). The area of the copper wire is 16,724 circular mills (d^2). Compute the length of each wire in meters.

$$R_t = \frac{(K) \cdot (L)}{d^2}$$

K = 10.8 (constant for copper)
L = length of wire in ft
d^2 = circular mills
1 m = 3.281 ft

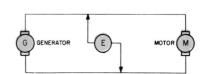

SECTION 12

RATIO AND PROPORTION

Unit 58 The Concept of Ratio

OBJECTIVES OF THE UNIT

After satisfactorily completing this unit, the student/trainee will be able to
* *Apply multiple and submultiple values to establish quantities for the comparison of first and second terms of a ratio.*
* *Solve practical ratio problems that require a comparison to be made between two related quantities.*
* *Reduce a ratio result to lowest terms.*

A ratio is a comparison of one quantity with another like quantity or value. In other words, ratio is simply a statement of the relationship between two things. The ratio of corresponding sides of two squares, like A and B (*figure* 58-1), may be stated as 4 to 6. Both objects are alike in that each is a square, but they differ in their lengths as 4 to 6.

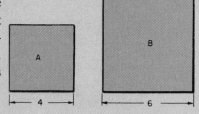

FIGURE 58-1

A. FORMS FOR EXPRESSING RATIO

In squares A and B (*figure* 58-1) the ratio is 4 to 6. This same ratio or comparison may be given in two other ways: (1) by the use of a colon, as 4:6; (2) by the use of a division sign, as 4 ÷ 6 or 4/6.

The two numbers 4 and 6 are the *terms* of the ratio: 4 is the *first term* and 6 is the *second term*. Since a ratio indicates division, the number may be reduced to lowest terms without changing the value.

RULE FOR APPLYING PRINCIPLES OF RATIO

* Read the problem to determine like quantities.
* Compare two quantities by writing the first and second terms of a ratio.

• Reduce ratio to lowest terms.
• Apply the ratio to other parts of the problem, if required, to determine other dimensions.

EXAMPLE: Express the relationship of sides *b* and *c* of triangle *ABC* as a ratio.

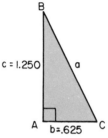

Step 1   Measure the lengths of sides *b* and *c*.
Step 2   Write as a ratio.   ***b:c* or .625:1.250**
Step 3   Reduce to lowest terms.   **1:2   Ans**

Review and Self-Test

ASSIGNMENT UNIT 58

Note. Problems may be solved by conventional mathematical processes or by using a calculator, or both.

A. Application of Ratio in Measurement

1. Measure the length of each line (a through e) to the nearest $\frac{1}{32}''$. Using these lengths, indicate the ratios given in the table.

   a. ├────────────────────┤
   b. ├──────────────┤
   c. ├──────────────────────┤
   d. ├────────┤
   e. ├─────────────────┤

Required Ratio
b:a
a:c
d:b
c:e

2. Determine the ratio between the first and second quantities of electrical measurements a through e. Reduce results to lowest terms. Round off decimal values, when required, to two decimal places.
   a. 10 A to 250 A
   b. 60 Ω to 2.4 mΩ
   c. 500 V to 75 V
   d. 90 mA to 25 μA
   e. 11 Ω to 3.41 kΩ

3. Measure the lengths indicated for each object Ⓐ through Ⓔ. Then, with these values, give the ratio of dimensions for the combinations given.

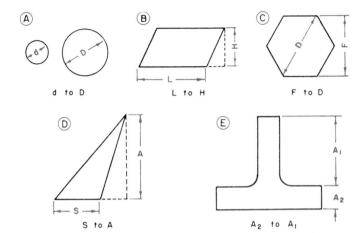

4. A pinion driver gear has 24 teeth; the driven gear, 84 teeth. What is the ratio of the driver gear to the driven gear?

5. Give the ratio of a primary winding of 350 turns to a secondary winding of $17\frac{1}{2}$ turns.

6. The rise and span of three roof layouts are given in the table. Determine the pitch in each case, rounded to whole numbers. $P = R/S$

Roof Layout	Rise (R)	Span (S)	Pitch (P
A	.91 m	3.66 m	
B	1.82 m	3.66 m	
C	.91 m	5.49 m	

7. Determine the ratio of the shaft diameter (D) to bearing length (L) in problems A through E.

Shaft	A	B	C	D	E
Shaft Diameter	1.000"	.750"	1.250"	.312"	.625"
Bearing Length	$2\frac{1}{2}''$	1.500"	3.750"	1"	$1\frac{7}{8}''$

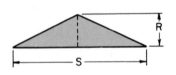

8. Calculate the following ratios of the four field rheostats as illustrated. Reduce to lowest terms. Round off decimals to nearest whole number for (c) and (d) values.

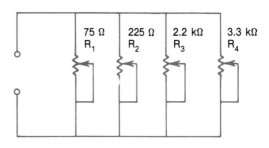

| 75 Ω R₁ | 225 Ω R₂ | 2.2 kΩ R₃ | 3.3 kΩ R₄ |

   a. R_1 to R_2
   b. R_3 to R_4
   c. R_2 to R_4
   d. $(R_1 + R_2)$ to $(R_3 + R_4)$

9. a. Determine the gear ratio of each driver gear to each driven gear for the five sets of gears in the table.

Gear Set	Gear Teeth	
	Driver	Driven
A	32	32
B	32	64
C	32	80
D	96	108
E	54	132

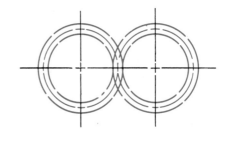

   b. Express the gear ratios a second way, this time in terms of the driven gear to the driver gear.

10. Give the compression ratio of each piston and cylinder combination for each compressed volume listed in the table. Express each ratio in terms of how many times the gas in cylinder (B) is compressed in relation to cylinder (A).

Combination	Cubic Inches	
	Full Cylinder (A)	Compressed Cylinder (B)
1	24	12
2	24	4.8
3	$12\frac{1}{2}$	$9\frac{3}{8}$
4	13.6	2.2

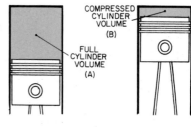

COMPRESSED CYLINDER VOLUME (B)

FULL CYLINDER VOLUME (A)

Unit 59 The Concept of Proportion

OBJECTIVES OF THE UNIT
After satisfactorily studying this unit, the student/trainee will be able to
- *Form and express a proportion from written data about two equal value ratios.*
- *Calculate the missing term (value) in a direct proportion.*
- *Solve practical problems in which the two terms in the first and second ratio are reserved and the ratios vary inversely (inverse proportion).*

Two ratios, equal in value and placed on opposite sides of an equal sign, make up a *proportion*. A proportion, therefore, is an equality of two ratios.

Expressing a Proportion

Two ratios, like 1:2 and 4:8, representing equal quantities, may be expressed as a proportion.

Terms of a Proportion

The two outside terms (1 and 8) are known as *extremes,* while the two inner terms (2 and 4) are called the *means.* In any proportion, the product of the means is equal to the product of the extremes.

A. APPLICATION OF DIRECT PROPORTION

This principle is important because if three of the four terms of a proportion are known, the fourth term may be computed easily.

RULE FOR FINDING MISSING TERM IN A PROPORTION

- Read the problem and set up two sets of ratios.
- Examine each ratio to make certain that there is an equality of the two ratios. Then write as a proportion.
- Determine whether the missing term is a mean or an extreme term.
- Multiply the means and divide the product by the known extreme to find the missing extreme; *or,*
- Multiply the extremes and divide the product by the known mean to get the missing mean.

EXAMPLE: A wire 1320 feet long has a resistance of .42 ohms. What resistance is there in one mile of the same wire?

1320:5280

Step 1   Set up two like ratios. X = the missing value.

.42 ohms: X

$1320{:}5280 = .42{:}X$

Step 2   Multiply the two means and place on one side of the (=) sign.

$5280 \times .42 = 2217.6$

Step 3   Multiply the two extremes and place on the opposite side of the (=) sign.

$1320 \times (X) = 1320X$

$1320X = 2217.6$

Step 4   Divide 2,217.6 by 1,320.

$X = \frac{2217.6}{1320} = 1.68$ ohms   Ans

Step 5   Check by substituting 1.68 ohms for X in the original proportion.

$1320{:}5280 = .42{:}1.68$

Step 6   Multiply extremes.

$1320 \times 1.68 = 2217.60$

Step 7   Multiply means.

$5280 \times .42 = 2217.60$

Step 8   Compare both products to make certain the values on both sides of the equation are equal. If they vary, start at the first step and check.

$2217.60 = 2217.60$ Check

B. APPLICATION OF INDIRECT PROPORTION

Proportion, thus far, has been covered as *direct* because the relationship of one ratio to another is direct. A proportion may also be *indirect*, in which case the ratios vary inversely. In writing an *indirect* or *inverse proportion*, the two terms in the first or second ratio are reversed.

Use is made of indirect proportion in velocity and speed problems. If the ratios of the speeds of two different-sized pulleys were written as a direct proportion, the larger pulley would have to rotate faster. Since just the opposite is true, the proportion is an inverse one.

To illustrate, determine what RPM an 8-inch diameter pulley, turning at 100 RPM, will drive a 4-inch diameter pulley.

Direct Proportion (Incorrect)

$8{:}4 = 100{:}X$

$8X = 400$

$X = 50$

Indirect Proportion (Correct)

$8{:}4 = X{:}100$ (terms 3 and 4 reversed)

$800 = 4X$

$200 = X$

It is readily apparent that by using indirect proportion the correct RPM of 200 is obtained.

RULE FOR APPLYING INDIRECT PROPORTION

* Determine the three values that are given and the fourth value that is to be computed.
* Set up the first ratio the same as for direct proportion.
* Set up the second ratio by arranging the terms in inverse order.
* Check by inserting the computed value in the original problem and reworking.

E XAMPLE: A larger gear of 96 teeth revolves at 250 RPM and drives a
smaller gear of 48 teeth. Determine the RPM of the smaller gear.

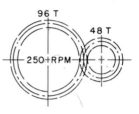

Step 1 Set up ratio of the number of teeth in the *driver* to the number of teeth in the *driven* gear.	**96:48**
Step 2 Set up ratio of RPM in *driven* gear to *driver* gear.	**X:250**
Note. This places the third and fourth terms in inverse order.	
Step 3 Multiply the extremes.	**96(250) = 24,000**
Step 4 Divide the product of the extremes by the known mean.	$\frac{24,000}{48}$ **= 500 RPM**
Step 5 Label answer and check by substituting computed value in the original equation.	**500 RPM** Ans

Review and Self-Test

ASSIGNMENT UNIT 59

Note. Final solutions or problem checks may be determined in the conventional manner or by using a four-function or scientific calculator.

A. Application of Direct Proportion (Practical Problems)

1. Determine how many milliliters of an 8.5 percent saline solution are needed to prepare 450 milliliters of a 3.5 percent solution.
2. A cylindrical container holds 1000 gallons of oil when filled to a depth of 8 feet. How many gallons are there when the level is $3\frac{1}{2}$ feet?
3. One cubic inch of water weighs .58 ounce. The specific gravity of cast iron is 7.82.
   a. What is the weight of 1 cubic foot, 864 cubic inches of cast iron?
   b. Convert the weight to its equivalent metric value.
4. What is the weight of tin, antimony, and copper in 40 pounds of Babbitt metal? Babbitt is made up of 96 parts tin, 8 parts antimony, and 4 parts copper.
5. A motor-driven pump discharges 560 liters of water in 2.5 minutes. How long does it take to discharge 10,900 liters?
6. A wire that is 925 feet long has a resistance of 2.92 ohms. Determine the length of copper wire of the same area that has a resistance of 4.24 ohms.

B. Application of Indirect Proportion (Practical Problems)

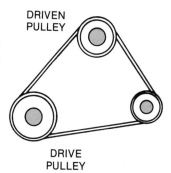

DRIVEN PULLEY

DRIVE PULLEY

1. In the direct drive illustrated, the 4″ drive pulley revolves at 1700 RPM. What size pulley is needed to increase the RPM of the driven pulley to 3200? Round off pulley diameter to nearest $\frac{1}{2}$″.

2. Compute the missing value for gears A through E.

	No. Teeth in Driver	No. Teeth in Driven	RPM of Driver	RPM of Driven
A	24		200	150
B	56	72		300
C	8	36	70	
D		144	50	32
E	72	96	60	

3. Determine the spindle speed for each step of the step-cone pulley and drive.

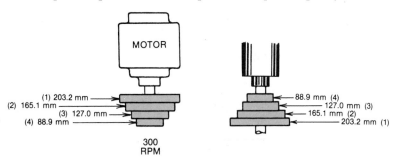

MOTOR

(1) 203.2 mm
(2) 165.1 mm
(3) 127.0 mm
(4) 88.9 mm

300 RPM

88.9 mm (4)
127.0 mm (3)
165.1 mm (2)
203.2 mm (1)

4. The pressure of a contained gas is indirectly proportional to its volume. Determine the compression at the four different stages of compression (A through D) given in the table. State the compression as pounds per square inch, rounded to the nearest pound. *Note.* The pressure on the original volume is 20 lb/□″.

Stage of Compression	Volume of Gas (Cubic Inches)
Original	20
A	16
B	12
C	9
D	7.5

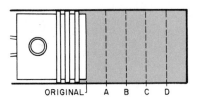

5. The applied force on a simple machine is indirectly proportional to the generated force. Calculate the weight that can be lifted (generated force) for the four lever combinations. Round off answers to the nearest pound.

Lever	Distances in Centimeters		Applied Force (kg)
	D_1	D_2	
A	8	4	120
B	$6\frac{1}{4}$	$4\frac{1}{2}$	160
C	15.2	16.8	88
D	19.6	24.9	2760

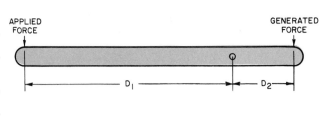

6. An electrical conductor wire of 34.221 mils diameter has a resistance of 6.032 ohms. Determine the resistance of a wire 42.610 mils in diameter of the same material and diameter. The resistance varies inversely as the square of the diameter.

Unit 60 Application of Relationships to Proportion

OBJECTIVES OF THE UNIT

After satisfactorily studying this unit, the student/trainee will be able to
- *Interpret the relationships that exist between quantities, particularly where a measurement scale is used in proportion problems.*
- *Solve problems in proportion that depend on relationships to an established scale.*

Two quantities that depend on one another for their values are related and any change in one quantity produces a change in the other.

One of the most widespread applications of relationships is in the use of scale drawings. A drawing that says, Scale 1″ = 1′-0″ means that for every measurement on the drawing, the actual size is 12 times larger. The scale, therefore, indicates a relationship between the size of the drawing and the actual size of the part.

Wherever a scale is indicated, it should be used in the order in which the terms or numbers are arranged. A scale 1:2 means that the drawing is one-half the size of the part.

RULE FOR APPLYING RELATIONSHIPS TO PROPORTION PROBLEMS

- Determine the relationship of the part to an established scale.
- Form a proportion with the scale ratio on one side of the equal sign and a ratio of known to unknown size on the other side.
- Multiply the extremes.
- Multiply the means.
- The product of the extremes equals the product of the means; solve for the unknown.

EXAMPLE: Determine dimensions *A*, *B*, *C*, and *D*.

Step 1   Measure the length of *A*.
Step 2   Form a proportion.
            1:2 = measured length : actual length of *A*.
Step 3   Multiply extremes.
Step 4   Multiply means, using the measured length of *A*.
Step 5   Solve for the actual length *A*.
            Note. The product of the means equals the product of the extremes.
Step 6   Repeat the same steps for *B*, *C*, and *D*.

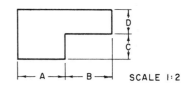

SCALE 1:2

Review and Self-Test

ASSIGNMENT UNIT 60

Note. The problems in relationships in proportion may be solved in the conventional manner or by using a basic or scientific calculator or a combination of both methods.

A. Relationships to Proportion

1. The specific weight of bronze is 8.3 as compared with a specific weight of 7.8 for cast iron. Determine how much castings A through E will weigh in cast iron.

Castings	A	B	C	D	E
Weight of Parts in Bronze	3 oz	1 lb	2 lb, 8 oz	8 lb, 4 oz	12 lb, 2 oz

2. Determine the cutting speed for the five diameters of work given in the table. Then give the rate of change in cutting speeds as a ratio.

$$CS = \frac{RPM \times \pi D}{12}$$

Work Pieces	A	B	C	D	E
Work Diameter	2″	$1\frac{1}{2}″$	1″	$\frac{3″}{4}$	$\frac{1″}{2}$
Spindle RPM	250	250	250	250	250
Required Rate of Change		A to B	A to C	B to D	A to E

3. Measure the various lengths and diameters (A through P) as shown on each drawing. Then, using the scale given for each drawing, compute the actual size.

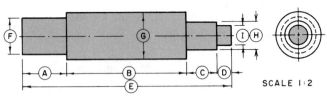

SCALE 1:2

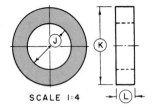

SCALE 1:4

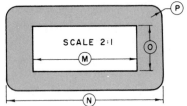

SCALE 2:1

Unit 61 Achievement Review on Ratio, Proportion, and Relationships

OBJECTIVES OF THE UNIT

This achievement review serves as an overall test for Section 12. The unit is designed to measure the student's/trainee's ability to
- *Express a ratio from written information and data.*
- *Apply rules related to ratio to direct measurement problems and solve for a required value.*
- *Express the equality of two ratios as a proportion.*
- *Find the missing term (quantity or value) in a proportion.*
- *Solve practical problems involving direct or indirect proportion.*
- *Apply measurement relationships to proportion problems.*

Note. Solutions for the ratio, proportion, and relationship problems that follow may be found by conventional mathematical processes, or by using a calculator, or by a combination of the two methods.

A. THE CONCEPT OF RATIO

1. Express in lowest terms the rise of each roof to the span (also known as *pitch*).

	Rise	Span
A	9′–0″	26′–0″
B	10′–0″	30′–0″
C	8′–0″	26′–6″
D	14′–6″	32′–6″
E	9′–6″	28′–8″

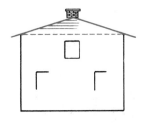

2. Determine what scale is used on drawings Ⓐ, Ⓑ, and Ⓒ.

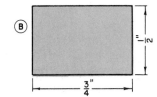

B. APPLICATION OF DIRECT PROPORTION (PRACTICAL PROBLEMS)

1. Find the value of each unknown quantity.

   a. $A:2 = 5:10$

   b. $4:B = 9:4\frac{1}{2}$

   c. $17\frac{1}{2}:8\frac{3}{4} = 32.5:C$ f. $22/F = 3.6/4.7$

   d. $8.25:10.50 = D:36.75$ g. $9.2/13.8 = G/6.50$

   e. $E/16 = 7/15$ h. $22.250/36.750 = 12.5/H$

2. Four actual measurements on a job were 30, 45, 68, and 104 inches. To what length will each line be drawn if a scale of $1:2$ is used for a drawing of the job?

3. A crew of 12 assemble 45 units a day. If the production is to be increased to 80 units a day, how many additional crew members will be needed?

4. The label of a large quantity of a medical solution reads: 1 mℓ = $\frac{1}{3}$ gr. A prescribed dose of $\frac{1}{2}$ gr is needed.

   Find the volume of the required medical solution.
   Prescribed dose: Required volume of medicine = Amount of medicine in preparation: Amount in stock solution

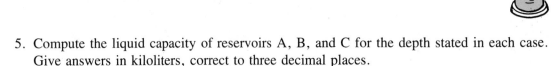

5. Compute the liquid capacity of reservoirs A, B, and C for the depth stated in each case. Give answers in kiloliters, correct to three decimal places.

| Reservoir | Total Capacity | | Required Capacity |
	Volume	Depth	Depth
A	11 358 ℓ	2.74 m	1.83 m
B	27 465 ℓ	4.69 m	2.47 m
C	92.397 kℓ	6.75 m	4.29 m

6. The number of primary turns (T_P) in a transformer is 180; the secondary turns (T_S), 720. The transformer delivers a secondary current (I_S) of 1.2 amperes. Required: Primary current (I_P) in amperes.
   Note. Use the proportion

$$I_P:I_s = T_P:T_S$$

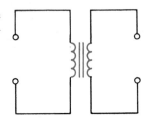

C. APPLICATION OF INVERSE PROPORTION (PRACTICAL PROBLEMS)

1. A crankshaft turns at 1250 RPM. Referring to the pulley diameters in the sketch, compute the RPM of
   a. the generator pulley and
   b. the fan pulley.

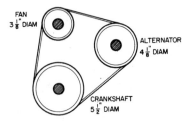

2. Find the values that are missing in the table.

	Driver Gear		Driven Gear	
	No. Teeth	RPM	No. Teeth	RPM
A	36	120	48	
B	44	200		160
C		320	96	240
D	82		104	140
E	18	400	80	

D. RELATIONSHIPS IN RATIO AND PROPORTION

1. Determine the circumferences and areas of circles A through F. Then express the relationship between the circumferences and areas as ratios.

	Diameters (Inches)	Required Ratio
A	5	Circumference A to C
B	8	Circumference D to B
C	3	Circumference E to C
D	4	Area F to D
E	9	Area E to A
F	6	Area F to B

13

EXPONENTS

Unit 62 The Concept of Exponents

OBJECTIVES OF THE UNIT

After satisfactorily studying this unit, the student/trainee will be able to
- *Interpret the functions of exponents.*
- *Raise literal and numerical values to an indicated power.*
- *Carry on the factoring of numerical terms and express the result in exponential form.*
- *Solve general and practical problems involving applications of exponents and factoring.*

A simplified method of stating mathematically that one quantity is to be multiplied by itself a number of times is to write a small number to the right of and slightly above the original quantity. This number, called an *exponent*, shows that the quantity is to be *raised* (multiplied) to the power as indicated.

A. EXPONENTS USED WITH NUMERICAL VALUES

If the number 3 is to be multiplied by itself four times, the mathematical operation may be stated a number of ways. In each case, the *factored* form indicates that (3) is to be taken as a factor four times. A more convenient way of expressing this problem is to use the exponent and the *exponential* form, 3^4. The quantity 3 is to be raised to the fourth power. The results of either the factored or exponential forms are equal, as the meanings are the same.

Method 1	$(3)(3)(3)(3)$ =
Method 2	$3 \cdot 3 \cdot 3 \cdot 3$ =
Method 3	$3 \times 3 \times 3 \times 3$ =

The area of a square .6 in. on a side is .36 sq in. This area may be determined by multiplying $.6'' \times .6''$. The problem may be stated as $A = .6'' \times .6''$. The same problem may be written more conveniently as $A = .6^2$. The exponent^2 shows that $.6''$ is to be raised to the second power or is to be taken as a factor two times.

Exponents, therefore, serve a twofold purpose.

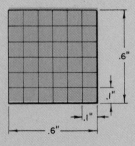

FIGURE 62-1

- To provide a simplified technique of expressing a problem mathematically.
- To simplify the mathematical processes required to solve a problem.

When exponents are used, attention must be paid to the unit of measurement in which the result is expressed. A dimension given in a linear unit, when multiplied by itself, becomes a unit in square measure. A dimension multiplied three times by itself is expressed in terms of cubic measure.

RULE FOR USING EXPONENTS (WITH SINGLE LITERAL OR NUMERICAL VALUES)

- Substitute the value given for a literal term.
- Multiply the numerical value by itself the number of times indicated by the exponent.
- Express the result in the appropriate unit of measure.

EXAMPLE: *Case 1.* Find the area of a square 8 inches on a side.   **The area = 8^2**

Step 1 Multiply 8 by itself the number of times indicated by the $8^2 = 8 \times 8 = 64$
exponent (^{2}).

Step 2 Express the area in square inches.   **64 sq in.**   Ans

EXAMPLE: *Case 2.* Determine the volume of a sphere 6 inches in
diameter, correct to two places. The volume is equal to $\frac{\pi D^3}{6}$ $V = \frac{3.1416(6)^3}{6}$

Step 1 Substitute the known diameter (6) for D and 3.1416 $= \frac{3.1416(6)(6)(6)}{6}$
for π.

Step 2 Cancel like factors in numerator and denominator. $= \frac{3.1416(6)(6)(\overset{1}{\cancel{6}})}{\underset{1}{\cancel{6}}}$

Step 3 Multiply and round off the result to two places. $= 113.0976$
                                          $= 113.10$

Step 4 Express in appropriate unit of measure. $= $ **113.10 cu in.**   Ans
('') $\times$ ('') $\times$ ('') $=$ cubic inches

B. FACTORING

The term *factoring* refers to the steps used in determining the series of smaller numbers that, when multiplied by each other, produce the original number. If the factors are all the same, they may be written in exponential form. In a problem like $3 \times 3 \times 3 \times 3 \times 3$, the 3 is a common factor. The exponential form 3^5 is the preferred method of stating that "three is to be raised to the fifth power."

In some instances, a product is given and it becomes necessary to find the factors. If the product is 27, this number may be divided by 3 to get 9. The nine may also be divided by 3. Therefore, when 3 is multiplied by itself three times (3^3), it is equal to 27.

RULE FOR FACTORING

- Divide the original number by any number that will divide into it exactly.
- Divide the quotient by the same number, if possible.
- Continue to divide until the last number may no longer be divided.
- Determine what factor, if any, is common.
- Simplify the result.

EXAMPLE: Factor 125.

Step 1 Divide original number by any number that will divide exactly.

Step 2 Divide the quotient by the same number.

Step 3 Determine the common factor (5).

$\frac{125}{5} = 25$

$\frac{25}{5} = 5$ **Ans**

$5 \cdot 5 \cdot 5 = 125$ **Check**

Review and Self-Test

ASSIGNMENT UNIT 62

A. Application of Exponents

1. Express the quantities a through h in exponential form.
   a. Two to the second power
   b. Four to the fourth power
   c. Ten to the seventh power
   d. Five to the fourth power
   e. Four-tenths to the second power
   f. One and two-tenths to the second power
   g. One-half to the third power
   h. Six and one-fourth to the second power

2. Find the numerical value of numbers (a through l) when raised to the power indicated in each case.

   a. 15^2
   b. 12^3
   c. 8^4
   d. 10^5
   e. 2^8
   f. $.5^2$
   g. 1.5^3
   h. $(\frac{1}{4})^2$
   i. $(2\frac{1}{2})^3$
   j. $(.25'')^2$
   k. $(.50'')^3$
   l. $(1.25')^3$

B. Factoring

1. Prepare a table similar to the one illustrated. Factor each value A through N and include the factors in the appropriate column. Then write in exponential form in the next column.

		Factors	Exponential Form			Factors	Exponential Form
A	4			H	$\frac{1}{16}$		
B	16			I	$\frac{1}{64}$		
C	25			J	125		
D	64			K	1000		
E	.09			L	512 cu in.		
F	.36			M	.729 cu in.		
G	.81 sq in.			N	1.728 cu ft		

C. Application of Exponents in Practical Problems

1. Compute the areas of circles Ⓐ through Ⓓ correct to three decimal places.

	Given
A	Diameter = 22.5 cm
B	Diameter = 18.75 cm
C	Radius = 2.4 cm
D	Radius = 10.32 mm

AREA = 0.7854 D^2

2. Find the area of squares A through H. Round off answers to three decimal places.

	Given		Given
A	3″	E	1.5 m
B	15″	F	5.625 cm
C	14′	G	6.35 mm
D	.8″	H	109.54 mm

A = L^2

3. Determine the power (P, in watts) required for each of three electrical circuits that draws 12 amperes (I) of current and has a resistance (R) of (a) 10 ohms, (b) $8\frac{1}{2}$ ohms, and (c) 17.62 ohms, respectively. $P = I^2 (R)$

4. Use the horsepower formula given and find the horsepower ratings of motors A through D, to the nearest whole number value.

$$\text{Horsepower ratings} = \frac{(\text{Diameter of cylinder})^2 \times (\text{Number of Cylinders})}{2.6}$$

$$H = \frac{(D)^2(N)}{2.6}$$

Motor	Number Cylinders (N)	Diameter (Bore) of Cylinders (D)
A	2	4″
B	4	4.2″
C	6	$3\frac{1}{4}''$
D	12	3.75″

Unit 63 Scientific Notation and Calculator Functions

OBJECTIVES OF THE UNIT

After satisfactorily completing this unit, the student/trainee will be able to
- *Build upon earlier experiences in establishing multiple and submultiple values for customary and SI metric units of measure.*
- *Express a power-of-ten quantity in a positive or negative exponential form.*
- *Convert a normal display format on a calculator to a scientific notation display format.*
- *Intermix data in scientific notation and standard formats.*
- *Program a scientific calculator to solve problems involving very large and very small quantities using a positive or negative mantissa or power-of-ten exponent and a scientific notation format.*

Prefixes, symbols, and power-of-ten multiples and submultiples in the scientific notation system were covered in earlier units. Measurements in the system are identified in terms of a quantity and a stated unit of measure. The different units were then related to the seven base units of SI metrics: length, mass, time, electric current, temperature, luminous intensity, and substance; the supplementary units of angular measure; and the commonly used derived units.

This unit deals with the use of a calculator, programming it to receive and to carry on mathematical processes with extremely large or small quantities using a scientific notation format. Powers-of-ten values are entered as positive or negative exponents.

A. EXPRESSING A POWER-OF-TEN VALUE IN EXPONENTIAL FORM

The use of exponents simplifies a problem statement and the actual processing of data. When the *scientific notation method* is employed, a given value is generally expressed by a number between 1 and 10. This number is followed by the multiplication (×) sign and the positive or negative power-of-ten exponent that applies. In other words, the numbers in the problem are expressed as a single digit followed by (a) a decimal point, (b) the remaining digits, (c) the (×) sign, (d) the power of ten, and (e) the exponent representing the positive or negative quantity.

For submultiple values, the power (exponent) is preceded by the minus (−) sign. It is assumed that the exponent is positive unless indicated by the (−) sign. The quantity in an answer is identified in terms of a specific unit of measure, for example: 2.75×10^2 kg; 17.03×10^{-3} mV; or 0.02×10^{-2} N/m^2.

RULE FOR EXPRESSING A POWER-OF-TEN VALUE IN EXPONENTIAL FORM

Whole Number Values: Positive Power-of-Ten Exponents
* Add a decimal point after the first whole number digit.
* Count the number of digits in the decimal portion of the number. This number expresses the original number as a power-of-ten multiple.

Decimal Values: Negative Power-of-Ten Exponents
* Move the decimal point to a position following the first digit which contains a numerical value.
  Note. This number of places which the decimal point is moved represents the submultiple value as a negative exponent.

EXAMPLE: *Case 1.* Positive Exponent
Express the following quantities as power-of-ten values in positive exponential form.

(1500), (22.7), and (65,320)

		①	②
Step 1	Place a decimal point following the first whole number digit.	(1500) 1.500	3
		(22.7) 2.27	2
		(65,320) 6.5320	4

Step 2 Count the number of places in the remaining decimal value (if any). This number represents the exponential value.

Step 3 Remove the excess zeros from the decimal value.

Step 4 Express the original value in exponential form.
  Note. Each quantity is generally followed by the appropriate unit of measure.

④
1.500 to 1.5×10^3
22.7×10^1
6.5320 to 6.532×10^4
③

EXAMPLE: *Case 2.* Negative Exponent: Submultiple values
Express the following numerical values as power-of-ten submultiples.

$$(0.987), (0.0265), \text{ and } (0.0078400)$$

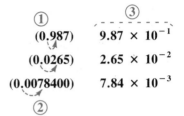

Step 1 Move the decimal point to the right, positioning it after the first digit.

Step 2 Count the number of digits the decimal point is moved to the right.
Note. This number represents the negative power-of-ten negative exponent value.

Step 3 Write the original numerical value in scientific notation by adding the multiplication sign and the submultiple power-of-ten number as a negative exponent.

Step 4 Label the answer in the unit of measure used in the problem.

B. CONVERTING A POWER-OF-TEN QUANTITY TO AN EQUIVALENT STANDARD VALUE

A quantity expressed as a power of ten (scientific notation) may be converted to its standard form by reversing the steps.

EXAMPLE: Express each of the following quantities in scientific notation form in standard form.

$$(122.6 \times 10^1 \text{ P/s}), (0.31415 \times 10^{-2} \text{ m}\Omega), (1.706 \times 10^5 \text{ lb})$$

Step 1 Check the sign of the exponent.

Step 2 Move the decimal point to the left the number of digits identified in a negative sign exponent.
Note. The decimal point is moved to the right for a positive sign exponent.

Step 3 Label each answer with the appropriate unit of measure.

$$122.6 \times 10^1 \quad + \quad 10^1$$
$$0.31415 \times 10^{-2} \quad - \quad 10^{-2}$$
$$1.706 \times 10^5 \quad + \quad 10^5$$

$$122.6 \times 10^1 = 1226 \text{ P/s}$$
$$0.31415 \times 10^{-2} = 0.0031415 \text{ m}\Omega$$
$$1.706 \times 10^5 = 170,600 \text{ lb}$$

C. PROGRAMMING A SCIENTIFIC CALCULATOR FOR PROCESSING SCIENTIFIC NOTATION PROBLEMS

The eight-digit display scientific calculator with eleven-digit calculator capacity is again used as a model. In scientific notation problems, the last two right digits on the calculator register display the exponent. The next digit on the left indicates the sign of the exponent. The numerical value and its sign (called *mantissa*) appear as the remaining digits. The mantissa consists of the *integer* (whole number) and decimal value and the sign of the number. A typical scientific notation display is illustrated in figure 63-1. In this example, the mantissa (-6.9078) is to be multiplied by ten twelve (12) times. Expressed algebraically, -6.9078×10^{-12} means -6.9078 is raised to the 12 power of -10. Thus, the value of -6.9078×10^{-12} is $-0.000\ 000\ 000\ 006\ 907\ 8$.

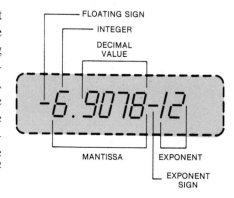

FIGURE 63-1

Converting a Normal (Standard) Display Format to a Scientific Notation Display Format

A normal (standard) display format may be converted to a scientific notation format on a calculator by using the $\boxed{\times}$ $\boxed{\text{EE}}$ (exponent entry) and $\boxed{=}$ keys.

E XAMPLE: Convert 78×929 to scientific notation.

Numerical Entry	Key Sequence Programming	Display	Notes and Functions
	$\boxed{\text{ON/C}}$ $\boxed{\text{ON/C}}$	0	• Power on • All registers are clear
78	$\boxed{\times}$	78	• (78 ×
929	$\boxed{=}$ $\boxed{\times}$	72462	• 929) = 72 462
	$\boxed{\text{EE}}$ $\boxed{=}$	7.2462 04	• Enters power-of-ten exponent value as 04 • Processes data • Displays answer of 7.2462×10^4

Intermixing Data in Scientific Notation and Standard Format

E XAMPLE: Solve 7.2×10^3 Pa $+11,682.412$ Pa.

Numerical Entry	Key Sequence Programming	Display
	ON/C	0
7.2	EE	7.2 00
3	+	7.2 03
11682.412	=	1.8882 04
	INV EE =	18 882.412 Pa

The answer indicates that the calculator carried eleven digits internally. The value in standard display in the final result is 18 882.412. This number includes the (.412) carried internally in the calculator.

The number of digits in a mantissa for scientific notation in the calculator used as a model is rounded to five. However, up to eight digits may be entered for a mantissa and used for calculations. If more whole numbers are added, the display will not go into the scientific notation format.

Attaching a Negative Sign to a Mantissa or a Power-of-Ten Exponent

The +/− *change sign key* is used to enter and display a negative sign for the mantissa and/or the exponent sign.

E XAMPLE: Enter (-62.14×10^{-5}) in scientific notation.

Numerical Entry	Key Sequence Programming	Display
	ON/C	0
62.14	+/− EE	−62.14 00
5	+/− =	−62.14-05

If the scientific notation format is to be changed to standard format, the keys INV (inverse key), EE, and = are pressed. The display answer is −0.00062.14.

Review and Self-Test

ASSIGNMENT UNIT 63

A. Expressing Standard Quantities as Power-of-Ten Values in Positive and Negative Exponential Form

1. Express quantities A through H dealing with light and sound physical properties in simplified scientific notation form. Label each answer with the appropriate unit of measure.

	Wavelength	Unit of Measure	Measurement
A	0.000 032 4 to 0.000 014 4	inch (″)	light
B	0.000 81 to 0.000 36	mm	
C	0.000 02	inch (″)	reflected
D	0.000 5	mm	light
E	0.000 016	inch (″)	color light
F	4000	angstrom (Å)	(indigo)
G	200 000	vibrations per second (vps)	sound
H	40 000 000		
		cycles per second (cps)	

B. Conversion of Scientific Notation Values to Standard Measurement Values

1. Convert the quantities expressed in scientific notation to standard form. Label each quantity in the stated unit of measure.

	Metal/Material	Electrical Resistivity	Unit of Measure
A	Aluminum	2.655×10^{-6}	ohms/cm^3
B	Copper	1.673×10^{-6}	
		Coefficient of Linear Expansion	″/°F
C	Tungsten	2.4×10^{-6}	″/°F
D	Silver	13×10^{-6}	
		Mass Density	
E	Concrete	2.3×10^{3}	kg/m^3
F	Nickel	5.5×10^{2}	lb/ft^3

C. Programming a Calculator for Scientific Notation

1. a. Identify scientific notation features Ⓐ through Ⓕ on the scientific calculator register.
   b. Interpret the two scientific notation readings as displayed.

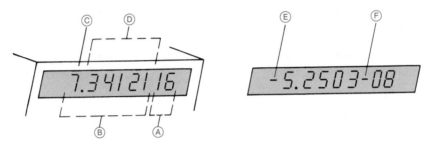

2. Convert quantities (a) through (d) to the required measurement as specified in each case. Answers are to be labeled with the appropriate unit of measure. Express each answer in (1) scientific notation and (2) standard formats.
   a. 2000 miles per hour to feet per second (according to the digit capacity of the calculator). Round the answer to four decimal places.
   b. 273.5 lb_m to kg_m (kilogram mass). Use 1 lb_m = 0.453 592 427 7 kg_m.
   c. 200.5 tons of coal to Btu of energy. Use 1 lb coal = 12 500 Btu.
   d. 16.5 lb uranium to kW · h of energy and the equivalent Btu. Use 1 lb uranium = 10 200 000 kW · h and 35 000 000 000 Btu.

D. Intermixing Data in Scientific Notation and Standard (Normal) Formats on a Calculator

1. The liquid flow in a pipeline is 3 × 10^5 gallons per 24-hour day.
   a. Compute the flow in ft^3/min.
   b. Convert the ft^3/min to m^3/min. Use 1 gal = 0.134 ft^3 and 1 ft^3/min = 2.83 × 10^{-2} m^3/min.
2. Calculate the electrical resistivity of 1275 feet of copper wire that is 250 mils in diameter. Use π = 3.1416; 1 in^3 = 16.39 cm^3, and the electrical resistivity = 2.655 × 10^{-6} ohms/cm^3. Round off the scientific notation answer to four decimal places.
3. A machine delivers 220 000 ft lbs of energy. Convert this quantity to equivalent electronvolts (eV). Use 1 joule (J) = 1.36 ft lb; also, 6.24 × 10^{18} eV.

Unit 64 Algebraic Multiplication of Numbers and Letters with Exponents

OBJECTIVES OF THE UNIT

After satisfactorily studying this unit, the student/trainee will be able to
- *Multiply like literal and numerical factors with exponents.*
- *Use the scientific notation system to multiply unlike exponent factors.*
- *Solve algebraic problems that require the multiplication of numbers and letters expressed in exponential form.*

The same principles of using exponents that apply to numerical values may be used with literal values or any combination of numbers and letters. If the quantity A is to be taken as a factor 5 times, it may be written in exponential form as A^5. The form A^5 has the same value as $(A \times A \times A \times A \times A)$.

A. MULTIPLICATION OF LIKE FACTORS WITH EXPONENTS

When the factors of a number are the same, the original number may be stated in exponential form. The number 9, when factored, is equal to 3×3. Written in exponential form, the value of 3^2 is the same as 9.

Occasionally, like factors are to be multiplied. A shortcut method of doing this is to add the exponents. Thus, $3^2 \times 3^4 = 3^{(2+4)}$ or 3^6. This process may be checked easily by multiplying each factor long hand and then comparing the results.

Step 1 $3^2 = 3 \times 3 = 9$
Step 2 $3^4 = 3 \times 3 \times 3 \times 3 = 81$
Step 3 $3^2 \times 3^4 = 9 \times 81 = 729$

Check by adding exponents.   Check
Step 4 $3^2 \times 3^4 = 3^{(2+4)} = 3^6$
Step 5 $3^6 = 3 \times 3 \times 3 \times 3 \times 3 \times 3 = 729$

These same steps are followed when letters are used in place of numerical values.

RULE FOR MULTIPLYING LIKE FACTORS

- Determine like numerical or literal factors.
- Add the exponents of all like factors.
- Show the products of all numerical and literal factors.
  Note. If no exponent is given, the exponent is understood to be ^{1}.

EXAMPLE: Multiply $S^3 \times S \times S^2$ and express the result in exponential form.
    Step 1   Add exponents of all like factors. $S^{(3\ +\ 1\ +\ 2)} = {}^6$
    Step 2   Express in exponential form. $S^3 \times S \times S^2 = S^6$ **Ans**

B. MULTIPLICATION OF UNLIKE FACTORS IN THE SCIENTIFIC NOTATION SYSTEM (SI METRICS)

Unlike factors expressed in the scientific notation system may be multiplied. The steps are similar to those followed for like factors. The answer may be a positive or negative quantity, depending on the quantities involved.

EXAMPLE: Multiply $6.6(10^{-2})$ hertz by $3.2(10^{-4})$ hertz.
    Step 1   Multiply the unlike factors. $(6.6) \cdot (3.2) = 21.12$
    Step 2   Add the exponents. $10^{(-2\ +\ -4)} = 10^{-6}$
    Step 3   Combine the results and label the answer. $21.12(10^{-6})$ **hertz (Hz)** **Ans**
            Note. The 10^{-6} value may be expressed by using the prefix ("micro-") or symbol (μ) with the unit of measure (hertz, Hz). The answer $21.12(10^{-6})$ hertz may be written 21.12 microhertz or 21.12 μHz.

The same steps are used for positive power-of-ten values. For instance, $6.6(10^3) \times 3.2\ (10^4) = 21.12(10^7)$. Positive and negative exponents may be combined in a similar manner.

EXAMPLE: Multiply $6.6(10^{-3})$ Hz by $3.2(10^5)$ Hz.
    Step 1   Multiply the unlike factors. $(6.6) \cdot (3.2) = 21.12$
    Step 2   Add the exponents. $(10^{-3}) + (10^5) = (10^2)$
    Step 3   Combine the values. $21.12(10^2)$
    Step 4   Add the appropriate unit of measure in the answer. $21.12(10^2)$ **Hz** **Ans**

Using these simple steps and the scientific notation system simplifies the multiplication process. This is particularly important where quantities involve a great number of digits.

C. EXPONENTS USED WITH NUMERICAL AND LITERAL VALUES COMBINED

Sometimes a quantity containing several terms may be included within parentheses. The exponent indicates the number of times the quantity is to be multiplied by itself. The expression $(A + 1)^2$ means that each number and letter in the parentheses is to be multiplied by every other number and letter. Further, $(A + 1)^2$ means that $(A + 1)$ must be multiplied by $(A + 1)$.

RULE FOR MULTIPLYING EXPONENTS (WITH QUANTITIES COMBINING LITERAL AND NUMERICAL VALUES)

- Determine the terms that are affected by the exponent.
- Multiply each literal and/or numerical term in the parentheses first by one term and then by the next.

- Add like terms.
- Express the product in the appropriate unit of measure.

EXAMPLE: *Case 1.* Solve the problem $(A + 1)^2$.

Step 1 Multiply each term by 1.
Step 2 Multiply each term by A.
Step 3 Add and combine like terms.
Step 4 Arrange the terms with the term having the highest power (exponent) first.

Note. This step is known as *arranging the terms in the descending order of power*.

The quantity $A^2 + 2A + 1$ is the same as $(A + 1)^2$.

①
$$\begin{array}{r} A + 1 \\ \times \quad 1 \\ \hline A + 1 \end{array}$$

$$\begin{array}{r} A + 1 \\ \times A \\ \hline A^2 + A \end{array}$$ ②

③ $(A + 1) + (A^2 + A) = 2A + A^2 + 1$
④ $= A^2 + 2A + 1$ Ans

EXAMPLE: *Case 2.* Multiply: $5x^3 \cdot 6y \cdot 4x^2 \cdot 3\,xy^2$

Step 1 Multiply numerical factors of all terms.
Step 2 Add exponents of all like literal factors.

Step 3 Show the product as the combination of all numerical and literal terms.

$(5) \cdot (6) \cdot (4) \cdot (3) = 360$
$(x^3 + x^2 + x^1) = x^6$
$(y^1 + y^2) = y^3$
$360(x^6)(y^3)$ Ans

Review and Self-Test

ASSIGNMENT UNIT 64

A. Multiplication of Like Numerical and Literal Factors

1. Write the results of each multiplication in exponential form.

a. $2^2 \times 2^2$
b. $3^2 \times 3 \times 3^3$
c. $10^5 \times 10^7$
d. $9^4 \times 9^7 \times 9^5$

e. $\left(\frac{1}{2}\right)^2 \times \left(\frac{1}{2}\right)^3$
f. $\left(\frac{1}{4}\right)^3 \times \left(\frac{1}{4}\right)^5$
g. $(1.2)^4 \times (1.2)^9$
h. $(3.02)^3 \times (3.02)^5 \times (3.02)^6$

2. Multiply the following literal values and express each answer in exponential form.

a. $A^3 \times A$
b. $C^5 \times C^2$
c. $X^6 \times X^3$

d. $f^2 \times f^{12} \times f^3$
e. $E^4 \times E^6 \times E^0 \times E$
f. $y^2 \times y^9 \times y^3 \times y^5$

g. $d^{10} \times d^{12}$
h. $A^3 \times A^2 \times A^5$
i. $L^2 \times L^4 \times L^6 \times L^{10}$

j. $D^4 \cdot D^6 \cdot D \cdot D^8$
k. $Z \cdot Z^3 \cdot Z^6 \cdot Z^2$
l. $P^4 \cdot P^2 \cdot P^{12} \cdot P$

B. Multiplication of Unlike Factors in the Scientific Notation System

1. Multiply each of the following quantities.
   a. $3.6(10^2)$ volts $\times$ $4.7(10^4)$ volts
   b. $17.92(10^6)$ grams $\times$ $12.8(10^3)$ grams
   c. $27.15(10^{-8})$ hertz $\times$ $14.6(10^{-4})$ hertz
   d. $422.7(10^{-18})$ meters $\times$ $6.5(10^{12})$ meters

2. Rewrite each answer in the problem B.1. Give the (a) computed quantity, numerical value, (b) appropriate prefix, (c) symbol, and (d) unit of measure. Then, simplify the writing of the answer (e).

Ans.	Computed Quantity (a)	Numerical Value	Prefix (b)	Symbol (c)	Unit of Measure (d)	Simplified Answer (e)
a.						
b.						
c.						
d.						

C. Multiplying Exponents with Quantities Combining Literal and Numerical Values

1. Multiply the number and letter values and simplify the results in each case.
   a. $(A + 2)^2$
   b. $(C + 5)^2$
   c. $(B + 8)^2$
   d. $(y + 3)^2$
   e. $(X + 4)^2$
   f. $(d + 3)^3$
   g. $(n + 1)^3$
   h. $(Z + 6)^2$
   i. $(g + 2)^3$
   j. $(m + 7)^3$
   k. $(R - 1)^2$
   l. $(S - 1)^3$

D. Applications of the Calculator in Solving Problems Involving Scientific Notation and Multiplication

Note. a. Express all literal and numerical terms in problems 1, 2, 3, and 4 in scientific notation format.
b. Use a hand calculator to solve each problem.
c. State each answer in exponential form, followed by the appropriate unit of measure.
d. Round off each final answer to two decimal places.

1. Required: Number of cycles per second (hertz, Hz)

$$(3.12 \ \mu Hz) \times (.425 \ mHz) \times (105^{10-2} Hz)$$

2. Required: Overall linear expansion

   7.192 μin. · 986 in. · 172°   7.192 represents linear expansion/in.
   986 in. represents overall assembly length
   172° represents change in temperature

3. Required: Surface speed (in./min)
   The gear ratio between a reduction gearbox and a drive shaft for an 8″ diameter form roller is 1:3.5. The gearbox RPM is 150. Use $\pi = 3.14159$.

4. Determine the power (watts, W) drawn by a diode that has a voltage (E) drop across it of 0.8 V and a current of 40 mA. Use the formula

$$\text{Power (W)} = \text{Current } (I) \times \text{Voltage } (E) \qquad W = I \cdot E$$

40 mA (I)

0.8 V (E)

Unit 65 Algebraic Division of Numbers and Letters with Exponents

OBJECTIVES OF THE UNIT

After satisfactorily studying this unit, the student/trainee will be able to

- *Simplify the division of like factors of literal and numerical terms by using exponents.*
- *Use the scientific notation system to divide unlike exponential factors.*
- *Program a calculator according to directions to perform basic algebraic processes using data that require translating multiple and submultiple values to exponential powers.*

Division has been described as a shortcut method of repeated subtraction of one quantity from another quantity in the same unit of measure. Where a number, followed by an exponent, is to be divided by another number with an exponent, the division process is the reverse of algebraic multiplication.

A problem like $3^4 \div 3^2$ may be solved by either one of two methods.

The second method is simplified because only a few simple subtraction and multiplication steps need to be fol-

Method 1 $\dfrac{3 \times 3 \times 3 \times 3}{3 \times 3} = \dfrac{81}{9} = \mathbf{9}$

Method 2 $\dfrac{3^4}{3^2} = 3^{(4-2)} = 3^2 = \mathbf{9}$

lowed. The exponent of the divisor $(^2)$ is subtracted from the exponent of the dividend $(^4)$. The difference between these exponents $^{(4\ -\ 2\ =\ 2)}$ becomes the exponent of the quotient factor (3), as 3^2.

A. DIVISION OF LIKE FACTORS

The division of letters, or numbers and letters in combination, in exponential form is possible only when the letter in both dividend and divisor is the same.

RULE FOR DIVIDING NUMBERS AND LETTERS WITH EXPONENTS

* Determine the factor common to both dividend and divisor. This becomes the quotient factor.
* Subtract the exponent of the divisor from the exponent of the dividend.
* Write the difference between the exponents as the exponent of the quotient factor. *Note*. If the exponent of the divisor is greater than that of the dividend, the difference will be a negative value.

EXAMPLE: *Case 1.* Divide B^6 by B^2.

Step 1 Determine the quotient factor.

Step 2 Subtract the exponents.

Step 3 Combine the quotient factor and exponent.

$$\frac{B^6}{B^2} = B$$
$$_{(6-2)} = \ ^4$$
$$\frac{B^6}{B^2} = B^4 \quad \text{Ans}$$

EXAMPLE: *Case 2.* Divide $\dfrac{C^6 \times C^b}{C^2 \times C^{2b}}$.

Step 1 Multiply the quantities in the numerator.

Step 2 Multiply the quantities in the denominator.

Step 3 Determine the common factor.

Step 4 Subtract the exponents of the divisor from the exponents of the dividend. *Note*. Change the signs and add.

Step 5 Write the result $^{(4\ -\ b)}$ as the exponent of the quotient factor (C).

①② $\dfrac{C^6 \times C^{\ b}}{C^2 \times C^{2b}} = \dfrac{C^{(6\ +\ b)}}{C^{(2\ +\ 2b)}} = C$ ③

$$
\begin{array}{ll}
(6 + b) \ \ \text{Dividend} = & 6 + b \\
- \ (2 + 2b) \ \ \text{Divisor} \ \ ④ & \\
& \underline{+(-2 - 2b)} \\
& (4 - b) \quad \text{Ans}
\end{array}
$$

⑤

$$\frac{C^6 \times C^b}{C^2 \times C^{2b}} = \boxed{C^{(4\ -\ b)}}$$

The same rule of division applies when the quotient factor is common to both dividend and divisor but the exponents are different.

EXAMPLE: *Case 3.* Divide 8^X by 8^Y.

Step 1 Write quotient factor.

Step 2 Subtract exponents. Since $(X - Y)$ cannot be reduced, the answer is $8^{(X - Y)}$.

$$\frac{8^x}{8^y} = 8^{(x - y)} \quad \textbf{Ans}$$

B. DIVISION OF UNLIKE FACTORS IN THE SCIENTIFIC NOTATION SYSTEM (SI METRICS)

Scientific and other occupational problems often require the division of unlike factors. In the scientific notation system these are expressed as powers of ten. Like multiplication, the division process is carried out first. The sign of the exponent in the denominator is changed. The value of the exponent is then added to the exponent in the numerator.

The answer may be expressed in a power-of-ten form. Otherwise, an appropriate prefix may be used. The prefix is followed by the unit of measure.

EXAMPLE: Divide $7.5 \times (10^3)$ pascal by $1.5 \times (10^{-6})$ pascal

Step 1 Divide the numerator by the denominator.   **7.5/1.5 = 5.0**

Step 2 Change the sign of the exponent in the denominator. $\mathbf{(10^{-6})}$ **to** $\mathbf{(10^{+6})}$

Step 3 Add the values of the exponents. $\mathbf{(10^{3+6}) = 10^9}$

Step 4 Combine the values.   **5.0 and (10^9) = 5.0 × (10^9)**
    5.0(10^9) pascal

Step 5 Express the answer in the appropriate unit of measure (pascal, Pa). Simplify the writing of the answer by using the symbol of the prefix and the abbreviation of unit of measure.

$$\overset{\text{G} \quad \text{Pa}}{5.0 \; \overbrace{\text{gigapascal}} =}$$
5.0 GPa   Ans

Exponential values may result in a submultiple value. In the case of $7.5(10^2) \div 1.5(10^5) = 1.5(10^{-3})$ pascal. The negative exponential value 10^{-3} is associated with the prefix "milli-." The value may be stated as 1.5 millipascal or 1.5 mPa.

Review and Self-Test

ASSIGNMENT UNIT 65

A. Algebraic Division of Like Numerical Factors

1. Divide each combination of numbers.

a. $\dfrac{4^4}{4^2}$ b. $\dfrac{3^6}{3^3}$ c. $\dfrac{12^8}{12^6}$ d. $\dfrac{132^7}{132^5}$

e. $\dfrac{2.5^4}{2.5^2}$
f. $\dfrac{.63^3}{.63^2}$
g. $\dfrac{\left(\dfrac{1}{2}\right)^4}{\left(\dfrac{1}{2}\right)^2}$
h. $\dfrac{\left(\dfrac{5}{8}\right)^6}{\left(\dfrac{5}{8}\right)^4}$

B. Algebraic Division of Like Literal Factors

1. Divide each combination of letters.

   a. $\dfrac{A^3}{A^2}$
   c. $\dfrac{X^{15}}{X^{12}}$
   e. $\dfrac{Y^{4m}}{Y^{2m}}$

   b. $\dfrac{B^4}{B}$
   d. $\dfrac{A^c}{A^d}$
   f. $\dfrac{C^{10a}}{C^{14a}}$

2. Perform the operations indicated in a through d.

   a. $\dfrac{c^2 \cdot c^4}{c^3}$
   c. $\dfrac{X^4 \cdot X^Y}{X^2 \cdot X^{2Y}}$

   b. $\dfrac{B^3 \cdot B^{3d}}{B^2 \cdot B^{2d}}$
   d. $\dfrac{A^7 \cdot A^3 \cdot A^{2m}}{A^2 \cdot A^4 \cdot A^{3m}}$

C. Algebraic Division of Unlike Factors in the Scientific Notation System

1. Divide quantities a through d.

   a. $\dfrac{6.9(10^{-6})\text{ meters}}{2.3(10^{-4})\text{ meters}}$
   c. $\dfrac{112.95(10^9)}{150.6(10^6)}$ meters

   b. $\dfrac{72.45(10^2)}{16.1(10^5)}$ liters
   d. $\dfrac{2.746(10^3)}{2.288(10^{-6})}$ volts

2. Express each answer (a through d) in the problem C.1 in terms of the appropriate
   a. Numerical value   c. Symbol
   b. Prefix   d. Unit of measure

   Then, restate as a simplified answer.

Ans	Computed Quantity	Numerical Value	Prefix	Symbol	Unit of Measure	Simplified Answer
a.						
b.						
c.						
d.						

D. Applications of Calculators in Algebraic Problems Involving Exponents and Basic Computing Processes

1. Calculate electrical quantities identified in items A through E according to the information provided in the table for the series circuit as illustrated. Round off computed values to three decimal places.

27.4 V E_S

R_1 1.6 kΩ R_2 6.4 kΩ

2.6 kΩ R_4 272 Ω R_3

   a. Write each electrical unit of measurement in exponential form (where applicable).
   b. Compute required measurements A, B, C, D, and E using exponents.
   c. Check the total of the computed source voltages (C) against the given E_s voltage on the circuit diagram.
   d. Express each answer in two forms: (1) scientific notation and (2) standard notation. Label each answer with the required unit of measurement.

	Required Quantity	Electrical Unit of Measure	Information for Programming
A	Total Resistance (R_t)	$kΩ(10^3\ Ω)$ (kiloohms)	$R_t = R_1 + R_2 + R_3 + R_4$
B	Current (I)	mA $(10^{-3}A)$ (milliamperes)	$I = \dfrac{E_s(\text{volts, V})}{R_t(\text{in } kΩ)}$
C	Source Voltage (E)	(V) volts	$E_{R_1} = E_S \times \dfrac{R_1(\text{in } kΩ)}{R_t(\text{in } kΩ)}$ $E_{R_2} = E_S \times \dfrac{R_2(\text{in } kΩ)}{R_t(\text{in } kΩ)}$ $E_{R_3} = E_S \times \dfrac{R_3(\text{in } kΩ)}{R_t(\text{in } kΩ)}$ $E_{R_4} = E_S \times \dfrac{R_4(\text{in } kΩ)}{R_t(\text{in } kΩ)}$
D	Check on Total Source Voltage	(V) volts	$E_{R_1} + E_{R_2} + E_{R_3} + E_{R_4} = E_S(27.4\ V)$
E	Total Circuit Power (P_t)	mW $(10^{-3}W)$ (milliwatts)	$P_t = I(\text{mA}) \times E_s(V)$

Unit 66 Achievement Review on Exponents

OBJECTIVES OF THE UNIT

This achievement review serves a an overall test for Section 13. The unit is designed to measure the student/trainee's ability to
- *Factor numerical terms and express the result in exponential form.*
- *Express positive or negative power-of-ten quantities in exponential form and convert a normal display format on a calculator to a scientific notation format.*
- *Multiply and divide like and unlike factors of literal and numerical terms, using exponents.*
- *Program a calculator from given data to perform basic algebraic processes using exponents and scientific notation.*

A. THE CONCEPT OF EXPONENTS

1. Raise each number to the power indicated by the exponent.

   a. 10^2 c. 3^5 e. $\left(\frac{1}{2}\right)^3$

   b. 8^3 d. $(.10)^4$ f. $(.15')^3$

2. Compute the volume of cubes A, B, C, and D in cubic inches correct to three decimal places.

	Given Length of Side
A	4"
B	5"
C	4.25"
D	$3\frac{1}{8}''$

   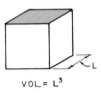

   VOL. = L³

3. Determine the area of circles A through D, correct to two decimal places. Use the value of 3.14 for π.

	Given Radius
A	4"
B	.8"
C	2.75"
D	$3\frac{5}{8}''$

   A = π r²

4. Compute the volume of spheres A through D, correct to three decimal places. Use the value of 3.14 for π.

	Given Diameter
A	5.08 cm
B	8.255 cm
C	31.75 mm
D	0.92 m

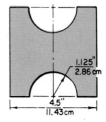

$$V = \frac{\pi D^3}{6}$$

5. Determine the area of the square part with the cutouts as illustrated. Give the areas in both customary and metric units. Round off answer correct to three places.

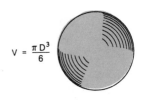

B. EXPRESSING STANDARD QUANTITIES IN POWER-OF-TEN SCIENTIFIC NOTATION FORMAT

1. Express quantities A through E in scientific notation format. State the unit of measure in each case.

	Application	Quantity
A	Force	445 000 dynes
B	Volume	0.003 785 m³
C	Pressure	6 900 N/m²
D	Time	526 000 min
E	Density (helium)	0.000 18 g/cm³

C. CONVERTING SCIENTIFIC NOTATION QUANTITIES TO STANDARD MEASUREMENT

1. a. Show how values A through D appear on the scientific notation display on a calculator.
   b. Express each scientific notation quantity as an equivalent standard notation.

	Application	Scientific Notation Quantity	Calculator Display	Standard Notation Equivalent
A	Mass (Density)	1.1×10^{-2} lb/ft³		
B	Reflected Light	3.78×10^{-4} mm		
C	Light	2.72×10^{-5} in.		
D	Linear Expansion	6.25×10^{-6}″/°F		

2. a. Compute the linear expansion of a chromium element coil that is 20' in diameter. The coil is heated from 120°F to 1560°F. Use $\pi = 3.14159$ and the linear expansion of chromium as $9.2 \times 10^{-6\prime\prime}/°F$. Give the answer in scientific notation form as displayed on a calculator.

   b. Convert the answer in scientific notation to standard measurement, correct to five decimal places. Label the quantity.

D. ALGEBRAIC MULTIPLICATION OF NUMBERS AND LETTERS

1. Multiply each set of factors and raise to the power indicated at a through d. Round off decimal values correct to two places.

   a. $2^2 \times 2^3$ c. $.8^3 \times .8$

   b. $\left(\frac{1}{6}\right)^2 \times \left(\frac{1}{6}\right)^4$ d. $(.15)^3 \times (.15) \times (.15)^2$

2. Multiply the number and letter values a through d and simplify.

   a. $(B + 3)^2$ c. $(X + 4)^2$

   b. $(5 + C)^3$ d. $(3 + AB)^3$

E. ALGEBRAIC MULTIPLICATION OF QUANTITIES IN SCIENTIFIC NOTATION

1. Multiply each of the four quantities. Express each answer in terms of the appropriate symbols of the prefix and unit of measure.

   a. $2.9(10^3)$ joules (J) $\times 6.2(10^6)$ joules

   b. $272.4(10^{-8})$ lumen (lm) $\times 4.8(10^{-10})$ lumen

   c. $16.72(10^{-12})$ amperes (A) $\times 7.56(10^{15})$ amperes

   d. $8.56(10^{-18})$ coulomb (C) $\times 12.97(10^{12})$ coulomb

F. ALGEBRAIC DIVISION OF NUMBERS AND LETTERS

1. Divide each number in problems a through d and raise the quotient factor to the resulting exponential value. Round off decimal values to two places.

   a. $\dfrac{242^5}{242^3}$ c. $\dfrac{1.125^5}{1.125^2}$

   b. $\dfrac{.625^4}{.625^2}$ d. $\dfrac{2.25^7}{2.25^3}$

2. Perform the operations indicated at a through d.

   a. $\dfrac{X^{7m}}{X^m}$ c. $\dfrac{P^7 \cdot P^4 \cdot P}{P^6 \cdot P^3}$

   b. $\dfrac{L^3 \times L^3}{L^5}$ d. $\dfrac{A^9 \cdot A^{12} \cdot A^{10} \cdot A}{A^8 \cdot A^2 \cdot A^3}$

G. ALGEBRAIC DIVISION OF QUANTITIES IN SCIENTIFIC NOTATION

1. Divide each of the four quantities. Express each answer in terms of the symbols for the appropriate prefix and unit of measure.

   a. $\dfrac{7.5(10^{-5})}{1.25(10^{-2})}$ amperes (A)

   b. $\dfrac{54.6(10^{-6})}{4.55(10^{-12})}$ watts (W)

   c. $\dfrac{242.8(10^{6})}{192.7(10^{12})}$ newtons (N)

   d. $\dfrac{5.426(10^{6})}{0.096(10^{9})}$ meters (m)

H. SOLVING ALGEBRAIC PROBLEMS ON A CALCULATOR INTERMIXING SCIENTIFIC NOTATION AND STANDARD FORMAT DATA

1. Calculate the total resistance (R_t) in kΩ of the series electrical circuit.
   Note. $R_t = R_1 + R_2 + R_3 + R_4$
   a. Express each value in scientific notation for kΩ.
   b. Calculate and state the answer as a standard unit of electrical measurement.

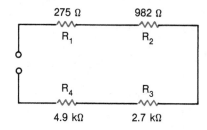

2. Calculate the total current (I_t in amperes, A) and the total circuit resistance (R_t in ohms, Ω) for the values given on the circuit diagram.
   Note. $I_t = I_1 + I_2 + I_3 + I_4$

$$R_t(\text{ohms}) = \frac{E_s(\text{volts})}{I_t(\text{amperes})}$$

   a. Express the milliampere values as amperes and the voltage values as millivolts for setting up the problem in scientific notation.
   b. Give the total current answer in scientific notation; state the circuit resistance as a conventional measurement.

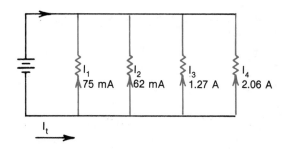

SECTION 14

RADICALS

Unit 67 Square Root of Numbers

OBJECTIVES OF THE UNIT

After satisfactorily studying this unit, the student/trainee will be able to
- *Understand the meaning of the terms root, square root, radical sign, and perfect squares.*
- *Calculate the square root of whole numbers, fractional numbers, and decimal values by conventional mathematical processes.*
- *Extract the square root of any number by approximation.*
- *Use a scientific calculator to determine the square root of any number.*

A working knowledge of how to find the root of whole numbers, fractions, decimals, literal terms, and algebraic numbers is important. The process is applied constantly in business, industry, health occupations, distribution and marketing, and in the home. The easiest way to compute the root of a number is to use a scientific calculator, which is described later in this unit. This unit also has an introduction to conventional methods used by technicians and craftspersons to find square root values.

A. PERFECT SQUARES OF WHOLE NUMBERS IN SQUARE ROOT

The phrase, *extracting the square root,* refers to the process of finding the two equal factors that, when multiplied together, give the original number. The process is identified by the use of the *radical symbol* ($\sqrt{}$). This *radical sign* is a shorthand way of saying, mathematically, that the equal factors of the number under the radical sign are to be determined.

The $\sqrt{16}$ is read "the square root of 16." This number consists of the two equal factors 4 and 4. Thus, when 4 is raised to the second power or *squared*, it is equal to 16. The term *squaring a number* merely means to multiply the number by itself.

The 16 is also referred to as a "perfect" square. Numbers that are perfect squares have whole numbers as square roots. For example, the square roots of perfect squares 4, 25, 36, 121, and 324 are all whole numbers.

RULE FOR EXTRACTING THE SQUARE ROOTS OF WHOLE NUMBERS THAT ARE PERFECT SQUARES

- Determine what number when multiplied by itself is equal to the whole number under the radical sign.
  Note. This number is known as the square root of the original number.
- If the factors cannot be determined readily:
  Break the original number into more than one factor.
  Extract the square root of each of the smaller numbers to get the factors.
- Multiply the factors. The product of these factors equals the square root of the original number. This square root multiplied by itself equals the original number that was a perfect square.

EXAMPLE: Find the square root of 1024.

Step 1 Break the original number into a series of smaller numbers.

$$\sqrt{1024} = \sqrt{4 \times 256} = \sqrt{4 \times 4 \times 64}$$

Step 2 Extract the square root of each smaller number.

$$\sqrt{4} \times \sqrt{4} \times \sqrt{64} = 2 \times 2 \times 8$$

Step 3 Multiply the factors. The product (32) is the square root of 1024.

$$2 \times 2 \times 8 = 32 \quad \text{Ans}$$

B. COMPUTING THE SQUARE ROOT BY APPROXIMATION

RULE FOR EXTRACTING THE SQUARE ROOT BY APPROXIMATION

All numbers under a radical sign are not necessarily "*perfect*" squares. In such instances, an approximate square root may be computed to a specified degree of accuracy. Problems of this type are simplified by first extracting the square root of the perfect square. Then the square root of the number remaining under the radical sign is determined. The product of the square root of the perfect square and the approximate square root of the other number is the approximate square root of the original number.

EXAMPLE: Find $\sqrt{32}$, correct to two decimal places.

Step 1 Break the original number into smaller numbers that may be perfect squares.

$$\sqrt{32} = \sqrt{16}\,\sqrt{2}$$

Step 2 Extract the square root of any perfect square. Place this number in front of the radical sign of remaining number.

$$\sqrt{16} = 4\sqrt{2}$$

Note. If there is more than one perfect square, multiply the roots of these to get the number to place in front of the radical sign.

Step 3 Examine the remaining number under the radical sign and determine the approximate number of the square root. In this case, the square root is closer to 1 than 2 because $1^2 = 1$ and $2^2 = 4$.

Step 4 Place the 1 outside the radical sign above the 2. Multiply the 1 by itself and place the product (1) under the 2.

Step 5 Subtract as in division to get the remainder.

Step 6 Add a decimal point following the 2, and one more set of zeros (00) than the required number of decimal places. *Note.* The zeros are added because the square root is a whole number and a decimal fraction.

Step 7 Carry down one set of zeros (00) and add to the remainder.

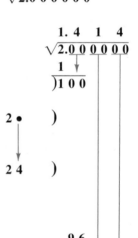

Step 8 Double the first number (1) and write outside the division frame in the tens column.

Step 9 Determine how many times the "*two tens*" plus a number in the units place may be divided into the 100. In this instance, 24 will go into 100 four times.

Step 10 Place the 4 above the radical sign following the decimal point. Multiply the 24 × 4 and place the product (96) under the 100.

Step 11 Subtract and bring down next set of zeros.

Step 12 Double the 1.4 above the radical sign and place the result in the hundreds and tens digits as a divisor into the 400.

Step 13 Determine how many times the 280 will divide into 400. Add the 1 to the 280 and place 1 above the radical sign.

Step 14 Multiply the 281 × 1 and place the product under the 400.

Step 15 Subtract and continue the same processes until the factor is determined correct to a specified number of places.

Step 16 Multiply the factor 4 as the square root of $\sqrt{16}$ and 1.414 as the approximate square root of $\sqrt{2}$.

$$4 \times 1.414 = 5.656$$

Step 17 The $\sqrt{32} = 5.66$, correct to two decimal places.

5.66 Ans

The four separate mathematical processes used in determining the approximate square root of $\sqrt{32}$ are illustrated as they are carried out in actual practice.

$$\sqrt{32} = \sqrt{16} \times \sqrt{2} = 4\sqrt{2}$$

②

①

$$\begin{array}{r} 1.\ 4\ \ 1\ \ 4 \\ \sqrt{2.0\ 0\ 0\ 0\ 0\ 0} \\ \hline 1 \end{array}$$

③ ⟶ 1

$$\begin{array}{r} 2\ 4\ \ )\overline{1\ 0\ 0} \\ 9\ 6 \\ \hline 2\ 8\ 1\ \ )\ \ \ \ 4\ 0\ 0 \\ 2\ 8\ 1 \\ \hline 2\ 8\ 2\ 4\ )\ \ \ 1\ 1\ 9\ 0\ 0 \\ 1\ 1\ 2\ 9\ 6 \\ \hline \end{array}$$

④ ⟶ **4(1.414) = 5.656 = 5.66** **Ans**

Note. The 5.66 is the approximate square root of 32.

C. EXTRACTING THE SQUARE ROOT OF A FRACTION (PERFECT SQUARE)

Extracting the square root of a fraction is a double process of first finding the square root of the numerator and then of the denominator.

RULE FOR EXTRACTING THE SQUARE ROOT OF A FRACTION (PERFECT SQUARE)

- Determine the factors of the numerator. If any factor is a perfect square, remove from under the radical sign and place its square root as a numerator.
- Determine the factors of the denominator and proceed as with the numerator.
- Combine the terms of the fraction and simplify.

EXAMPLE: Find the square root of $\frac{4}{9}$.

Step 1 Determine the square root of the numerator.

Step 2 Determine the square root of the denominator.

Step 3 Place the numerator (2) over the denominator (3). The result is $\frac{2}{3}$.

$$\sqrt{\frac{4}{9}} = \frac{\sqrt{4}}{\sqrt{9}}$$
$$\sqrt{4} = 2$$
$$\sqrt{9} = 3$$
$$\sqrt{\frac{4}{9}} = \frac{2}{3}\ \ \text{Ans}$$

D. APPROXIMATING THE SQUARE ROOT OF FRACTIONS

When neither the numerator nor the denominator of a fraction (or both) can be factored as perfect squares, then the approximate method is used to compute the square root of the fraction under the radical sign.

RULE FOR DETERMINING SQUARE ROOT OF A FRACTION (APPROXIMATE METHOD)

- Place the numerator under a radical sign.
- Place the denominator under a radical sign.
- Extract the square root of the numerator by the approximate method the same as for a whole number.
- Perform the same operation for the denominator.
- Express the resulting fraction in lowest terms.

EXAMPLE: Solve $\sqrt{\dfrac{2}{3}}$, correct to two decimal places.

Step 1 Place the numerator and denominator under separate radical signs. $\sqrt{\dfrac{2}{3}} = \dfrac{\sqrt{2}}{\sqrt{3}}$

Step 2 Extract the square root of the numerator; then of the denominator. $\dfrac{\sqrt{2}}{\sqrt{3}} = \dfrac{1.414}{1.732}$

Note. Carry out to three decimal places. $= \dfrac{.707}{.866}$

Step 3 Express in lowest terms and round off to two decimal places. $= \dfrac{.71}{.87}$ Ans

E. CALCULATING SQUARE ROOT USING A SCIENTIFIC CALCULATOR

The square root process is simplified by using a calculator and the second function key $\boxed{\text{2nd}}$ and the square root key $\boxed{\sqrt{x}}$ sequence. The calculator is cleared of any preceding operation. The number from which the square root is to be extracted is then entered. This entry is followed by pressing the $\boxed{\text{2nd}}$ $\boxed{\sqrt{x}}$ key sequence. The number that appears in the display represents the square root value. The x value entry must be positive in all cases.

EXAMPLE: Determine the $\sqrt{12.96}$.

Numerical Entry	Key Sequence	Display	
12.96	$\boxed{\text{2nd}}$ $\boxed{\sqrt{x}}$	3.6	Ans

F. PERFORMING SQUARE ROOT AND MULTIPLE MATHEMATICAL PROCESSES

A scientific calculator may be programmed to permit numerical entries for other basic mathematical processes. These may be combined with square root values.

EXAMPLE: Calculate the value of $[\sqrt{7.044\ 78} - 12.392 \times (2.6)^2 + 79.841]$. Round the final result to three decimal places.

Numerical Entry	Key Sequence	Display
7.044 78	2nd $\sqrt{x}$ −	2.6542
12.392	x	−9.7378
2.6	x^2	6.76
79.841	+	−65.8276
	+ =	14.0134 **14.013** Ans

Review and Self-Test

ASSIGNMENT UNIT 67

A. Determining Square Roots of Whole Numbers that Are Perfect Squares

1. Determine, by observation, the square root of each perfect square under a radical sign.
   a. $\sqrt{36}$ d. $\sqrt{625}$ g. $\sqrt{.25}$ j. $\sqrt{2.25}$
   b. $\sqrt{64}$ e. $\sqrt{144}$ h. $\sqrt{1.21}$ k. $\sqrt{3600}$
   c. $\sqrt{100}$ f. $\sqrt{256}$ i. $\sqrt{1.44}$ l. $\sqrt{2.89}$

2. Compute the square root of each perfect square in (a) through (i). If necessary, factor the original number to simplify a problem.
   a. $\sqrt{1024}$ d. $\sqrt{5184}$ g. $\sqrt{12.25}$
   b. $\sqrt{4096}$ e. $\sqrt{6.25}$ h. $\sqrt{.2025}$
   c. $\sqrt{1296}$ f. $\sqrt{9.61}$ i. $\sqrt{.5625}$

B. Extracting Square Root by Approximation

1. Extract the square root of (a) through (i) by the approximation method. Give each answer correct to two decimal places and check the accuracy of each result.
   a. $\sqrt{2}$ d. $\sqrt{5}$ g. $\sqrt{24}$
   b. $\sqrt{3}$ e. $\sqrt{15}$ h. $\sqrt{320}$
   c. $\sqrt{7}$ f. $\sqrt{17}$ i. $\sqrt{431}$

2. Find the value of (a) through (e).

   a. $\sqrt{(3)^2 + (6)^2 + (8)^2 + (4)^2}$

   b. $\sqrt{(10)^2 + (5.2)^2 + (6.4)^2 + (3.1)}$

   c. $\sqrt{(9.3)^2 + (4.8)^2 + (5.2) + (7.4^2)}$

   d. $\sqrt{(2)^2 + (7)^2 + (9)^2 + (4)^2}$

   e. $\sqrt{(1.2)^2 + (2.5)^2 + (5.5)^2 + (3.9)^2}$

C. Determining Square Roots of Fractions (Perfect Squares)

1. Determine, by observation, the square root of each fraction (perfect square) under a radical sign.

   a. $\sqrt{\dfrac{4}{25}}$

   b. $\sqrt{\dfrac{16}{25}}$

   c. $\sqrt{\dfrac{9}{16}}$

   d. $\sqrt{\dfrac{25}{36}}$

   e. $\sqrt{\dfrac{36}{49}}$

   f. $\sqrt{\dfrac{9}{25}}$

   g. $\sqrt{\dfrac{64}{81}}$

   h. $\sqrt{\dfrac{100}{121}}$

   i. $\sqrt{\dfrac{.16}{.25}}$

   j. $\sqrt{\dfrac{.25}{.36}}$

   k. $\sqrt{\dfrac{1.00}{1.21}}$

   l. $\sqrt{\dfrac{.64}{1.44}}$

2. A heating coil in a dryer uses 16 watts (W). The heat resistance is 25 ohms (Ω). How many amperes (A) are there in the circuit when $A = \sqrt{W/\Omega}$?

3. Compute the square root of each fraction. Wherever practical, factor to simplify.

   a. $\sqrt{\dfrac{1600}{2025}}$

   b. $\sqrt{\dfrac{1764}{2704}}$

   c. $\sqrt{\dfrac{10.89}{42.25}}$

   d. $\sqrt{\dfrac{110.00}{129.6}}$

D. Extracting the Square Root of Fractions by Approximation

1. Extract the square root of fractions (a) through (f). Reduce and/or factor, where practical. Also, reduce to lowest terms and round off answers correct to two decimal places.

   a. $\sqrt{\dfrac{5}{7}}$

   b. $\sqrt{\dfrac{10}{24}}$

   c. $\sqrt{\dfrac{44}{35}}$

   d. $\sqrt{\dfrac{3}{32}}$

   e. $\sqrt{\dfrac{5.0}{12.5}}$

   f. $\sqrt{\dfrac{7.84}{15.36}}$

E. Practical Problems in Square Root

1. Determine the length of side (L) for triangles A through E. Substitute the values given in the equation.

$$L = \sqrt{X^2 + Y^2}$$

	A	B	C	D	E
X =	3"	6.6"	1'	10"	5.6"
Y =	4"	8.8"	2'	16"	7.8"

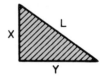

2. Compute the value of D correct to two decimal places for each value of S. Use 3.142 for π.

$$D = \sqrt{S/\pi}$$

	A	B	C	D
Value of S	9.426	31.42	38.704	62.84

3. Find dimension (S) for each of the four square sheet metal ducts, according to the cross-sectional areas given in the table. Round answers to two decimal places.

Duct	Cross-Sectional Area (X)
A	3715.2 cm²
B	7901.25 cm²
C	0.93 m²
D	2.51 m²

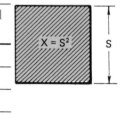

4. The voltage diagram shows voltage E_R is 82.2 volts. E_X is 66.4 volts. Compute the line voltage E_L correct to two decimal places.

$$E_L = \sqrt{(E_R)^2 + (E_X)^2}$$

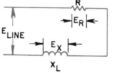

VOLTAGE DIAGRAM

F. Use of Scientific Calculator to Find Square Root Values

1. Determine the diameters of the individual strands of wire within cables A and B. Give the diameters in circular mils, rounded to two decimal places. Use

$$D_S = \frac{\sqrt{A(\text{in.}^2)}}{\#\ \text{Strands}}$$

Cable	Area (□")	# Strands	Diameter of Individual Strands in Circular Mils
A	62 500$^{10-6}$	38	
B	1.000	94	

2. Calculate the voltage (E) used in a heater that requires 17.64 watts of power (P) and has a resistance (R) of 0.5625 ohms. Use

$$E = \sqrt{P \times R}$$

3. Find the length of the sides of the square integrated circuit chips A, B, and C. Round each final value to three decimal places. Use

$$S = \sqrt{A}$$

	Area (A)	Required (S)
A	0.015 63 □″	
B	10.24 □mm	
C	$1.7024^{10'}$ □mm	

Unit 68 Roots of Algebraic Numbers

OBJECTIVES OF THE UNIT
After satisfactorily completing this unit, the student/trainee will be able to
- *Calculate the square root of algebraic numbers (numerals and literal factors) by addition, subtraction, multiplication, and division.*
- *Determine the cube root of algebraic numbers by observation and by using a table of roots.*
- *Use a scientific calculator to find the xth root of a y factor.*

An algebraic number contains letters and numbers used in combination. The square root of an algebraic number is the product of the square root of the number and the square root of the letter or letters.

A. SQUARE ROOT OF ALGEBRAIC NUMBERS

RULE FOR DETERMINING THE SQUARE ROOT OF ALGEBRAIC NUMBERS

- Determine the square root of the numerical value. See if the number may be factored as a perfect square. Otherwise, use the approximate method.
- Place the result outside the radical sign.
- Factor the literal value.
- Multiply the numerical and literal values outside the radical sign.

EXAMPLE: Find $\sqrt{9Y^2}$.

Step 1 Break up $\sqrt{9Y^2}$ into numerical and literal terms.

Step 2 Determine the square root of the numerical term.

Step 3 Determine the square root of the literal term.

Step 4 Multiply the square roots of the terms.

$\sqrt{9Y^2} = \sqrt{9} \times \sqrt{Y^2}$
$\sqrt{9} = 3$
$\sqrt{Y^2} = Y^{2\div2} = Y$
$3 \times Y = 3Y$ **Ans**

This same procedure is followed for both the numerator and denominator of algebraic fractions.

B. CUBE ROOT OF ALGEBRAIC NUMBERS

The volume of a solid shape like a cube is computed by multiplying the dimensions of length, width, and height. Since each dimension is equal to each other, the volume of the cube in figure 68-1 $= 2 \cdot 2 \cdot 2$ or 8 units. This relationship is simplified by the expression $\sqrt[3]{8}$. The radical symbol $\sqrt{}$ and the placement of the index number three (3) in the symbol indicates that the *third (^{3}) or cube root* is to be found in the measurement (quantity) within the symbol.

The index number gives the number of times the algebraic root number is taken (multiplied by itself) as an equal factor to produce the cube root value.

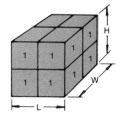

FIGURE 68-1

Determining Cube Root Values by Using Tables

Handbook and other manufacturers' tables are often used to obtain the cube root of a number. Figure 68-2 (abstracted from Table 8 in the Appendix) displays *cube power* and *cube root values*. The three-place decimal table provides the degree of accuracy needed in general applications. Cube roots of additional decimal quantities may be computed or determined by using a calculator.

Number	Powers		Roots		Number	Powers		Roots	
	Square	Cube	Square	Cube		Square	Cube	Square	Cube
1	1	1	1.000	1.000	51	2,601	132,651	7.141	3.708
2	4	8	1.414	1.260	90	8,100	729,000	9.487	4.481
3	9	27	1.732	1.442	91	8,281	753,571	9.539	4.498
4	16	64	2.000	1.587	92	8,464	778,688	9.592	4.514
5	25	125	2.236	1.710	93	8,649	804,357	9.644	4.531
6	36	216	2.449	1.817	94	8,836	830,584	9.695	4.547
7	49	343	2.646	1.913	95	9,025	857,375	9.747	4.563
8	64	512	2.828	2.000	96	9,216	884,736	9.798	4.579
9	81	729	3.000	2.080	97	9,409	912,673	9.849	4.595
10	100	1,000	3.162	2.154	98	9,604	941,192	9.900	4.610
11	121	1,331	3.317	2.224	99	9,801	970,299	9.950	4.626
12	144	1,728	3.464	2.289	100	10,000	1,000,000	10.000	4.642

FIGURE 68-2 Partial table of square and cube powers and roots

EXAMPLE: Use a table of cube roots and solve $\sqrt[3]{96}$ Pa.

Step 1	Locate the given value in the table.	**96 =**
Step 2	Read the cube root directly from the values given in this column.	**4.579**
Step 3	Express the answer in terms of the required unit of measurement.	**4.579 Pa** Ans

C. DETERMINING THE CUBE ROOT OF AN ALGEBRAIC NUMBER

RULE FOR DETERMINING CUBE ROOT FOR ALBEGRAIC NUMBERS

* Separate the numerical and literal values of the terms.
* Find the $\sqrt[3]{}$ of the numerical terms using a table or calculator.
* Find the $\sqrt[3]{}$ of the literal term by dividing the index number into the exponent of the literal term.
* Combine the cube root values of the numerical and literal terms.

EXAMPLE: *Case 1.* Find $\sqrt[3]{99a^3}$.

Step 1 Separate literal and numerical terms. $\sqrt[3]{99} \cdot \sqrt[3]{a^3}$

Step 2 Look up the cube root of the numerical term. $\sqrt[3]{99} = 4.626$

Step 3 Find the cube root of the literal term. $\sqrt[3]{a^3} = a^{3 \div 3} = a^1$
Step 4 Combine the numerical and literal roots.   **$4.636 \cdot a = 4.626a$** Ans

EXAMPLE: *Case 2.* Find $\sqrt[3]{86z^6y^3}$.

Step 1 Separate numerical and literal terms $\sqrt[3]{86} \cdot \sqrt[3]{z^6} \cdot \sqrt[3]{y^3}$

Step 2 Look up the cube root of the numerical term. $\sqrt[3]{86} = 4.414$

Step 3 Find the cube root of the literal term. $\sqrt[3]{z^6} = z^{6 \div 3} = z^2$
Step 4 Combine the numerical and literal roots.   **$4.414 \cdot z^2 \cdot y = 4.414z^2y$** Ans

D. ADDING AND SUBTRACTING LIKE RADICALS

RULE FOR ADDING AND SUBTRACTING LIKE RADICALS

* Add or subtract the coefficient of the radical.
  Note. Only like radicals can be added or subtracted.

E XAMPLES: Add or subtract like radicals.
- $a\sqrt{d} + c\sqrt{d} = a + c\sqrt{d}$
- $3\sqrt{5} + 2\sqrt{5} = 3 + 2\sqrt{5} = 5\sqrt{5}$
- $\sqrt[4]{16} - \sqrt[4]{16} = \sqrt[0]{16} = 0$
- $2\sqrt{2} - \sqrt{3} =$ **No subtraction can take place**
- $10\sqrt[3]{X} + 6\sqrt[3]{X} = 16\sqrt[3]{x}$

E. MULTIPLYING ROOT RADICALS

RULE FOR MULTIPLYING ROOT RADICALS

- Multiply the coefficient of each term.
- Multiply the literal or numerical quantities under the radical sign.
- Combine the coefficient product and the radical product for the answer.

E XAMPLE: Multiply each of the following radicals.
- $\sqrt{4} \cdot \sqrt{5} = \sqrt{20}$
- $3\sqrt{2} \cdot 5\sqrt{3} = 15\sqrt{6}$
- $\sqrt{\frac{7}{8}} \cdot \sqrt{\frac{2}{3}} = \sqrt{\frac{14}{24}} = \sqrt{\frac{7}{12}}$
- $a\sqrt{b} \cdot b\sqrt{x} = ab\sqrt{x\,b}$
- $(3\sqrt{2})(a\sqrt{3}) = 3a\sqrt{6}$
- $9\sqrt[3]{E\,A} \cdot 6\sqrt[3]{E\,A} = 54\sqrt[3]{E^2A^2}$

F. DIVIDING RADICALS

RULE FOR DIVIDING ROOT RADICALS

- Extract the common factor from the numerator and denominator.
  Note. This step applies to numerical and literal factors within and outside the radical sign.
- Reduce the remaining numerator and denominator to lowest terms.

E XAMPLE: *Case 1.* Divide $\dfrac{20\sqrt{12}}{4\sqrt{4}}$.

$\dfrac{20}{4} = 5$

Step 1 Extract the common numerical factor outside the radical sign.

Step 2 Extract the common factor from the values inside the radical sign.

$\dfrac{\sqrt{12}}{\sqrt{4}} = 3$

Step 3 Combine the values using the radical sign.

$5\sqrt{3}$ **Ans**

Note. If the numerical value is required, the square root value of 3 of 1.732 is multiplied by 5 for an answer of 8.66.

EXAMPLE: *Case 2.* Divide $\dfrac{120\sqrt[3]{Y}}{40\sqrt[3]{Y}}$.

Step 1 Extract the common numerical factor. $\qquad\qquad \dfrac{120}{40} = 3$

Step 2 Extract the common root factor and literal factors. $\qquad \dfrac{\sqrt[3]{Y}}{\sqrt[3]{Y}} = 1$

Step 3 Combine the values. $\qquad\qquad\qquad\qquad (3) \cdot (1) = 3 \quad$ Ans

G. DETERMINING THE *x*TH ROOT OF A *y* VALUE USING A SCIENTIFIC CALCULATOR

The $\boxed{\text{INV}}$ *inverse key* and the $\boxed{y^x}$ *y to the xth power key* are used in combination *to take the xth root of a y value* (y^x). The *y* value is entered followed by pressing the $\boxed{\text{INV}}$ $\boxed{y^x}$ keys. The display shows the entry value. The numerical value of the required root is entered. The $\boxed{=}$ key is pressed. The display identifies the required *x*th root of the *y* value.

EXAMPLE: Solve $\sqrt[6]{87\,648.83}$.

Numerical Entry	Key Sequence	Display	
87 648.83	$\boxed{\text{INV}}$ $\boxed{y^x}$	87648.83	
6	$\boxed{=}$	2.4	Ans

It must be noted that the variable *y* must be a positive value. Otherwise, an *ERROR* is displayed after *x* and an operation are pressed. The accuracy of the *x*th root is within ± in the eighth significant digit for general applications of *x* and *y* values.

H. POWER FUNCTIONS PERFORMED ON A SCIENTIFIC CALCULATOR

A given number (*y*) may be raised to any positive or negative (*x*) power by using the $\boxed{y^x}$ *xth root of y key,* the $\boxed{+/-}$ key for negative powers, and the $\boxed{=}$ key.

EXAMPLE: *Case 1.* Raise 14.26 to the fourth power (14.26^4).

Numerical Entry	Key Sequence	Display	
14.26	$\boxed{y^x}$	14.26	
4	$\boxed{=}$	589654.49	Ans

EXAMPLE: *Case 2.* Solve $4.74^{-.36}$.

Numerical Entry	Key Sequence	Display	
4.74	y^x	4.74	
.36	$+/-$	-0.36	
	$=$	0.571111	Ans

Review and Self-Test

ASSIGNMENT UNIT 68

A. Extracting the Square Root of Algebraic Numbers

1. Find the square root of each algebraic number and fraction.

   a. $\sqrt{4a^2}$

   b. $\sqrt{9b^2}$

   c. $\sqrt{36c^2}$

   d. $\sqrt{.25X^2}$

   e. $\sqrt{1.44Y^2}$

   f. $\sqrt{2.56Z^2}$

   g. $\sqrt{\dfrac{16A^2}{25B^2}}$

   h. $\sqrt{\dfrac{27b^2}{32c^2}}$

   i. $\sqrt{\dfrac{2.42X^2}{7.2Y^2}}$

   j. $\sqrt[3]{212}$

   k. $\sqrt[3]{51.90a^3}$

   l. $\sqrt[3]{19.03c^3}$

2. Determine length (b) of $\triangle ABC$ when $b = \sqrt{c^2 - a^2}$.

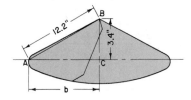

B. Adding and Subtracting Root Radicals

Add or subtract the following algebraic numbers.

1. $\sqrt{9} - \sqrt{2}$

2. $3.5\sqrt{8} - 1.5\sqrt{8}$

3. $X\sqrt{Y} + 2\sqrt{Y}$

4. $\frac{3}{4}\sqrt[3]{a} + \frac{3}{8}\sqrt[3]{a}$

5. $6.4\sqrt{b} - 6.4\sqrt{b}$

6. $12\frac{1}{2}\sqrt[3]{Z} - 9\frac{1}{4}\sqrt[3]{Z}$

C. Multiplying or Dividing Radicals

1. $z\sqrt{a} \cdot y\sqrt{b}$

2. $\dfrac{\sqrt{18}}{\sqrt{9}}$

3. $.5\sqrt[3]{12a} \cdot 6\sqrt[3]{4a}$

4. $\frac{1}{2}\sqrt{3z} \times \frac{1}{4}\sqrt{5y}$

5. $\dfrac{14\sqrt{9b}}{7\sqrt{3b}}$

D. Radicals (Practical Problems)

Note. Problems 1, 2, 3, and 4 may be solved by any combination of algebraic and scientific calculator processes, including the use of tables.

1. The design diameter (*D*) of an automotive crankshaft is established by using the formula

$$D^3 = \frac{T}{0.2S}.$$

Using values of 35 000 for (*S*) and 133 000 for *T*, compute the diameter correct to three decimal places.

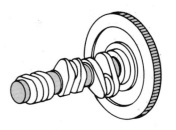

2. The length of a beam for a steel truss is equal to the square root of the sum of the rise squared and the span divided by two, squared.
   a. Compute the length of trusses A and B according to the specifications for rise and span.

Truss	Rise	Span
A	4'–0"	20'–0"
B	6'–6"	32'–0"

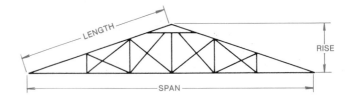

   b. Convert each answer to feet and inches, rounded to the closest $\frac{1''}{4}$.

3. Five cylindrical support columns require 308 cubic feet (11.4 cubic yards) of poured concrete. The height of each column is 10 feet. Calculate the diameter of the columns in feet and inches to the nearest $\frac{1''}{8}$. Use $\pi = \frac{22}{7}$.

4. Determine the required measurements for programming a machine for distances L_1 and L_2. Convert measurement L_2 to a customary inch value. Round the measurements to the nearest 0.000″.

Part	Side (a)	Side (b)
A	4.195″	6.800″
B	127 mm	152.4 mm

$$L = \sqrt{a^2 + b^2}$$

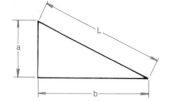

Unit 69 Application of Square Root Tables

OBJECTIVES OF THE UNIT

After satisfactorily studying this unit, the student/trainee will be able to
- *Read square root values directly from selected tables within a specified range of accuracy.*
- *Find the square root of any number by factoring, using tables, or by the approximate method.*
- *Check the accuracy of a square root against an original number.*

The simplest method of finding the square root of a number is to use a *table of square roots*. This practice is widespread in the business and industrial world. It conserves time and, what is more important, it ensures accuracy.

While the form in which tables are prepared varies, the factual information they contain is the same. All square root tables list in sequence and in one column all whole numbers. The square root of each of these numbers is given in another column. The square root values, except in the case of perfect squares, are approximate. The values are accurate to the degree indicated for the entire table. The tables are called, for example, *two-point,* or *three-point*. The *point* indicates that the approximate values are correct to two or three decimal places, respectively.

Larger square root tables are accurate to four and five places. Also, these tables contain the approximate square roots of whole numbers and tenths.

A. READING TABLES OF SQUARE ROOTS

Portions of these tables are illustrated in figures 69-1 and 69-2. Figure 69-1 lists whole numbers from 1 to 10 in one column. The square root of each number is given in a second column. Note that the square root is correct to two decimal places. Figure 69-2 contains the square root values of numbers from 1.0 through 9.9, correct to two decimal places. Figure 69-3 shows only a small part of a larger table of square root values for numbers ranging from 1.00 to 9.99 in steps of .01. This three-point table is accurate to the nearest thousandth.

Square Root Values 1 through 10	
No.	**Square Root**
1	1.00
2	1.41
3	1.73
4	2.00
5	2.24
6	2.45
7	2.65
8	2.83
9	3.00
10	3.16

FIGURE 69-1 Table of square root values (1 through 10)

Square Root Values

1.0 through 9.9

No.	.0	.1	.2	.3	.4	.5	.6	.7	.8	.9
1	1.00	$1.0\overline{5}$	1.10	1.14	1.18	1.22	1.26	1.30	1.34	1.38
2	1.41	$1.4\overline{5}$	1.48	1.52	$1.5\overline{5}$	1.58	1.61	1.64	1.67	1.70
3	1.73	1.76	1.79	1.82	1.84	1.87	1.90	1.92	$1.9\overline{5}$	1.97
4	2.00	2.02	$2.0\overline{5}$	2.07	2.10	2.12	2.14	2.17	2.19	2.21
5	2.24	2.26	2.28	2.30	2.32	$2.3\overline{5}$	2.37	2.39	2.41	2.43
6	$2.4\overline{5}$	2.47	2.49	2.51	2.53	2.55	2.57	2.59	2.61	2.63
7	$2.6\overline{5}$	2.67	2.68	2.70	2.72	2.74	2.76	2.77	2.79	2.81
8	2.83	$2.8\overline{5}$	2.86	2.88	2.90	2.92	2.93	2.95	2.97	2.98
9	3.00	3.02	3.03	3.05	3.07	3.08	3.10	3.11	3.13	3.15

Note: Wherever a line appears over the second place digit of 5, like $1.5\overline{5}$, drop the last digit in rounding off. The $1.5\overline{5}$ correct to one decimal place is 1.5.

FIGURE 69–2 Table of square root values (1.0 through 9.9)

Square Root Values of Numbers from 1.00 through 9.99										
No.	0	1	2	3	4	5	6	7	8	9
1.0	1.000	1.00$\overline{5}$	1.010	1.01$\overline{5}$	1.020	1.02$\overline{5}$	1.030	1.034	1.039	1.044
1.1	1.049	1.054	1.058	1.063	1.068	1.072	1.077	1.082	1.086	1.091
1.2	1.095	1.100	1.10$\overline{5}$	1.109	1.114	1.118	1.122	1.127	1.131	1.136
1.3	1.140	1.14$\overline{5}$	1.149	1.153	1.158	1.162	1.166	1.170	1.175	1.179

FIGURE 69-3 Table of square root values (1.00 through 9.99)

RULE FOR READING A TABLE OF SQUARE ROOTS

- Determine the degree of accuracy required. Then select a table of square roots that will provide the desired accuracy.
  Note. If tables are not available, it may be necessary to compute by the approximate method.
- Find the given number in the column that is so marked.
- Locate the square root value of the given number in the column in which the square roots are given.
  Note. If the number contains a decimal, look for the square root value in the column to the right of the whole number and below the decimal value.
- Round off answer to as many places as may be required.

EXAMPLE: *Case 1.* Use a table to find $\sqrt{2}$ correct to one decimal place.

			No.	Sq. Root
Step 1	Locate the whole number 2 in the first vertical column of a table of square roots.		1	**1.00**
Step 2	Read the square root to the right of the given number in the column of square roots.		2	**1.41**
			3	**1.73**
Step 3	Round off to one decimal place. **1.41 = 1.4 Ans**		4	**2.00**

EXAMPLE: *Case 2.* Find in a table of square
roots the $\sqrt{3.1}$ correct to one decimal place.

> *Step 1* Select a table of square roots which
> gives values of numbers in steps of
> .1.
>
> *Step 2* Locate the 3 in the vertical column
> of whole numbers.
>
> *Step 3* Read the square root value in the
> column to the right of 3 and below
> .1. $\sqrt{3.1} = 1.76$
>
> *Step 4* Round off the square root value of
> 1.76 to one decimal place. **1.76 = 1.8** Ans

	.0	.1	.2
1	1.00	1.05	1.10
2	1.41	1.45	1.48
3	1.73	1.76	1.79
4	2.00		

B. COMPUTING THE SQUARE ROOT OF NUMBERS OUTSIDE THE RANGE OF TABLES

Sometimes, the range of a table of square roots is not large enough to take care of all numbers. When this is the case, the number may be factored so the perfect square is removed. The number remaining under the radical sign may then fall within the range of the table.

RULE FOR COMPUTING THE SQUARE ROOT OF ANY NUMBER (FACTORING, APPROXIMATE METHOD, AND USE OF TABLES)

- Factor the given number as far as possible.
- Locate the square root value of the remainder under the radical sign in a table of square roots.
  Note. If no table is available, compute this value by the approximate method.
- Multiply the first factor (or combination of factors) by the square root obtained from a table or by the approximate method.
- Round off the result, if required.

EXAMPLE: *Case 1.* Find $\sqrt{192}$ to two decimal places.

> *Step 1* Factor.
>
> *Step 2* Determine $\sqrt{64}$.
>
> *Step 3* Find $\sqrt{3}$ from a table.
> *Note.* Compute this value if no table is available.
>
> *Step 4* Multiply both factors 8 and 1.73.

$$\sqrt{192} = \sqrt{64} \times \sqrt{3}$$
$$\sqrt{64} = 8$$
$$\sqrt{3} = 1.73$$
$$\sqrt{192} = (8) \times (1.73)$$
$$= 13.84 \quad \text{Ans}$$

EXAMPLE: *Case 2.* Find $\sqrt{.24}$ to two decimal places.
  Step 1 Express $\sqrt{.24}$ in terms of a fraction.

$$\sqrt{.24} = \sqrt{\frac{24}{100}} = \frac{\sqrt{24}}{\sqrt{100}}$$

  Step 2 Factor the numerator.

$$\sqrt{24} = \sqrt{4} \times \sqrt{6} = 2\sqrt{6}$$

  Step 3 Find the value of $\sqrt{6}$.

$$\sqrt{6} = 2.45$$

  Step 4 Multiply both factors in the numerator.

$$\sqrt{24} = (2) \times (2.45) = 4.90$$

  Step 5 Factor the denominator.

$$\sqrt{100} = 10$$

  Step 6 Place the square root of the numerator (4.90)
    over the square root of the denominator (10)
    and divide.

$$\frac{4.90}{10} = .49$$

$$\sqrt{.24} = .49 \quad \textbf{Ans}$$

  The square root values of numbers may be checked readily by simply multiplying the square root of the original number by itself. When the square root value has been determined by the approximate method, the result should fall within the range of the required degree of accuracy. The square roots, particularly of decimal numbers, should be checked for accurate placement of the decimal point.

Review and Self-Test

ASSIGNMENT UNIT 69

  Note. Refer to table 8 in the Appendix (Powers and Roots of Numbers) to solve the problems in this unit that require the use of a table.

A. Reading Tables of Square Roots

  1. Find the square root of each number correct to two decimal places.
    a. $\sqrt{2}$ b. $\sqrt{3}$ c. $\sqrt{5}$ d. $\sqrt{6}$ e. $\sqrt{7}$ f. $\sqrt{10}$
  2. Find the square root of each number correct to one decimal place.
    a. $\sqrt{1.1}$ d. $\sqrt{2.3}$ g. $\sqrt{9.3}$
    b. $\sqrt{1.7}$ e. $\sqrt{5.2}$ h. $\sqrt{8.7}$
    c. $\sqrt{3.3}$ f. $\sqrt{6.3}$

B. Computing Square Root Values

  1. Compute the square root of each number by factoring, approximation, or by using a table. Give answers correct to two decimal places.
    a. $\sqrt{32}$ d. $\sqrt{324}$ g. $\sqrt{3.2}$
    b. $\sqrt{80}$ e. $\sqrt{1694}$ h. $\sqrt{2501.7}$
    c. $\sqrt{63}$ f. $\sqrt{27.5}$

C. Application of Square Root (Practical Problems)

1. Compute dimension Z of the part illustrated, for each dimension given for X and Y, correct to three decimal places. Wherever practical, use a table of square root values.

	Given Values	
	X	**Y**
A	3	4
B	6	7
C	12.5	15
D	8.6	10.5

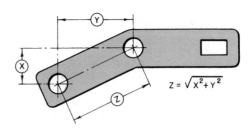

$z = \sqrt{x^2 + y^2}$

2. A series of rectangular pipes is to be constructed. The width of each pipe is to be five times the height. Find the dimensions of sheet metal pipes A through E for the area given in each instance.

Pipe	Area
A	80 cm²
B	100 cm²
C	245 mm²
D	366.25 mm²
E	5.5 m²

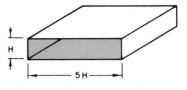

Note. Problems 3 through 7 may be solved by conventional algebraic processes or by using a hand calculator.

3. What amperage is flowing through an electrical circuit when the wattage is 440 and the resistance is 200?

$$\text{Amperage} = \sqrt{\frac{\text{Watts}}{\text{Resistance}}}$$

4. An electrical heating unit has a 20-ohm resistance and uses power at the rate of 2420 watts. Determine the voltage that the heater works on when

$$\text{Voltage} = \sqrt{(\text{Watts}) \times (\text{Ohms Resistance})}.$$

5. Calculate the cylinder diameter (c) in centimeters of a V-6 cylinder engine (N) rated at 96.4 horsepower (hp) when

$$c = \sqrt{\frac{\text{hp} \times 2.5}{N}}.$$

6. The resistance of an electric heating device is 672 ohms. It uses 34 watts of power. Determine the amperes required by the resistance. Round off the answer to three decimal places.

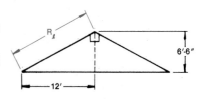

$$A = \sqrt{\frac{\text{watts}}{\text{ohms}}}$$

7. Carpenters often apply square root to determine rafter length.
   a. Find the rafter length (R_l) for a wooden roof truss with a rise of 6′–6″ and a run of 12′.
   b. Convert the final answer to feet and inches, rounded to the nearest $\frac{1}{4}''$.

$$R_l = \sqrt{(rise)^2 + (run)^2}$$

Unit 70 Achievement Review on Radicals

OBJECTIVES OF THE UNIT

This achievement review serves as an overall test for Section 14. The unit is designed to measure the student's/trainee's ability to
* *Calculate square root, cube root, and other roots of whole numbers, fractional numbers, and decimal values by conventional algebraic processes.*
* *Determine the square roots by perfect squares and fractional parts by approximation.*
* *Read tables of square and cube root values within a required range of accuracy.*
* *Use a scientific calculator to find square, cube, and xth root of a y factor.*

A. DETERMINING SQUARE ROOTS OF WHOLE NUMBERS (PERFECT SQUARES)

1. Determine the square root for a through d.
   a. $\sqrt{729}$ b. $\sqrt{361}$ c. $\sqrt{18.49}$ d. $\sqrt{30.25}$

B. EXTRACTING SQUARE ROOT BY APPROXIMATION

1. Find the square root of (a) through (d) by the approximation method. Round off each answer to two decimal places.
   a. $\sqrt{7}$ b. $\sqrt{30}$ c. $\sqrt{28}$ d. $\sqrt{5.6}$

2. Determine the diameter (d) in mils of a #8 wire having a cross-sectional area of 16,510 circular mils ($C.M.$), when $d = \sqrt{C.M.}$

C. DETERMINING SQUARE ROOTS OF FRACTIONS (PERFECT SQUARES)

1. Determine the square roots of each fraction.

  a. $\sqrt{\dfrac{2209}{3364}}$ b. $\sqrt{\dfrac{50.41}{8.41}}$ c. $\sqrt{\dfrac{.1521}{.3959}}$ d. $\sqrt{\dfrac{.0625}{.0124}}$

D. EXTRACTING SQUARE ROOT OF FRACTIONS BY APPROXIMATION

1. The amperes (A) in a circuit $= \sqrt{\dfrac{\text{watts (W)}}{\text{ohms(O)}}}$.

$W =$	1	2	8	1
$O =$	3	5	11	7

Determine the amperes for the four circuits whose values appear in the table.

2. Find the square root of each fraction by approximation.

  a. $\sqrt{\dfrac{1.7}{3.2}}$ b. $\sqrt{\dfrac{.75}{.25}}$ c. $\sqrt{\dfrac{30.2}{37.3}}$ d. $\sqrt{\dfrac{2.15}{1.25}}$

E. DETERMINING THE SQUARE ROOT OF ALGEBRAIC NUMBERS

1. The largest side (hypotenuse) of a 90° triangle equals $\sqrt{a^2 + b^2}$. Find the hypotenuse for triangles A, B, and C.

Triangle	A	B	C
Side $a =$	$6\frac{1}{2}''$	$7\frac{1}{2}''$	$11.5''$
Side $b =$	$2''$	$3\frac{1}{2}''$	$4.5''$

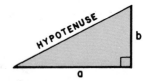

2. Find the value of each algebraic fraction.

  a. $\sqrt{\dfrac{A^2}{3}}$ b. $\sqrt{\dfrac{2b^2}{5}}$ c. $\sqrt{\dfrac{a^2b^2}{11}}$ d. $\sqrt{\dfrac{X^2Y^2}{7}}$

3. The vector diagram shows a voltage of 27 between L_3 and L_2 and 35 volts between L_1 and L_2. What is the voltage across L_1 and L_3? This voltage is equal to the square root of the sum of the squares of the other two sides of the triangle.

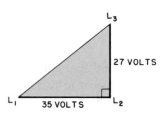

F. READING TABLES OF SQUARE ROOTS

a	3	e	1.1	i	7.01
b	5	f	2.1	j	1.02
c	7	g	3.9	k	3.03
d	8	h	5.5	l	1.38

1. Refer to a table of square root values. Find the square root of numbers (a) through (h) correct to two decimal places and (i) through (l) to three decimal places.

G. APPLICATION OF SQUARE ROOT TABLES, FACTORING, AND APPROXIMATION

1. Determine the square root of the following numbers that are outside the range of a table. Solve each number by factoring, approximating, or using tables, whichever method or combination is appropriate.
   a. $\sqrt{200}$ b. $\sqrt{156}$ c. $\sqrt{.32}$ d. $\sqrt{224.8}$
2. The diameter of a wire in mils is equal to the square root of the circular mils area. Find the diameter of each of three wires where the cross-sectional areas equal 5.232, 3.109, and 4.378 circular mils, respectively.
3. Find distance D, when $D = \sqrt{X^2} - \sqrt{Y^2}$.

H. CALCULATING RADICAL VALUES USING A SCIENTIFIC CALCULATOR

1. Calculate the length of each side of air conditioning ducts A and B and the radius of ducts C and D. Label each dimension and round off the decimal values to two places.

	Duct Form	Area (A)
A	Square	0.0625 □″
B		53.582 □ mm
C	Round	120.73 in.^{2}
D		2025.835 mm^2

$$L = \sqrt{A}$$

$$r = \sqrt{\frac{A}{3.1416}}$$

2. Find the root value of measurements A, B, C, and D.
   Note. Use a three-place decimal table to establish the square root value for quantities A and B and cube root for C and D. Label each measurement.

A	B	C	D
$\sqrt{5}$ in.^{2}	$\sqrt{27.5}$ m^2	$\sqrt[3]{99}$ ft^3	$\sqrt[3]{81}$ dm^3

3. Use a scientific calculator to calculate the root values for measurements a, b, and c.
   a. $\sqrt[6]{10\,587.224}$ Pa   c. $\sqrt[32]{161^{10.4}}$ g

   b. $\sqrt[8]{\dfrac{(97.656)^2}{(16)^2}}$ N

FORMULAS

Unit 71 The Concept and Use of Algebraic Formulas and Complex Equations

---

OBJECTIVES OF THE UNIT

After satisfactorily studying this unit, the student/trainee will be able to
- *Substitute known values in a formula and then solve for missing values.*
- *Change a formula to a more usable form by rearranging the variables.*
- *Analyze given data, rearrange a formula to accommodate changes, substitute data, and solve for unknowns.*
- *Translate word statements and measurements into a specific formula and then solve for required quantities.*

A. CONCEPT OF FORMULAS

A *formula* is a set equation. The equation is used to solve problems in which a relationship exists between two or more variables. A formula is an abbreviated way of expressing a combination of mathematical processes that, when solved, always give the same result. Formulas are widely used in the shop, laboratory, and clinic to obtain dimensions and other data. Formulas are applied in banking, business, marketing, agriculture, and in the home.

A formula also provides a technique of simplifying the solution of a problem. It gives very specific directions for computing unknown quantities. A formula is an *equality*. Solution of a formula requires the application of the principles covered thus far. These relate to the use of symbols, exponents, radicals, and the four basic processes of addition, subtraction, multiplication, and division of numerical and literal values. Formulas, therefore, express basic mathematical truths in a simplified form. Values may be substituted in formulas and comparatively simple processes followed to solve for a required dimension or quantity.

Formulas are found in farm journals, industrial publications, technical handbooks, and household publications. Formulas are used wherever it is practical to take a shortcut in solving a problem. If an orderly procedure is followed for writing the formula, substituting values in the formula, and computing for missing values, the solution of the problem is simplified and the chance for error is lessened. It should be noted that as a formula is written, the unknown quantity does not have to be written for each step. The unknown quantity is understood to be there before the equal sign.

B. FORMULA EVALUATION

Formula evaluation is the process of (1) substituting all known numerical and literal factors in a formula and (2) solving for a particular variable by performing the indicated operations.

Before data are substituted in a formula, they must be in the same unit of measure. If linear measurements are required, the quantities must be expressed in customary inch or SI metric units. Conversion tables and factors are used to change measurements to appropriate units.

RULE FOR APPLYING FORMULAS

- Analyze the problem by reading it carefully.
- Determine what quantities are given and what values are required.
- Review the formula for the meaning of each literal and numerical value.
- Write the formula to one side with sufficient room around it to work the problem.
  Note. Where no formula is available, it is practical to develop one for solving the whole problem or parts of the problem.
- Substitute the given quantities in the formula.
- Perform the mathematical operations as indicated.
- Simplify the answer.
- Label the answer with the correct unit.
- Check the final result by estimating a reasonable answer and reworking the problem in another form or order.

EXAMPLE: The acceleration (a) of an object is determined by the formula

$$a = \frac{V_f - V_i}{t}.$$

If the final velocity (V_f) of the object is 65 m/s, the initial velocity (V_i) is 30 m/s, and the time interval (t) is 8 seconds, compute the acceleration (a) in m/s.

Step 1 Write the formula. $a = \frac{V_f - V_i}{t}$

Step 2 Substitute the known data.

Step 3 Perform the mathematical processes. $= \frac{65 - 30}{8}$

Step 4 Label the answer in the correct unit. $= \frac{35}{8} = 4.375$

Step 5 Check. Substitute the value of (a) in the formula for equality. $a = 4.375$ **m/s**   Ans

C. REARRANGING FORMULA VARIABLES

The sequence and/or placement of the variables in a formula do not always appear in the most usable form. Known and unknown quantities *(variables)* may appear anywhere in a formula. It is therefore important to be able to rewrite a formula to simplify the solution for different variables. The process of changing the position and sequence of the variables is called *formula rearranging*.

RULE FOR REARRANGING FORMULA VARIABLES

* Determine the given quantities and the required quantities.
* Read the formula to determine the required unknown quantity.
* Determine the mathematical processes needed to isolate the unknown quantity.
* Perform the mathematical processes needed to rearrange the variables into an equivalent new formula (equation).
* Follow the rules for applying formulas and solve for the required variable.
* Check the answer and correctness of the unit of measure.

EXAMPLE: Rearrange the following volume formula to solve for (r).

$$V = \frac{\pi \cdot r^2 \cdot h}{3}$$

V = volume
r = radius
h = height

Step 1 Determine the quantity sought (r). $\qquad V = \frac{\pi \cdot r^2 \cdot h}{3}$

Step 2 Multiply both sides of the equation by 3. $\qquad (3)V = \frac{\pi \cdot r^2 \cdot h \cancel{(3)}}{\cancel{3}}$

Step 3 Divide both sides of the equation by (π) and (h). $\qquad \frac{(3)V}{(\pi)(h)} = \frac{\cancel{\pi} \cdot r^2 \cdot \cancel{h}}{\cancel{(\pi)}\cancel{(h)}}$

Step 4 Take the square root of r^2. $\qquad \sqrt{\frac{3V}{(\pi)(h)}} = \sqrt{r^2}$

Step 5 Rearrange the formula with the unknown on the left. $\qquad r = \sqrt{\frac{3V}{\pi h}}$ **Ans**

Review and Self-Test

ASSIGNMENT UNIT 71

A. Introduction to and Use of Formulas

1. Develop a formula from the information furnished in each rule (a through e).
   a. The diameter (D) of a circle is equal to twice the radius (R).
   b. The area of a square (A) is found by squaring a side (S).

c. The list price (L) is equal to the cost of an article (C) plus 25% markup.

d. The length of one side of a right triangle (A) is equal to the square root of the sum of squares of the other two sides, (B and C).

e. The cutting speed (CS) of a drill is the product of the circumference of the drill times the revolutions per minute.

2. Interpret formulas A through E by writing the rule which each one expresses.

	Formula	Designation of Letter Symbols		
A	$A = \dfrac{\pi D^2}{4}$	A	=	Area of circle
		D	=	Diameter
B	$V = .785\ D^2 H$	V	=	Volume of cylinder
		D	=	Diameter
		H	=	Height
C	$\dfrac{OD}{N + 2} = P$	P	=	Diametral pitch
		OD	=	Outside diameter
		N	=	Number of teeth
D	$CS = \dfrac{\pi D\ (RPM)}{12}$	CS	=	Cutting speed in feet per minute
		D	=	Diameter of cutter
		RPM	=	Revolutions per minute
E	$L = \sqrt{r^2 + \left(\frac{S}{2}\right)^2}$	L	=	Length of roof slope
		r	=	Rise
		S	=	Span

3. Rule three vertical columns (I, II, III) on a sheet of paper.

   a. Write five formulas from trade handbooks, technical journals, catalogs, or any other printed material in column I.

   b. Identify each letter or symbol by writing in column II what each stands for.

   c. Translate each formula into a rule in column III.

   Use an appropriate formula or rearrange the variables within a formula for parts B and C of this assignment. The problems may be solved by algebraic processes or in combination with a scientific calculator.

B. Rearranging Formula Variables

1. Determine the H_o value for an engine that operates at an efficiency (E) of 83.8%. The H_1 input is 26,000 calories.

   Note. Rearrange the formula $E = \dfrac{H_1 - H_o}{H_1}$ to solve for H_o.

2. Rearrange the given formula of $V = \frac{1}{2}(l) \cdot (w) \cdot (D + d)$ to solve for variable (d). Use values of $l = 6.24''$, $w = 5.86''$, $D = 2.05''$, and $V = 106.04\ \text{in}^3$.

3. Calculate the body diameter of the oval head rivet, correct to three decimal places. The given area (A) is 0.01227 in². The formula for $A = 0.7854 \cdot d^2$.

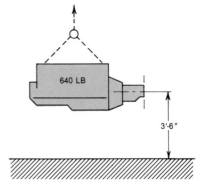

4. Set up a simple formula to determine the joules (J) of energy required to raise the 640-lb engine 3'–6" from the floor. Note that work is the product of the weight and distance and 1 J = 0.738 ft lb. Round the answer to two decimal places.

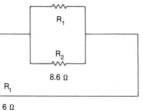

5. An electrician must find the resistance (R_1) in a parallel circuit that has a total resistance (R_t) of 6 ohms (Ω).
   a. Rearrange the formula:

$$R_t = \frac{R_1 + R_2}{(R_1 + R_2)} \text{ to solve for } R_1.$$

   b. Determine the value of R_1 when $R_2 = 8.6$ ohms (Ω).

C. Formulas with Applications to Practical Problems

1. Translate each of the following three mathematical statements into formulas.
   a. The root diameter (R) of a thread is equal to the outside diameter (o) minus twice the depth of the thread (d) and twice the clearance (c).
   b. The length of metal (L) required to make a right-angle bend is equal to the length of the two legs (L_1 and L_2) plus one-half the thickness of the metal (t).
   c. The approximate length (L) of a flat belt is equal to twice the distance between the centerlines (c) of the two pulleys (D_1 and D_2) plus 3.25 times the sum of the pulley diameters divided by two.
   Note. All dimensions used in the formula must be in the same unit of measurement.
2. Use the appropriate formula and solve each of the three following problems.
   a. Determine the length of metal required to produce right-angle parts A, B, and C.

Right-Angle Parts	Dimensions of Legs		Metal Thickness (t)
	L_1	L_2	
A	10.2 cm	15.2 cm	0.32 cm
B	13.98 cm	22.72 cm	0.24 cm
C	0.165 m	0.244 m	0.005 m

b. Find the root diameter of threaded parts A, B, and C.

Threaded Part	Outside Diameter (O)	Depth of Thread (D)	Clearance (C)
A	1.125"	.125"	.015"
B	3.3125"	.1875"	.022"
C	$4\frac{1}{16}''$	$\frac{5}{32}''$	$\frac{1}{64}''$

c. Calculate the approximate lengths of flat belt that are required to fill the three conditions given in the table. Round off answers to the nearest $\frac{1}{8}''$ and reduce to feet and inches.

Condition	Center-to-Center Distance (c)	Pulley Diameters	
		(D_1)	(D_2)
A	36"	10"	14"
B	4'-6"	9"	13"
C	8'-7"	15"	$18\frac{1}{2}''$

Unit 72 Complex Equations

OBJECTIVES OF THE UNIT

After satisfactorily studying the unit, the student/trainee will be able to
- *Eliminate one unknown from theoretical and applied complex equations comprising two unknowns by processes of addition and substitution.*
- *Solve complex equations and formulas that have been reduced to one unknown by using algebraic rules and basic mathematical processes.*
- *Set up and work through complex equations that are derived from word statements involving two unknown variables.*

Up to this point, equations and formulas have been solved that involve only one unknown numerical or literal quantity. However, technicians and craftspersons need to solve many occupational problems for two unknown variables. Such equations are called *complex equations* or a system of equations. In the system, there are two equations, each consisting of two variables. This system of equations has only one solution set in common. The solution set is often referred to as an ordered pair such as *X, Y*. Generally, complex equations are said to be solved simultaneously (with no reference to time) by developing a set of equations (system of equations), which is solved by a pair of values (ordered pair). Complex equations can be solved *simultaneously* by developing a set of equations that are solved by a pair of values.

Several techniques may be used to solve complex equations. In one instance, one unknown is eliminated. Once this is done, the equation is solved by using standard algebraic processes. Two other common methods are used in this text to solve simultaneous equations for two unknowns. The methods deal with *addition* and *substitution*.

A. SOLVING SIMULTANEOUS EQUATIONS BY ADDITION

This method of solving pairs of complex equations for two unknown variables also involves applications of basic algebraic processes of adding, subtracting, multiplying, and dividing quantities.

RULE FOR SOLVING COMPLEX EQUATIONS BY ADDITION

- Identify two equations with the same two variables in each equation.
- Make the numerical factor of one of the variables the negative of the same variable in the second equation. Apply the appropriate basic algebraic processes to both sides of one or both equations.
- Add the two equations. Eliminate one variable.
- Solve the remaining equation for the unknown.
- Take the value for one unknown and substitute it in either one of the two equations.
- Solve for the second unknown.
- Check the answers for the unknowns by substituting the computed value for each variable in the remaining original equation.

EXAMPLE: *Case 1.* Solve the following pair of complex equations by addition. $X - Y = 6; X + Y = 9$.

Step 1 Identify the two equations.

Step 2 Add the equations.

Note. In this case, the signs do not need to be changed.

$$\begin{array}{ll} \text{(a)} & X - Y = 6 \\ \text{(b)} & \underline{X + Y = 9} \\ & 2X \quad\;\; = 15 \\ & X \quad\;\;\; = 7.5 \;\; \text{Ans} \end{array}$$

Step 3 Substitute the value of the first unknown variable in the original equation.

Step 4 Solve for the second unknown variable.

$$\text{(a)} \; (7.5) - Y = 6$$

$$\begin{array}{l} -Y = 6 - (7.5) \\ -Y = -1.5 \\ Y = 1.5 \;\; \text{Ans} \end{array}$$

Step 5 Check the answers by substituting both unknown values in the remaining original equation.

$$\text{(b)} \; X + \quad Y = 9$$
$$(7.5) + (1.5) = 9$$
$$9 = 9 \;\; \text{Check}$$

E XAMPLE: *Case 2.* Solve the following pair of complex
equations by addition. $2X + 4Y = 6$; $4X + 3Y = 17$.

(a) $\qquad 2X + 4Y = 6$
(b) $\qquad 4X + 3Y = 17$

Step 1   Identify two equations.

(a) $-2(2X) + -2(4Y) = (6) \cdot (-2)$
(b) $\qquad -4X - 8Y = -12$

Step 2   Multiply both sides of the equation by a numerical factor that will eliminate one unknown.
Note. Multiplying both sides of an equation by the same number keeps the equality.

Step 3   Add the equations to eliminate one variable.

(a) $\qquad -4X - 8Y = -12$
(b) $\qquad \underline{\quad 4X + 3Y = 17\quad}$
$\qquad\qquad -5Y = 5$

Step 4   Solve the new equation.

$-Y = 1$
$Y = -1$ Ans

Step 5   Substitute the value of the one unknown $(Y = -1)$ in the original equation.

(a) $2X + 4(-1) = 6$
$2X - 4 = 6$
$2X = 10$
$X = 5$ Ans

Step 6   Check the answers by substituting both values in the remaining equation.

(b) $\quad 4X + 3Y = 17$
$(4) \cdot (5) + (3) \cdot (-1) = 17$
$(20) \quad + \quad (-3) = 17$
$17 = 17$ Check

B. SOLVING SIMULTANEOUS EQUATIONS BY SUBSTITUTION

An easier method than addition of finding an exact solution may be used if one equation of a pair of simultaneous equations has a numerical coefficient of one. This method requires one unknown to be eliminated by substitution.

RULE FOR SOLVING COMPLEX EQUATIONS BY SUBSTITUTION

- Identify two equations with the same two variables in each equation.
- Write the equation for the variable with coefficient on one side; the second variable on the other side.
- Solve for the numerical value of the second unknown.
- Substitute the numerical value of the second unknown in one of the equations. Solve for the numerical value of the first unknown.
- Check the solutions. Substitute the numerical value of the variables in one equation. The answer must be an equality for the correct solution.
- Label the answers with the correct units of measure.

EXAMPLE: Solve the following pair of complex equations and check the answers.

$Y - X = 6; Y = \frac{1}{2}X + 2.$

Step 1 Identify the two equations.	**(a)** $Y - X = 6$
	(b) $Y = \frac{1}{2}X + 2$
Step 2 Write the equation for the variable with the coefficient of (1) on one side.	**(a)** $Y = 6 + X$
Step 3 Substitute the value of the first unknown in the other equation.	**(b)** $Y = \frac{1}{2}X + 2$
	$(6 + X) = \frac{1}{2}X + 2$
Step 4 Solve for the second unknown.	**(b)** $6 + X = \frac{1}{2}X + 2$
	$\frac{1}{2}X = -4$
	$X = -8$ Ans
Step 5 Solve for Y by substituting the value for $X(-8)$.	**(a)** $Y - X = 6$
	$Y - (-8) = 6$
	$Y = 6 - 8$
	$Y = -2$ Ans
Step 6 Check the answers by substituting the values of $X(-8)$ and $Y(-2)$ in the second equation.	**(b)** $Y = \frac{1}{2}X + 2$
	$(-2) = \frac{1}{2}(-8) + 2$
	$-2 = -4 + 2$
	$-2 = -2$ Check

C. PREPARING SIMULTANEOUS EQUATIONS FROM WORD STATEMENTS

Formulas are represented by equations that express an equality of relationship. Although most occupational formulas are given in technical handbooks, there are times when the technician needs to translate problem statements into a pair of simultaneous equations. Then, once the simultaneous equations are established, the unknown quantities are determined by addition or substitution.

EXAMPLE: The federal and state taxes on a business operation total $3,150,000. The ratio of federal to state tax is $1:6 + \$350,000$. Determine the separate federal and state taxes.

Step 1 Set up a pair of simultaneous equations. Use $X =$ federal tax; $Y =$ state tax.	**(a)** $X + Y = 3\ 150\ 000$
	(b) $Y = 6X + 350\ 000$
Step 2 Substitute the value of one unknown (Y) in the first equation.	**(a)** $X + (6X + 350\ 000) = 3\ 150\ 000$
	$7X = 2\ 800\ 000$
Step 3 Solve for the unknown (X).	$X = \$400\ 000$ Ans
Step 4 Substitute the value for X in the second equation.	**(b)** $Y = 6(400\ 000) + 350\ 000$
	$= \$2\ 750\ 000$ Ans

Step 5   Check the values of X and Y.   **(b)**

$$Y = 6X + 350\ 000$$
$$2\ 750\ 000 = 6(400\ 000) + 350\ 000$$
$$2\ 750\ 000 = 2\ 750\ 000 \quad \text{Check}$$

Review and Self-Test

ASSIGNMENT UNIT 72

A. Solving Complex Equations by Addition and/or Substitution

1. Solve simultaneous equations a, b, c, and d by either addition or substitution. Check each answer. Label each method.

   a. $X + 4Y = 21$
       $3X - Y = 11$

   b. $4A - B = 12$
       $3A + B = 66$

   c. $6z - 5q = -1$
       $10z + 35q = 33$

   d. $5a + 2b + 8 = 0$
       $4a = 3b + 2$

B. Changing Word Statements to Equations and Solving for Unknown Values

1. It requires a 2.50″ length of stock to stamp out a part and provide waste allowance between stampings. The waste allowance equals one-fourth the length of the stamped part. Set up the necessary equations to determine the length of the part and the waste allowance. Check each answer.

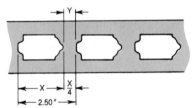

2. Refer to the part drawing for the relationship of dimensions. Set up the necessary equations to solve for center-to-center distances X and Y. Check answers.

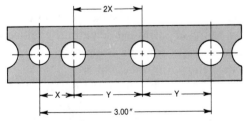

3. The combined capacity of two cylinders is 32 cubic feet. The capacity of cylinder X is three times that of cylinder Y. Set up the necessary equations and determine the capacity of each cylinder. Label the answers. Check for accuracy.

4. A tile mason paid $792.00 for 840 ceramic tiles of two different sizes. The cost per tile was either $1.10 or $.72, depending on size.
   a. Set up two simultaneous equations.
   b. Determine the number of tiles of each size. *Note.* Round off the number of tiles to the next whole number.
   c. Check the answers, making allowance for any variation due to whole number rounding.

Unit 73 Applications of Special Formulas and Handbook Data

OBJECTIVES OF THE UNIT

After satisfactorily completing the unit, the student/trainee will be able to

• *Understand the broad range of formulas that are used to express quantity relationships between a number of variables in science, industry, business, and consumer mathematics.*

• *Apply letters, symbols, and constant values that designate elements, features, and quantities commonly contained in formulas.*

• *Take a scientific or industrial formula, substitute known values, and solve for a missing dimension or factor.*

• *Apply the required units of measurement in a formula and express the answer in the appropriate measurement unit.*

A. HANDBOOK TABLES AND CONVERSION FACTORS

Examination of a trade or technical handbook reveals that engineering, scientific, and other occupational standards are based on data that are compared in a special fixed relationship. The relationship of each quantity of an element, of a feature, or of conditions within a system is expressed as a formula.

The relationships may deal with mechanical, electrical, electronic, aerospace, or other design features or with such common elements as liquids, gases, and solids. Formulas are based on truths about tested scientific, engineering, business, and production experiences. In simplified form, a formula is used to find unknown values from given dimensions, quantities, and other known information.

Some of the simplest formulas incorporate the values used in conversion between customary and British units of measure and SI metric units. In review, for instance, to convert 1 Btu to joules of energy, multiply by $1.054\ 3 \times 10^3$. A force given in lb/in.^{2} is changed to its metric equivalent of kg/m^2 by multiplying by $7.030\ 7 \times 10^{-4}$. In other cases, a known value is used in an equality such as 1 Btu = 1 J $(1.054\ 3 \times 10^3)$.

In physical science one group of tables relates to conversion factors. Values are converted into units of measurement that may be used to deal with linear, surface, flow, and solid measurements. Other measurements relate to acceleration, energy, force, heat, power, mechanics, and every other quantity as related to a material, properties, or physical phenomenon.

B. FORMULAS RELATED TO PHYSICAL PHENOMENA

Certain laws in mathematics, physics, chemistry, biology, and other scientific fields are expressed in simplified form as formulas. In solving problems dealing with the basic properties of gases, there are fixed relationships. For example, volume and pressure are affected by *temperature*, by *density*, and by the *specific weight* of a gas.

In mechanics, there is an equality between an *effort force* and the *distance through which it moves* and the *resistance force* and its *movement*. This statement appears as the simple formula

$$E \times ED = R \times RD.$$

The symbols or letters used in formulas tend to be standardized and are representative of specific quantities. In the example, *E* is *effort force; R, resistance force*. The *D* indicates *distance*, like *ED* for *effort distance*. A code or log usually appears with a formula to clarify the designation of each symbol.

In uniform acceleration or deceleration problems, known values are inserted in the appropriate formula.

$V_g = (A) \cdot (T)$ $V_l = (D) \cdot (T)$	$V_g = $ Velocity gained $V_l = $ Velocity lost $D = $ Deceleration $A = $ Acceleration $T = $ Time

C. APPROPRIATE USE OF MEASUREMENT UNITS

Importantly, every value computed by using a formula is identified with a particular unit of measure. The pitch of a customary-inch standard screw thread equals one (1) divided by the number of threads per inch.

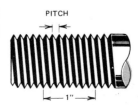

PITCH

$$P = \frac{1}{\textbf{Number Threads per Inch}}$$

FIGURE 73-1 (Reprinted from *Fundamentals of Applied Physics*, 3rd ed., Olivo and Olivo, figure 14-1, © 1984 by Delmar Publishers Inc.)

The resulting answer is expressed as a definite linear measurement. In this example there are four threads per inch. The pitch $(P) = \frac{1}{4} = 0.250''$.

It must be understood that when using a formula more than one unit of measure may be involved. For instance, in pressure and volume relationships in fluid power problems, pressure may be expressed in pounds (lb) or in newtons (N); volume (V) in cubic feet (ft^3) or cubic meters (m^3). According to *Boyle's law* (formula), $P_1 \times V_1 = P_2 \times V_2$. If $P = 44.5$ N, $V_1 = 0.5$ m, $P_2 = 89$ N, then V_2 may be computed by rearranging the formula (without changing any values) so

$$V_2 = \frac{P_1 \times V_1}{P_2}$$

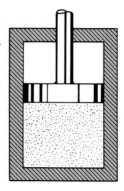

FIGURE 73-2

D. RANGE OF TECHNICAL INFORMATION

Handbooks and trade and technical journals provide latest standards information that is derived by agreement among engineering, scientific, and other responsible organizations and individuals. Standards for the composition, sizes, forms, fabrication, and other important data are organized according to needs and demands within major occupational groupings. There are relevant tables for machine and metal products manufacturing occupations; welding, structural, civil, transportation, and other industries; industrial and communications electronics industries, and cutting through all other industries. Similarly, technical data are available with specialized measurement requirements in agriculture, business, health, and related industries.

Tables may relate to basic raw materials, compositions, properties, and treatments. The whole ANSI, SAE, aluminum industry, and similar tables provide important design information. Characteristics of machine elements and fasteners like screws, bolts, bearings, and drives are covered by tables and formulas. Machining processes, such as thread cutting, gear cutting, turning, milling, and grinding all require computations for required dimensions. So, too, are cutting speeds, feeds, and surface finishes often computed.

Two items are self-evident from this brief overview on tables and formulas. First, handbooks provide a reliable source of latest information and formulas relating to standards within each occupation. Second, the exacting nature of formulas requires the continual use of trade and technical handbooks.

E. FORMULAS IN MACHINE, TOOL, AND PRODUCT DESIGN

Designers are familiar with appropriate formulas that relate properties of materials to the performance of particular machining operations. For example, in the design of single or multiple-point cutting tools, attention is directed to such factors as *coefficient of friction* (μ) to measure the resistance between a sliding chip and the face of the cutting tool over which it travels. Other design considerations include *friction force, main shear strength, work done in shear,* and other fractors. From this mass of important mechanical quantities, design handbooks provide such formulas as

$$F_c = (W_n) \cdot (A_o)$$

where F_c represents a cutting force (lb); A_o, a cross-sectional area of a chip (in.); and F_c the cutting force component acting in the direction of the tool head (lb).

Strength is built into tools by considering *force (F), overhang (L), modulus of elasticity* of the material *(E)*, the *moment of inertia (I)* and the *tool deflection (D)*. The rigidity of the tool may then be determined by the formula:

$$D = \frac{F \times L^3}{3E \times I}$$

RULE FOR USING FORMULAS

- Study the given and required quantities in a problem.
- Determine the appropriate formula(s) that will produce the unknown quantity.
- Substitute the known data for the value of each symbol in the formula.
  Note. Pay particular attention to the unit of measurement associated with each quantity.
- Carry out the basic algebraic processes as identified in the formula to the specified degree of accuracy.
- Label the answer according to the required unit of measure. Check the answer mathematically.
  Note. During production, other regular precision measurement and quality control checks are made. These provide additional checks against the mathematical computations.

F. APPLICATION OF HANDBOOK DATA

The number and variety of formulas are limitless, as their use cuts across many activities in life like science, industry, and the home. Most simple formulas that are used daily may be developed and applied with the knowledge obtained through the mastery of all the units on applied algebra. Essentially, the parts of a formula are similar, as each formula represents an equality between values that are given and those that are required.

The spur gear is a common example of how handbook data are supplied as a series of formulas.

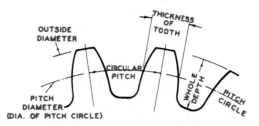

FIGURE 73-3

These may be used in combination to compute all the required dimensions for machining the object. Some of the essential parts are illustrated and identified. A partial table of the kind of information furnished in trade handbooks is given in figure 73-4. Note the designation of letter symbols in the formulas.

	Formulas for Spur Gear Teeth			
Required Dimension	Outside Diameter (*O*)	Pitch Diameter (*D*)	Whole Depth of Tooth (*W*)	Thickness of Tooth (*T*)
Formula	$O = \dfrac{N + 2}{P}$	$D = \dfrac{N}{P}$	$W = \dfrac{2.157}{P}$	$T = \dfrac{1.5708}{P}$

N = Number of Teeth in Gear
P = Diametral Pitch

FIGURE 73-4

In this particular table, four major dimensions may be computed. Simply substitute given values for letters in the formulas and perform the mathematical operations indicated in each case. In like manner, other formulas may be used to solve for missing dimensions.

Review and Self-Test

ASSIGNMENT UNIT 73

A. Applications of Formulas in Science

1. State two major functions that are served by formulas.
2. Translate formulas a through e into word statements.

	Area of Application	Formula	Values Represented
a	Materials testing	$e = \dfrac{\Delta L}{L}$	e = Strain ("/" or m/m) ΔL = Total deformation (" ") L = Original length (" ")
b	Temperature	$T_c = \frac{5}{9}(T_f - 32)$	T_c = °T Celsius T_f = °T Fahrenheit
c	Heat	$\Delta L = \propto \times L \times \Delta T$	ΔL = Change in length $\propto$ = Coefficient linear expansion L = Original length ΔT = Change in temperature

	Area of Application	Formula	Values Represented
d	Fluid flow	$R_f = V \times A$	R_f = Rate of flow (ft³/min, CFM) V = Fluid velocity (ft/min) A = Area (ft²)
		Note. Convert CFM to GPM by multiplying by 7.48	
e	Electric current	$Z = \sqrt{R^2 + X_1^2}$	Z = Impedance R = Resistance X_1 = Inductive reactance

Use the formulas given in problem A.2 to solve problems A.3 to A.7 in science.

3. Determine the strain on a machine member, correct to six decimal places. Given: $\triangle L = 0.000\ 25''$, $L = 3.500''$.
4. Calculate the equivalent temperature readings for processes (a) and (b).
   a. Annealing temperature of 1830°F in Celsius degrees.
   b. Red heat working temperature of 1300°C in Fahrenheit degrees.
5. Determine the change in overall length of a 4,000-foot steel cable during temperature changes from −50°F to 112°F. The thermal coefficient of linear expansion is 0.000 006 5″/°F.
6. Calculate the rate of flow to the nearest CFM and GPM through a 12″ diameter pipeline. The velocity of flow is 200 ft/min.
7. Find the impedance (ohms) of a current within a coil that has an inductive reactance of 4.0 ohms and a resistance of 5.5 ohms.

B. Applications of Industrial Formulas and Handbook Data

Formulas for establishing certain gear tooth dimensions are reproduced in abstracted form from a complete table.

Required Dimension	Formula	Identification of Symbol
Addendum	$a = \dfrac{1}{P}$	a = Addendum
Dedendum	$d = \dfrac{1.157}{P}$	P_c = Circular pitch d = Dedendum
Circular pitch	$P_c = \dfrac{3.1416}{P}$	D = Pitch diameter
Whole depth	$WD = \dfrac{2.157}{P}$	WD = Whole depth of tooth
Working depth	$W_d = .6366\ P_c$	W_d = Working depth of tooth P = Diametral pitch

1. Use the appropriate formula to solve for each missing dimension as indicated for spur gears A, B, and C, rounded to three decimal places.

Gear	Number of Teeth	Diametral Pitch (P)*	Addendum (a)	Dedendum (d)	Whole Depth (WD)	Working Depth (W_d)	Circular Pitch (P_c)
A	36	12					
B	24		0.167	0.193			
C	48				0.120		0.175

*Rounded to whole number value

2. A 96-tooth webbed spur gear of 10 diametral pitch (P) to fit a $1\frac{1}{4}''$ diameter shaft (B) is needed. The handbook data relating to the six dimensions required for machining the gear are given in table form. Use the formulas in the table to compute the dimensions. Label each dimension with the appropriate unit of measure.

Required Dimensions	Solution
Outside diameter	$O = \dfrac{N + 2}{P}$
Width of face	$F = \dfrac{3\pi}{P}$
Length of hub	$L = 1\frac{1}{4}F$
Diameter of hub	$H = 1\frac{3}{4}B$
Thickness of web	$G = \dfrac{F}{3}$
Thickness of rim	$R = \dfrac{\pi}{2P}$

N = Number of Teeth in Gear
P = Diametral Pitch B = Bore

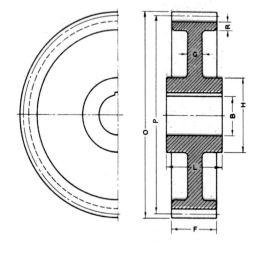

Unit 74 Achievement Review on Formulas and Handbook Data

OBJECTIVES OF THE UNIT

This achievement review serves as an overall test for Section 15. The unit is designed to measure the student's/trainee's ability to

- *Write a formula from word statements of existing equality conditions between known and unknown factors.*
- *Convert values obtained by simple formulas to equivalent units of measure in other measurement systems.*
- *Solve complex equations for unknown quantities.*
- *Interpret the meaning of each term or symbol in a formula and substitute values.*
- *Solve practical problems requiring the application of complex equations and formulas in science, business, and industry.*

A. CONCEPT OF FORMULAS

1. State briefly two characteristics of formulas.
2. Give three main reasons why formulas are used.
3. Interpret what is meant by each formula.
   a. $C = \pi D$ (Circle)
   b. $A = S^2$ (Square)
   c. $P = 2(L + H)$ (Rectangle)
   d. $A = \pi r^2$ (Circle)
4. State each rule as a formula.
   a. To convert a Celsius reading to a Fahrenheit (F) reading, multiply the Celsius (C) reading by 1.8 and add 32.
   b. The area of a sector (A) is equal to the sector angle (n) over 360° times the area of the circle.
   c. The cutting speed of a drill is the product of the circumference of the drill (πD) times the revolutions per minute (RPM) divided by 12.
   d. The mechanical efficiency (E) is equal to the output (O) divided by the input (I).
   e. The SAE horsepower rating formula states that the horsepower (hp) is equal to the cylinder bore (D) squared times the number of cylinders, divided by 2.5.

B. GENERAL PROBLEMS INVOLVING COMPLEX EQUATIONS

Solve the following complex equations. Check each solution.

1. $X + Y = 5$
   $2X - Y = 1$
2. $3X + Y = -6$
   $2X - Y = 1$
3. $3m + 5n = 21$
   $10n = 38 - 4m$
4. $-9X + 8Y = -3$
   $5Y = 12X + 4$

C. APPLICATIONS OF FORMULAS IN SCIENCE AND SPECIAL PROBLEMS

1. The mechanical advantage of force (MA_f) of the hydraulic system as illustrated equals the ratio of the force exerted on the large piston (F_2) by the force applied by the small piston (F_1).

   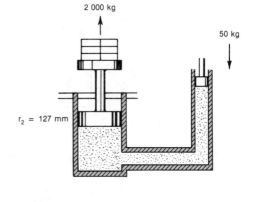

   a. Express this relationship as a formula.
   b. Use the formula to compute the MA_f of the system.
   c. Determine the force that needs to be applied on the small piston to raise a load of 2,722 kg on the large piston.
   d. Compute the radius of the small piston using the formula:
   $$MA_f = \frac{(r_2)^2}{(r_1)^2}$$

2. What length of copper wire 2.743 mm in diameter gives a resistance (R) of 4.2 ohms?

$$R = \frac{KL}{D^2}$$

   R = Resistance
   K = 10.8 (for copper)
   D = Diameter in mils = Diameter (mm) $\times$ 0.03937

3. Determine the SAE horsepower rating of an 8-cylinder (N) automobile engine with a bore (D) of 3.375 inches.

$$\text{hp} = \frac{D^2 N}{2.5}$$

4. Find the value of the missing letter in each formula. Use $\pi = 3.1416$ and label each answer, where possible.

A	$C = \frac{5}{9}(F - 32)$	$F = 77°$
B	$E = \frac{O}{I}$	$O = 22{,}000$ Btu, $I = 32{,}000$ Btu
C	$A = .7854 D^2$	$D = 75$ cm
D	$I = \frac{E}{R}$	$E = 12$ volts, $R = 4.5$ ohms
E	$S = \frac{n}{2}(a + L)$	$n = 72, a = 9, L = 165$
F	$A = ab + \frac{d}{2}(a + c)$	$a = 6.4, b = 5.2, c = 12, d = 8.6$
G	$V = \frac{\pi H}{3}(3A^2 + 3B^2 + H^2)$	$A = 9.4, B = 7.6, H = 8.4$
H	$A = \sqrt{B^2 + C^2}$	$B = 27.2$ m $\quad C = 24.6$ m
I	$S = \sqrt{\frac{B}{6}}$	$B = 612$
J	$Z = 5 + \sqrt{\frac{25 - (4 \times Y)}{2X}}$	$X = 1.5, \quad Y = 3.4$

D. FORMULAS IN INDUSTRY (PRACTICAL PROBLEMS)

1. Determine the pitch (P) of threads A through E, the single depth (d) for screw threads F through J, and the major diameter (D) of numbered screws K through O. Round off answers to three decimal places.

$P = \frac{1}{N}$		$d = \frac{.6495}{N}$		$D = W(.013) + .060$	
	Threads per Inch (N)		Threads per Inch (N)		Wire Size (W)
A	10	F	10	K	2
B	20	G	20	L	3
C	8	H	8	M	4
D	16	I	16	N	6
E	24	J	24	O	8

2. The revolutions per minute that the work, or a drill, or a cutter turns, is equal to the cutting speed in feet per minute divided by πD. Find the RPM for each operation indicated in table form. Use $\pi = 22/7$.

$$\text{RPM} = \frac{CS \text{ in ft} \times 12}{\pi D}$$

	Cutting Speed Ft/Min		Given Diameter	Required
A	100		$\frac{7''}{8}$	RPM of Lathe Spindle
B	80	Work Diameter	$2\frac{1''}{8}$	
C	60		$3\frac{1''}{2}$	
D	55		$\frac{1''}{2}$	RPM of Drill Press Spindle
E	42	Drill Diameter	$1''$	
F	80		$\frac{7''}{8}$	
G	90		$2\frac{3''}{4}$	RPM of Milling Machine Spindle
H	60	Cutter Diameter	$5\frac{1''}{2}$	
I	40		$5''$	

3. Use the necessary formula to find the dimensions that are not supplied in the table for spur gears A, B, and C. Decimal dimensions must be correct to four places.

$$\text{Whole Tooth Depth} = \frac{2.157}{\text{Diametral Pitch}}$$

$$\frac{\text{Pitch}}{\text{Diameter}} = \frac{\text{No. of Teeth}}{\text{Diametral Pitch}}$$

Gear	Number of Teeth	Pitch Diameter	Diametral Pitch	Whole Tooth Depth
A	48	6"		
B	32		6	
C		7.6666"		.1798"

PART FIVE

Fundamentals of Applied Geometry (with Calculator Applications)

16

GEOMETRIC LINES AND SHAPES

Unit 75 The Concepts of Lines, Angles, and Circles

OBJECTIVES OF THE UNIT

After satisfactorily completing the unit, the student/trainee will be able to
- *Apply theorems and corollaries in geometry that relate to lines, angles, and circles.*
- *Interpret geometric statements.*
- *Identify and solve problems involving adjacent, alternate interior, corresponding, vertical, and other angles formed with transversal and parallel and nonparallel lines.*
- *Describe different lines and angles in relation to circles and tangents and solve circle problems.*
- *Solve geometric problems requiring the addition, subtraction, multiplication, and division of line measurements, angle measurements, and measurements of circular forms.*

Geometry is one of the oldest branches of mathematics. Applications were made of geometric constructions centuries before the mathematical principles on which the constructions were based were recorded.

Geometry is a mathematical study of points, lines, planes, closed flat shapes, and solids. Using any one of these alone, or in combination with others, it is possible to describe, design, and construct every visible object.

The purpose of this section is to provide a foundation of geometric principles and constructions on which many practical problems depend for solution.

Plane geometry deals with objects that have two dimensions where all features lie within a *plane*. *Solid geometry* relates to objects that are formed and measured by three or more dimensions that lie in two or more planes.

A. GEOMETRIC THEOREMS AND COROLLARIES

Theorems and *corollaries* in geometry are mathematical truths that can be proved. Statements based on theorems are called *corollaries*. Many of these truths deal with such properties as quantities and constructions. Under a given set of conditions, the result remains *constant* (true). Theorems are used in problem solving and product development and manufacture.

General Theorems Relating to Quantities

Relationship of a Whole and Its Parts

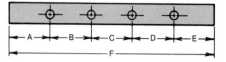

- The whole is equal to the sum of all of its parts.

$$F = A + B + C + D + E$$

- The whole is greater than any one or combination of its parts.

$$F > A, B, C, D, \text{ or } E \text{ or } AB, ABC, \text{ or } ABCD$$

Equality of Quantities

- Quantities that are equal to the same quantity or to an equal quantity are equal to each other.

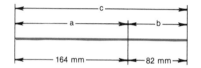 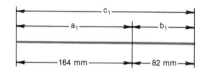

$$a + b = 266 \text{ mm}$$
$$a_1 + b_1 = 266 \text{ mm}$$
$$\therefore c = c_1$$

- When equals are added to equals, the sums are equal.

  Example. $d + e = d_1 + e_1 \therefore f = f_1$

- *When equals are subtracted from equals, the remainders are equal.*

  Example. $f = f_1, d = d_1, f - d = f_1 - d_1 \therefore e = e_1$

- When equals are multiplied by equals, the products are equal.

  Example. $a \times b = a_1 \times b_1. \ a = a_1 \therefore c = c_1$

- When equals are divided by equals, the quotients are equal.

  Example. $3a = 3b \therefore a = b$

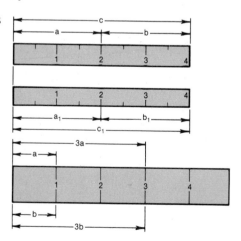

B. PROPERTIES OF LINES

Lines are basic to all geometric shapes. There are two kinds of lines, the straight line and the curved line. A line is said to be of indefinite length when there are no fixed points to indicate a specific size.

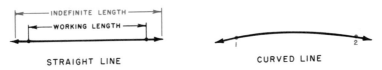

STRAIGHT LINE   CURVED LINE

Points are usually used to mark the *end points* of a line to show that it is a working line from which construction is started. The points are marked with letters or numbers to identify the working length of the line. This working length is sometimes referred to as a *line segment*.

The points lettered *A* and *B* on the straight line and numbered 1 and 2 on the curved line denote the end points. The line segment for the straight line is *AB*.

Direction and Types of Lines

There are three general directions to lines.

- *Horizontal,* in which a level line goes from left to right or right to left.
- *Vertical,* in which the straight line is vertical or at 90° to a horizontal line or surface.
- *Slanted,* in which the line is inclined.

Parallel and Tangent Lines

Lines may be parallel to one another, may touch at one or more points, or may intersect. In the first instance, when two lines are parallel, it means they are always the same distance apart.

Straight lines *AB* and *CD* are parallel because they are equidistant no matter how far they are extended. Measurements taken at lines 1–2 and 3–4, or at any other points, are equal.

When lines touch at one point, they are said to be *tangent* to one another. Two lines that cross each other are said to *intersect* or cut one another at a given point.

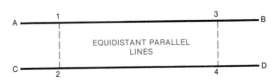

LINE TANGENT TO ARC   TANGENT ARCS   INTERSECTING LINES

C. TYPES OF ANGLES

As two straight lines come together (intersect) at a point, an angle is formed. In general, there are five common types of angles: straight, right, obtuse, acute, and reflex.

| STRAIGHT ANGLE | RIGHT ANGLE | OBTUSE ANGLE | ACUTE ANGLE | REFLEX ANGLE |

In each illustration, three letters are used to describe each angle. One letter (O in each case) is placed at the point where the two lines intersect. This point is called the *vertex*.

The first illustration shows a *straight* angle of 180°. This angle may be identified as angle *AOB*. The symbol ∠ is sometimes used in place of the word angle. ∠*AOB* has the same meaning as angle *AOB*.

A *right* angle has two sides, *CO* and *OD*, that form an angle of 90°. Occasionally, a 90° angle is denoted on a drawing by a square symbol (└) placed in the corner of the 90° angle. The right angle in the illustration is └*COD*.

When an angle formed by two intersecting lines is greater than 90° and smaller than 180°, the angle is known as an *obtuse* angle. Angle *EOF* is an obtuse angle. An *acute* angle is less than 90°, as shown by ∠*GOH*; ∠*IOJ* represents a *reflex* angle between 180° and 360°.

D. MATHEMATICAL OPERATIONS INVOLVING EQUAL ANGLE MEASUREMENTS

The importance of precise angular measurements and principles of reading vernier bevel protractors were covered in earlier units. The addition, subtraction, multiplication, and division of angle measurements often require the exchanging of equivalent values. This also happens when an answer is expressed in lowest terms. For instance, an answer like 126° 106′ 72″ is reduced to 127° 47′ 12″. This value is obtained by using the equivalent value of 60′ for 1° and 60″ for 1′, exchanging these values, and *carrying over* (adding) equivalent values. In this case, since 72″ is equivalent to 1′ 12″, the 72″ is reduced to 12″ and the 1′ is carried over. The same process continues with the 106′ (+ 1′ carried over). The 107′ value = 1° 47′. The 1° is carried over to the 126°, making it 127°. Thus, by exchanging and carrying over equivalent values, the 126° 106′ 72″ is reduced to lower terms (127° 47′ 12″).

Addition, Subtraction, Multiplication, and Division of Angles

When subtracting angles where the value of the subtrahend is greater than the minuend, values need to be exchanged. In the angles as illustrated, ∠B is found by subtracting ∠A from 89° 12′ 48″.

$$
\begin{array}{r}
89°\ 12'\ 48'' \\
-\ 45\ \ 18\ \ 56 \\
\hline
\end{array}
$$

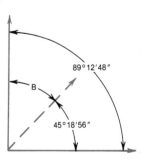

Since the 56″ cannot be subtracted from 48″ nor the 18′ from 12′, equivalent units need to be exchanged.

$$89° \; 12' \; 48'' = \quad 88° + (60 + 11) + (60 + 48)$$

$$
\begin{array}{llll}
= & 88° & 71' & 108'' \\
 & -45 & 18 & 56 \\
\hline
 & 43° & 53'' & 52'' & \textbf{Ans}
\end{array}
$$

In the multiplication of angles, any answer in minutes that exceeds 60′ is reduced to equivalent degree and minute value. Similarly, any second (″) value that is greater than 60″ is reduced to the equivalent ′ and ° value.

In the division of angles, it is often necessary to add a remainder by changing an equivalent value. For example, if $165^{1'}$ is to be divided by 4, the answer is 41′ and a remainder 1′ divided by 4. The remainder of 1′ is converted to its equivalent 60″. By further dividing, 60 ÷ 4 = 15″. Combining values, 165′ ÷ 4 = 41′ 15″ **Ans**

Angles Formed by an Intersecting Line

A line that intersects two or more lines is called a *transversal*. The angles formed by such an intersection are either *alternate interior, corresponding,* or *supplemental*. In the illustration, lines *AB* and *CD* are cut by transversal line *EF*. Angles 1 and $\angle 2$ and $\angle 3$ and $\angle 4$ are *alternate interior angles*. These pairs of interior angles are on alternate (opposite) sides of the transversal; angles 5 and $\angle 6$ and $\angle 7$ and $\angle 8$ are *alternate exterior angles*.

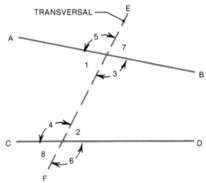

Corresponding angles consist of one inside and one outside angle on the same side of the transversal. Angles 4 and $\angle 5$, $\angle 1$ and $\angle 8$, $\angle 2$ and $\angle 7$, and $\angle 3$ and $\angle 6$ are corresponding angles.

Supplemental angles are interior angles on the same side of the transversal. Angles 1 and $\angle 4$ and $\angle 2$ and $\angle 3$ are supplemental angles.

Theorems on Angles Formed by Cutting Parallel Lines with a Transversal

- The alternate interior angles and alternate exterior angles formed by cutting parallel lines with a transversal are equal.
- The sum of the interior angles on the same side of a transversal that cuts through two parallel lines is equal to 180°.

E. CIRCLES AND PARTS OF CIRCLES

While many objects are represented by straight lines, others require the use of curved lines. Such parts of the circle as the center, diameter, radius, and chord are used.

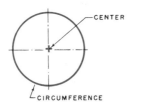 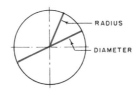 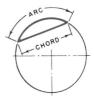

PARTS OF A CIRCLE

Each of these parts may be redefined in simple geometric terms. A *circle* is a closed curve on which all points are the same distance from a point (center) inside the circle. A straight line extending from the center to the closed curve line is the *radius*.

The straight line that passes through the center of the circle and ends at opposite sides of the closed curved line is the *diameter*.

An *arc* is a portion of the curved line or circle. A *chord* refers to the straight line segment between the end points of an arc. These are the terms relating to the circle that are most generally used in solving practical problems.

Angles Formed within Circles

A *central angle* is the included angle formed between two radii. The angle extends from the vertex at the center of the circle to the periphery (circumference). An *inscribed angle* is formed by two chords that extend from a common vertex and terminate on the circle.

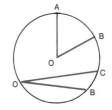

∠*AOB*   is a central angle.
∠*COB*   is an inscribed angle.

Angles in SI metrics are related to one of two supplementary units of angular measure: *radian* (r) or *steradian* (sr). The *radian* is a *plane angle* that represents an angle formed between a vertex at the center of a circle and an arc that is equal in length to the radius. Mathematically, one radian equals 180° divided by π, or 57.2956°. For most computations, the value of 1 r = 59° 30′ or 57.5°.

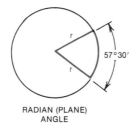

RADIAN (PLANE) ANGLE

STERADIAN (SOLID) ANGLE

A *solid angle* is measured in a *steradian unit* of measure. A steradian represents an enclosed area that is equal to a square whose sides are equal in length to the radius of a sphere (solid).

Theorems Related to Circles and Angles Formed

- The circumference of a circle is equal to π*D* (diameter) or 2π*r* (radius).
- The length of an arc is equal to the number of degrees in the arc divided by 360 and multiplied by the circumference (π · diameter or 2π*r*).

- Equal chords cut off *(subtend)* equal arcs on the same circle or equal diameter circles.
- The *perpendicular bisector of a chord* cuts through the center of the circle.
- A line that is perpendicular to a radius and touches the periphery of the circle is *tangent to the circle*.
  Line *A-B* is tangent to the circle at point *X*.

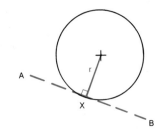

- A line that connects the centers of two tangent circles, passes through the tangent point, and is perpendicular to a tangent line.
  Center line *A-B* passes through tangent line *C-D* at the point of tangency of the circles.

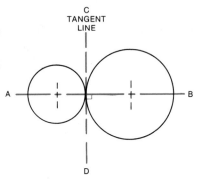

Review and Self-Test

ASSIGNMENT UNIT 75

A. The Concept of Lines

1. Lay out the lines to the lengths and in the directions indicated in the table. Label the end points of each line.

Line	Direction and Length		
	Horizontal	**Vertical**	**Inclined 45°**
AB	3″		
AC		4″	
AD		$3\frac{1}{2}″$	
XY			152 mm
XZ			130 mm

2. Identify each pair or combination of lines by name. Then indicate the characteristics of each.

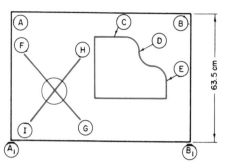

3. Calculate the peak voltages of (*a*) and (*a₁*), when $a = a_1$.

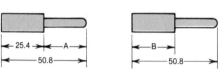

4. Determine the lengths *A* and *B*, when $A = B$.

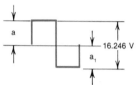

5. Find the voltage (*A*) in the circuit when both peaks are equal. The voltage $= 0.707 \times$ the peak voltage.

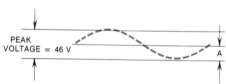

B. Addition, Subtraction, Multiplication, and Division of Angles (General Applications)

1. Add each combination of angle values.

   a.   27° 18'
       +42° 28'

   b.   72° 47'
       +41° 59'

   c.   37° 29' 39"
       +104° 24' 57"

   d.   185° 36' 58"
       +104° 59' 54"

2. Compute angles 1, 2, and 3 as illustrated at (a), (b), and (c).

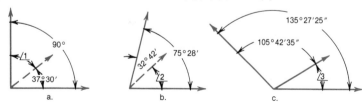

3. Compute included angles A, B, and C. Express values within 60′ and 60″, where applicable.

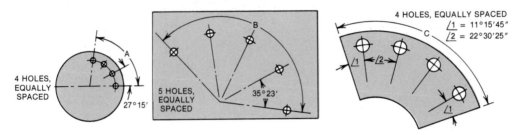

4. Determine the included angle from the data given for problems (a) through (e).
   a. 76° 24′ 16″ divided by 4.
   b. 315° 28′ 3″ divided by 9.
   c.

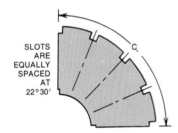

   d.

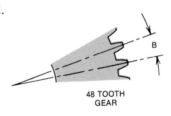

   e.

C. Properties of Angles

1. Name the type of angle used in the cases shown.

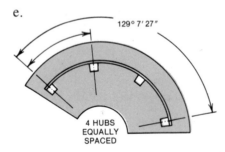

2. Draw the angles with sides and included angles equal to the sizes indicated in the table. Use a flat protractor to measure the angle.
   a. Label each angle for identification.
   b. Name the type of angle in each case.

Angle	Length of Sides				Included $\angle$
AOB	AO	2″	OB	3″	180°
AOC	AO	2″	OC	2″	90°
AOD	AO	$3\frac{1}{2}″$	OD	3″	60°
XOY	XO	4″	OY	$3\frac{1}{4}″$	$28\frac{1}{2}°$
YOZ	YO	2″	OZ	$1\frac{7}{8}″$	127°

3. Give the number of degrees in the layout angles *A* through *E*.

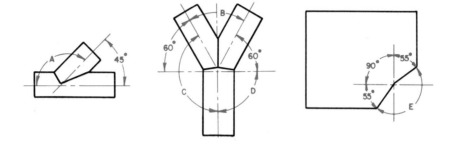

D. Properties of Circles and Parts of a Circle

1. Name and determine the value of each missing dimension.

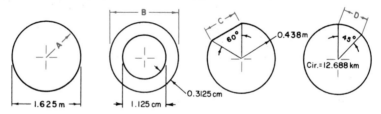

2. Compute the depths for parts A and B (correct to three decimal places) and the *X* or *Y* dimension for parts C and D (correct to two decimal places).

Part	Dimension	
	X	*Y*
C	25.4 cm	
D		27.68 cm

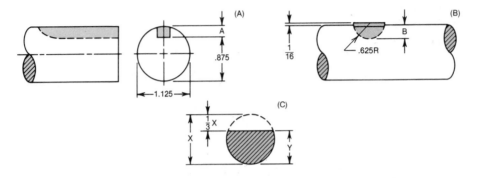

Unit 76 Basic Flat Shapes

OBJECTIVES OF THE UNIT
After satisfactorily studying this unit, the student/trainee will be able to
• *Use appropriate terms to identify characteristic features of triangular, rectangular, circular, and other common flat shapes.*
• *Make sketches and other drawings of round, rectangle, and angle forms according to given dimensional information.*

The terms and properties of lines, angles, and circles may be applied in the layout, design, development, and construction of closed flat shapes. A new term, *plane,* must be understood in order to visualize a closed flat shape accurately. A *plane* refers to a flat surface on which a straight line connecting any two points lies. In figure 76-1, the flat shape is in one plane because any straight line connecting any two lines or points on the lines lies in the plane.

Fundamental to most design and construction are the three flat shapes of the triangle, rectangle, and circle.

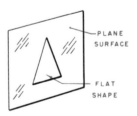

FIGURE 76-1

A. CHARACTERISTICS OF TRIANGLES

A triangle consists of three line segments which are joined at the end points to form a closed flat shape. The line segments are called *sides,* and the angles formed, *inside angles.*

There are three broad classes of triangles: *equilateral, isosceles,* and *scalene.* As the name equilateral implies, the three sides of this type of triangle are the same length. The isosceles triangle has two sides of the same length. In the scalene triangle, all three sides are of different lengths.

Each of these triangles is illustrated in figure 76-2. Note that a short line appears (△) on those sides of the triangle that are equal. A short curved line indicates the angles of the triangle that are equal (△).

TYPES ⇨	EQUILATERAL	ISOSCELES	SCALENE
⇩ RIGHT TRIANGLE			
ACUTE TRIANGLE			
OBTUSE TRIANGLE			

FIGURE 76-2

In each type of triangle the word *base* usually refers to the horizontal or lower side. Opposite this base is the *vertex* angle. The point at which the two sides of an angle come together is the *vertex.*

Triangles are further described by their included angles. The most commonly used triangle is the *right* triangle in which one angle is a right angle. An *acute* triangle has three acute angles. An *obtuse* triangle has one angle greater than 90°.

VERTEX ANGLE

BASE

FIGURE 76-3

B. CHARACTERISTICS OF RECTANGLES

The second basic flat shape is the *rectangle*. A rectangle has four closed sides with four right angles. When the two pairs of sides are of equal length, the object is a rectangle. When all four sides are equal and the four angles are 90°, the shape is called *square.*

The base of the rectangle is usually the horizontal side at the bottom. The altitude or height is the vertical side. The diagonal of a rectangle is a straight line that connects the opposite corners.

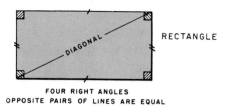

RECTANGLE

DIAGONAL

FOUR RIGHT ANGLES
OPPOSITE PAIRS OF LINES ARE EQUAL

SQUARE

FOUR RIGHT ANGLES
FOUR EQUAL SIDES

C. CHARACTERISTICS OF CIRCLES

The third basic flat shape is the circle. Three types of circles are used constantly in design. The circle, where all parts on the curve are equidistant from a fixed center, is the type used most widely. Round parts that have more than one diameter are usually machined or designed *concentric*. Concentric means that all points on each circle are equidistant from the same center. Occasionally, the centers of the circles on a part or mechanism are off-center. Circles drawn in the same plane from the different centers of a part are *eccentric*. The characteristics of the regular, concentric, and eccentric circles are shown graphically.

REGULAR CIRCLE   CONCENTRIC CIRCLES   ECCENTRIC CIRCLES

All three of the basic flat shapes may be modified, thus making possible a limitless number of layout constructions.

D. OTHER COMMON FLAT SHAPES

The next group of flat shapes that is used by craftspersons and others includes the parallelogram, trapezoid, and hexagon. Sometimes, these shapes are broadly classified as *polygons*. A *polygon* is any closed flat shape of three or more sides. When all the sides are the same size, the polygon is *regular*.

A *parallelogram* has four sides consisting of two pairs of lines that are parallel and the same length and two opposite pairs of angles that are equal. The altitude of the parallelogram is the distance from the base to the opposite parallel line.

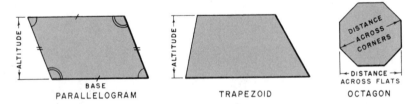

PARALLELOGRAM   TRAPEZOID   OCTAGON

A *trapezoid* has four sides, two of which are parallel. The altitude, as illustrated, is the vertical distance between the two parallel sides.

The *hexagon* and *octagon* are common. As each name implies, the hexagon is a six-sided figure and the octagon, an eight-sided figure. The hexagon and octagon are regular polygons when each side of the hexagon is equal to every other side and the same holds true for the octagon. Two dimensions are referred to with the hexagon and octagon. The *distance across flats* is the measurement across any set of parallel sides. The *distance across corners* is the measurement from the vertex of one pair of sides to the vertex of the opposite pair of sides.

Review and Self-Test

ASSIGNMENT UNIT 76

A. Properties of Triangles

1. Classify each triangle as to the group in which it falls.

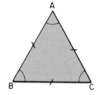 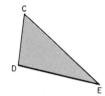

2. Sketch triangles 1, 2, 3, and 4, freehand.
   a. Label the sides and angles.
   b. Mark the sides or angles in each triangle that may be equal.

Triangle	Type
1	Right Triangle
2	Acute Triangle
3	Equilateral Triangle
4	Obtuse Triangle

B. Properties of Rectangles

1. Sketch rectangles ①, ②, ③, and ④ and mark each line and angle for identification.

 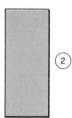 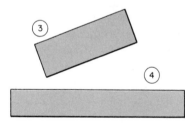

2. Identify all sides and angles in each rectangle that should be equal.
3. Draw a diagonal in each rectangle.

C. Properties of Circles

1. Identify the types of circles used in parts ①, ②, and ③.

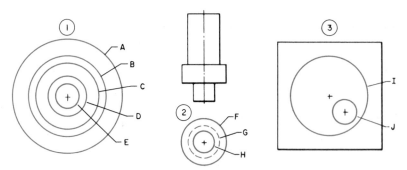

D. Properties of Other Common Flat Shapes

1. Name parts ①, ②, ③, and ④.

2. Sketch each part (1, 2, 3, and 4) freehand. Show either the altitude and base or the distance across flats, as may apply.
3. Use a drafting protractor and lay out the four shapes A, B, C, and D in the table.

Required Shape		Description
A	Square	Sides are 2″
B	Parallelogram	1. Inclined sides are 2″ long 2. Horizontal sides are 3″ long 3. Included angle is 60°
C	Hexagon	Distance across flats is $2\frac{1}{4}''$
D	Octagon	Distance across corners is $2\frac{1}{2}''$

Unit 77 Basic Solid Shapes

OBJECTIVES OF THE UNIT
After satisfactorily studying this unit, the student/trainee will be able to
* *Identify and letter the features of each basic solid geometric shape on drawings.*
* *Sketch prisms, regular solid forms, and cylinders and pyramids and other solids of revolution that are common in business and industry.*

A. THE THREE BASIC SOLIDS

The three flat shapes of the triangle, rectangle, and circle may be changed into solids by adding the third dimension of depth. The triangle becomes a prism; the rectangle, a rectangular solid; and the circle, a cylinder.

The word *base* has a somewhat different meaning with solids in that it usually refers to the surface on which the solid rests. Each circular end of

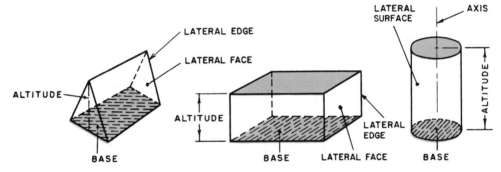

	TRIANGLE	RECTANGLE	CIRCLE
FLAT SHAPES	△	▭	○
SOLID SHAPES	PRISM	RECTANGULAR SOLID	CYLINDER

the cylinder may be referred to as its base. The *altitude of the prism* is the perpendicular line segment that joins the point at which the triangular faces come together with the base. The *altitude of the rectangular solid* is the same as for the rectangle. The *altitude or height of the cylinder* is the distance between the two bases measured along the axis. The *axis* is the imaginary line that passes through the center of each base circle. A solid object that is cut by a plane at an angle to the base is said to be *truncated*.

The term *lateral edge* refers to the line (segment) where the lateral faces (sides) meet. The faces or sides of solid forms are called *lateral faces*.

B. SOLIDS OF REVOLUTION

Solids are also formed by revolving a flat form and tracing the shape that the outside line or lines take during revolution. Such solid shapes are known as *solids of revolution*. The simplest solid of revolution to visualize is the sphere that is formed by revolving a circle completely around an axis.

In like manner, if an isosceles triangle is revolved about its axis, a cone is formed. The shape of the solid of revolution is governed by the original flat shape.

C. PRISMS, PYRAMIDS, AND OTHER COMMON SOLID GEOMETRIC SHAPES

Right prisms, pyramids, and segments of spheres are used in science and industry. A *right prism* is a solid formed by visualizing the shape that a flat form takes as it moves perpendicular to its base to a predetermined height known as the altitude. Right prisms are named according to the shape of the bases. A triangular base produces a *triangular prism;* an octagonal base, an *octagonal prism*.

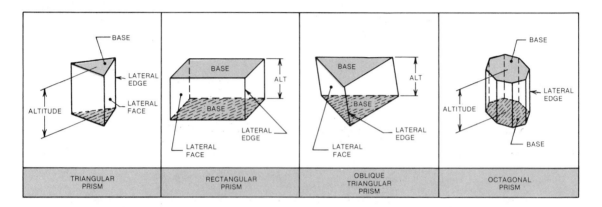

A *pyramid* is formed connecting each corner of the base of a flat shape with a point outside the base. Each triangular face thus formed meets at the *vertex.* The name of the pyramid as formed depends on the shape of the base. A three-sided (triangular) base produces a *triangular pyramid;* a four-sided base, a *quadrangular* (or *square*) *pyramid;* a six-sided base, a *hexagonal pyramid*.

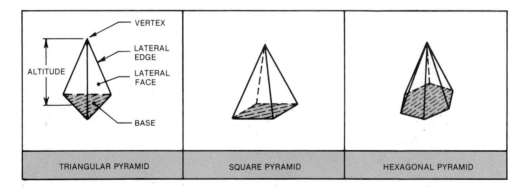

If a sphere is cut in two, or the circle is revolved through only 180°, half a sphere, or a *hemisphere,* is formed.

While there are many other basic and common solid shapes, those that have been described are the ones that have greatest daily application in the science, occupational and technical fields, and in everyday living.

Review and Self-Test

ASSIGNMENT UNIT 77

A. Properties of Prisms, Rectangular Solids, and Cylinders

1. Identify each solid shape or object as lettered on the pictorial drawings.

 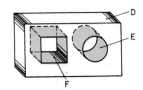 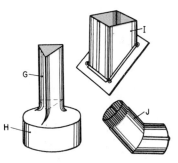

2. Name the type of line or lines that describe the two parts shown on the mechanical drawings (A_1, B_1, E_1, F_1).

3. Sketch each of the following solid shapes.
   a. A cube
   b. A rectangular solid
   c. A triangular prism
   d. A cylinder
4. Compare each solid shape with the flat shape given in each case.
   a. A cube with a square
   b. A rectangular solid with a rectangle
   c. A triangular prism with a triangle
   d. A cylinder with a circle

B. Properties of Solids of Revolution

1. Sketch and name the solid shape that develops as each basic flat shape is revolved on the given axis.

  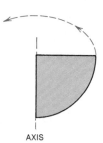

AXIS   AXIS

AXIS

C. Other Common Solid Geometric Shapes

1. Sketch a square and a triangular pyramid. Identify the base, altitude, a lateral edge, and a lateral face.
2. Name the two solid geometric shapes that are combined in Ⓐ and Ⓑ to form the two different types of bolt heads.

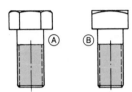

Unit 78 Congruent and Symmetrical Plane and Solid Geometric Shapes

---

OBJECTIVES OF THE UNIT
After satisfactorily studying this unit, the student/trainee will be able to
- *Interpret the factors of size and shape as each affects the symmetry and/or congruence of plane and solid geometric forms.*
- *Apply the rules for establishing the congruence and symmetry of parts.*
- *Make sketches and dimension congruent and symmetrical objects.*
- *Translate congruence and symmetry in relation to actual industrially produced parts.*

The production of identical shapes and interchangeable parts and mechanisms in modern manufacturing is based on congruence. *Congruence* is a geometric term that means that all the physical properties of size and shape in one part are duplicated in another. All bolts of the same size in a box or the patterns in a cloth are congruent when each one is an exact copy of the other.

The characteristics of congruent plane (flat) and solid shapes are covered in this unit. In proving the congruence of two shapes, the sides that fit, or are compared, or are constructed, are called *corresponding sides*. In the case of angles, they are known as *corresponding angles*. When a side is equal to its corresponding side, the sides are said to *coincide*. The same holds true when an angle is identical with its corresponding angle. The corresponding sides are located opposite the angles to which they correspond.

A. CONGRUENCE OF FLAT GEOMETRIC SHAPES

RULE FOR DETERMINING CONGRUENCE OF TRIANGLES

- Draw or place the triangles to be compared on a plane or flat surface.
- Compare the length of each corresponding side.

- Compare each corresponding angle to see if they coincide.
- Triangles are congruent:
  - When two corresponding angles and a corresponding side are equal.
  - When two corresponding sides and the corresponding included angle are equal.
  - When the three sides are equal.

EXAMPLE: Determine which of the three triangles are congruent.

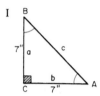 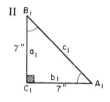 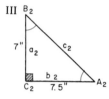

Step 1 Check the length of the corresponding sides a, a_1, and a_2. Note that all are equal.

$a = 7''$
$a_1 = 7''$
$a_2 = 7''$

Step 2 Compare the largest angle in triangles I, II, and III.

$\angle BCA = 90°$
$\angle B_1C_1A_1 = 90°$
$\angle B_2C_2A_2 = 90°$

Step 3 Determine the length of corresponding sides b, b_1, and b_2.

$b = 7''$
$b_1 = 7''$
$b_2 = 7.5''$

Step 4 Compare the values of the two sets of corresponding sides and the corresponding included angle. Triangles I and II are congruent. Triangle III is not because the side b_2 is larger than both b and b_1.

This same problem may be checked by taking any other combination of corresponding sides and corresponding angles.

RULE FOR DETERMINING THE CONGRUENCE OF POLYGONS

- Check each side of the polygons for congruence.
- Check each angle of the polygons against each corresponding angle.
  Note. Plane polygons are similar when all the corresponding angles are equal.

RULE FOR DETERMINING CONGRUENCE OF PLANE CIRCLES

- Check the diameter of each circle. When the diameters are equal, the circles are congruent.

B. CONGRUENCE OF SOLID GEOMETRIC SHAPES

Solids are congruent when every point, face, edge, and vertex of one solid are equal to the corresponding part of the other solid.

CONGRUENT SOLIDS

RULE FOR DETERMINING THE CONGRUENCE OF GEOMETRIC SOLIDS

- Place solids in the same relative position to each other.
- Measure or test the length of each corresponding edge.
- Check each corresponding angle.
  Note. When every part of one solid is equal to the corresponding part of the other, the solids are congruent.

C. SYMMETRICAL PLANE GEOMETRIC SHAPES

Symmetry is important in production and design. Symmetrical parts are balanced, prevent undue wear and destructive vibration, and simplify manufacturing. A flat shape is symmetrical when corresponding points on each side of an axis are equidistant from the axis.

The axis that cuts the part in two symmetrical halves is known as the *axis of symmetry* or the *line of symmetry*.

SYMMETRICAL FLAT SHAPES

D. SYMMETRICAL SOLID GEOMETRIC SHAPES

A solid shape is symmetrical when all corresponding parts are equal but in an opposite position. This idea of symmetry is extremely valuable in drafting and sketching. It is the geometric principle on which many types of section drawings are made. On symmetrical objects fewer views are needed and considerable drafting time is saved.

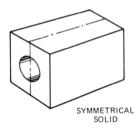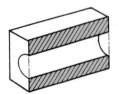

SYMMETRICAL
SOLID

ONE HALF SYMMETRICAL SOLID

Review and Self-Test

ASSIGNMENT UNIT 78

A. Congruent Flat (Plane) Shapes

1. Prepare a table similar to the one with places for the two sides and the included angle of a triangle. Complete the missing information by referring to the triangle.

Side	Included Angle	Side
AB		
	$\angle C$	
		BA
		BC
	$\angle A$	
BC		

2. Use a protractor, rule, and compass to lay out triangles *A, B,* and *C.*

Triangle	Side	Angle	Side
A	3"	45°	$3\frac{1}{2}''$
B	4"	60°	$2\frac{3}{4}''$
C	$2\frac{1}{2}''$	75°	$4\frac{1}{8}''$

B. Congruent Solid Geometric Shapes

1. Ten pieces of hexagon stock are all cut 2 cm long from a bar of stock. Are the ten pieces congruent to each other? Why?

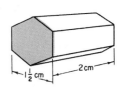

2. A plastic strip, triangular in shape, is formed by extrusion. If the strip is cut into $\frac{1}{2}''$ lengths, are the pieces congruent? Why?

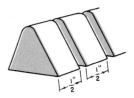

3. Pentagon-shaped plugs are to be stamped from 3.175 mm thick brass. If stock 2.778 mm is used instead of 3.175 mm, will the parts be congruent with those 3.175 mm thick? Why?

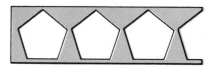

C. Symmetrical Plane Shapes

1. Give three reasons why symmetry is essential in designing for production.
2. Sketch three flat shapes that are symmetrical. Label the axis in each case.

D. Symmetrical Solid Geometric Shapes

1. Make a sketch of a simple part that is symmetrical.
   a. Pass an imaginary plane through the part.
   b. Sketch how the exterior and interior shapes of the part look.
2. Explain the conditions under which each of the following machined parts is congruent.
   a. A pair of spur gears
   b. A set of pistons
   c. Precision-ground threaded shafts

Unit 79 Achievement Review on Geometric Lines and Shapes

OBJECTIVES OF THE UNIT

This achievement review serves as an overall test for Section 16. The unit is designed to measure the student's/trainee's ability to

- *Recognize the properties and characteristics of such common plane (flat) shapes as triangles, rectangles, circles, parallelograms, and trapezoids.*
- *Interpret the features of basic solid geometric forms including prisms, rectangular solids, cylinders, spheres, and pyramids.*
- *Make drawings, label features, and dimension plane and solid geometric parts and products.*
- *Prove the congruence and symmetry of plane and solid geometric forms and workpieces.*

A. THE CONCEPTS OF LINES, ANGLES, AND CIRCLES

1. Draw horizontal lines to the lengths indicated in the table.
   a. Connect the end point of each line with an inclined line drawn to the given length at the given angle.
   b. Name the type of angle formed in each case.

Line	Length of Sides		Included Angle
	Horizontal	Inclined	
AB	4"	$3\frac{1}{2}''$	90°
BC	$3\frac{1}{2}''$	$3\frac{1}{2}''$	29°
CD	$3\frac{1}{2}''$	$3\frac{1}{4}''$	120°
DE	$3\frac{1}{4}''$	3"	135°
EF	$2\frac{7}{8}''$	$3\frac{1}{8}''$	$59\frac{1}{2}°$

2. Refer to the sketch and name lines Ⓐ, Ⓑ, Ⓒ, Ⓓ, and Ⓔ.

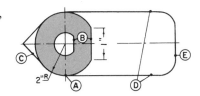

a. Give the diameter of the outside cylindrical shape.

b. What is the chordal dimension for the larger circle?

B. BASIC FLAT GEOMETRIC (PLANE) SHAPES

1. Draw the triangles indicated in each table. Label the base. Then show the altitude of each triangle by a dotted line.

△	Three Equal Sides	△	Two Equal Sides	Incl. Angle	△	Base	Side	Side
A	4″	D	3″	90°	G	4″	$2\frac{1}{2}''$	$3\frac{1}{2}''$
B	$3\frac{1}{4}''$	E	2″	120°	H	$3\frac{1}{2}''$	$2\frac{1}{4}''$	$2\frac{3}{4}''$
C	$2\frac{7}{8}''$	F	$2\frac{1}{8}''$	60°	I	$2\frac{11}{16}''$	$1\frac{5}{8}''$	$3\frac{1}{16}''$

2. Name each flat shape (a through f) and mark the sides and angles that are equal.

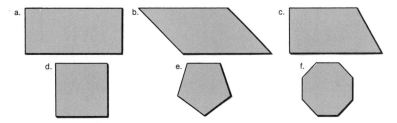

3. Sketch and label two simple round parts. The first should have all surfaces concentric; the second, eccentric.

C. BASIC SOLID GEOMETRIC SHAPES

1. Sketch the rectangular, cylindrical, and triangular parts to the sizes (approximately) given in the table. Draw either full size or to a reduced scale, if preferred. Use a ruled or graph sheet if available and indicate the scale to which the parts are sketched.

	Rectangular			Cylindrical		Triangular
	Length	Height	Depth	Diameter	Height	3 Sides
A	4″	2″	3″	2″	2″	3″
B	3″	$1\frac{1}{2}''$	2″	$1\frac{1}{2}''$	$2\frac{1}{2}''$	$2\frac{1}{2}''$
C	$2\frac{1}{8}''$	3″	$\frac{5}{8}''$	$2\frac{1}{2}''$	$\frac{3}{4}''$	$1\frac{1}{4}''$

2. Sketch freehand a cylindrical screw head that has a hexagonal prism shape formed in the head to receive a wrench.

D. CONGRUENT AND SIMILAR PLANE AND SOLID GEOMETRIC SHAPES

1. Lay out two congruent equilateral triangles and two congruent isosceles triangles. Mark the sides and the angles of each pair of corresponding sides or angles that are equal.

2. Determine which of the four pins are congruent. Explain.

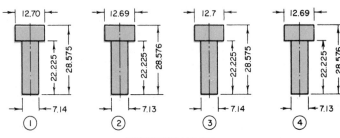

DIMENSIONS ARE IN mm

3. Determine whether parts Ⓐ and Ⓑ are similar or congruent. Explain.

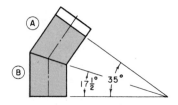

FORMULAS APPLIED TO GEOMETRIC FORMS

Unit 80 Application of Formulas to Plane and Solid Geometric Forms (Regular Polygons)

OBJECTIVES OF THE UNIT

After satisfactorily completing the unit, the student/trainee will be able to
- *Apply formulas for square polygons to find the perimeter, area, and volume and pattern stretchout measurements.*
- *Solve rectangular polygon problems relating to perimeter, area, and volume.*
- *Apply the Pythagorean theorem to establish dimensions for the sides of right triangles.*
- *Use combination mathematical processes in solving problems involving materials conservation practices in pattern layout work.*

The *polygons* that are applied in everyday business, industrial, and consumer problems include squares, rectangles, triangles, and combinations of these straight-line forms. The computations involve formulas dealing with perimeters, areas, and volumes. Unit 81 relates to formulas for circles, sectors, and segments in both plane and solid geometric forms.

A. DEVELOPING FORMULAS FOR SQUARE OBJECTS

Formula for Perimeter of a Square

Considering the square first, the distance around it (called *perimeter*) is equal to the sum of the four sides. Since each one of the four sides is equal, the perimeter (P) is equal to four times the length of a side (S). $P = 4S$

Formula for Area of a Square

The *area of a square* is equal to the width times the height. Here again, since each side is equal, the area (*A*) is found by multiplying a side (*S*) by itself. Expressed as a formula, $A = S^2$.

Formula for Volume of a Cube

Where the length of each side of a regular solid is equal, the object is called a cube. The *volume of a cube* is equal to its length multiplied by its width multiplied by its height. Since each of these sides is the same length, $V = S^3$.

Formula for Area of a Stretchout

The *area of a stretchout* (*A*) of a square surface is equal to the length of the four sides (*S*) multiplied by the height of the object. The formula $A = 4S(H)$ is the abbreviated way of saying the same thing. Note the selection of letters in the formula: *A* for area, *S* for the length of one side, and *H* for height.

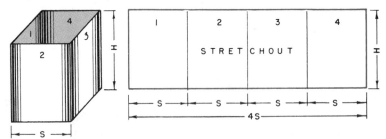

If the object has a seam or overlap, the overall length is equal to 4*S* plus the seam allowance.

B. DEVELOPING FORMULAS FOR RECTANGULAR POLYGONS

Rectangles differ from squares in that in place of four sides of equal length, the two sets of opposite sides are the same length.

Formula for Perimeter of a Rectangle

The *perimeter of a rectangle* is equal to $2W + 2H$. *W* is the width of one side and *H* the height or length of the other. $P = 2(W + H)$ is a simpler way of stating the formula.

An allowance is usually added to this perimeter for overlapping the ends or for seams.

Formula for Area of a Rectangle

The *area of a rectangle* is the product of its two adjacent sides. The formula is $A = (S_1) \times (S_2)$. *A* denotes area; (S_1) the length of one side; and (S_2) the length of the other side.

Formula for Volume of a Rectangular Solid

As the rectangle takes on height, it becomes a rectangular solid. The *volume,* then, is equal to the area of the base $\times$ the height. $V = (S_1) \times (S_2) \times (H)$

C. FORMULA FOR SIDES OF A RIGHT TRIANGLE

The dimension of a third side of a right triangle may be computed by a simple formula, when the two remaining sides are given. The formula is often referred to as the *Pythagorean theorem.* The theorem is a proven mathematical law. The formula expresses the relationship among the three sides of a right triangle. The square of the hypotenuse of a right triangle is equal to the sum of the squares of the remaining two sides. Expressed as a formula, $H^2 = (S_1)^2 + (S_2)^2$. The sides are generally identified as *A, B,* and *C.* The formula is stated as $C^2 = A^2 + B^2$.

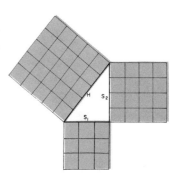

D. CONSERVING MATERIALS IN STRETCHOUTS

Once the sizes of a stretchout are determined, the next step is to determine the most economical use of the material from which one or more of the same object may be made. This process calls for simple calculations.

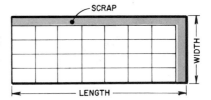

RULE FOR CONSERVING MATERIAL FOR STRETCHOUTS

- Make a simple sketch of the shape of the material and dimension.
- Lay out the pattern in one of the extreme corners.
- Divide the length of the material by the length of the stretchout.
- Divide the width in the same manner.
- Multiply quotients in both cases.
- Turn the pattern around 90°.
- Repeat the same steps.

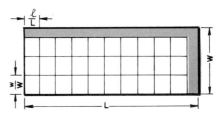

Compare the number of pieces that may be cut from the material with the pattern in both positions.

EXAMPLE: How many parts 7 inches in diameter by 12 inches high may be cut from a sheet 36″ × 120″? Add $\frac{1}{2}$ inch to the length of each part for a lap seam and use $\frac{22}{7}$ for π.

Step 1   Find the length of the stretchout, using the circumference formula and adding $\frac{1''}{2}$.

$$C = \pi D + \tfrac{1}{2}$$
$$= \tfrac{22}{7} (7) + \tfrac{1}{2}$$
$$= 22\tfrac{1}{2}$$

Step 2   Make a simple sketch of the materials with dimensions.

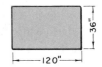

Step 3   Divide the length of the sheet by the length of the stretchout.

$$\frac{120}{22\tfrac{1}{2}} = 5\tfrac{1}{3}$$

Step 4   Divide the width of the sheet by the height of the part.

$$\frac{36}{12} = 3 \text{ pieces}$$

Step 5   Multiply the whole numbers only, to determine the number of pieces that may be cut.

$$5 \times 3 = 15 \text{ pieces}\quad \text{Ans}$$

Step 6   Turn the pattern around 90° and repeat the same steps.

Step 7   Compare both products to determine the position in which to place the pattern to make the most economical use of the material.

Review and Self-Test

ASSIGNMENT UNIT 80

A. Application of Formulas to Square Objects

1. Determine the number of cubic yards of topsoil required for an area 48 feet square. The topsoil is to be 4 inches deep.

   Note. Problems may be solved by conventional mathematical processes or by using an electronic calculator.

2. How much paint is needed to cover five rooms with two coats? The room sizes are shown in the table. Deduct 25 square feet for each opening. One gallon of paint covers 350 square feet of surface.

	Room Size			No. of
	Width	Length	Height	Openings
A	12′	12′	8′	2
B	16′	16′	8′	3
C	8′	8′	8′	4
D	10′	10′	7′–6″	3
E	12′	12′	7′–6″	4

3. Find the area of each of the stretchouts for square parts A through E. Add $\frac{1}{2}$ inch to the length of the seam allowance.

	A	B	C	D	E
Length of Side	8″	$9\frac{1}{2}″$	3′–3″	5′–4″	$4′–9\frac{1}{2}″$

B. Application of Formulas to Rectangular Polygons

1. What is the total area in square yards of two sections of a roof, 32′ × 28′ and 24′–8″ × 30′?

2. A bathroom 7′–6″ wide × 8′–3″ long has four sidewalls to be covered with metal lath to a height of 4′–6″. Allow 20 square feet for a door and window opening and determine to the nearest whole number the square yards of wall surface to be covered.

3. Determine the capacities of the containers at A through E.
   Use:
   1 cubic foot = 7.48 gallons
   1 gallon = 231 cubic inches
   1 barrel = $31\frac{1}{2}$ gallons

	Given Dimension			Required Capacity
	L	*W*	*H*	
A	24″	36″	30″	Gal
B	$12\frac{1}{2}″$	25″	28″	Gal
C	2′-4″	3′-3″	2′	Gal
D	5′-6″	6′-4″	10′-6″	Bbl
E	20′-8″	15′-6″	12′-6″	Bbl

C. Application of Formulas to Right Triangle Sides

1. Find the length of the crossbraces for structural trusses A, B, and C. State each length correct to two decimal places.

Truss	Length (*L₁*)	Height (*H*)	Crossbrace (*B*)
A	2.3 m	1.98 m	
B	2.84 m		3.69 m
C		2.92 m	4.86 m

2. The 20.93 kilometer power line shown in the sketch is to be shortened by a new right of way. Determine the length of the new line, correct to two decimal places.

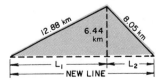

D. Application of Formulas to Stretchouts

1. Find the number of rectangular pieces that may be cut from each of the material sizes given in the table for parts A through E.

	Sizes of Rectangular Parts			Allowance on Length and Height for Seams	Dimension of Materials
	L	*W*	*H*		
A	6″	12″	12″	$\frac{1}{4}$″	24″ × 60″
B	8″	10″	12″	$\frac{1}{4}$″	30″ × 96″
C	$8\frac{1}{2}$″	12″	12″	$\frac{1}{4}$″	36″ × 96″
D	1′–4″	1′	1′	$\frac{1}{2}$″	36″ × 96″
E	1′–6″	10″	2′	$\frac{1}{2}$″	48″ × 120″

2. Determine the number of pieces that can be cut from a sheet 36″ × 96″ to make the rectangular part as illustrated.

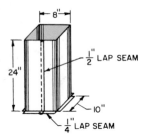

3. Three square ducts having cross-sectional areas of (a) 930 square centimeters, (b) 1,648 square centimeters, and (c) 3,721 square centimeters, respectively, are needed. A lap seam of 1.3 cm is added. What length of metal is required to make each square duct?

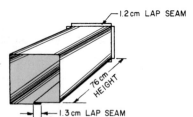

Unit 81 Application of Formulas to Plane and Solid Circular Forms

OBJECTIVES OF THE UNIT

After satisfactorily completing the unit, the student/trainee will be able to
- *Interpret problem requirements of plane and solid geometric shapes related to circles, sectors, segments, and circular solids.*
- *Solve typical shop, laboratory, and consumer problems involving plane and solid forms based on or generated from a circle or parts of a circle.*

One of the most widely known and used formulas deals with the circumference and diameter of a circle. The formula, $C = \pi D$, like all other formulas, serves more than one purpose as it may be used to find the value of any letter or missing quantity. In this case, when the circumference is given and substituted in the formula, the diameter may be computed.

A. APPLICATION OF FORMULAS TO CIRCLES

EXAMPLE: Determine the diameter of a circle with a 22-inch circumference.

Step 1 Analyze the problem. The diameter (D) is required. The circumference (C) is given.

Step 2 Select the formula. $C = \pi D$

Step 3 Substitute given values for letters in the formula. Use the value of $\frac{22}{7}$ for π.
$$22 = \pi D$$
$$22 = \frac{22}{7}D$$

Step 4 Perform mathematical operations as indicated.
$$\overset{1}{22}\left(\frac{7}{\underset{1}{22}}\right) = D$$

Step 5 Label answer with correct unit of measure.
$$7'' = D \quad \text{Ans}$$
$$C = \pi D$$

Step 6 Check answer by substituting the computed values in the formula.
$$22 = \tfrac{22}{7} \times 7$$
$$22 = 22 \quad \text{Check}$$

Formula for Stretchout of Circular Parts

The application of the circumference formula is especially helpful in finding the length of the *stretchout* of a circular job. In figure 81-1, it is apparent that the length of the stretchout is the same as the circumference; the height, the same as H.

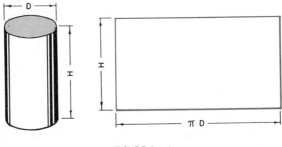

FIGURE 81-1

B. APPLICATION OF FORMULAS TO PARTS OF CIRCLES

Formula for Lengths of Arcs

An *arc* is part of a circle, and a circle contains 360°. The ratio of the length of the arc to the circumference is the same as the ratio of the angle of the arc to 360°. With this information, a formula may be derived for computing arc length.

RULE FOR DEVELOPING FORMULA FOR ARC LENGTH

- Analyze the problem.
- Select appropriate letters for given values and values to be computed.
  Let N = Number of degrees in angle of arc
  D = Diameter of circle
  C = Circumference
  L = Length of arc

- Write in abbreviated form the statement that explains what the length of arc is equal to, using letters and numbers.

Required		**Formula Derived**
▼		▼

$$L = \frac{N}{360}(C) = \frac{N}{360} \times (\pi D)$$

EXAMPLE: Find the length of arc in an angle of 45° in a 7-inch diameter circle. Use $\frac{22}{7}$ for π.

Step 1 Read problem for what is given and what is required. $N = 45°, D = 7''$

Step 2 Select formula and substitute given values for letters. $L = \frac{N}{360} \times (\pi D)$

Step 3 Perform the operations as indicated.

$$= \frac{\overset{1}{\cancel{45}}}{\underset{8}{\cancel{360}}} \times \left(\frac{22}{7} \cdot \overset{1}{\cancel{7}}\right)$$

Step 4 Label answer with correct unit of measure.

$$= \frac{22}{8} = 2\frac{3}{4}'' \quad \text{Ans}$$

C. APPLICATION OF FORMULAS TO CIRCULAR GEOMETRIC SOLIDS

As a third dimension of height is added to a circle, the object becomes a *circular solid*. The volume of this cylinder is equal to the area of the base × the height. In abbreviated form, $V = \pi r^2 H$. Note that three dimensions are multiplied so that the product is in cubic measure.

EXAMPLE: Find the volume in cubic inches of a cylindrical container having a radius of 7 inches and a height of 10 inches.

$$V = \pi r^2 H$$

Step 1 Read problem for given and required values.

Step 2 Select formula and substitute values.

$$= \tfrac{22}{7}(7 \cdot 7 \cdot 10)$$

Step 3 Perform mathematical operations.

Step 4 Label answer.

$$= 1540 \text{ cu in.} \quad \text{Ans}$$

D. APPLICATION OF FORMULAS TO SEMICIRCULAR-SIDED OBJECTS

Formula for Area

A semicircular-sided object is an example of how the information about rounds and rectangles may be combined to solve a modified shape. The area of a semicircular-sided object is a composite of the area of a circle plus the area of a rectangle.

Where A = Area

Expressed as a formula

$\pi = \tfrac{22}{7}$ or the decimal equivalent

$$A = \pi r^2 + (L \times W)$$

r = Radius of semicircular end
L = Length of rectangle
W = Width of rectangle

$$A = \frac{\pi r^2}{2} + (L \times W) + \frac{\pi r^2}{2} = \pi r^2 + (L \times W)$$

Formula for Volume

As the sides of the semicircular-sided object are extended, a solid is formed. The volume of this solid is equal to the area of the base times the height (H) of the object.

$$V = [\pi r^2 + (L \times W)]H$$

EXAMPLE: Find the capacity of a semicircular-sided tank in cubic feet. The end radius is 7 inches, the length of the rectangular center section is 12 inches, and the height is 10 inches.

Step 1   Select or develop formula. $V = [\pi r^2 + (L \times W)]H$

Step 2   Substitute given values for letters. $V = \left[\frac{22}{7}(7)(7) + (12 \times 14)\right]10$

Step 3   Perform mathematical operations. $V = \left[\frac{22}{\overset{1}{\cancel{7}}}(\overset{1}{\cancel{7}})(7) + (12 \times 14)\right]10$

$$= [154 + 168]10$$
$$= 322 \times 10 = 3220$$
$$\frac{3220}{1728} = 1 \text{ cu ft } 1492 \text{ cu in.} \quad \text{Ans}$$

Step 4   Divide 3220 cu in. by 1728 to get the answer in cu ft.

Review and Self-Test

ASSIGNMENT UNIT 81

A. Application of Formulas to Circles

1. Determine the circumference of blanks A through E for each diameter given in the table. Use $\frac{22}{7}$ for the value of π.

Blanks	A	B	C	D	E
Given Diameter	14″	70″	$3\frac{1}{2}''$	10.5′	9″

2. Compute the circumference of pipes A through E for the diameters and radii given. Use $\pi = 3.1416$ and give each answer correct to three decimal places.
Note. D = Diameter, R = Radius.

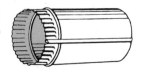

Pipes	A	B	C	D	E
Given Diameter	$D = 8\frac{1}{7}''$	$D = 5\frac{1}{2}''$	$R = 3''$	$R = 4.2'$	$R = 4.10''$

3. Find the diameter of pulleys A through E for each given circumference. Use the formula $C = \pi D$ and the value of 3.1416 for π. Give each diameter to three decimal places.

Pulleys	A	B	C	D	E
Given Circumference	63.84 cm	35.92 cm	88.9 mm	142.88 mm	1.524 m

B. Application of Formulas to Parts of Circles

1. Compute the circular length at Ⓐ, Ⓑ, Ⓒ, using the dimensions given on the sketch of the cam plate.

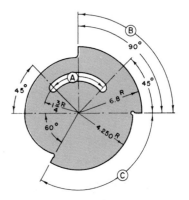

2. Determine the length of stretchout for (a) the throat and (b) the heel of sheet metal elbows A, B, C, and D. Use the dimensions given in the table.

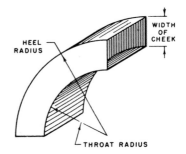

	A	B	C	D
Throat Radius	9″	8″	30″	6″
Central Angle	90°	60°	45°	135°
Width of Cheek	4″	5″	15″	$3\frac{1}{2}″$

3. Compute the circular length for each central angle and dimension as given in the accompanying table. Use $\pi = 3.142$.

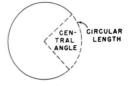

	A	B	C	D	E
Central Angle	180°	90°	60°	135°	72°
	Diameter	Diameter	Diameter	Radius	Radius
Dimension	10″	$12\frac{1}{4}″$	12.6″	8″	6.250″
	25.4 cm	31.12 cm	0.320 m	20.32 cm	158.75 mm

C. Application of Formulas to Cylindrical Solids

1. Find the volumes of containers A, B, and C and the capacities of D and E in the units of measure indicated in the table. Use $\frac{22}{7}$ for π and the formula $V = \pi r^2 H$.

	Diameter	Height	Unit of Measure
A	14″	10″	Cubic inch
B	20″	32″	Cubic inch
C	$14\frac{1}{2}″$	18″	Cubic inch
D	28′	15′	Barrel
E	$20\frac{1}{2}′$	$12\frac{1}{2}′$	Barrel

2. Determine the liquid capacities of containers A, B, and C. Give the volumes in the measure indicated in each case. Use $\pi = \frac{22}{7}$ and the formula $V = \pi r^2 H$ for both sections which make up each container. Round off decimal answers to two places.

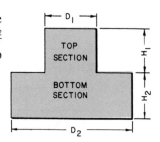

Container	Top Section		Bottom Section		Unit of Measure
	Diameter$_1$	Height$_1$	Diameter$_2$	Height$_2$	
A	14″	10″	28″	20″	Cubic inch
B	21″	1′–8″	35″	2′–4″	Cubic foot
C	3′–8″	8″	4′–10″	1′–9″	Cubic yard

3. Determine the quantity of concrete required for the concrete pipes A and B. Use $\pi = 3.1416$. Give the answers in cubic meters, correct to the nearest two decimal places.

Concrete Pipe	Inside Diameter (ID)	Outside Diameter (OD)	Length (L)
A	71.12 cm	99.06 cm	1.8 m
B	88.9 cm	124.46 cm	2.4 m

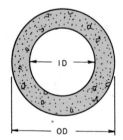

Unit 82 Achievement Review on Formula Applications to Common Geometric Forms

OBJECTIVES OF THE UNIT

This achievement review serves as an overall test for Section 17. The unit is designed to measure the student's/trainee's ability to
- *Apply formulas in solving typical shop and laboratory problems involving square and rectangular geometric forms.*
- *Use formulas in handling circle, surface area, and volumes of round parts, sections of circular forms, and weights.*
- *Develop formulas and compute dimensional and quantity values for plane and solid regular geometric forms and combinations of different shapes.*
- *Compute surface area and major and minor diameters of elliptical forms.*
- *Calculate lateral surface areas, total surface areas, and volumes of cone- and pyramid-shaped objects.*

A. APPLICATION OF FORMULAS TO SQUARES AND RECTANGLES

1. Determine the greatest number of pieces (A through E) that may be cut from the sheets given in the table.

	Shape	Dimension			Allowance in Length for Seams	Sheet Size
		Length	Width	Height		
A	Square	4'	4"	6"	$\frac{1}{8}''$	24" × 72"
B	Square	1'–3"	1'–3"	8"	$\frac{1}{2}''$	36" × 96"
C	Rectangle	5"	8"	12"	$\frac{1}{4}''$	30" × 72"
D	Rectangle	7.6"	9.8"	11.2"	.4"	36" × 96"
E	Rectangle	2'–9"	1'–3"	10"	$\frac{3}{8}''$	48" × 120"

B. APPLICATION OF FORMULAS TO CIRCLES AND PARTS OF CIRCLES

1. The diameters of a series of parts are given in table form. Compute each circumference accurate to two decimal places.

	A	B	C	D	E
Diameter	7 mm	5.5 mm	9.25 cm	3.25 m	5.438 m

2. The accompanying table gives the circumference of pulleys A through E. Determine the diameter in each case to three places, using $\pi = 3.1416$.

	A	B	C	D	E
Circumference	28.274 mm	20.422 mm	25.918 cm	10.603 cm	9.623 m

3. Determine the weight of A and B for the part as illustrated. Use the tabular data and $\pi = 3.1416$. Round off the answer to one place.

	Outside Diameter	Inside Diameter	Height	Sector Angle	Weight per cu in.
A	6.750"	3.125"	$8\frac{1}{4}''$	135°	.33 lb
B	6.750"	3.125"	$8\frac{1}{4}''$	225°	.15 lb

C. APPLICATION OF FORMULAS TO MODIFIED REGULAR SHAPES

1. Develop three formulas for volumes of square, rectangular, and cylindrical solids.

2. Develop a formula and determine the weight of the bronze casting from the tabular data given.

Part Specification		
1	Outside Diameter	$3\frac{1}{4}''$
2	Inside Diameter	$1\frac{3}{8}''$
3	Sector Angle	135°
4	Height	4"
5	Number Required	24
6	Weight	.28 lb/cu in.

3. Compute the required measurement as identified in the table for ducts A, B, C, and D. Use $\pi = \frac{22}{7}$; 1 m³ = 264.17 gal.

Duct	Given Dimensions			Required Dimension
	Diameter	Width	Length	
A	7 ft	8 ft		Perimeter
B	6.50 ft	3.75 ft		Area of end
C	1.54 m	2.50 m	4.62 m	Area of stretchout
D	0.88 m (radius)	176.2 cm	1.46 m	Liquid capacity (gallons)

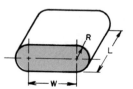

D. APPLICATIONS OF FORMULAS TO AREAS OF ELLIPTICAL FORMS

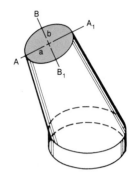

1. Compute the area of the elliptical section of the duct. Use the formula $A = \pi (a) \cdot (b)$

A–A_1 (major axis) = 12 m

B–B_1 (minor axis) = 8 m

$a = \dfrac{\text{major axis}}{2}$

$b = \dfrac{\text{minor axis}}{2}$

$\pi = 3.1416$

2. The elliptical face of a structure is designed to have 957 square feet of surface area. The major axis is to be two times the minor axis. Calculate the major and minor axis dimensions.

   Note. Use the formula

$$A = \pi \left(\frac{\text{major axis}}{2}\right) \cdot \left(\frac{\text{minor axis}}{2}\right).$$

Use $\pi = 3.14$

E. APPLICATIONS OF FORMULAS TO LATERAL AND SURFACE AREAS AND VOLUMES OF REGULAR CONES AND PYRAMIDS

1. a. Find the lateral area of the right circular cone. Use $\pi. = 3.14$.

   b. Determine the total surface area.

   $$\text{Note. } A_l = \frac{(C_b) \cdot (S)}{2}$$

   $$A_t = A_l + A_b$$

A_l = lateral area

C_b = circumference of base = D

A_b = area of base

S = slant height

A_t = total surface area

2. a. Use the Pythagorean theorem to compute the slant height of the
      square pyramid.
      Note. $S^2 = X^2 + Y^2$
   b. Determine the lateral area.

      Note. $A_l = \dfrac{\text{perimeter of base}}{2} \times \text{slant height}$

   c. Find the total surface area.

      $A_t = A_l + A_b$

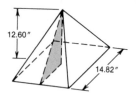

3. a. Calculate the cubic yards of concrete needed to
      pour the pyramid-shaped footing. Use the formula,

      $$\text{Volume} = \frac{\text{Altitude}}{3}\left[A_B + A_b + \sqrt{A_B \cdot A_b} \right].$$

      Round off the answer to the next cubic yard.
   b. Determine the slant height of the steel footing form
      to the nearest inch.

COMMON GEOMETRIC CONSTRUCTIONS

Unit 83 Basic Geometric Constructions

OBJECTIVES OF THE UNIT

After satisfactorily studying this unit, the student/trainee will be able to
- *Apply geometric theorems relating to design features of parallel lines, perpendicular lines, straight lines, angles, circles, and tangent lines.*
- *Use simple shop, drafting room, or laboratory instruments to construct parallel lines, perpendicular lines, angles, tangents to circles, and to modify each of these plane geometric forms according to specified dimensions.*
- *Make a layout that requires applications of basic geometric constructions.*

The most practical applications of geometric principles are in the construction of parallel, perpendicular, and tangent lines, the dividing of straight and curved lines, and the bisecting of angles. These fundamental constructions are applied to closed flat shapes such as squares, rectangles, regular polygons, and circles. In advanced work, the constructions are used in drawing solid shapes that include cylinders, triangular prisms, and rectangular solids.

A. CONSTRUCTING PARALLEL LINES

Parallel lines are usually drawn at a specified distance apart. Regardless of the position of the parallel line, the steps used are the same.

RULE FOR CONSTRUCTING PARALLEL LINES

- Draw the first line.
- Set the compass the specified distance that the lines are to be drawn apart.
- Place the point end of the compass (or divider on metal layouts) on the first line near one end of it.
- Swing an arc on the side on which the parallel line is required.

- Leave the compass (or divider) set at the same distance and swing an arc at the other end of the line.
- Draw a line that touches the two arcs (is tangent). This line is parallel to the first line.

E XAMPLE: Draw a line parallel to the horizontal line *AB* and 1 inch away from it.

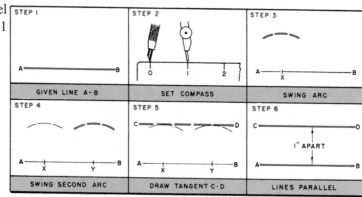

B. CONSTRUCTING ANGLES AND TRIANGLES

RULE FOR CONSTRUCTING AN ANGLE EQUAL TO A GIVEN ANGLE

- Draw the given angle.
- Lay out a line to be one side of the angle.
- Swing an arc, with the vertex of the angle as center, that intersects both sides.
- Transfer this arc, using the end point of the line as a center.
- Set the compass or divider to the chordal distance between the two intersecting points of the arc and the sides of the original angle.
- Swing this same chordal distance (arc) from the point where the arc intersects the side of the new angle.
- Draw a line from the end point through the point of intersection of the arcs.
- The new angle thus formed is equal to the given angle.

EXAMPLE: Draw an angle equal to angle *ABC*.

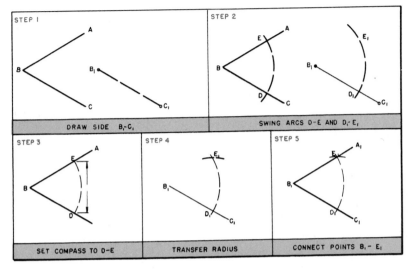

When constructing the sides of one angle equal to the sides of another angle and repeating the same steps to construct equal angles, it is possible to construct congruent angles, triangles, and other polygons.

C. DIVIDING A LINE

When a line of given length is bisected, it is divided into two equal parts. These are produced by swinging the same arc from both ends and drawing a line through the two points where the arcs intersect.

RULE FOR BISECTING A GIVEN LINE

- Set the compass at any radius greater than half the length of the given line.
- Place the point end of the compass at each of the end points of the line and swing an arc.
- Draw a light line through the points of intersection of the two arcs. The point where this line cuts the given line is the midpoint.

EXAMPLE: Lay out hole C the same distance from centers A and B in the plate illustrated.

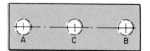

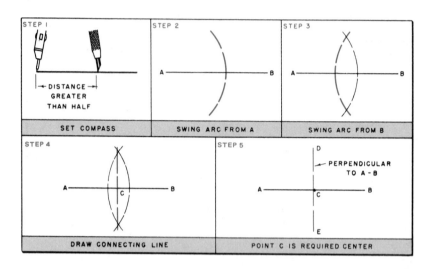

RULE FOR DIVIDING A LINE INTO A GIVEN NUMBER OF EQUAL PARTS

Method I
- Draw a line of given length.
- Draw another line at any angle and length to the given line from one end point.
- Step off this angular line the number of equal spaces into which the line is to be divided.
- Connect the last point with the second end point on the given line.
- Draw lines parallel to this connecting line. Lay out angles equal to the last angle at each point on the angular line. These lines are parallel to the first line drawn.

EXAMPLE: Divide a line $3\frac{5}{16}''$ long into three equal parts.

Method I

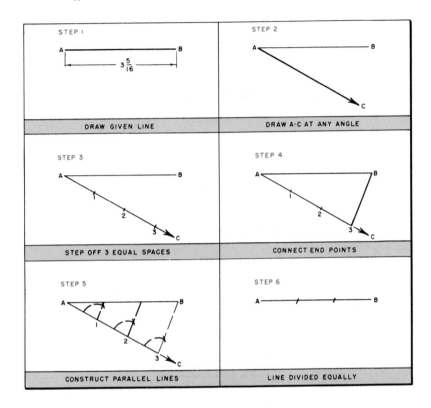

RULE FOR DIVIDING A LINE INTO A GIVEN NUMBER OF EQUAL PARTS

Method II

- Draw a line of given length.
- From each end point, on the opposite sides of the line, construct equal angles to the given line.
- Step off on each angular line the number of equal spaces into which the original line is to be divided.
- Connect each end point of the original line with the last step-off point on the angular lines.
- Draw lines parallel to these connecting lines from the step-off point.

Method II

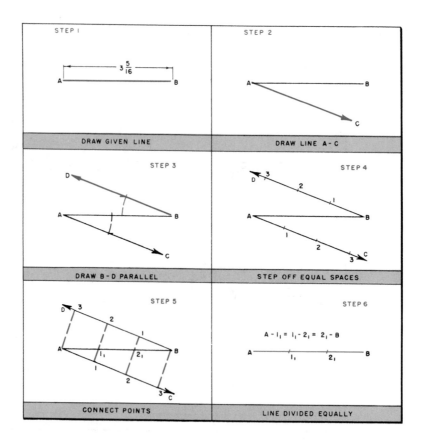

STEP 1

$3\frac{5}{16}$

A   B

DRAW GIVEN LINE

STEP 2

A   B

C

DRAW LINE A-C

STEP 3

D

A   B

C

DRAW B-D PARALLEL

STEP 4

D 3 2 1

A   B

1 2 3 C

STEP OFF EQUAL SPACES

STEP 5

D 3 2 1

A 1_1 2_1 B

1 2 3 C

CONNECT POINTS

STEP 6

$A - 1_1 = 1_1 - 2_1 = 2_1 - B$

A 1_1 2_1 B

LINE DIVIDED EQUALLY

D. ERECTING A PERPENDICULAR

Two lines are perpendicular when they form an angle of 90° with each other. When a perpendicular is to be erected at the center of a line of given length, the same steps are followed as for bisecting a line. If a perpendicular is to be erected from any other point on a line, another procedure is used.

RULE FOR ERECTING A PERPENDICULAR (AT A GIVEN POINT ON A LINE)

* Swing equal arcs from the given point to intersect the given line.
* Increase the size of the radius and swing equal arcs on the side of the line on which the perpendicular is to be erected. Use the intersecting points on the given line as centers.
* Draw a line from the point at which the two arcs intersect to the given point.
* This line is perpendicular at the given point to the given line.

EXAMPLE: Erect a perpendicular on line *AB* at *C*.

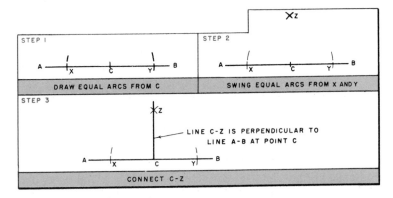

E. BISECTING A GIVEN ANGLE

RULE FOR BISECTING A GIVEN ANGLE

- Swing an arc (from the vertex of the angle) that intersects both sides.
- Use the intersecting points as centers and swing two more arcs to intersect.
- Draw a line that passes through the vertex and the point where the two arcs intersect.
- This line bisects the given angle into two equal angles.

EXAMPLE: Bisect angle *ABC*.

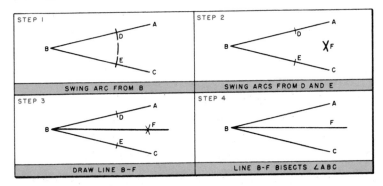

F. CONSTRUCTING TANGENTS TO A CIRCLE

The construction of tangents to circles, arcs, and straight lines is common in layout work. The drawing shows straight lines that are tangent to parts of circles and arcs that are tangent to other arcs. Two constructions are used widely in drawing tangents.

1. Constructing a tangent to a given circle at a given point.

2. Constructing a tangent to a circle from a given point outside the circle.

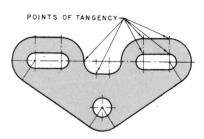

RULE FOR CONSTRUCTING A TANGENT (TO A GIVEN CIRCLE AT A GIVEN POINT)

- Draw the given circle and locate the given point on the circumference.
- Draw the radius from the center to the given point and extend it beyond the circle.
- Erect a perpendicular at the given point.
- The perpendicular is the tangent to the given circle at the given point.

EXAMPLE: Construct a tangent to circle O at point C.

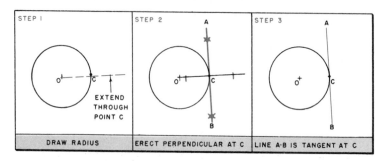

RULE FOR CONSTRUCTING A TANGENT (TO A GIVEN CIRCLE FROM A POINT OUTSIDE THE CIRCLE)

- Draw a line connecting the given point outside the circle with the center.
- Bisect this line.
- Use the center point of the bisected line as a center. With a radius equal to the distance from this center point to the center of the circle, draw an arc that intersects the circle.
- Draw a line between the given point outside the circle and the point of intersection of the arc on the circle.
- This line is tangent to the given circle from a point outside the circle.

EXAMPLE: Construct a tangent to circle O from point X.

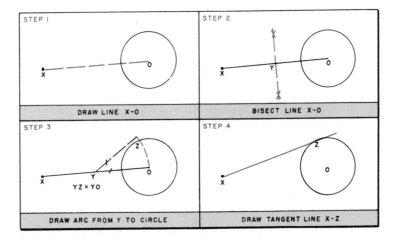

These basic constructions may be used alone or in combination as shown in the sketch.

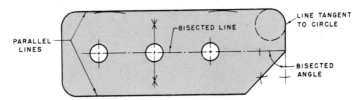

Review and Self-Test

ASSIGNMENT UNIT 83

Note. Show all construction lines for all problems.

A. Constructing Parallel Lines

1. Draw a parallel line for each line given in the table at the distance shown in each case.

Line	Length of Line	Distance
AB	6"	2"
CD	5"	$1\frac{1}{2}''$
EF	4"	$1\frac{1}{4}''$
GH	3"	$2\frac{1}{8}''$
IJ	2"	$1\frac{7}{16}''$

2. Draw two vertical lines of different lengths. Then draw the perpendicular bisector of each line. The perpendicular bisectors should be parallel to each other.

B. Dividing Lines

1. Bisect each line in the table.

Line	Length
AB	$4\frac{1}{2}''$
CD	$3\frac{3}{4}''$
EF	$2\frac{7}{8}''$
GH	10.5 cm
IJ	120 mm

2. Lay out the centers for each of the five holes on the Drill Plate. Solve by geometric construction and draw to full size.

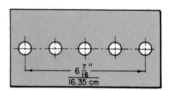

$6\frac{7}{16}''$
16.35 cm

C. Erecting Perpendiculars

1. Erect a perpendicular to each line given in the table at the point in the line as specified in each case.

Line	Length	Distance from End of Line
AB	5"	2"
CD	$4\frac{1}{2}''$	$1\frac{1}{2}''$
EF	$3\frac{3}{4}''$	$2\frac{3}{8}''$
GH	$5\frac{5}{8}''$	$2\frac{3}{8}''$
IJ	$4\frac{15}{16}''$	$1\frac{11}{16}''$

2. Erect a perpendicular to each given line from a point outside the line as indicated for each line in the table.

Line	Length and Point Outside Line
A – B	✗ A⊢————————————⊣B
C – D	C ＼———————＼ D ✗

D. Bisecting Angles

1. Lay out the five angles given in the table with a protractor. Then bisect each angle.
2. Lay out an angle of 90° with a protractor.
   a. Bisect the angle.
   b. Construct two 22 1/2° angles by bisecting the 45° angle.

Angle	Number of Degrees
ABC	90°
CED	120°
DEF	75°
FGH	140°
IJK	29°

3. Show the construction used to lay out the centers of the $\frac{1}{4}''$ circular slot.

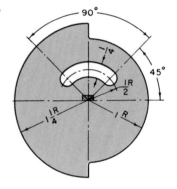

E. Constructing Tangents

1. Construct tangents to circle O at points (A), (B), and (C).

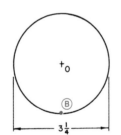
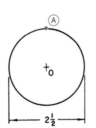
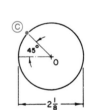

2. Construct tangents to circle (X) from points outside the circle at the distances given in the table.

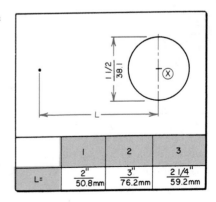

	1	2	3
L=	$\frac{2''}{50.8mm}$	$\frac{3''}{76.2mm}$	$\frac{2\ 1/4''}{59.2mm}$

F. Combining Basic Plane Geometric Constructions

1. a. Make a full-size layout of the Slotted Plate. Use basic geometric constructions for parallel and perpendicular lines, bisecting angles, and constructing tangent lines, circles, and arcs.

   b. Dimension the layout drawing.

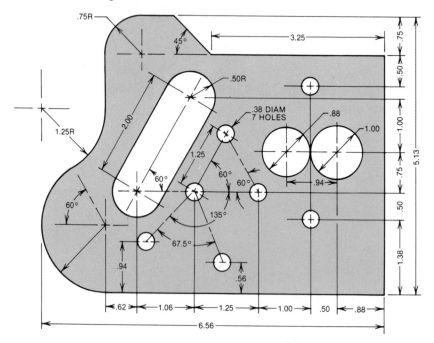

Unit 84 Constructions Applied to Geometric Shapes

OBJECTIVES OF THE UNIT

After satisfactorily studying this unit, the student/trainee will be able to
- *Apply principles relating to geometric constructions of parallel and perpendicular lines, dividing of lines and angles, and constructing tangents.*
- *Inscribe squares and hexagonal shapes in circles.*
- *Circumscribe regular polygons about circles.*

Geometry takes on increased value as more and more applications are made of the basic constructions in laying out geometric shapes. While the number of applications becomes limitless, only a few of the most widely used constructions are described in this unit.

A. INSCRIBING SQUARES AND HEXAGONS IN A CIRCLE

A square is *inscribed* in a circle when all corners of the square lie within the circumference of the circle.

RULE FOR INSCRIBING A SQUARE IN A CIRCLE

- Draw any diameter in the given circle.
- Construct a perpendicular to this diameter that passes through the center of the circle.
- Extend the perpendicular so it becomes a second diameter.
- Connect the point at which the diameter intersects the circumference with the next corresponding point.
- Repeat this process to draw the next three sides of the square.
- The square thus formed is inscribed in a circle of given diameter.

EXAMPLE: Inscribe a square in a 2″ diameter circle.

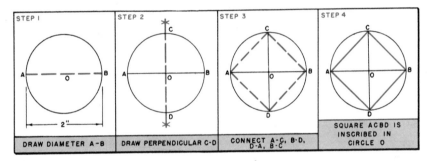

RULE FOR INSCRIBING A REGULAR HEXAGON IN A CIRCLE

- Take the radius of the circle and start at any point on the circumference to step off an arc.
- Use the point of intersection as a center and step off another arc of the same radius.
- Continue in the same way until the circle is divided into six equal parts.
- Connect each point of intersection with the next successive point until all points are connected.
- The six-sided figure inscribed in the circle is the required regular hexagon.

EXAMPLE: Construct a regular hexagon in a $1\frac{1}{2}''$ diameter circle.

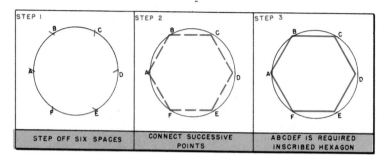

STEP 1	STEP 2	STEP 3
STEP OFF SIX SPACES	CONNECT SUCCESSIVE POINTS	ABCDEF IS REQUIRED INSCRIBED HEXAGON

B. CIRCUMSCRIBING SQUARES AND HEXAGONS

Circumscribing a polygon about a given circle is a geometric expression. It means that a specially shaped figure is constructed outside a circle so that each line of the figure is tangent to the circle.

RULE FOR CIRCUMSCRIBING REGULAR POLYGONS

- Divide the circle into the number of desired parts the same as for inscribed shapes.
- Draw radii from the point of intersection to the center of the circle.
- Draw a tangent line perpendicular to each of the radii.
- The object formed by the intersecting lines is the circumscribed polygon.

EXAMPLE: Circumscribe a regular hexagon about a 2″ diameter circle.

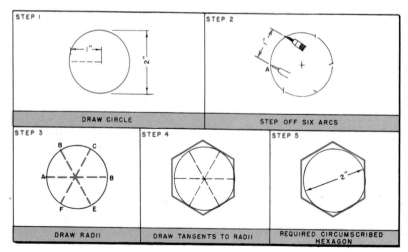

STEP 1	STEP 2
DRAW CIRCLE	STEP OFF SIX ARCS
STEP 3	STEP 4	STEP 5
DRAW RADII	DRAW TANGENTS TO RADII	REQUIRED CIRCUMSCRIBED HEXAGON

Steps similar to these may be followed in circumscribing other regular polygons about circles of known diameters.

C. GEOMETRIC CALCULATIONS WITH THE FOUR-DIGIT CALCULATOR

The importance of geometry is self-evident in the construction of every visible part, component, mechanism, and object. Some descriptions and measurements relate to closed flat surfaces on a plane. Triangles, circles, squares, parallelograms, trapezoids, and combinations of these shapes are examples of everyday applications.

Adaptations of these flat shapes are found in cones, pyramids, prisms, spheres, and other combinations. These shapes are identified as solids. Measurements apply to both plane and solid shapes. Direct measurements are taken with measuring instruments; others are computed.

An analysis of the formulas used to compute quantities such as linear, circular, area, and volume measurements reveals that they incorporate the four basic mathematical processes, higher powers, and roots of numerical values. Table 10 in the Appendix lists calculator processes that relate to measurements of common geometric forms. Examples of calculator processes, in figure 84-1, abstracted from the full table, show how formulas and a calculator are used to simplify mathematical computations.

Geometric Calculation		Formula	Calculator Processes	Measurement Unit Values
Areas	Triangle	$A = s^2$	$(s) \times (s)$	square units
Perimeter	Rectangle	$P = 2(l \times w)$	$(l \times w) \times 2$	base unit
Degrees and Radians	Angle (as a radian value)	$\angle A = \dfrac{n°}{57.295\ 78°}$	$n°\ \ 57.295\ 78$	radian
Volume	Sphere	$V = \dfrac{4\pi r^3}{3}$	$(4) \cdot (\pi) \cdot (r) \cdot (r) \cdot (r) \div 3$	cubic units

FIGURE 84-1 Samples of calculator processes for common geometric constructions.

Review and Self-Test

ASSIGNMENT UNIT 84
Note. Show all construction, where used, for all problems.

A. Inscribing Polygons in Circles

1. Construct each of the three polygons (a, b, c). Use a straight edge and compass.

 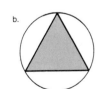

2. A piece of round stock is $2\frac{1}{2}''$ in diameter. Construct the largest square that can be machined at one end.

3. What is the distance across flats for the hexagon that can be planed on a 6″ diameter post?

B. Circumscribing Regular Polygons about Circles

1. Circumscribe the square, equilateral triangle, and regular hexagon (a, b, c) about the circles for the diameter given for each shape.

a.

b.

c.

2. Circumscribe a regular hexagon about a 2″ diameter circle.
   a. Determine the distance (X) across corners by measuring.

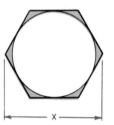

3. a. Lay out the Sheet Metal Template by using geometric constructions. Draw to a scale of 1:2.
   b. Show all construction details.
   c. Dimension the drawing.

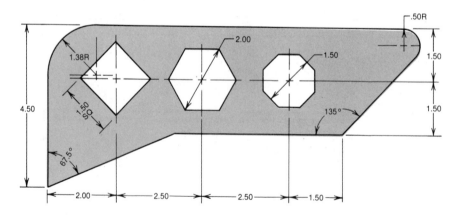

Unit 85 Achievement Review on Geometric Constructions

OBJECTIVES OF THE UNIT

This achievement review serves as an overall test for Section 18. The unit is designed to measure the student's/trainee's ability to

- *Visualize the relationship among all the basic geometric terms that apply to lines, angles, circles, and regular polygons and tangents.*
- *Draw parallel, perpendicular, and straight-line tangents; divide lines into any given number of equal parts; bisect angles and construct tangent arcs.*
- *Apply geometric construction principles and practices to parts drawings. Make line drawings of parts and mechanisms that involve combinations of regular polygon forms, triangular features, circular areas, and contoured surfaces.*

A. BASIC CONSTRUCTIONS

1. Determine whether or not the dotted lines constructed with the framing square are parallel. Give reasons.

2. The value of angle *A* is 110° 30′. Lines 1-2 and 3-4 are parallel. Find the value of each angle (*B* through *G*).

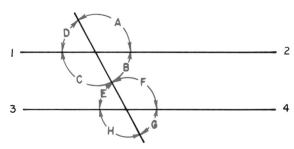

3. Lay out the 45° elbow. Start with a straight angle.
   a. Erect a perpendicular to the straight angle.
   b. Bisect the 90° angle formed.
   c. Draw lines parallel to the center lines to form the sides, and perpendicular lines to form the ends.

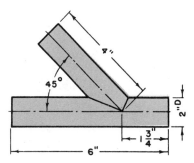

B. BASIC CONSTRUCTIONS APPLIED TO GEOMETRIC SHAPES

1. Construct the template according to the dimensions as given.

   Note. Show all lines used for constructing parallel lines, erecting a perpendicular, bisecting an angle, and drawing tangents.

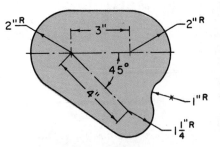

2. Construct the special plate with the circumscribed square and the inscribed portion of the regular hexagon. Show all construction lines.

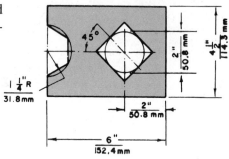

PART SIX

Fundamentals of Applied Trigonometry (with Calculator Applications)

19

RIGHT, ACUTE, AND OBLIQUE TRIANGLES

Unit 86 The Concept of Right Triangle Trigonometry

OBJECTIVES OF THE UNIT

After satisfactorily studying this unit, the student/trainee will be able to

• *Diagram, label, and interpret relationships in trigonometry of the angles and sides of right triangles.*

• *Identify and apply the six basic trigonometric functions of right triangles in a variety of problem situations.*

• *Solve practical problems involving linear and angular measurements and calculations involving the basic functions, including the cofunctions of cotangent, secant, and cosecant.*

• *Apply the Pythagorean theorem and, in combination with other trigonometric processes, solve technical word problems associated with vectors and other measurements dealing with right triangles.*

• *Set up vector diagrams and solve representative physical science and business and industry problems.*

• *Determine precision angle gage-block combinations to set up for tooling, measurement, or inspection of angles within ±1″.*

The word trigonometry, as derived from the Greek, means two things: *triangle* and *measurement*. Trigonometry is a branch of mathematics that deals with the measurement of angles, triangles, and distances. A working knowledge of trigonometry requires an understanding of fundamental principles and constructions in geometry. Also, since symbols are used and algebraic equations are constructed, problems in trigonometry are solved by using algebra.

Trigonometry is preferred by many workers to other branches of mathematics. Trigonometry conserves time and effort and simplifies the solution of common technical problems in different occupations and in consumer applications.

A. PARTS OF THE RIGHT TRIANGLE

While trigonometry includes both the right triangle and the oblique triangle, the right triangle is covered in this unit. A *right triangle,* as the name implies, is a triangle with one right angle. In the illustration of the right triangle *ABC,* the side opposite the right angle *C* is the *hypotenuse* (*c*). The side opposite the acute angle *A* is called the *side opposite* (*a*). Adjacent to angle *A* is the *side adjacent* (*b*). Note in figure 86-1 and in the explanation that the angles are indicated by capital letters. The sides are denoted by either lowercase letters or the words opposite, adjacent, or hypotenuse.

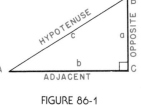

FIGURE 86-1

B. TRIGONOMETRIC FUNCTIONS

Six terms are widely used in trigonometry to express the ratios between the sides. The terms and the abbreviations of each are *sine* (sin), *cosine* (cos), *tangent* (tan), *cotangent* (cot), *secant* (sec), and *cosecant* (csc).

The sine (sin) is the ratio of the side opposite to the hypotenuse. In triangle *ABC,* the

$$\sin A = \frac{\text{side opposite}}{\text{hypotenuse}} = \frac{a}{c}$$

$$SIN\ A = \frac{a}{c}$$

The cosine (cos) is the ratio of the side adjacent to the hypotenuse.

$$\cos A = \frac{\text{side adjacent}}{\text{hypotenuse}} = \frac{b}{c}$$

$$COS\ A = \frac{b}{c}$$

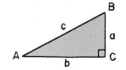

The tangent (tan) is the ratio of the side opposite to the side adjacent.

$$\tan A = \frac{\text{side opposite}}{\text{side adjacent}} = \frac{a}{b}$$

$$TAN\ A = \frac{a}{b}$$

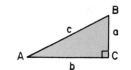

The cotangent (cot) is the ratio of the side adjacent to the side opposite.

$$\cot A = \frac{\text{side adjacent}}{\text{side opposite}} = \frac{b}{a}$$

The secant and cosecant of angle *A* may be expressed in algebraic form.

$$\sec A = \frac{\text{hypotenuse}}{\text{side adjacent}} = \frac{c}{b} \qquad \csc A = \frac{\text{hypotenuse}}{\text{side opposite}} = \frac{c}{a}$$

A comparison of the values of the six trigonometric functions shows two things. First, the sine and cosine, tangent and cotangent, and secant and cosecant may be grouped in pairs. Second, three of the trigonometric functions are reciprocals of the other three. Starting with the sine:

$$\sin A = \frac{a}{c} \text{ and the cosecant reciprocal, } \csc A = \frac{c}{a}$$

$$\cos A = \frac{b}{c} \text{ and the secant reciprocal, } \sec A = \frac{c}{b}$$

$$\tan A = \frac{a}{b} \text{ and the cotangent reciprocal, } \cot A = \frac{b}{a}$$

C. FUNCTIONS AND COFUNCTIONS

In the right triangle *ABC*, angle *A* and angle *B* are *complementary* to each other and the sum of the two angles is 90%. The term *cofunction* refers to the function of the complementary angle.

The cofunction of the sine is the cosine, which is a contraction of the phrase ''the sine of the complement.'' The cofunction of the tangent is the cotangent; that of the secant, the cosecant.

- The sine of an acute angle is equal to the cosine of its complement.
- The tangent of an acute angle is equal to the cotangent of its complement.
- The secant of an acute angle is equal to the cosecant of its complement.

With the knowledge that the function of one acute angle in a right triangle is equal to the cofunction of the complement, it is possible to work with either acute angle. For instance, the tan of 30° = cot of 60°; the sin of 25° = cos 65°; the sec 75° = csc 15°.

D. VECTORS, DIAGRAMS, AND MEASUREMENT OF FORCES

Many physical science and trade, technical, and business occupations involve trigonometric applications that deal with *vectors*. (Vectors are called *phasors* in electrical engineering and related fields.) A *vector* is a force that has a *known direction* and *known magnitude*. Magnitude refers to the size, or amount, or quantity.

The term *scalar quantity* is used with vectors to represent a physical quantity. Scalar quantity means a *quantity (number)* and a *unit of measurement*. A cutting speed of 4.75"/min, a speed of 70 km/hr, and a force of 60 pascals per square meter (60 Pa/m^2) are examples of scalar quantities.

Characteristics of Vectors and Representation

Vector quantities, having both magnitude and direction, are widely used to represent force, velocity, or displacement. *Displacement* refers to a *change* in position. Vectors are stated in terms of direction and magnitude for *force problems;* speed of a body and direction for *velocity problems;* and by speed and direction for *displacement problems.*

Sketches, called diagrams, are used to show vectors. The magnitude is represented by the length of a vector line segment. Vectors are usually drawn to scale. An arrowhead at the end of a vector line shows the

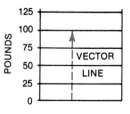

FIGURE 86-2

direction of force. An upward force of 100 pounds may be represented graphically as shown by the vector line (*figure* 86-2).

When two forces are applied at an angle to each other, the tail of each successive vector is placed at the termination point of the preceding vector. The *resultant force,* in the case of forces acting at a right angle, is the hypotenuse of the triangle (*figure* 86-3).

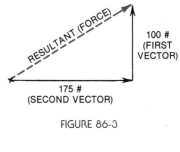

FIGURE 86-3

The resultant is drawn as a vector that connects the tail of the first vector with the vertex of the arrowhead of the last vector quantity. Figure 86-4 shows three forces and the resultant force.

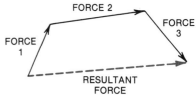

FIGURE 86-4

Solving Vector Problems by Graphic Methods

Vector problems are often solved by graphing the vector quantities and labeling the scalar quantities. If measurements with coarse limits of accuracy are required and the quantities are represented to a scale on grid sheets, the resultant may be read directly for differing conditions. More accurate measurements may require mathematical computations by conventional algebra, geometry, or trigonometry or the use of these processes in combination with certain calculator functions.

Laws Relating to Vectors

Vector problems usually require the scalar quantity of a *resultant vector.* Guidelines for determining the resultant vector are found in three laws.

- The resultant vector of *parallel forces* moving in the same direction is equal to the *sum of the forces.*
- The resultant vector of parallel forces acting in opposite directions is equal to the *difference between the forces.*
- The resultant vector of *angular forces* at right angles is equal to the scalar quantity of the hypotenuse.
- The resultant vector of angular forces acting at angles other than 90° is equal to the diagonal drawn after a parallelogram is formed. The method of determining the resultant force is called the *composition of forces.*

Measurement of Opposite and Equal Forces: Equilibrant

Some applications require an equal and opposite force to the resultant in order to *balance* the vector forces. The new force is known as the *equilibrant.* Refer to figure 86-5. Two forces acting at a right angle are graphed to a scale of one square for each 10 pascals (10 Pa).

A parallelogram is formed. The resultant force is represented by the diagonal. Using the scale, the resultant is measured to be equal to 72 Pa. The equilibrant, as an equal and opposite force, is also 72 Pa. The equilibrant may also be calculated mathematically as the hypotenuse of a right triangle whose legs represent 40 Pa and 60 Pa.

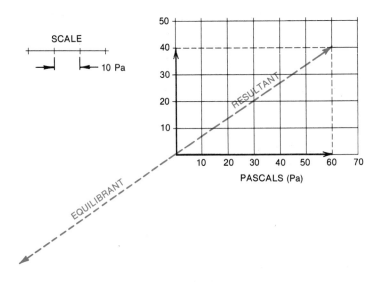

FIGURE 86-5

E. APPLYING THE PYTHAGOREAN THEOREM IN COMBINATION WITH TRIGONOMETRIC FUNCTIONS

This theorem was discovered by the Greek mathematician Pythagoras. As stated before, this theorem is used to determine the length of the hypotenuse or a leg of a right triangle when two of the three legs (sides) are known. Once the value of the sides is computed, any of the trigonometric functions may be applied to find the remaining angles of a right triangle.

Pythagorean Theorem Relating to Right Triangles

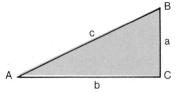

- The square of the hypotenuse equals the sum of the squares of the two other sides (legs).

$$\text{Hyp}(a)^2 = b^2 + c^2; \qquad a = \sqrt{b^2 + c^2}$$

- The square of a side (leg), other than the hypotenuse, is equal to the difference between the squares of the two remaining sides.

$$\text{Side } (b)^2 = (a)^2 - (c)^2 \qquad \text{Side } (c)^2 = (a)^2 - (b)^2$$

$$(b) = \sqrt{(a)^2 - (c)^2} \qquad (c) = \sqrt{(a)^2 - (b)^2}$$

Practical Applications of Vectors, Pythagorean Theorem, and Trigonometric Functions

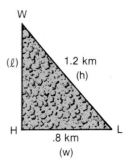

W

(ℓ) 1.2 km (h)

H .8 km L
(w)

EXAMPLE: *Case 1.* Determine the length of the wooded land area (*l*) and angles (*W*) and (*L*).

Note. Use the Pythagorean theorem to find (*l*). Then, calculate the angle values by the laws of sines and cosines. Round off the length to four decimal places and the angles to ±15″.

Step 1 Calculate the length.

$$(h)^2 = (l)^2 + (W)^2$$

$$(l)^2 = (h)^2 - (W)^2$$
$$l = \sqrt{(h)^2 - (W)^2}$$
$$= \sqrt{(1.2) - (.8)}$$
$$= \sqrt{.80} = 0.8944 \text{ km} \quad \text{Ans}$$

Step 2 Use the law of sines to calculate the trigonometric function.

$$\sin \angle l = \frac{.8944}{1.2} = 0.74533 \quad \text{Ans}$$

Step 3 Refer to a table of trigonometric functions and establish the angle.

Note. Since the value lies between 48° 11′ and 48° 12′, the exact value is interpolated. The computed sine value of 0.74533 is $\frac{5}{20}$ of 60 seconds or 15″.

Table Values
sin 0.74548 = 48° 12′
sin 0.74528 = 48° 11′
  0.00020
sin 0.74533 = 48° 11′ 15″ Ans
      Angle (*l*)

Step 4 Subtract the computed angle from 90° to find ∠*W*.

89° 59′ 60″
−48° 11′ 15″
=41° 48′ 45″ for ∠*W* Ans

Many problems in the electrical, electronics, instrumentation, aerospace, and other high technology fields are solved with vector processes, by applying the Pythagorean theorem and with trigonometric functions.

The problems are usually centered around a schematic drawing like the illustration (*figure* 86-6) of a resistance (*R*), inductance (*L*), and capacitance (*C*) circuit, called an RLC circuit.

The effective voltage in an AC circuit consists of (*R*), (*L*), and (*C*). The two voltage drops of V_L and V_C are represented in the vector diagram (*figure* 86-7) as being 180° apart. The voltage drop (V_R) across the resistor is 90° out-of-phase.

The *resultant voltage* is equal to the vector sum of the three voltages. Since the resultant of parallel forces (V_L and V_C)

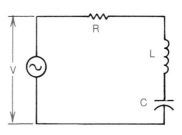

FIGURE 86-6

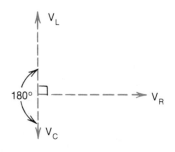

FIGURE 86-7

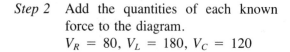

FIGURE 86-8

represents the difference, the new vector diagram (*figure* 86-8) shows this relationship. By adding the V_R force, drawing the parallelogram and then the diagonal, the resultant angle formed is called the phase angle or $\angle\theta$. This angles measures how much the voltage leads ($+$) or lags ($-$) the current.

EXAMPLE: *Case 1.* Determine the total voltage (V_T) of the circuit specified in the diagram.
Note. Solve for V_T by using the Pythagorean thorem.

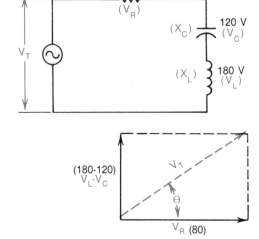

Step 1   Draw a vector diagram. Label the forces.

Step 2   Add the quantities of each known force to the diagram.
$V_R = 80$, $V_L = 180$, $V_C = 120$

Step 3   Substitute the industry symbols for the conventional symbols used with the Pythagorean formula.

Step 4   Substitute the voltage values in the formula.

Step 5   Carry out the mathematical processes. The total voltage (V_T) of the circuit is 100 volts.

From $a^2 = b^2 + c^2$
To $(V_T)^2 = (V_R)^2 + (V_L - V_C)^2$
$V_T = \sqrt{(V_R)^2 + (V_L - V_C)^2}$
$V_T = \sqrt{(80)^2 + (180 - 120)^2}$

$= \sqrt{6400 + 3600}$
$= \sqrt{10\,000} = 100$ volts   Ans

EXAMPLE: *Case 2*. Determine the V_T of the preceding circuit by using trigonometric functions.

Step 1 Use the same vector diagram as in Case 1.

Step 2 Apply the tangent function to the phase angle ($\angle\theta$).

Step 3 Substitute the voltage values for V_L, V_C, and V_R.

Step 4 Refer to a trigonometry table for the phase angle ($\angle\theta$) value of the trigonometric function: tan 0.7500.

From $\tan A = \dfrac{\text{opposite side } (a)}{\text{adjacent side } (b)}$

To $\tan C = \dfrac{(V_L - V_C)}{V_R}$

$= \dfrac{180 - 120}{80}$

$= 0.7500$

$= 36°\ 52'$ **Phase angle**   Ans

Step 5 Use the sine of 36° 52′ to calculate the total voltage (V_T).

$$\sin\theta = \frac{\text{side opposite}}{\text{hypotenuse}}$$

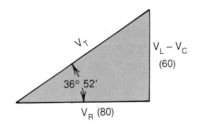

Step 6 Substitute the values for volts and degrees in the sine formula.

Step 7 Label the answer.

$$\sin 36°\ 52' = \frac{60}{V_T}$$

$$V_T = \frac{60}{0.59995}$$

$$= 100 \text{ volts}\quad\text{Ans}$$

F. PRECISION ANGULAR GAGE-BLOCK CALCULATIONS AND SETUPS

Parallel gage blocks were applied in earlier units to tooling setups, inspection, and linear measurements. The measurements were computed in either customary inch or SI metric units of measure. Accuracies fell within precision limits as fine as $\pm 0.000\ 001''$ (one microinch, μin) or $\pm 0.000\ 002$ mm (two micrometers, μm).

Similarly, angle gage-block sets are available for setups involving angular measurements in increments of one second and to accuracies of $\pm\frac{1}{4}$ second ($\pm 0.25''$). The table provides an example of the number and different angles of gage blocks that are included in a popular 16-block set.

Number of Blocks	Angle Series	Gage Block Sizes
6	Degree	1, 3, 5, 15, 30, 45
5	Minute	1, 3, 5, 20, 30
5	Second	1, 3, 5, 20, 30

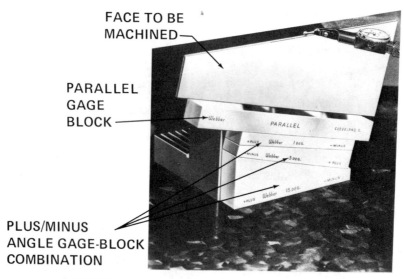

FIGURE 86-9 Angle gage-block setup (*Courtesy of Webber Division, the L. S. Starrett Company*)

The gage blocks may be positioned in *plus* and *minus* relationships. The photograph (*figure 86-9*) shows a typical angle gage-block setup. Angles may be established with this set in one-second increments and ranging from 0° to 99°.

RULE FOR SETTING UP ANGLE GAGE BLOCKS

- Select the angle gage block or combination of blocks that, when added (plus) or subtracted (minus), produce the required number of *degrees*.
- Repeat the same process to select the blocks for the required number of *minutes*.
- Repeat the selection process to meet the required number of *seconds* in the angle setup.
- Clean, slide, and wring the selected combinations of degree, minute, and second blocks to form the required angle.

EXAMPLE: Select the combination of angular gage blocks necessary to position the workpiece so that the 14° 20′ 32″ ± 1″ angle may be produced.

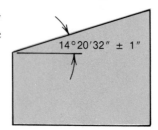

14°20′32″ ± 1″

Step 1   Select the 15° and 1° angle blocks.

15°
− 1°
14°   combination

Step 2   Clean and wring the −*1*° face and the *plus 15*° side together to produce the 14°.

Step 3   Select a 20′ gage block and wring it to the 14° combination.

+ 20′ block = 14° 20′

Step 4   Select 30″, 5″, and 3″ gage blocks. Wring the plus 30″ and plus 5″ faces together. Match the 35″ combination with the minus 3″ face to produce the 32″ measurement.

30″ angle block
+ 5″   ″   ″
35″ combination
− 3″ angle block
= 32″ combination

Step 5   Wring this last combination of blocks to the 14° 20′ combination.

Step 6   Check the angles and plus and minus combinations against the required 14° 20′ 32″.

14° + 20′ + 32″ =
14° 20′ 32″   Check

Review and Self-Test

ASSIGNMENT UNIT 86

A. Parts of the Right Triangle

1. Letter the angles and name the side opposite, side adjacent and hypotenuse of each right triangle.

2. Draw each of the right triangles according to the dimensions given in the table. Label the angles and dimension the sides.

Triangle	Right Angle	Length of Side	
ABC	B	A	C
		3″	3″
LMN	N	L	M
		$2\frac{1}{4}''$	3″
XYZ	X	Y	Z
		5.25″	4.50″

B. Trigonometric Functions (Practical Problems)

1. Prepare a table similar to the one shown. Complete the information called for in each column.

Function	Abbreviation	Trigonometry Ratios	
		Expressed as Sides	Sides Represented by Letters
Sine			
Cosine			
Tangent			
Cotangent			
Secant			
Cosecant			

2. The included angle of the National Form thread is 60°. The crest and the root of the thread are each made flat for a depth of $\frac{1}{8}$ the pitch. Sketch a single thread and show by dotted lines the trapezoid at the crest of the thread and the triangle at the root that are allowed for strength and clearance.

3. Determine the decimal value of each one of the six trigonometric ratios as they apply to angle D.

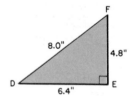

4. Guidelines for laying up ceiling tiles on a job are checked for squareness. A measurement of 1.62 meters is marked on line AB and 2.46 meters on line AC. Determine measurement BC, correct to three decimal places.

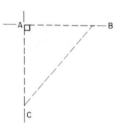

5. A derrick rig is 19.51 meters high. A guy wire is attached 1.83 meters from the top and 7.62 meters out from the base. Compute the length of the guy wire to two decimal places.

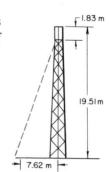

C. Functions and Cofunctions (General Applications)

1. Give the cofunction of each angle in a right triangle.

Function	Cofunction
sin 45°	
tan 20°	
sec $29\frac{1}{2}°$	

2. a. Make a freehand sketch of each of the six right triangles given in the table. Label the angles with the given letters and the sides with corresponding lowercase letters.

b. State the ratio of the sides for each function A through F.

c. Give the cofunction of each of the six functions.

	Right Triangle	Function
A	BAC	sin∡A
B	DEF	tan∡E
C	JKL	sec∡K
D	MNO	sin 30°
E	STU	tan 25°
F	XYZ	sec $37\frac{1}{2}°$

D. Applications of the Pythagorean Theorem in Combination with Trigonometric Functions

1. Use the Pythagorean theorem to calculate the length to which rafters are to be cut to produce the rise as specified on the drawing.

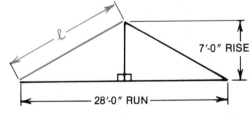

2. Determine the offset according to the given dimensions for the piping system.

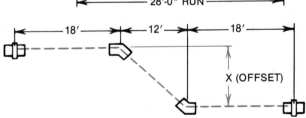

3. Check the accuracy of the 90° angle (±5′) from the dimensions given for the three reamed holes in the Die Block.

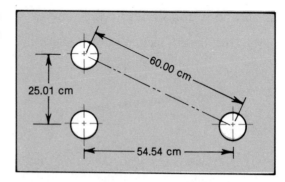

4. Compute the magnitude of the instantaneous voltage (e) in an alternating current application. The alternating current has reached 60° of its cycle. The maximum voltage produced by the circuit is 720 volts.

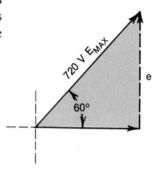

Note that the force diagram shows the voltage at 60°, the resultant E_{max} of 72 volts. The instantaneous voltage (e) is represented by the vertical vector.

E. Vector Measurements

1. a. Solve for impedance (Z) of the circuit as described by the diagram.
   b. Use the force diagram and the computed value of the impedance of the circuit to determine the negative value phase angle $\angle\theta$.

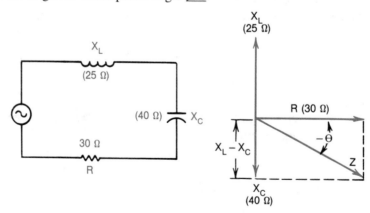

Note. In the ac circuit, the resistance in the inductor is represented by X_L; resistance of the capacitor by X_C. The formula for determining the impedance is

$$Z = \sqrt{R^2 + (X_L - X_C)^2}.$$

The tangent formula is used to find the phase angle ($\angle\theta$).

2. a. Compute the resultant force of two forces that are acting at 90° to each other. Force 1 of 22.5 kg/m² is exerted horizontally. Force 2 of 72.6 kg/m² is exerted vertically. *Note.* Round the answer to two decimal places.
   b. Determine the angle of the resultant force to the horizontal plane. State the angle as a decimal-degree value, correct to two decimal places.
   c. Convert the decimal-degree value to degrees, minutes, and seconds with the seconds rounded to one decimal place.

F. Determining Gage-Block Combinations to Produce Required Angles

1. a. Select the precision angle gage-block combinations that will produce the two required angles shown on the drawing.
   b. Check the combinations against the required settings.

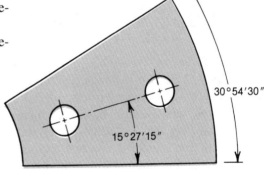

30°54'30"

15°27'15"

Unit 87 Applications of Tables of Trigonometric Functions

OBJECTIVES OF THE UNIT

After satisfactorily studying this unit, the student/trainee will be able to
- *Determine the sign (+ or −) of any trigonometric function according to the quadrant in which the angle falls.*
- *Read the numerical value of any trigonometric function or cofunction from 0° to 360° that appears in tables.*
- *Determine the angles of trigonometric function values directly from tables using English units of degrees and minutes or the decimal-degree system as required in SI metric problems.*
- *Interpolate numerical values for precise angles that require accuracies in the second of a degree range or two or more decimal places.*
- *Solve general and applied technology problems by trigonometry that involve direct reading of angle measurements or the interpolation of numerical values of trigonometric functions from handbook tables.*

The values of trigonometric functions may be found by construction and measurement or by tables. Of the two methods, the tables are more reliable where precision is necessary. The average accuracy where measuring rules and instruments are used is within two or three decimal places. Most tables of trigonometric functions are carried to four or more decimal places.

As long as the two acute angles in a right triangle remain the same, the trigonometric function of each angle is constant and does not change regardless of the size of the triangle. The sine of a 45° angle of a right triangle with 4-inch sides is the same as for one with 12-inch sides or any other length sides.

A. CONSTRUCTION AND MEASUREMENT OF TRIGONOMETRIC FUNCTIONS

RULE FOR FINDING TRIGONOMETRIC FUNCTIONS (BY CONSTRUCTION AND MEASUREMENT)

Method I (Given an Angle)
- Construct the desired angle.
- Measure a convenient even distance on the hypotenuse.
- Drop a perpendicular from the end point of the hypotenuse to the side adjacent.
- Measure the side opposite.
- Divide the two measurements. The quotient is the value of the sine of the angle.

The value of the other functions may be found by using the same steps and the actual measurements of the sides used.

Method II (Given the Value of a Function)
- Draw a horizontal line.
- Erect a perpendicular at the end of the line.
- Measure the given distance on the perpendicular.
- Swing a 1-inch arc from the end point of the perpendicular to intersect the horizontal line.
- Draw the hypotenuse through the end point and the intersecting point.
- Measure the acute angle formed with the horizontal line.

EXAMPLE: Given the sin D = .7071, find the angle by construction.

Step 1 Draw a horizontal line AB.

Step 2 Erect a perpendicular at B.

Step 3 Measure .7071 inch on BC as accurately as a measuring tool or instrument permits.

Step 4 Swing a 1-inch arc from the .7071-inch mark to intersect line A-B.

Step 5 Connect points D and C with a straight line.

Step 6 Measure angle D (= 45°). Thus, the angle with the sine of .7071 is 45°.

B. READING TABLES OF NATURAL TRIGONOMETRIC FUNCTIONS

It is apparent that while the construction methods are interesting and practical for the layout of many jobs, precision is not possible. Since the value of the function of an angle is constant (once the value of angles is computed accurately) the results may be combined in table form. One of the simplest tables of natural trigonometric functions is illustrated. The more complicated tables give five or more decimal places for each degree and each minute in a degree.

Angle	Sine	Cosine	Tangent	Angle	Sine	Cosine	Tangent	Angle	Sine	Cosine	Tangent
1°	.0175	.9998	.0175	31°	.5150	.8572	.6009	61°	.8746	.4848	1.8040
2°	.0349	.9994	.0349	32°	.5299	.8480	.6249	62°	.8829	.4695	1.8807
3°	.0523	.9986	.0524	33°	.5446	.8387	.6494	63°	.8910	.4540	1.9626
4°	.0698	.9976	.0699	34°	.5592	.8290	.6745	64°	.8988	.4384	2.0503
5°	.0872	.9962	.0875	35°	.5736	.8192	.7002	65°	.9063	.4226	2.1445
6°	.1045	.9945	.1051	36°	.5878	.8090	.7265	66°	.9135	.4067	2.2460
7°	.1219	.9925	.1228	37°	.6018	.7986	.7536	67°	.9205	.3907	2.3559
8°	.1392	.9903	.1405	38°	.6157	.7880	.7813	68°	.9272	.3746	2.4751
9°	.1564	.9877	.1584	39°	.6293	.7771	.8098	69°	.9336	.3584	2.6051
10°	.1736	.9848	.1763	40°	.6428	.7660	.8391	70°	.9397	.3420	2.7475
11°	.1908	.9816	.1944	41°	.6561	.7547	.8693	71°	.9455	.3256	2.9042
12°	.2079	.9781	.2126	42°	.6691	.7431	.9004	72°	.9511	.3090	3.0777
13°	.2250	.9744	.2309	43°	.6820	.7314	.9325	73°	.9563	.2924	3.2709
14°	.2419	.9703	.2493	44°	.6947	.7193	.9657	74°	.9613	.2756	3.4874
15°	.2588	.9659	.2679	45°	.7071	.7071	1.0000	75°	.9659	.2588	3.7321
16°	.2756	.9613	.2867	46°	.7193	.6947	1.0355	76°	.9703	.2419	4.0108
17°	.2924	.9563	.3057	47°	.7314	.6820	1.0724	77°	.9744	.2250	4.3315
18°	.3090	.9511	.3249	48°	.7431	.6691	1.1106	78°	.9781	.2079	4.7046
19°	.3256	.9455	.3443	49°	.7547	.6561	1.1504	79°	.9816	.1908	5.1446
20°	.3420	.9397	.3640	50°	.7660	.6428	1.1918	80°	.9848	.1736	5.6713
21°	.3584	.9336	.3839	51°	.7771	.6293	1.2349	81°	.9877	.1564	6.3138
22°	.3746	.9272	.4040	52°	.7880	.6157	1.2799	82°	.9903	.1392	7.1154
23°	.3907	.9205	.4245	53°	.7986	.6018	1.3270	83°	.9925	.1219	8.1443
24°	.4067	.9135	.4452	54°	.8090	.5878	1.3764	84°	.9945	.1045	9.5144
25°	.4226	.9063	.4663	55°	.8192	.5736	1.4281	85°	.9962	.0872	11.4301
26°	.4384	.8988	.4877	56°	.8290	.5592	1.4826	86°	.9976	.0698	14.3007
27°	.4540	.8910	.5095	57°	.8387	.5446	1.5399	87°	.9986	.0523	19.0811
28°	.4695	.8829	.5317	58°	.8480	.5299	1.6003	88°	.9994	.0349	28.6363
29°	.4848	.8746	.5543	59°	.8572	.5150	1.6643	89°	.9998	.0175	57.2900
30°	.5000	.8660	.5774	60°	.8660	.5000	1.7321	90°	1.0000	.0000	

FIGURE 87-1 Table of natural trigonometric functions (four place, 0° through 90°)

RULE FOR READING A TABLE

* Solve for one of the trigonometric functions.
* Locate the function in the appropriate column of a table of trigonometric functions.
* Read the degree equivalent to the closest value of the function.
  Note. Where a larger table is used, the reading of the angle may be in degrees and minutes.

EXAMPLE: Find the two acute angles in triangle *ABC* whose sides are 3″, 4″, and 5″.

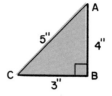

Step 1 Determine the value of the functions sin *C* and sin *A*. $\sin C = \frac{4}{5} = .8000$

$\sin A = \frac{3}{5} = .6000$

Step 2 Find the angle closest to .8000 in the sine column of a $\sin 53° = .7986$
table of trigonometric functions. $\sin 37° = .6018$

Step 3 Repeat the step for the .6000 value. $\angle C = 53°$

Step 4 Check to see that the complementary angles equal 90°. $\angle A = \dfrac{37°}{90°}$ Check

C. INTERPOLATION OF TABLES OF TRIGONOMETRIC FUNCTIONS

Angles that are given in degrees or trigonometric ratios of whole degrees may be read directly on all tables of trigonometric functions. The longer tables make it possible to read degrees and minutes directly from the table.

When such tables are not available, or when degrees, minutes, and seconds are needed and direct values are not accessible in table form, the values are *interpolated. Interpolation* is the process of finding an exact value between two consecutive values in a series. For instance, if the sine of an angle of 45° 30′ 25″ is needed, and a table is available reading in minutes, the desired reading falls between sin 45° 30′ and sin 45° 31′. To be exact, the numerical value is $\frac{25}{60}$ of the difference between 30′ and 31′ more than the sin 45° 30′ value.

RULE FOR INTERPOLATING TABLES OF TRIGONOMETRIC FUNCTIONS

- Find the numerical difference in a table of trigonometric functions between the function of the given number of minutes and the next larger number.
- Place the given number of seconds over 60.
- Multiply the numerical difference by the fraction.

Sine, Tangent, and Secant Functions
- Add the product to the function of the given number of minutes.
The sum is the value of the sine, tangent, or secant function of an angle in degrees, minutes, and seconds.

Cosine, Cotangent, and Cosecant Functions
- Subtract the product from the function of the given number of minutes.
The difference is the value of the cosine, cotangent, and cosecant functions.

E XAMPLE: Determine the sine value of angle A.

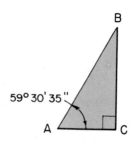

Step 1   Locate two numbers in the sine column of a table
of trigonometric functions, one $1'$ smaller and
one $1'$ larger.

Step 2   Subtract. Then take $\frac{35}{60}$ of the difference (.00015).

Step 3   Add the difference to the function of the smaller
angle.

The sum .86172 is the numerical value of the sin
59° 30′ 35″.

Note. If the value of the function is decreasing,
subtract rather than add the computed amount.

$$\textbf{sin 59°31}' = \textbf{.86178}$$
$$\textbf{sin 59°30}' = \textbf{.86163}$$
$$\textbf{Difference} = \textbf{.00015}$$
$$\textbf{35/60 (.00015)} =$$

$$.00009$$
$$+ \ .86163$$
$$.86172 \quad \text{Ans}$$

Alternative Method of Interpolating to Determine Angle Measurements

An alternative method of interpolation requires the use of a proportional equation to find
minute and/or second values.

E XAMPLE: Determine angle A within $\pm 1'$. Use the sine function.

Step 1   Set up the ratio for the angle.

Step 2   Refer to the table of Natural Trigonometric
Functions (*figure* 87-1). Locate the two
sine function values closest to the computed value.

Step 3   Make a simple listing of the angles in a
left column; trigonometric function values
in a right column.

Step 4   Subtract the sine function value for the
40° 00′ − 39° 00′ values.

Step 5   Determine the difference between the
computed sine value of 0.6383 and the
larger angle value.

$$\text{sin}\ \angle\text{A} = \tfrac{30}{47} = 0.6383$$

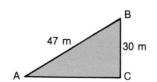

$$1° = 0.0135$$

$$= 0.0045$$

Step 6   Set up a proportional equation for the unknown angle.

$$X:1 = 0.0045:0.0135$$

$$X = \frac{0.0045}{0.0135} = \frac{1}{3}$$

Step 7   Convert the $X = \frac{1}{3}$ into equivalent minutes.

$$1° = 60'$$

$$\tfrac{1}{3}(60) = 20'$$

Step 8   Subtract the 20′ from the 40° 00′ angle. The result is the answer.

$$(40° \ 00') - 0° \ 20' =$$

$$39° \ 40' \quad \text{Ans}$$

Additional steps are taken if the problem involves degrees, minutes, and seconds. Fractional minute values are converted to second values by setting up a proportion and following the same steps. However, instead of using $60' = 1°$, equivalent second values in a minute are used $(60'' = 1')$.

D. READING PRECISE ENGINEERING/SCIENTIFIC HANDBOOK TRIGONOMETRIC FUNCTION TABLES

More precise tables of trigonometric functions are provided in engineering handbooks. Such tables contain five and six decimal-place numerical values of sine, cosine, and tangent functions and the cofunctions of secant, cosecant, and cotangent from 0° to 360°.

The tables are designed so that certain degree and function values are read downward; others, upward. Engineering tables of angular measurements in the British system are expressed in degrees and minutes. Measurements of angular values in seconds are interpolated.

Column Arrangements for Angle and Trigonometric Function Values

The contents of a partial table of values within a one-degree range is reproduced in figure 87-2. English units of measure are used. The 16° or 73° relate to quadrant I; 106° or 163°, to quadrant II; 196° or 253°, to quadrant III; and 286° or 343°, to quadrant IV.

Trigonometric function values for angles between 16° and 17° or 196° and 197° are read downward. The left column is used for minute values.

Function values for angles between 163° and 164° or 343° and 344° are also read downward but with reference to the right column for minutes.

For example, the trigonometric function value for the sin 16° 30′ is found in the 16° to 17° sine column to the right of the 30′ minute column.

The left minute column is used with angle values between 106° and 105°, 286° and 285°, 73° and 72°, or 253° and 252°. The values for the minute increments are read in the appropriate function column according to the minute value in the right column.

Similar practices are followed for the angles shown at the top and bottom of the table on the right side. Function values are read downward from 163° to 162° or 343° to 342°. Function values are read upward from 73° to 74° and 253° to 254°. Note, however, that the values in minute increments are read according to the functions shown at the bottom of the table. The minutes in the right column are used for the 73° to 74° and 253° to 254° values as well as the 163° to 164° and 343° to 344° values.

Read Downward

16° →

Min.	Sine	Cosine	Tan.	Cotan.	Secant	Cosec.	Min.
0	0.27564	0.96126	0.28675	3.4874	1.0403	3.6280	60
1	.27592	.96118	.28706	.4836	.0404	.6243	59
2	.27620	.96110	.28738	.4798	.0405	.6206	58
3	.27648	.96102	.28769	.4760	.0406	.6169	57
4	.27676	.96094	.28801	.4722	.0406	.6133	56
5	0.27704	0.96086	0.28832	3.4684	1.0407	3.6097	55
6	.27731	.96078	.28864	.4646	.0408	.6060	54
7	.27759	.96070	.28895	.4608	.0409	.6024	53
8	.27787	.96062	.28927	.4570	.0410	.5988	52
9	.27815	.96054	.28958	.4533	.0411	.5951	51
10	0.27843	0.96046	0.28990	3.4495	1.0412	3.5915	50
11	.27871	.96037	.29021	.4458	.0413	.5879	49
12	.27899	.96029	.29053	.4420	.0413	.5843	48
13	.27927	.96021	.29084	.4383	.0414	.5808	47
14	.27955	.96013	.29116	.4346	.0415	.5772	46
15	0.27983	0.96005	0.29147	3.4308	1.0416	3.5736	45
16	.28011	.95997	.29179	.4271	.0417	.5700	44
17	.28039	.95989	.29210	.4234	.0418	.5665	43
18	.28067	.95981	.29242	.4197	.0419	.5629	42
19	.28095	.95972	.29274	.4160	.0420	.5594	41
20	0.28123	0.95964	0.29305	3.4124	1.0421	3.5559	40
21	.28150	.95950	.29337	.4087	.0421	.5523	39
22	.28178	.95948	.29368	.4050	.0422	.5488	38
23	.28206	.95940	.29400	.4014	.0423	.5453	37
24	.28234	.95931	.29432	.3977	.0424	.5418	36
25	0.28262	0.95923	0.29463	3.3941	1.0425	3.5383	35
26	.28290	.95915	.29495	.3904	.0426	.5348	34
27	.28318	.95907	.29526	.3868	.0427	.5313	33
28	.28346	.95898	.29558	.3832	.0428	.5279	32
29	.28374	.95890	.29590	.3796	.0429	.5244	31
30	0.28402	0.95882	0.29621	3.3759	1.0429	3.5209	30
31	.28429	.95874	.29653	.3723	.0430	.5175	29
32	.28457	.95865	.29685	.3687	.0431	.5140	28
33	.28485	.95857	.29716	.3652	.0432	.5106	27
34	.28513	.95849	.29748	.3616	.0433	.5072	26
35	0.28541	0.95841	0.29780	3.3580	1.0434	3.5037	25
36	.28569	.95832	.29811	.3544	.0435	.5003	24
45	0.28820	0.95757	0.30097	3.3226	1.0443	3.4699	15
46	.28847	.95749	.30128	.3191	.0444	.4665	14
47	.28875	.95740	.30160	.3156	.0445	.4632	13
48	.28903	.95732	.30192	.3122	.0446	.4598	12
49	.28931	.95724	.30224	.3087	.0447	.4565	11
50	0.28959	0.95715	0.30255	3.3052	1.0448	3.4532	10
51	.28987	.95707	.30287	.3017	.0449	.4499	9
52	.29015	.95698	.30319	.2983	.0450	.4465	8
53	.29042	.95690	.30351	.2948	.0450	.4432	7
54	.29070	.95681	.30382	.2914	.0451	.4399	6
55	0.29098	0.95673	0.30414	3.2879	1.0452	3.4367	5
56	.29126	.95664	.30446	.2845	.0453	.4334	4
57	.29154	.95656	.30478	.2811	.0454	.4301	3
58	.29182	.95647	.30509	.2777	.0455	.4268	2
59	.29209	.95639	.30541	.2743	.0456	.4236	1
60	0.29237	0.95630	0.30573	3.2709	1.0457	3.4203	0
Min.	Cosine	Sine	Cotan.	Tan.	Cosec.	Secant	Min.

30′ →

Read Upward

FIGURE 87-2 Trigonometric function values of degree-minute measurements (partial table)

As an example, the numerical value for tan 73° 10′ is found by reading upward in the tangent column to the left of the 10′ displayed in the right column. The numerical value of tan 73° 10′ is 3.3052.

Handbook tables generally follow this pattern. The position of the degree listing at the top or bottom of the table provides direction for reading function values upward or downward and whether to use the right or left minute column.

E. DECIMAL-DEGREE TRIGONOMETRIC FUNCTION VALUE TABLES FOR SI METRIC ANGLE MEASUREMENTS

While precision measurements on drawings and in specifications in the SI metric system may be expressed in degrees, minutes, and seconds, the preferred unit of measure is in the *decimal degree*. Tables of numerical values for trigonometric functions are similar in layout and application to degree-minute-second tables. The main difference is that each degree is divided into ten equal parts and values are given for each one-tenth of a degree.

An example of this breakdown is provided by a small portion of a *decimal-degree table* (*figure* 87-3). The table is read by locating the degree and decimal or by the trigonometric function value. Note that the complementary degree or function value is also read from the opposite direction. For instance, a measurement like 17.5°, reading downward, has a sine value of 0.3007. The complementary cosine angle function value, reading upward, is 72.5°.

Interpolation is used when two- and three-place decimal values of angles are required. An angle measurement like 72.58° is found by taking $\frac{8}{10}$ of the difference between the 72.5° and the 72.6° trigonometric function value. In this case, if cos 72.58 is required:

$$\cos 72.5° = 0.3007$$
$$\cos 72.6 = \underline{0.2990}$$
$$(\text{Difference}) \ 0.0017 \times 0.8 = 0.00136$$

Subtracting:

$$\cos 72.5° = 0.3007$$
$$\text{and} \quad \underline{0.00136}$$
$$= 0.30184 \text{ as the numerical value of cos } 72.58°$$

16.0° To 24.0°

Dec/Deg	Sin	Cos	Tan	Cot	Sec	Csc	Dec/Deg
16.0	0.2756	0.9613	0.2867	3.4874	1.0403	3.6280	74.0
.1	0.2773	0.9608	0.2886	3.4646	1.0408	3.6060	.9
.2	0.2790	0.9603	0.2905	3.4420	1.0413	3.5843	.8
.3	0.2807	0.9598	0.2924	3.4197	1.0419	3.5629	.7
.4	0.2823	0.9593	0.2943	3.3977	1.0424	3.5418	.6
.5	0.2840	0.9588	0.2962	3.3959	1.0429	3.5209	.5
17.0	0.2924	0.9563	0.3057	3.2706	1.0457	3.4203	73.0
.1	0.2940	0.9558	0.3076	3.2506	1.0463	3.4009	.9
.2	0.2957	0.9553	0.3096	3.2305	1.0468	3.3817	.8
.3	0.2974	0.9548	0.3115	3.2106	1.0474	3.3628	.7
.4	0.2990	0.9542	0.3134	3.1910	1.0480	3.3440	.6
.5	0.3007	0.9537	0.3153	3.1716	1.0485	3.3255	.5
.6	0.3024	0.9532	0.3172	3.1524	1.0491	3.3072	.4
.7	0.3040	0.9527	0.3191	3.1334	1.0497	3.2891	.3
.8	0.3057	0.9521	0.3211	3.1146	1.0503	3.2712	.2
.9	0.3074	0.9516	0.3230	3.0961	1.0509	3.2535	.1
18.0	0.3090	0.9511	0.3249	3.0777	1.0515	3.2361	72.0
.1	0.3107	0.9505	0.3269	3.0593	1.0521	3.2188	.9
.2	0.3123	0.9500	0.3288	3.0413	1.0527	3.2017	.8
.3	0.3140	0.9494	0.3307	3.0237	1.0533	3.1848	.7
.4	0.3156	0.9489	0.3327	3.0061	1.0539	3.1681	.6
.5	0.3173	0.9483	0.3346	2.9887	1.0545	3.1515	.5
.6	0.3190	0.9478	0.3365	2.9714	1.0551	3.1352	.4
.7	0.3206	0.9472	0.3385	2.9544	1.0557	3.1190	.3
.8	0.3223	0.9466	0.3404	2.9375	1.0564	3.1030	.2
.9	0.3239	0.9461	0.3424	2.9208	1.0570	3.0872	.1
19.0	0.3256	0.9455	0.3443	2.9042	1.0576	3.0716	71.0
Dec/Deg	Cos	Sin	Cot	Tan	Csc	Sec	Dec/Deg

66.0° To 74.0°

Note. See Appendix table 7 for values from 0.0° to 90.0°

FIGURE 87-3 Numerical values of trigonometric functions for decimal-degree angles

F. ESTABLISHING THE SIGNS OF TRIGONOMETRIC FUNCTIONS OF ANGLES IN ANY QUADRANT

The signs of trigonometric functions of angles depend on the quadrant in which the angle forms. There are four quadrants in a complete circle: I, II, III, and IV. The (+) or (−) sign of each of the three basic functions and the cofunctions are identified in figure 87-4 for each quadrant.

Note that the sign of all trigonometric functions in quadrant I is positive (+). The signs of the functions change, as, for instance, the tangent of 60° (quadrant I) = 1.7321; tan 120° (quadrant II) = −1.7321; tan 240° (quadrant III) = 1.7321; and tan 300° (quadrant IV) = −1.7321. The sign is understood to be positive unless the negative (−) sign appears before the numerical value of the function.

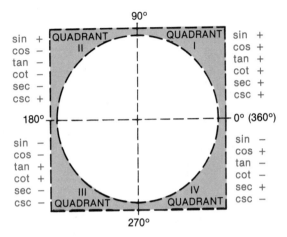

FIGURE 87-4

Review and Self-Test

ASSIGNMENT UNIT 87

A. Construction and Measurement of Trigonometric Functions

1. Find sin A by constructing the given angle, measuring the given distance on the hypotenuse, dropping and measuring a perpendicular, and applying the sine ratio.

Angle	Length of Hypotenuse
30°	4″
60°	$2\frac{1}{2}″$
$29\frac{1}{2}°$	$3\frac{1}{4}″$

2. Find by construction each angle whose sine value is given in the table.

Value of Sine (Angle D)			
.5000	.8660	.4226	.6947

B. Signs of Functions and Trigonometric Function Tables

1. a. Refer to the partial five-decimal-place degree-minute table of trigonometric function values.

16° or 196°   **163° or 343°**

Min	Sine	Cosine	Tangent	Cotangent	Secant	Cosecant	Min
0	0.27564	0.96126	0.28675	3.4874	1.0403	3.6280	60
1	.27592	.96118	.28706	.4836	.0404	.6243	59
2	.27620	.96110	.28738	.4798	.0405	.6206	58
3	.27648	.96102	.28769	.4760	.0406	.6169	57
4	.27676	.96094	.28801	.4722	.0406	.6133	56
5	0.27704	0.96086	0.28832	3.4684	1.0407	3.6097	55
6	.27731	.96078	.28864	.4646	.0408	.6060	54
50	0.28959	0.95715	0.30255	3.3052	1.0448	3.4532	10
51	.28987	.95707	.30287	.3017	.0449	.4499	9
52	.29015	.95698	.30319	.2983	.0450	.4465	8
53	.29042	.95690	.30351	.2948	.0450	.4432	7
54	.29070	.95681	.30382	.2914	.0451	.4399	6
55	0.29098	0.95673	0.30414	3.2879	1.0452	3.4367	5
56	.29126	.95664	.30446	.2845	.0453	.4334	4
57	.29154	.95656	.30478	.2811	.0454	.4301	3
58	.29182	.95647	.30509	.2777	.0455	.4268	2
59	.29209	.95639	.30541	.2743	.0456	.4236	1
60	0.29237	0.95630	0.30573	3.2709	1.0457	3.4203	0
Min	Cosine	Sine	Cotangent	Tangent	Cosecant	Secant	Min

106° or 286°   **73° or 253°**

b. Indicate the direction from which the table is read and which minute column is used to find the numerical values of the following functions.
Note. Using the sin 73° 10′ as an example, the table is read from the bottom upward, using the right minute column. The symbol is .

(1) sin 73° 10′

(2) cos 106° 42′

(3) tan 163° 39′

(4) cot 196° 58′

(5) sec 253° 24′

(6) csc 286° 17′

(7) tan 343° 22′

┌── READ DOWNWARD
┊
▼
USE LEFT
COLUMN OF
MINUTES

USE RIGHT
COLUMN OF
MINUTES
▲
┊
READ UPWARD ───┘

2. Identify the (+) or (−) sign for the numerical values of each function in each quadrant.

Quadrant	Sin	Cos	Tan	Cot	Sec	Csc
I						
II						
III						
IV						

C. Determining Angle Measurements and Trigonometric Values from Tables and by Interpolation

Establishing Trigonometric Function Values of Angle Measurements

1. a. Use the four-decimal-place degree table of trigonometric functions in the appendix (table 7). Determine the numerical trigonometric function value for functions (A) through (F). Interpolate where needed.
   b. State (1) the complementary angle of each trigonometric function and (2) the value of each angle in degrees and minutes.

Given Angle		Required Trigonometric Function Value	Name and Angle Measurement of Complementary Angle
A	1° 30′	sin	
B	10° 30′	cos	
C	25° 30′	tan	
D	50° 40′	sin	
E	60° 36′	cos	
F	75° 25′	tan	

Determining Angle Measurements of Trigonometric Functions Using Tables for Degree Values and Degree-Minute Values

2. Refer to an appropriate four-place decimal table in a handbook. Determine the angle measurements that correspond to trigonometric function values (a) through (f).
   a. sin 0.0436
   b. cos 0.9863
   c. tan 0.5206
   d. sin 0.7167
   e. cos 0.4636
   f. tan 23.532

3. Use an engineering degree-minute table.
   a. Determine the quadrant from the plus or minus signs before the trigonometric functions (A) through (G).

b. Refer to the table to establish the angle measurement in each case, correct to the nearest minute.

Trigonometric Function Value		Quadrant	Angle Measurement
A	sin 0.30043		
B	cos 0.30071		
C	cos −0.95372		
D	tan −0.51283		
E	sec −1.29160		
F	cot −0.60921		
G	sin −0.63383		

D. Applications of Decimal-Degree Metric Trigonometric Table Values

Determining SI Metric Angle Measurements from Function Values

1. Use a Table of Decimal-Degree Trigonometric Functions. Read the angles in metric units of measure that correspond to the function values for A through H in the quadrants indicated.

Trigonometric Function Value		Quadrant	Angle Measurement
A	sin A = 0.30071	I	
B	cos A = 0.25038	I	
C	tan A = −0.98270	II	
D	cot A = −0.04891	II	
E	sec A = −1.00490	III	
F	csc A = −1.72628	III	
G	cos A = 0.18567	IV	
H	cot A = −4.27374	IV	

Note. These values may be rounded off if a four-place decimal table is used.

2. Use a Table of Decimal-Degree Trigonometric Functions. Determine the trigonometric function value as stated for each angle measurement (a) through (f). Mark the value with a (−) sign, where applicable.

	Required Function Value				Required Function Value
a. 22.8°	sin		d. 45.85°	sin
b. 96.4°	cos		e. 135.85°	cos
c. 184.7°	tan		f. 312.75°	tan

3. Use a scientific calculator. Determine the sign and numerical value of trigonometric functions (a) through (d). Round off the calculator display register values to five decimal places.
   a. sin of 22.75°
   b. tan of 137.26°
   c. sec of 197.45°
   d. csc of 275.64°

4. Use a scientific calculator. Determine the angles in metric units of measure for function values (a) through (d), according to the stated quadrants. Round the angle measurements to two places.
   a. sin 0.6275 (quadrant I)
   b. cos −0.10192 (quadrant II)
   c. cos −0.99479 (quadrant III)
   d. tan −7.47474 (quadrant IV)

E. Reading and Interpolating Tables of Trigonometric Functions

1. Find dimensions A and B, correct to three decimal places. Solve by using trigonometric ratios and a table of trigonometric functions.

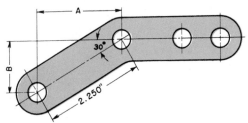

2. Determine each missing dimension in the table. Use trigonometric ratios and a table of trigonometric functions. Round off linear dimensions to three places and each angular dimension to degrees, minutes, and/or seconds.

Triangle	Angles			Sides		
	A	B	C	a	b	c
ABC	45°	90°		1.414		1.414
	D	E	F	d	e	f
DEF	90°	30°			2.500	
	X	Y	Z	x	y	z
XYZ			90°	1.420		4.280

3. Compute the value of the included angle for the tapered cylinder. Round off the angle to the nearest minute.

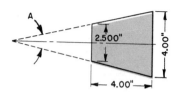

4. Compute the power output of the electrical circuit to a two-decimal-place accuracy.

$$E = 208 \text{ volts}$$
$$I = 8.25 \text{ amperes}$$

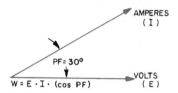

5. Solve for angle A of the cowl cover.

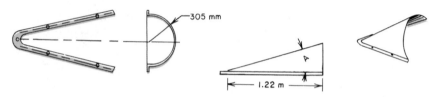

6. Determine the power factor angle from the values identified on the sketch. Express the angle in degrees and minutes.

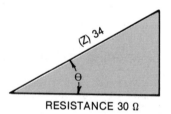

Unit 88 Four-Function and Scientific Calculator Programming: Trigonometric Applications

OBJECTIVES OF THE UNIT

After satisfactorily completing the unit, the student/trainee will be able to
- *Interpret algorithms for four-function calculators applied to trigonometry problems.*
- *Calculate sine, cosine, tangent, and cotangent values using algorithms.*
- *Use a scientific calculator to determine sine, cosine, and tangent and smallest angle values in degrees, minutes, and seconds and calculate inverse trigonometric functions.*
- *Obtain common or natural logarithms of a given trigonometric value.*
- *Determine reciprocal trigonometric function values for given angles.*
- *Use degree, radian, and grad mode functions to compute angles within $\pm 90°$ and $>90°$.*

In review, it may be stated that trigonometry is a study of the properties, applications, and solutions related to three-sided (triangular) shapes. Trigonometry incorporates basic mathematical processes that establish the relationships (ratio) of the sides and angles. The names and ratios between the sides and angles of right, acute, and oblique triangles, and the application of trigonometric function tables were covered in earlier units. This unit starts with further uses of *algorithms* for calculating sine, cosine, tangent, and cotangent values. It should be remembered that the term algorithm relates to a system of designated mathematical steps that are used with the four-function calculator to compute specific values.

A. TRIGONOMETRIC FUNCTIONS WITH THE FOUR-DIGIT CALCULATOR

Calculating Sine Values

Problem. The length (H) of a laminated timber truss must be determined to meet the construction requirements specified on the drawing.

The sine of angle A may be found be using the algorithm:

$$\text{Sin } A = 1.745\alpha[1 + 0.508\alpha^2(0.15\alpha^2 - 1)]$$

$$\alpha = \frac{\angle A}{100} = \frac{30}{100} = 0.30$$

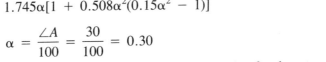

FIGURE 88-1

To eliminate the need for setting down an intermediate value for the square of angle A, the sequence in the chart may be used.

	Keyboard Entry	
Step	Numerical Value	Process
1	0.30	$\times$
2	0.30	$\times$
3	0.15	$-$
4	1	$\times$
5	0.30	$\times$
6	0.30	$\times$
7	0.508	$+$
8	1	$\times$
9	0.30	$\times$
10	1.745	$=$
Display readout 0.4998885 $=$ sin A		

FIGURE 88-2

If the sine value of 30° as produced by the calculator (0.499 888 5) is checked against a table of natural trigonometric functions (0.500 000 0), the calculator result is in error. Thus, a word of caution is in order. When this algorithm is used, a possible small error is introduced.

Continuing with the problem, the length of the laminated wooden truss is found by the sine formula:

$$\sin A = \frac{\text{opposite side (60)}}{\text{hypotenuse } (H)}; \text{ transposing, } H = \frac{60}{\sin A = (.499\ 888\ 5)}$$

$$= 120.026 \text{ meters long} \quad \textbf{Ans}$$

When calculated by conventional trigonometric methods, the truss would be exactly 120 meters long. Using this algorithm and a calculator, an error of +0.026 meters or 1.024″ is introduced. This degree of tolerance may not be acceptable. With engineering problems such as this one, the calculator may be used to supplement the conventional computation method (using table values for natural trigonometric functions). In this instance, the table value of the 30° angle of 0.500 000 would have produced the exact 120-meter measurement.

Calculating Cosine Values

A different formula and algorithm are used for determining cosine values.

$$\cos A = 1 - 1.523\ \alpha^2\ [1 + 0.254\ \alpha^2\ (0.1\ \alpha^2 - 1)]$$

The numerical value and process entries on the keyboard are similar to the steps followed with the sine calculations. Again α = angle $A \div 100$.

Problem. Determine the distance (D) a surveyor is to locate a construction bench mark, according to the drawing dimensions.

$$\cos 40° = D/3$$
$$D = \cos 40° \times 3$$

$$\cos A = \frac{\text{side adjacent}}{\text{hypotenuse}}$$

Using rules of order for solving internal parentheses values first and substituting known values,

$$\cos A = 1 - 1.523 \, (.4)^2 \, [1 + (0.254 \times .4^2)(0.1 \times (.4^2) - 1)] = 0.766 \, 064 \, 8$$

Substituting this value for cos A, $D = 0.766 \, 064 \, 8 \times 3 = 2.298 \, 194 \, 4$ kilometers. Again, it should be noted that this quantity varies from the exact dimension of 2.298 kilometers.

Calculating Secant Values

The secant is the ratio of the hypotenuse to the side adjacent. The value of the secant of an angle is the reciprocal of its cosine. Thus, the secant may be found by computing the reciprocal.

$$\sec A = \frac{1}{\cos A}$$

Calculating Tangent Values

A tangent value may be computed with a calculator using the algorithm:

$$\tan A = 1.745 \, \alpha \left(\frac{1.015 \, \alpha^2 - 5}{6.092 \, \alpha^2 - 5} \right) \qquad \alpha = \frac{\angle A}{100}$$

In setting up the procedure to follow, it is necessary to carry on the numerator and denominator computations first and to record the display readouts of the intermediate values.

Problem. The distance from a vertical beam to a mooring post is 40 meters. The angle of a connecting brace is 50 degrees. Determine the height (h) of the vertical beam.

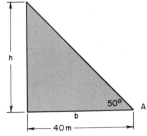

Using the formula and substituting the value 0.50 for α,

$$\tan A = 1.745\,\alpha\left(\frac{1.015\,\alpha^2 - 5}{6.092\,\alpha^2 - 5}\right)$$

$\boxed{= -4.746\ 25}$
STEP 1

$\underbrace{= 1.745\ (.5)}$ $\left(\dfrac{\boxed{1.015\ (.25) - 5}}{\boxed{6.092\ (.25) - 5}}\right)$
STEP 4   STEP 2

$\boxed{= -3.477}$

STEP 3

$\boxed{0.8725}$

$\boxed{\dfrac{-4.746\ 25}{-3.477}}$ ----- $\boxed{= 1.365\ 041\ 7}$

$= \underbrace{0.8725 \times 1.365\ 041\ 7} = 1.190\ 998\ 8$

STEP 5

$\tan A = \dfrac{\text{side opposite } (h)}{\text{side adjacent } (b)}$

$(h) = \tan A(b)$
$\ \ \ \ = 1.190\ 998\ 8\ (40)$
$\ \ \ \ = 47.639\ 952$ meters (m)   **Ans**

Calculating Cotangent Values

The cotangent denotes the relationship between the side adjacent to an angle and the side opposite to the same angle. Thus, the cotangent is the inverted tangent ratio. The cotangent may be determined by calculating the reciprocal of the tangent.

$$\cot A = \frac{1}{\tan A}$$

Note. The conventional trigonometric method of solving this problem is simpler and does not have the built-in error of this algorithm. The correct (error-free) height of the vertical beam is 47.62 meters. This height is calculated by using the tangent formula, a four-place-decimal natural trigonometric function value for the 60° tangent (1.1918), and substituting quantities. Since

$$(h) = \tan A\ (b)$$
$$= 1.1918\ (40)$$
$$= 47.62 \text{ meters (m).} \quad \textbf{Ans}$$

Solving this type of problem *by* the *conventional method, supplemented by* the *calculator* for carrying out the multiplication process, is simple, conserves time, and produces an accurate measurement.

B. APPLICATIONS OF THE SCIENTIFIC CALCULATOR WITH TRIGONOMETRIC FUNCTIONS

The scientific calculator is designed so that sequential steps, such as the algorithms used with the four-function calculator, are automatically performed within the electronic circuitry of the instrument. A trigonometric function is obtained on the scientific calculator by entering the angle and then pressing the appropriate sin , cos , or tan *trigonometric key.*

EXAMPLE: Use a scientific calculator to obtain the sine, cosine, and tangent functions for the angles:

$$\sin 44° 48', \cos 21° 32' 45'', \tan 43° 46' 35''$$

Note. Convert minute and second values to equivalent degrees before entering the angle values in the calculator.

Given Angle	Numerical Entry (Converted Angle Value)	Calculator Key Sequences	Display Register
sin 44° 48′	44.8°		44.8
		sin	0.7046342 sin function value Ans
cos 21° 32′ 45″	21.54°		21.54
		cos	0.9301615 cos function value Ans
tan 43° 46′ 35″	43.78°		43.78
		tan	0.9582957 tan function value Ans

Calculating Reciprocal Trigonometric Functions

As stated before, the cosecant (csc) has the reciprocal value of the sine (sin).

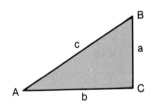

$$\text{Sin } A = \frac{a}{c}. \qquad \text{Csc } A = \frac{c}{a}.$$

The secant is the reciprocal of the cosine.

$$\text{Cos } A = \frac{b}{c}. \qquad \text{Sec } A = \frac{c}{b}.$$

The cotangent is the reciprocal of the tangent.

$$\text{Tan } A = \frac{a}{b}. \qquad \text{Cot } A = \frac{b}{a}.$$

EXAMPLE: Use the scientific calculator to find the first function and then the reciprocal function values for angles A, B, C, and D.

Given Angle		First Function	Reciprocal Function
A	19.333°	sin A = 0.3311	csc A = 3.0206
B	31.25°	cos A = 0.8549	sec A = 1.1697
C	70.2°	tan A = 2.7776	cot A = 0.3600
D	86.9°	sec A = 18.492	cos A = 0.0541

Step 1 Enter the given angle value.

Step 2 Press the required trigonometric function key.

Note. The display register is the trigonometric function value of the angle.

Step 3 Press the $\boxed{1/x}$ *reciprocal key.*

Note. The display register is the reciprocal function value.

C. OBTAINING COMMON LOGARITHMS USING TABLES

Logarithms are used principally in problems requiring multiplication and division, raising a number to a power, or extracting roots. One of the terms that is used to identify particular features of a logarithm is called a *characteristic*. The characteristic of a logarithm is a number that is related to the number of digits to the right or left of a decimal point in a given number. The second term, *mantissa*, is the decimal part and is found in logarithm tables. The logarithm of a number is a combination of the characteristic and the mantissa values.

RULE FOR DETERMINING THE CHARACTERISTIC OF A LOGARITHM

Numbers Greater than One
• Subtract (1) from the number of digits to the left of the decimal point. The remainder is the characteristic.

Numbers Less than One
• Add (1) to the number of zeros between the decimal point and the first nonzero digit.

Note. A negative sign is placed above the characteristic to show that the logarithm has a value of less than one.

EXAMPLE: *Case 1.* Determine the characteristic of the logarithm 2768.4

Step 1 Subtract (1) from the number of whole number digits in the given number.

Step 2 Identify the remainder as the characteristic.

2 7 6 8   (4) − (1)
(3)   Ans

EXAMPLE: *Case 2*. Determine the characteristic of the number.   **0.000 724**

Step 1   Count the number of zeros between the decimal point and   **0.0 0 0   (3)**
          the first nonzero digit.
Step 2   Add (1) to the number of zeros.   (3) + (1) = (4)
Step 3   Place a negative sign above this number. $\overline{4}$
Step 4   Locate the value of the mantissa 724 in a table of loga-
          rithms in a handbook.
Step 5   Combine the characteristic and the mantissa values.   **7 2 4   = .8597**
          Note. The logarithm of 0.000 724 is $\overline{4}$.8597   **$\overline{4}$.8597   Ans**

D. USING THE SCIENTIFIC CALCULATOR TO OBTAIN A COMMON OR A NATURAL LOGARITHM

Obtaining a Common Logarithm

The *common logarithm* is obtained by entering the given number and then pressing the ⎡log⎤ *common log key*

EXAMPLE: Determine the common logarithm of 32.01.

Numerical Entry	Key Sequence	Display Register
32.01	⎡log⎤	32.01 Common
		1.5052857 Logarithm Ans

Obtaining a Natural Logarithm

The *natural logarithm* (log e) is obtained by entering the given number on the display register and then depressing the ⎡lnx⎤ *natural logarithm key*.

Numerical Entry	Key Sequence	Display Register
1.026	⎡lnx⎤	1.026 Natural
		0.0256677 Logarithm Ans

E. COMPUTING ARCSINE, ARCCOSINE, AND ARCTANGENT VALUES

Angle values may be determined from trigonometric function values by pressing the ⎡INV⎤ *inverse key* and then the appropriate ⎡sin⎤, ⎡cos⎤, or ⎡tan⎤ function keys. The answer is identified as the *arcsine* ($\sin^{-1}$), *arccosine* ($\cos^{-1}$), or the *arctangent* ($\tan^{-1}$), depending on the function key that is pressed.

The arcsine represents the *smallest angle* for the sine value that is displayed in the register.

E XAMPLE: Determine the smallest angle for the sine value of 0.7071.

Numerical Entry	Key Sequence	Display Register
0.7071	INV sin	0.7071
		44.999451* Arcsine **Ans**

*In general practice, the angle is considered as 45°.

Similarly, the arcosine and the arctangent represent the smallest angles for the cosine and tangent, respectively, that are displayed on the register.

F. ANGLE MODE INDICATORS AND DEGREE, RADIAN, AND GRAD CONVERSIONS

Many problems require the conversion of angular measurements from one unit of measure to another. Angles are changed to degrees, to radians, or to grads, depending on the application. The DRG *degree, radian, grad key* is used to select the unit of measurement, called the *angular mode* of the calculator. The mode appears on the display: DEG for *degree mode*, RAD for the *radian mode*, and GRAD for the *grad mode*. When the instrument is turned on, the calculator is in the degree mode. The mode changes in a rotary pattern from the degree mode to the radian mode when the DRG key is pressed once. Another key press changes the mode from radian to grad. The grad mode is changed to degree mode by pressing the DRG key.

It must be noted that the range of angular conversions is limited to the first and fourth quadrants to $\pm 90°$, $\pm \frac{\pi}{2}$ radians $\left[\frac{\pi \cdot 57.2958}{2} \right]$, and ± 100 grads.

Converting Degree, Radian, and Grad Values within ±90°

Numerical Entry	Key Sequences	Display Register (Required Angular Mode)
Degrees	sin DRG INV sin	RAD
Degrees	sin DRG DRG INV sin	GRAD
Radians	sin DRG DRG INV sin	DEG
Radians	sin DRG INV sin	GRAD
Grads	sin DRG INV sin	DEG
Grads	sin DRG DRG INV sin	RAD

EXAMPLE: Convert 60 degrees to radians, to grads, and visa versa.

Numerical Entry	Key Sequences	Display Register (Required Angular Mode and Value)
60° (degrees)	[sin] [DRG] [INV] [sin]	1.047 1976 **RAD** Ans
60° (degrees)	[sin] [DRG] [DRG] [INV] [sin]	66.666667 **GRAD** Ans
66.6667 (grads)	[sin] [DRG] [INV] [sin]	60 **DEG** Ans
1.047 1976 (radians)	[sin] [DRG] [INV] [sin]	66.666667 **GRAD** Ans

Converting Angles in Any Quadrant

To review, a radian is equal to 57.295 8°. A grad is equal to 0.9° as there are 100 grads in 90°. Angles may be converted from one unit of measurement in one quadrant to another unit by using the table of conversion factors (*figure* 88-3).

Given Measurement	Conversion Unit of Measure and Factors		
	Degrees	**Radians**	**Grads**
Degrees		$\times \frac{\pi}{180}$	$\div .9$
Radians	$\times \frac{180}{\pi}$		$\times \frac{200}{\pi}$
Grads	$\times .9$	$\times \frac{\pi}{200}$	

FIGURE 88-3

EXAMPLE: Convert 135° to radians and then to grads.

Numerical Entry	Key Sequence	Display Register
135	×	135
3.1416	÷	424.116
180		2.35 62 Radians **Ans**
	ON/C	0
135	÷	135
.9		150 Grads **Ans**

EXAMPLE: Convert 2.3562 radians to degrees and grads.

Numerical Entry	Key Sequence	Display Register
2.3562	×	2.3562
180	÷	424.11 6
3.1416		135 Degrees **Ans**
2.3562	ON/C	0
200	×	2.3562
3.1416	÷	471.24
		150 Grads **Ans**

G. ACCURACY OF ALGORITHMS AND CALCULATOR COMPUTATIONS

Algorithms were presented in this unit to extend the capacity of the four-function calculator to solve certain trigonometric functions. Each algorithm introduces the possibility of a small error. Therefore, consideration should be given to using conventional computational methods as previously described. The calculations may then be checked or the mathematical processes carried out by calculator instead of long hand or with a slide rule.

Consideration must also be given to the additional internal capacity of scientific calculators to handle numbers with a greater number of digits than those displayed on the register. This is important in precise scientific and engineering calculations where extremely fine limits of accuracy are required.

Review and Self-Test

ASSIGNMENT UNIT 88

A. Applications of Four-Function Calculators and Algorithms

1. Describe briefly the function of algorithms in trigonometry.
2. Identify a precaution that must be taken in using algorithms.
3. Cite one advantage of using conventional methods of solving for trigonometric values supplemented by the calculator.
4. Calculate the structural steel beam lengths for four different trusses, according to the dimensions given in the table.
   a. Give the display readout of each measurement.
   b. Convert the display readouts to feet-inch dimensions. Use the sine algorithm and a calculator.

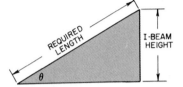

			Required Beam Length				
	Truss Angle	I-Beam Height	1. Calculator/ Algorithm Method		2. Conventional Method		3. Dimensional Variation
			Display Readout	Dimension (feet/inches)	Computed Value	Dimension (feet/inches)	
A	30°	20'					
B	45°	24'					
C	60°	32'					
D	22° 30'	16'-6"					

5. Check the structural steel length dimensions in problem A.4 by the conventional method. Use four-place decimal values for the natural trigonometric functions.
   a. Compute and record the lengths, rounding off the measurements to two decimal places.
   b. Convert these values to the closest $\frac{1}{8}''$ dimension.
6. Compare each structural steel beam length as computed by the two methods. Indicate any dimensional variation in the last right column of the table.

B. Applications of the Scientific Calculator in Trigonometry

Converting Angles and Trigonometric Functions
1. Calculate the natural trigonometric function values of angles *A, B,* and *C* using a scientific calculator.

	Angles	sin	cos	tan	csc
A	45°				
B	32° 30′				
C	85° 49′ 12″				

2. a. Determine the smallest angles (arcsine, arcosine, or arctangent) represented by trigonometric function values A, B, and C. Round off each angle value to two decimal places.

	Trigonometric Function Value	Calculator Angle Value	Converted Angle Value (°/′/″)
A	sin 0.05321		
B	cos 0.55291		
C	tan 16.087		

b. Convert the calculator angle value in degrees and a decimal to equivalent degrees, minutes, and/or seconds.

3. a. Convert angles B and C in the table to numerical entries for a calculator.
   b. Calculate the specified trigonometric function values for angles A, B, and C.
   c. Name and calculate the reciprocal function of each angle.

	Given Measurement	Numerical Entry (a)	Trigonometric Function Value (b)	Reciprocal Function and Value (c)
A	sin 60°		sin	
B	tan 85° 43′		tan	
C	sec 58° 35′30″		sec	

C. Using Tables and Calculator to Determine Natural and Common Logarithms

1. Give the characteristic of the logarithm for the following numbers (a through f).
   a. 0.625   c. 0.000 372   e. 192.0067
   b. 0.042   d. 27.6745   f. 26 827.61
2. Use a logarithm table to find the mantissa for numbers (a), (b), and (c).
   a. 2582.762   b. 3.2246   c. 0.008225

3. Refer again to a logarithm table and determine the logarithms for the numbers in problem C.2.
4. Find the number corresponding to logarithms a, b, and c.
   a. 2.7818   b. 2.3118   c. $\overline{4}$.8404

5. Use a scientific calculator to obtain the common logarithms for values a and b and natural logarithms for c and d. Round the answers to four decimal places.
   a. 64
   b. 1.7246
   c. 2.0765
   d. 0.00146

D. Converting Degree, Radian, and Grad Measurements in Any Quadrant

1. a. Program a scientific calculator to convert the $<90°$ angles A and B and the $>90°$ angles B and C to degrees, radians, and/or grads, as required.

Angle	Degrees	Radians	Grads
A	45°		
B	67.5°		
C		2.366	
D			300

   b. Express the degree measurement for angle C as displayed on the calculator to its equivalent degree, minute, and grad value.

Unit 89 Acute Triangles

OBJECTIVES OF THE UNIT

After satisfactorily studying this unit, the student/trainee will be able to
- *Apply trigonometric ratios to sides and angles of equilateral triangles to compute the value of a side.*
- *Solve either for a side or unknown angle in problems involving isosceles triangles.*

The sides or angles of isosceles and equilateral triangles may be computed by right triangle trigonometry. In these triangles, a perpendicular is erected to form two right triangles. The ratios that apply to sides and angles are used to solve problems for a missing angle or dimension.

A. APPLICATION OF TRIGONOMETRIC RATIOS TO EQUILATERAL TRIANGLES

By definition, the equilateral triangle has three equal sides and three equal angles. Thus, if the total of all angles in a triangle is 180°, each angle equals 60°. The altitude of an equilateral triangle divides it into two congruent triangles having angles of 30°, 60°, and 90°.

RULE FOR FINDING THE SIDES OF EQUILATERAL TRIANGLES (GIVEN THE ALTITUDE)

- Sketch an equilateral triangle. Label the sides and angles.
- Drop a perpendicular from the vertex to the base for the altitude. Thus, two triangles are formed having angles of 30°, 60°, and 90°.
- Substitute the given value of the altitude as the side opposite the 60° angle.
- Determine the value of sin 60° in a table of trigonometric functions. Then substitute the value in the sine ratio.
- Use a letter to represent the hypotenuse as the missing side and solve the equation for the hypotenuse.

EXAMPLE: The altitude of an equilateral triangle is 20 inches. What is the length of each side?

Step 1   Sketch an equilateral triangle and label the sides, angles, and altitude.

Step 2   Write the sine ratio.

Step 3   Substitute known values in the equation.

Step 4   Determine the value of sin 60° in a table of natural trigonometric functions. Then substitute this value in the equation.

Step 5   Solve the equation. The value of d is the length of the hypotenuse.

$$\sin B = \frac{\text{side opposite}}{\text{hypotenuse}} = \frac{b}{d}$$

$$\sin 60° = \frac{20}{d}$$

$$\sin 60° = 0.8660$$

$$0.866 = \frac{20}{d}$$

$$d = 23.09'' \quad \text{Ans}$$

B. APPLICATION OF TRIGONOMETRIC RATIOS TO ISOSCELES TRIANGLES

The isosceles triangle has two equal sides and two equal angles. Like the equilateral triangle, the value of any side or angle may be found by erecting a perpendicular and applying any one of the six trigonometric ratios. The perpendicular divides the triangle into two congruent triangles.

RULE FOR COMPUTING SIDES OR ANGLES OF ISOSCELES TRIANGLES

- Make a rough sketch of the triangle.
- Erect a perpendicular to the base from the vertex. The perpendicular divides the triangle into two congruent right triangles.
- Determine what the known values are and which of the six trigonometric ratios is easiest to apply.

- Substitute the given values in the selected ratio.
  Note. When either the vertex angle and/or the angle opposite the altitude is given, locate the numerical value in a table of trigonometric functions. Then, substitute this value in the ratio.
- Solve the missing value. Where an angle is to be computed, determine the number of degrees by locating the ratio in a table of trigonometric functions.

E XAMPLE: Find the number of degrees in angle B when line AA_1 is parallel to BC and angle $B =$ angle C.

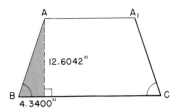

Step 1   Make a sketch of the isosceles triangle and label the given parts.

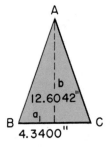

Step 2   Determine which trigonometry ratio may be used.

$$\tan B = \frac{\text{side opposite}}{\text{side adjacent}} = \frac{b}{a_1}$$

Step 3   Substitute the known value of the altitude ($b = 12.6042''$) and side adjacent ($a_1 = 4.3400''$) in the equation.

$$\tan B = \frac{b}{a_1}$$

$$= \frac{12.6042}{4.3400}$$

$$= 2.9042$$

Step 4   Determine from a table of trigonometric functions how many degrees are in an angle whose tangent is 2.9042. The table gives 71° as the angle.

$\angle B = 71°$ **Ans**

In a similar manner, by selecting one of the six trigonometric ratios it is possible to determine the values of the other angles or sides.

Review and Self-Test

ASSIGNMENT UNIT 89

A. Application of Trigonometric Ratios to Equilateral Triangles

1. Find dimensions O and altitude A for the plate illustrated.

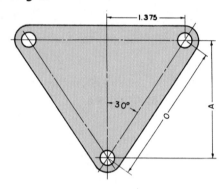

2. Compute the depth of a National Form (60°) screw thread according to the dimensions given on the drawing.

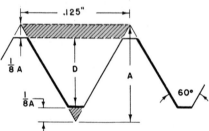

3. Calculate the dimensions of the cylindrical shell.
   a. Height of cone
   b. Volume of cone
   c. Overall length
   Round off each dimension to two decimal places.

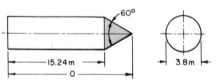

B. Application of Trigonometric Ratios to Isosceles Triangles

1. Determine the length of the chord for two consecutive end points on a bolt circle $4\frac{3}{8}$ inches in diameter on which the centers of 14 holes are to be laid out.

2. Find the length of the perpendicular from the center to the chord connecting two consecutive centers.

Unit 90 Oblique Triangles

OBJECTIVES OF THE UNIT

After satisfactorily studying this unit, the student/trainee will be able to
- *Interpret the law of sines and the law of cosines in relation to triangles with an angle greater than 90°.*
- *Solve industry and business problems involving oblique triangles using the law of sines or law of cosines for calculating side and angle measurements.*
- *Apply the law of supplements to simplify the solution of problems in trigonometry.*

Oblique triangles may be solved by drawing perpendiculars to make right triangles and using any combination of the six basic ratios of sides and angles. However, such a practice is time consuming and there are easier ways to solve such problems. Two new laws, the sine law and the cosine law, simplify the solution of measurements of sides and angles in oblique triangles.

A. THE LAW OF SINES

The *Law of Sines* is a short way of saying that the sides of any triangle are proportional to the sines of the opposite angles. For instance, in triangle ABC, the acute angles are A and B; the altitude, h; and the sides, a, b, and c. In triangle ADC,

$$\sin \angle A = \frac{\text{side opposite}}{\text{hypotenuse}} = \frac{h}{b}$$

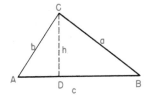

In triangle BCD,

$$\sin \angle B = \frac{side\ opposite}{hypotenuse} = \frac{h}{a}$$

If the values of the two sines are written as a proportion, then

$$\sin A : \sin B = \frac{h}{b} : \frac{h}{a} \text{ or } \frac{\sin A}{\sin B} = \frac{\dfrac{h}{b}}{\dfrac{h}{a}} = \frac{a}{b}$$

In a like manner, the relationship of any two sides is the same as that of the angles opposite the two sides. In the statement of such a relationship, there are four members. If three of these are known, the fourth may be found.

B. THE LAW OF COSINES

If, in setting up a proportion by the law of sines, two of the members are unknown, this law cannot be immediately applied. In such cases, the *Law of Cosines* can be used. This law

states that in *any triangle the square of any side is equal to the sum of the squares of the other sides minus twice the product of the two sides and the cosine of their included angle.*
Hence,

$$a^2 = b^2 + c^2 - 2(bc \cos A) \qquad b^2 = a^2 + c^2 - 2(ac \cos B)$$

$$c^2 = a^2 + b^2 - 2(ab \cos C).$$

Transposing these equations, the cosine values become

$$\cos A = \frac{b^2 + c^2 - a^2}{2bc} \qquad \cos B = \frac{a^2 + c^2 - b^2}{2ac}$$

$$\cos C = \frac{a^2 + b^2 - c^2}{2ab}.$$

C. APPLYING THE LAWS OF SINES AND COSINES

Problems in oblique triangles involve four types, depending on the data given.

Type I   Any two angles and any one side are given.
Type II   Any two sides and the angle opposite one of the sides are given.
Type III   Any two sides and the included angle are given.
Type IV   The three sides are given.

On type I and type II, the law of sines can be applied immediately. On type III and type IV, the law of cosines must be applied first. When the additional data are found, the law of sines can be used to complete the problem.

Type I

E XAMPLE: Find the missing sides and the third angle in triangle *XYZ*.

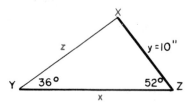

Step 1   The third angle can be found by using the fact that the sum of the angles of any triangle is 180°.

Step 2   Apply the law of sines.

$$\frac{\sin Y}{\sin Z} = \frac{y}{z} \qquad \frac{\sin 36°}{\sin 52°} = \frac{10}{z} \qquad \frac{.5878}{.7880} = \frac{10}{z} \qquad z = 13.4''$$

Also, $\qquad \dfrac{\sin X}{\sin Y} = \dfrac{x}{y} \qquad \dfrac{\sin 92°}{\sin 36°} = \dfrac{x}{10} \qquad \dfrac{.9994}{.5878} = \dfrac{x}{10} \qquad x = 17''$ Ans

Type II

<u>E XAMPLE</u>: Find the missing side and the two missing angles.

Step 1 Select the sides and angle where three members will be known in the proportion.

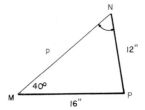

$$\frac{\sin M}{\sin N} = \frac{12}{16} \qquad \sin 40° = .6428 \qquad \text{Therefore,} \ \frac{.6428}{\sin N} = \frac{12}{16}$$

$N = 58° \ 59'$ **(to nearest minute)**

Step 2 $180° - (40° + 58°59') = 81°1' = \angle P$

Step 3 $\dfrac{\sin M}{\sin P} = \dfrac{12}{p} \qquad \dfrac{.6428}{.9877} = \dfrac{12}{p} \qquad p = 18.44''$ **Ans**

Type III

<u>E XAMPLE</u>: Find the missing side and the two missing angles.

Step 1 Since a proportion cannot be established in which three members are known, it is first necessary to apply the law of cosines.

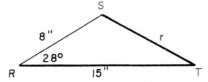

$$r^2 = t^2 + s^2 - 2 \ ts \cos \angle R$$
$$r^2 = 64 + 225 - 2 \times 8 \times 15 \times .8829$$
$$r^2 = 289 - 211.896 = 77.104$$
$$r = \sqrt{77.104} = 8.781''$$

Step 2 Now apply the law of sines.

$$\frac{\sin 28°}{\sin T} = \frac{8.781}{8}; \ \frac{.4695}{\sin T} = \frac{8.781}{8}; \ \sin T = .4277; \ \angle T = 25° \ 19' \ \text{(to nearest minute)}$$

Step 3 $\angle S = 180° - (28° + 25° \ 19') = 126° \ 41'$ **Ans**

Type IV

<u>E XAMPLE</u>: Find the three missing angles.

Step 1 Use the cosine form.

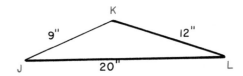

$$\cos J = \frac{l^2 + k^2 - j^2}{2lk}$$

$$\cos J = \frac{81 + 400 - 144}{2 \times 9 \times 20}$$

$$\cos J = .9361 = 20° \ 36' \ \text{(to nearest minute)}$$

Step 2 Now, apply the law of sines.

$$\frac{\sin 20°36'}{\sin L} = \frac{12}{9}; \qquad \frac{.3518}{\sin L} = \frac{12}{9}; \qquad \begin{array}{l} \sin L = .2638 \\ \angle L = 15° \ 18' \end{array}$$

Step 3 $\angle K = 180° - (15° \ 18' + 20° \ 36') = 144° \ 6'$ **Ans**

D. LAW OF SUPPLEMENTS

Problems in trigonometry are simplified by using supplements of angles. The *sine or cosine of an angle between 90° and 180° is the sine or cosine of the difference between the angle and 180°*. For cosines between 90° and 180°, the values are negative. When applied to the portion of the formula calling for $-2ab \cos C$, if C is negative, the two negatives will make this value positive and necessitate adding it to the first part of the formula.

Review and Self-Test

ASSIGNMENT UNIT 90

A. Application of the Laws of Sines and Cosines

1. Find the missing sides and angles for the values given in the table. Round off decimal values to one place.
2. Check the values of sides and angles of triangles A through E by first constructing the triangles and then measuring the angles with a protractor.

	Sides			Angles		
	m	n	o	M	N	O
A	4"		3"		100°	
B		$3\frac{1}{2}''$	2"	55°		
C	5.6"		7.8"		61°	
D	4.2"		1.8"		95°	
E		4.6"	2.5"	57°		

3. Determine the angle of slope (X) and the length (y) for the roof span as shown in drawing.

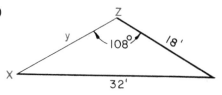

4. Find the number of degrees and minutes in all angles of triangles A through E.

	Sides		
	X	Y	Z
A	5"	12"	9"
B	$8\frac{1}{2}''$	8"	$14\frac{1}{2}''$
C	17.6"	12.8"	8.8"
D	8.75"	12.25"	5.75"
E	9.82"	3.76"	7.54"

5. Determine angles *X*, *Y*, and *Z* with the law of cosines. Give angles correct to seconds.

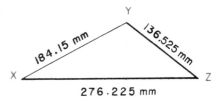

6. Find angles *A*, *B*, and *C* correct to minutes.

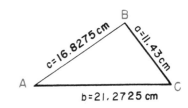

Unit 91 Achievement Review on Applied Trigonometry

OBJECTIVES OF THE UNIT

This achievement review serves as an overall test for Section 19. The unit is designed to measure the student's/trainee's ability to

- *Translate word statements relating to trigonometric problems into dimensioned sketches that show the relationship of known sides and angles and measurements that are to be computed.*
- *Use data from handbook tables of trigonometric functions expressed in English degrees-minutes-seconds or SI metric decimal-degrees and convert values between these systems.*
- *Determine the signs of trigonometric functions in any quadrant.*
- *Apply algorithms with four-function calculators and solve practical problems.*
- *Convert degree, radian, and grad values using a scientific calculator. Enter natural trigonometric function values in a calculator to establish angle measurements, interpolating as required.*
- *Determine natural and common logarithms using handbook tables and a scientific calculator.*
- *Solve industrial problems using trigonometry functions, the Pythagorean theorem, and interpolation.*
- *Apply trigonometric laws, principles, and practices to solve typical problems involving right, acute, and oblique triangles.*

A. THE CONCEPT OF RIGHT TRIANGLE TRIGONOMETRY

1. Sketch and label the sides and angles for the three right triangles given in the table.

Right Triangles	Sides			Angles			Trigonometric Function
ABC	a	b	c	A	B	C	$\dfrac{\text{side opposite}}{\text{hypotenuse}} = \underline{\hspace{1cm}} A$
LMN	l	m	n	L	M	N	$\dfrac{\text{side adjacent}}{\text{side opposite}} = \underline{\hspace{1cm}} L$
ZXY	z	x	y	Z	X	Y	$\dfrac{\text{side adjacent}}{\text{hypotenuse}} = \underline{\hspace{1cm}} Y$

2. Indicate the trigonometric function that each ratio of sides represents in the table.
3. Write each equation in the table using the letters for the sides as indicated in each case.

B. APPLICATIONS OF ENGLISH AND METRIC TABLES OF TRIGONOMETRIC FUNCTIONS AND INTERPOLATION OF NUMERICAL VALUES AND ANGLE MEASUREMENTS

1. Convert angle measurements (a) through (d) from English units of measure to metric decimal-degree units.
   a. 27° 30′   b. 27° 30′ 30″   c. 95° 27′ 48″   d. 212° 58′ 43″
2. Convert angle measurements (a), (b) and (c) from metric decimal-degree units to equivalent English degree, minute, and second units.
   a. 65.85°   b. 353.62°   c. 184.96°
3. Indicate the quadrant and the sign of trigonometric functions (a) through (e).
   a. sin 95° 22′   c. tan 282.94°   e. csc 343° 49′ 28″
   b. cos 245.75°   d. sec 65° 27′ 30″
4. Use a table of trigonometric functions in English units of measure to determine the numerical value of functions a, b, and c.
   Note. Interpolate to find the second values, rounded to four decimal places.
   a. sin 22° 10′   b. tan 87° 30′ 15″   c. csc 127° 42′ 28″
5. Determine the trigonometric function value of decimal-degree angles (a) through (e).
   a. sin 86.5°   c. tan 335.75°   e. cot 275.625°
   b. cos 227.62°   d. sec 3.592°

6. Determine either the height of gage blocks B or C, or angle X, whichever dimension is missing in the table. Give dimensions B and C to the nearest ten-thousandth; angle X, to the nearest second.

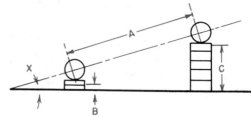

Sine Bar Length A (inches)	Heights of Blocks (in.)		Angle X
	B	C	
5.0000	1.2500	3.5000	
5.0000	1.0000	2.9375	
10.0000		4.3125	14° 30'
10.0000	1.5000		29° 1' 30"

C. APPLICATION OF THE FOUR-FUNCTION CALCULATOR IN TRIGONOMETRY

1. Compute the lengths of highway that must be paved for roads A, B, and C. The table indicates the angles of the feeder roads and the distance (X). Use the following formula and a calculator.

$$\tan \theta = 1.745\,\alpha \left(\frac{1.015\,\alpha^2 - 5}{6.092\,\alpha^2 - 5} \right)$$

$\alpha = \theta \div 100$ side $A = (X) \tan \theta$

a. Show the display readout for the value of tan θ and (b) the new highway distances in kilometers.

Road	Apex to			Calculator-Algorithm Method				Trigonometric Function Method			Accuracy of Computations (Variations)	
	Angle θ	Crossroad Distance (km)		Display Readout tan θ 1(a)	New Highway Distances		tan θ 3(a)	New Highway Distances		(km) 4(a)	(miles) (b)
					(km) 1(b)	(miles) (2)		(km) (b)	(miles) (c)		
A	20°	5.5									
B	25°	12.25									
C	32° 30'	16.375									

2. Convert each kilometer distance to its mile equivalent. Use 1 km = 0.6214 mi. Round the mileage values to four decimal places.
3. Check the highway lengths by conventional trigonometric methods, using natural trigonometric functions. Record (a) the natural function of tan θ and (b) the kilometer and (c) equivalent mile quantities.
4. Compare the calculator and the conventional method results. Indicate any variations in computations for road lengths A, B, or C. Express the variations (a) in decimal kilometer quantities and (b) equivalent mileage quantities.

D. CONVERSION OF DEGREE, RADIAN, AND GRAD MEASUREMENTS

1. a. Convert angle measurements A and B to equivalent values by direct calculator processes.
   b. Use formulas to change decimal-degree measurements C and D that are greater than 90°.
      Note. Round off degree values to two decimal places; radians to three decimal places, and grads to one decimal place.

	Degrees	Radians	Grads
A	60° 30′ 30″		
B		0.875	
C		2.625	
D			324.8

E. CALCULATOR APPLICATIONS: NATURAL TRIGONOMETRIC FUNCTIONS AND SMALLEST ANGLE FUNCTIONS

1. Enter angles (a) and (b) on a scientific calculator and read the trigonometric function values for sine, cosine, and tangent. Round off the display register value to four decimal places.
   a. Angle of 32°
   b. Angle of 72.46°
2. a. Enter the natural trigonometric function values of angles A and B on a scientific calculator.
   b. Determine the smallest angle in each case.
   c. Convert the display register value to English units of angle measure.

Function Value		Calculator Display	Converted Angle
A	sin 0.86646		
B	tan 4.8430		

F. APPLICATIONS OF NATURAL AND COMMON LOGARITHM TABLES AND CALCULATOR FUNCTIONS

1. Give the characteristic of the logarithm of numbers (a), (b), and (c).
   a. 0.0022
   b. 7.6493
   c. 746.2246
2. Use a handbook Table of Natural Logarithms and determine the mantissa of the logarithm for values (a) and (b).
   a. 0.0001903
   b. 8246.8678

3. Use a logarithm table and determine the logarithm for the same values (a) and (b).
   a. 0.0001903
   b. 8246.8678
4. Find the numbers that correspond to logarithms a, b, and c.
   a. 176.2553
   b. $\overline{2}$.7042
   c. $\overline{3}$.8782
5. Calculate the common logarithm values for (a) and the natural logarithms for (b) and (c). Round off the scientific calculator display values to three decimal places.
   a. 32.4444
   b. 0.012754
   c. 1.742

G. APPLICATIONS OF TRIGONOMETRIC FUNCTIONS, PYTHAGOREAN THEOREM, AND INTERPOLATION IN SOLVING INDUSTRIAL PROBLEMS

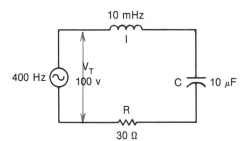

The series circuit as diagrammed consists of a 10 mHz inductor, a 10 μF capacitor, a 30 ohm resistor (R), and a voltage source of 100 volts with a frequency of 400 Hz.

1. Determine the reactance (X_L) of the inductor. Use the formula:

$$X_L = (2) \times (\pi) \times (F) \times (I).$$

2. Calculate the reactance of the capacitor (X_C). Use the formula:

$$X_C = \frac{1}{(2) \cdot (\pi) \cdot (F) \cdot (C)}$$

3. Determine the impedance (Z) using the Pythagorean theorem:

$$Z = \sqrt{R^2 + (X_L - X_C)^2}$$

4. Find the impedance phase angle $\angle\theta$. Use the tangent formula:

$$\tan \theta = \frac{X_L - X_C}{R}$$

H. APPLICATION OF ACUTE TRIANGLES

1. Determine the distance across flats on hexagons inscribed in circles having diameters of 1.5 inches, 2 inches, and 3 inches.
2. Find chordal distance C, included angle A and distance H for the three circles given in the table.

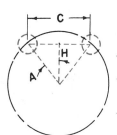

Diameter of Pitch Circle	No. of Holes on Pitch Circle
4″	6
$3\frac{1}{2}″$	7
6.250″	5

I. APPLICATION OF OBLIQUE TRIANGLES

1. Find the missing sides and angles of the triangles that are given in the table, using the law of sines. Dimensions for sides are to be rounded off to two decimal places; dimensions of angles, to the nearest minute.

Triangle	Angles			Sides		
	A	B	C	a	b	c
ABC	45°		30°	12"		
	D	E	F	d	e	f
DEF		100°	60°		$10\frac{1}{2}"$	
	K	L	M	k	l	m
KLM	$29\frac{1}{2}°$	$112\frac{1}{2}°$			20"	
	X	Y	Z	x	y	z
XYZ	59° 30'		14° 30'			12.50"

2. Determine the angle of slope (X) for the three roof spans indicated in the table. Use either the customary or the metric units.

S	$\dfrac{32'}{9.754 \text{ m}}$	$\dfrac{24'}{7.315 \text{ m}}$	$\dfrac{36'}{10.973 \text{ m}}$
R	$\dfrac{24'}{7.315 \text{ m}}$	$\dfrac{16'\text{-}8"}{5.08 \text{ m}}$	$\dfrac{22'}{6.706 \text{ m}}$

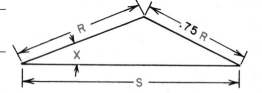

PART SEVEN

High Technology Applications of Mathematics

SYSTEMS AND PROGRAMMING FOR ADVANCED APPLICATIONS

Unit 92 Numerical Control (NC): Cartesian Coordinate and Binary Systems

OBJECTIVES OF THE UNIT

After satisfactorily completing the unit, the student/trainee will be able to

- *Understand the unique input of the Cartesian (rectangular) coordinate system in identifying any reference point in relation to coordinate dimensions that are perpendicular to the X, Y, and Z axes.*
- *Interpret basic numerical control commands to perform a specific block of work.*
- *Apply positional principles for the binary numbering system and change numerical values between the decimal and binary systems.*
- *Relate binary values and dimensional information for programming into numerical control.*

Numerical control (NC) refers to the operation of machine tools and other processing machines by a series of coded instructions. The instructions (NC commands) are in *alphanumeric code* that consists of numbers, letters, and other symbols. The commands are logically organized to make up a *program*. The program may be used continuously, interrupted, stored, and reused producing the same results each time.

NC programming and commands range from the positioning of a workpiece and spindle (or cutting tool) to cutter selection, tool wear compensation, speed and feed controls and direction, and machining processes, to the on/off control of coolant flow.

The foundations of numerical control include the Cartesian or rectangular coordinate system and the binary numbering system. This unit interlocks the relationship of these systems for later applications to programming, tooling up, and manufacturing, using sophisticated computer-aided equipment.

A. NUMERICAL CONTROL FOUNDED ON THE RECTANGULAR (CARTESIAN) COORDINATE SYSTEM

The old Cartesian coordinate system made it possible to describe any absolute point in mathematical terms from any other point in space along three coordinate axes that are mutually perpendicular.

The rectangular coordinate system may be explained simply by marking and folding a sheet of graph paper. A line drawn horizontally is marked as the *X axis*. A line drawn vertically at the midpoint represents the *Y axis*. The intersecting point of the two axes represents the *reference*, or *origin*, or *zero dimension*.

Numbers or other values may be added at regular intervals along the X and Y axes. Starting at the zero origin, all values along the X axis to the right of the zero reference point are positive (+). Similarly, all values that are up from the horizontal X axis line are positive (+). Values to the left of the index line are negative (−). Also, values below the X axis are negative (−).

The graph sheet is now divided into four parts along the axes. The parts are called *quadrants*. The graph sheet (*figure 92-1*) shows the positive and negative values of X and Y in relation to quadrants I, II, III, and IV. This type of graph is widely used for both plotting lines and curves and solving problems. It should be noted that the X and Y axes lie in the same plane.

The graph may also be considered as representing a machine table as, for example, a two-axis drilling machine. The table may be moved longitudinally (X axis) and transversely (Y axis). When the table

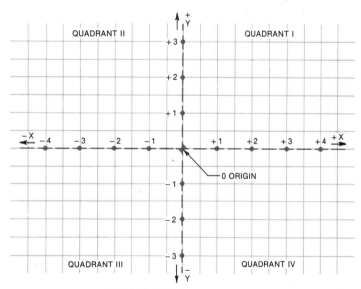

TWO-AXIS CARTESIAN COORDINATES

FIGURE 92-1

and spindle are positioned at a zero starting point, operations may be carried on according to X and Y *signed dimensions*. These are coordinate dimensions within a two-dimensioning system.

Third Major NC Axis or Z Coordinate

While a great deal of NC work relates to two-axis dimensioning, additional and advanced processes require dimensioning from a third major NC axis. This axis is perpendicular to the plane established by the *X* and *Y* axes. The new *Z* axis permits measurements to be made in three dimensions: length, width, and depth.

The line drawing (*figure* 92-2) shows that values along the *Z* axis above the *XY* axis (plane) are (+). Values below the axis line are (−). Thus, it is possible mathematically to describe the exact position of a cutting tool, process, or product anywhere in space.

To restate this fundamental principle of NC operation, any point may be located with respect to any other point on a workpiece or in relation to it by using three mutually perpendicular dimensions. Practically all numerical control motions and processes are based on and dimensioned according to rectangular coordinates. *X, Y,* and *Z* coordinates provide a mathematical approach for establishing reference points, specific dimensions, and other precise locations.

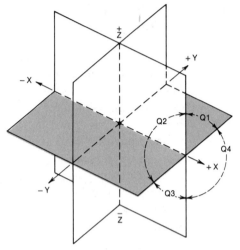

X, Y, AND Z THREE-AXIS CARTESIAN COORDINATES

FIGURE 92-2

B. POINT-TO-POINT AND CONTINUOUS-PATH PROGRAMMING

Point-to-Point Programming

There are two conceptual categories of numerical control programming: (1) point-to-point positioning and (2) continuous path. Point-to-point programming directs a machine spindle and cutter to move to a particular location (relationship to the workpiece) and to perform programmed events. Operations such as spot welding, drilling, punching, and tapping, are completed at fixed locations from a two-axis coordinate position. Before positioning, the workpiece and clamping device are cleared so the cutter may be advanced uninterrupted.

Continuous-Path Machining

Continuous-path machining indicates that a cutting tool is in constant contact with the workpiece. The coordinated path taken by the cutter or the contour produced is controlled by *interpolation*. Interpolation refers to the method of advancing from one point in a program to the next to produce a contoured workpiece according to the programmed design.

Contour profiling means that forms may be accurately generated that connect circles and arcs with straight or curved lines as illustrated in figure 92-3.

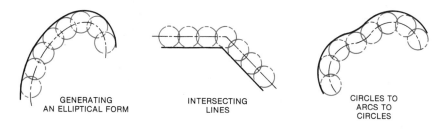

GENERATING
AN ELLIPTICAL FORM

INTERSECTING
LINES

CIRCLES TO
ARCS TO
CIRCLES

CONTINUOUS-PATH PROGRAMMING

FIGURE 92-3

C. METHODS OF LISTING COORDINATE POSITIONS

There are two *modes* (dimensioning techniques) for listing coordinate positions: *absolute* and *incremental*. These may be used separately or as a combination of the two. *All absolute dimensioning positions are identified from the fixed zero (reference) point*. This point represents the origin of the coordinate axes. The zero point may be established as a corner of a workpiece or as a particular location on a worktable. Absolute dimensioning is similar to what is called coordinate or base-line dimensioning on drawings. Each dimension has as its reference the point of origin. Thus, positioning errors are not cumulative from one dimension to the next. By contrast, *all incremental program locations are related in terms of distance and direction to the immediately preceding point.*

While the expression *positioning the spindle* is used, in actual practice the spindle location remains fixed. It is the workpiece and machine table that are positioned.

Figure 92-4 is an example of a workpiece where the coordinate locations of the three holes may be controlled by absolute or incremental dimensioning. In each instance, hole #1 is pro-

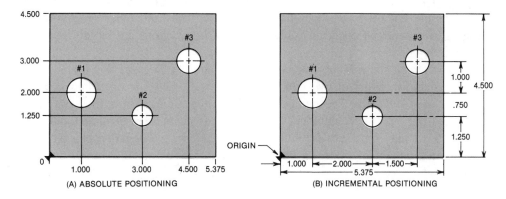

(A) ABSOLUTE POSITIONING

(B) INCREMENTAL POSITIONING

FIGURE 92-4

grammed by positioning the table and workpiece +1.000″ along the X axis and +2.000″ along the Y axis. After drilling and retracting the drill, hole #2 is positioned at +3.000″ (X axis) and +1.250″ (Y axis) in absolute programming.

In incremental programming, the table is moved +2.000″ (X) and −.750″ (Y) in relation to hole #1. Hole #3 at (A) is positioned at +4.500″ (X) and +3.000″ (Y). The same hole is located at (B) by moving the table +1.500″ (X) and +1.750″ (Y) in relation to hole #2.

Floating Zero

While the origin point has been described in a corner position, the zero may be floated to any position. Where operations and dimensions are related to a central location, the reference point is often moved as shown by the line drawing (*figure* 92-5). The operations are then programmed in relation to the floated zero.

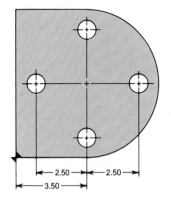

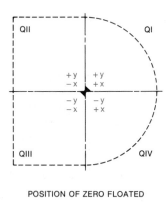

POSITION OF ZERO FLOATED

FIGURE 92-5

D. MATHEMATICAL COMMANDS FOR NUMERICAL CONTROL MACHINES

Thus far, a relationship has been established for representing signed numbers along X, Y, and Z axes. The Cartesian (rectangular) coordinate system provides the base for identifying any exact point in space. Movements of the machine worktable and spindle relate to absolute and floating reference points and point-to-point and continuous-path programming.

The heart of numerical control is the *numerical language command*. It is the command that provides a signal from a machine control unit to a machine to initiate one step in a complete program. Numerical control is the operation of machine tools and other processing machines and equipment by a series of coded instructions. These instructions are largely numbers, some letters, and other symbols.

Electronic machine control units and computers depend upon *cores* in a *logic* or *processing system* that are stable in either of two conditions: *on* or *off*, *charged* or *discharged*, ± or −, or *conductor* or *nonconductor*. The numbering system that uniquely fits these requirements is called the *binary system*. Binary numbers use only the two numbers of *binary one* and *binary zero*. A binary zero may represent a + charge; a binary one, a negative charge. By feeding binary numbers for numerical values, a binary notation may be added, subtracted, multiplied, and divided.

The computer instantaneously gathers a multitude of electronic ''yes'' and ''no'' signals that are either present or missing (on or off) within electronic circuits. The command is then fed to an actuating device to control the direction and quantity of movement. In other words, a quantity represented as a binary notation on a tape is transmitted through a command signal to a servomotor on a particular machine feed screw. When the command is executed, the signal instantly changes.

E. POSITIONAL PRINCIPLE OF BINARY NUMBERS

The binary number system requires only two *numeric symbols:* **0** and **1**. The base for the system is called *base two*. This means the symbol represents the number of distinct numerals (**0** and **1**) that are required before the binary numbers are used in combination to produce higher numbers.

In counting in the binary system, when the number in the last right digit is *exhausted,* the column returns to **0** and a value of **1** is carried over to the next left column. Most numbering systems are built on this *positional principle*. In the decimal system, the whole number columns are identified as the units, tens, hundreds, or thousands. When the number **9** is exhausted in the first right column, the column number returns to **0** and a value of **1** is carried over to the tens column. This same principle continues; for example, when **99** is exhausted, the units column returns to **0**, the tens column returns to zero, and **1** is carried over to the hundreds column.

The right-hand column in the binary number system represents values of ones. The next left column represents values of twos; the third column, fours; the fourth column, eights. In other words, the positional (place) values begin in the units column and progress through the twos, fours, eights, and sixteens columns.

F. COUNTING WHOLE NUMBERS IN THE BINARY SYSTEM

Counting whole numbers in the binary system begins with **0**. The binary numeral 0_2 is equal to **0** in the decimal system. The binary numeral 1_2 is equal to **1** in the decimal system. Examples are provided in figure 92-6 to further show the value of numbers in the positional binary system in relation to equivalent decimal values.

2^6	2^5	2^4	2^3	2^2	2^1	2^0	Decimal Number Equivalent
64	32	16	8	4	2	1	
						0	0
						1	1
					1	0	2
					1	1	3
				1	0	0	4
				1	0	1	5
				1	1	0	6
				1	1	1	7
			1	0	0	0	8
			1	0	0	1	9
			1	0	1	0	10

2^6	2^5	2^4	2^3	2^2	2^1	2^0	Decimal Number Equivalent
64	32	16	8	4	2	1	
			1	0	1	1	11
			1	1	0	0	12
			1	1	0	1	13
			1	1	1	0	14
			1	1	1	1	15
		1	0	0	0	0	16
		1	0	0	0	1	17
		1	0	0	1	0	18
		1	0	0	1	1	19
		1	0	1	0	0	20
	1	0	0	0	0	0	32
1	0	0	0	0	0	0	64
1	1	0	0	1	0	0	100

FIGURE 92-6

POSITIONAL RULE FOR BINARY NUMBERS

* Arrange the binary numbers in columns.
* Start to add the binary numbers in the last right column.
* Apply the three basic rules of addition.

$$
\begin{array}{ccc}
0_2 & 0_2 & 1_2 \\
+0_2 & +1_2 & +1_2 \\
\hline
0_2 & 1_2 & 0_2
\end{array}
$$

Note. When the binary numbers in the column are exhausted, enter a zero in the column. Carry over **1** to the next left column.

* Proceed to the next column on the left. Add the binary values.
Note. Again, when the last binary number for the column is exhausted, enter a **0** in the column and add **1** to the next left column.
* Continue these same steps until all values are added.

EXAMPLE: *Case 1.* Express the binary number 1001_2 plus one additional count.

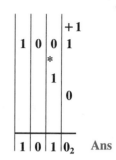

Step 1 Enter the additional count of **1** in the one unit column.

Step 2 Add the two binary numerals in the right position.

Note. The **1** returns to **0** and generates a carry (*) of **1** to the next left column.

Step 3 List the resulting binary number. 1010_2 **Ans**

EXAMPLE: *Case 2.* Record a second additional count and express the resulting binary number.

Step 1 Start with the 1010_2 and again enter the count of **1** in the one unit column.

Step 2 Add the binary numerals. 1011_2 **Ans**

EXAMPLE: *Case 3.* Record a third additional count and express the resulting binary number.

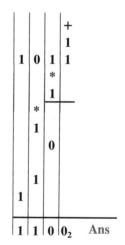

Step 1 Start with 1011_2 and again enter the count of **1** in the one unit column.

Step 2 Add the binary numerals in the ones column (= **0**). Carry the ***1*** over to the twos column.

Step 3 Add the binary numbers in the twos column (= **0**) and carry the ***1*** over to the fours column.

Step 4 Add the fours column (= **1**).

Step 5 Add the eights column.

Step 6 Write the total binary number resulting from all the additions and carryovers. 1100_2 **Ans**

Review and Self-Test

ASSIGNMENT UNIT 92

A. Point-to-Point and Continuous-Path Programming

1. a. Explain briefly how point-to-point positioning differs from continuous-path programming.

   b. Differentiate between absolute dimensioning and incremental dimensioning.

B. NC Rectangular Coordinate Dimensioning

1. a. Float the reference point to the center of the Mounting Plate.
    b. Determine the X, Y, and Z coordinate dimensions for holes #1, 2, 3, and 4.

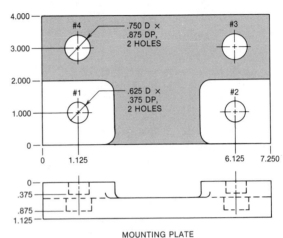

Hole	X	Y	Z
(0.001")			
1			
2			
3			
4			

MOUNTING PLATE

C. Mathematical Processing of Binary Numbers for NC Programming

1. Cite three major reasons why the binary notation system is especially adapted for numerical control of machine tool and other computer applications.

2. Find the binary sums of a, b, c, and d. Convert the answers to decimal numbers of equivalent values.

    a. 1 0 1 1 0
        1 0 0 1

    b. 1 0 1 1 1
        1 1 1 1 0

    c. 1 0 1 0 1 0
        1 1 0 0 1

    d. 1 1 1 1 1
        1 0 0 0 1

3. Translate decimal numbers a through e to equivalent binary numbers.
    a. 12   b. 36   c. 155   d. 477   e. 1298

Unit 93 Numerical Control Applications: Computer Numerical Control, Computer-aided Design, and Computer-aided Manufacturing

OBJECTIVES OF THE UNIT

After satisfactorily completing the unit, the student/trainee will be able to

- *Describe the functions of the channels and major features of standard Electronics Industries Association (EIA) tapes.*
- *Discuss the basic functions of numerical control.*
- *Interpret postprocessor, geometry, motion, and auxiliary statements in dimensional and nondimensioned words.*
- *Visualize programming features of computer-assisted design, (CAD), computer-assisted manufacturing (CAM), and computer numerical control (CNC), and other systems.*
- *Apply basic words in the APT (automatically programmed tools) language and interlock the programming of a part using postprocessor, geometry, motion, and auxiliary statements.*

Regardless of the numerical control system or equipment that it controls, it is evident that the programming, processes, and products depend upon the capability to precisely define and make measurements and to perform specific movements on command. Cartesian coordinates provide the system for the precise definition of motion and quantity. Binary notation and numbers are uniquely suited to generate, store, retrieve, modify, and control commands.

This unit advances binary-coded information through the numerical control unit to machine tools. Functions of CNC, CAD, CAM and other equipment are applied through widely used APT procedures.

A. TRANSFERRING BINARY-CODED INFORMATION TO AN ELECTRONIC COMPUTER

Commands that are binary coded through a computer (machine control unit) are communicated by the presence or absence of a hole or magnetism in a preselected area of a tape. Information is punched in rows (columns) across a one-inch wide tape. Successive rows along the length permit the entry of additional information.

The rows are called *channels*. A binary number is expressed in a column by the presence or absence of a signal. The nomenclature used, the tape width, hole number sequences, spacings between holes across and down the tape, feed holes, and other dimensions and sizes and colors have been standardized. Standards have been set by the Electronics Industries Association (EIA), Aerospace Industries Association of America, and National Aerospace Standards bodies.

A small section of a one-inch perforated tape is used in figure 93-1 to show the standard EIA character codes and other representative features. There are four separate codes:

- Numeric (0 through 9)
- Alphabetic (a through z)
- Symbols
- Parentheses, ampersand (&), and other non-standard symbols.

The tracks (channels) are numbered from **1** through **8**. The channel for the alignment sprocket holes is not numbered. Channels **1, 2, 3,** and **4** represent the binary notation codes of 2^0, 2^1, 2^2, and 2^3. The numeric values are 1, 2, 4, and 8, respectively. These values by horizontal and vertical rows are used for the numerical control of machines and processes.

Tape Tracks for the Binary-Coded Decimal System

Track **6** alone indicates **0** when holes are punched in the tape. The sum of the decimal equivalent is determined by the position and number of marked or punched holes in a row. The addition of the values in a single row represents all the decimal numerals from **0** through **9**. The line drawing shows the hole or marking combinations on a tape.

In operation, each row of holes is read in binary-code sequence in a tape reader. The data are then converted through the circuitry of the computer from binary codes to decimal equivalents. This numeric information is then stored until required for positioning or controlling particular processes. The perforated or otherwise marked tape communicates commands.

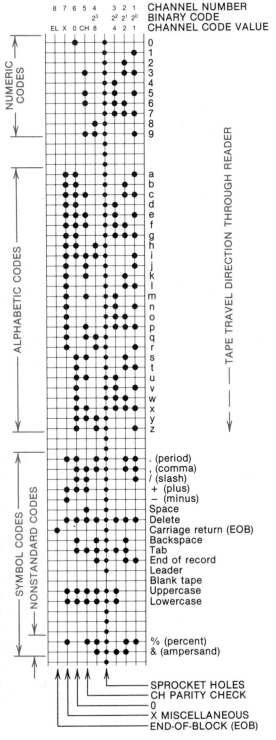

FIGURE 93-1 EIA character codes and channels for NC machine tools

It must be remembered that the manner and sequence in which a part is programmed depends on the type of machine tool or machining center, the range of processes, and other factors. The tape code must be known by the designer, programmer, and machine operator.

Sample Binary-coded Decimal (BCD) Punched Tape

The short section of tape in figure 93-1 also illustrates the nature and extent of information that is manipulated through a computer (MCU). The fifth channel is called the *parity check* (*CH*). A parity check verifies accuracy in preparing the tape. The *characters* (entries in a row) are either all-odd or all-even, depending on the coding system. A punched or marked CH channel entry provides this check.

The *X*-value *channel* designates *miscellaneous functions* or it may be used with alphabetic codes. The *EOB* channel represents the *end-of-block*.

B. BASIC FUNCTIONS OF A NUMERICAL CONTROL SYSTEM

The design, tooling, processing, inspection, assembly, and disassembly of manufactured parts are controlled by eight basic functions of numerical control.

- Positioning and controlling machine and workpiece movements. Using Cartesian coordinates, every dimensioned location may be accurately charted by coordinate dimensions.
- Controlling tool movements with respect to machining within prescribed dimensional limitations.
- Establishing tool storage, retrieval, and other sequences.
- Controlling cutting speeds and feeds.
- Compensating for tool wear; monitoring performance, surface quality, and accuracy.
- Producing digital display and other production control readouts.
- Interrupting one sequence of processes to change, store, or retrieve other processes and/or products.
- Controlling shutdown and start-up of machine components or systems.

All of these functions and others are programmed. Particular commands for dimensions, as well as nondimensioned words, are all part of programming. A *program manuscript* is prepared. *Statements* are made in a special sequence using words that have a definite meaning in the programming language. A *system dictionary* is used.

Commands for a machine tool to perform any combination of the eight basic functions are usually contained in four types of statements.

1. *Postprocessor statements*. This part of the computer program takes the control system data output and adapts it to control machine, tooling, other systems, and processes.
2. *Geometry statements* that describe precise features about the part.
3. *Motion statements* relating to positioning, retracting, and other positions and rotating directions of cutting tools.
4. *Auxiliary statements* that provide commands for functions other than those relating to coordinates of the workpiece, geometry, or cutter functions.

C. NONDIMENSION NC WORDS AND FUNCTIONS

Nondimension words are direction (command) words. Each word is identified by a category or function that follows in this order: (1) sequence, (2) preparation, (3) feed rate, (4) spindle speed, (5) tool selection, and (6) miscellaneous. Each function is designated by a letter or what is often called an *address character*. Each address character is accompanied by one or more numeric digits. The following characters and number of numeric digits are used to identify nondimension words.

Nondimension Word Category	Character	Designation
		Number of Numeric Digits
(1) Sequence Number	N	Three
(2) Preparatory Function	G	Two
(3) Feed Rate	F	Maximum Eight
(4) Spindle Speed	S	Three
(5) Tool Selection	T	Maximum Five
(6) Miscellaneous Function	M	Two

FIGURE 93-2

The character in each case indicates the start of a programming sequence within a *block*. A block consists of a word grouping that provides complete information for a particular process or function. The grouping may be one or more rows of characters separated from other words by an EOB character. The nondimension letter and coded words prepare the numerical control unit for a specified mode of operation.

For example, in preparing the NC unit for preparatory functions, the word **G00** indicates that the mode of machine operation is to be point-to-point positioning. A **G01** command directs positioning by linear interpolation. The commands **G17, G18,** and **G19** relate to *XY, YZ,* or *XZ* plane selections.

D. CNC, CAD, AND CAM SYSTEMS

With computer-aided programming (CAP), laborious, time-consuming numerical coordinate and other data calculations may be made cost-efficiently and error-free. CAP permits editing by running a sample part through a programmed cycle or by visually plotting a cutter path or displaying it on a *cathode-ray tube* (CRT).

The CRT is capable of visually showing all part numbers of all programs entered on computer memory; cutter radius, length, offsets, and other features for which compensation is to be made; special messages to the operator at specified times; and diagnostic information the CNC units isolate as malfunctions in the numerical control system.

Selected Design Features of a CNC System

Examples of CNC system programming features include

- Tape controlled.
    —Feed rates in inches per minute (ipm) or (mm/m);
    —Spindle speeds in rpm;
    —Rotary table indexing and simultaneous movement with any x, y, or z plane movements;
    —Random tool selection.
- Keyboards of CNC systems provide a cutter diameter compensation (CDC) feature for oversized and undersized cutters and workpiece irregularities.
- Tool length and tool depth compensation in increments of 0.0001″.
- Electronic tool gages and digital readouts.

Computer-aided Design (CAD) and Manufacturing (CAM) Systems

Former functions of computers are now locked into what are called hierarchial levels of computers. The computer's capability to design, evaluate, and accurately take information from an original source, manipulate it, and transmit it to a required point makes it possible to move into a computer-aided manufacturing system.

The ultimate of CAM is an interlocked system of design and development functions with manufacturing, marketing, inventory controls, financing, sales promotion, management, and all other functions within an industry. The CAM system requires coupling the output of mainframe computers to serve multiple functions at one time, minicomputer input, and microprocessor capability to program terminals and other machine control units.

E. COMPUTER-AIDED PROGRAMMING (CAP)

The automatically programmed tools (APT) system is one of the most widely used processor languages. The APT system is described briefly because it contains programming activities that are common in computer-aided tooling and manufacturing.

At the outset, a conventional drawing and specifications may be used to provide full information from which a single part or a large quantity of components may be produced. The dimensioning may be changed to absolute (base line) or incremental dimensioning, as required. A complementary sketch is generally prepared showing the positioning or tool path. A programming sheet with special notes is then developed giving the sequence of every event from setup to machine shutdown. All functions are described in statements that are commands for every movement and activity performed on the machine tool.

In APT language there is a complete vocabulary of allowable words. The words are spelled in a particular way. The English-like form must be entered into the computer according to the dictionary spelling. Punctuation marks appear between every two words, symbols, or numbers. An incorrectly spelled word produces an *error message*.

APT statements consist of two parts. These are separated by a slash (/). For example, **GØLFT/BASLIN** directs the tool to go left from its present position until its periphery is tangent to the base line. APT statements follow a logical outline, producing a check for the programmer. The parts of a particular program begin with descriptive information about the part number and identifying part name and remarks. Machine control and cutter identification are recorded. Tolerances and other computations and setting point information are programmed. Geometric and other part features are described on the *X-Y* plane.

The programmer's guide includes speed, rotation, and spindle conditions, feed rates, coolant requirements, machining motions, and, finally, spindle and motion ''off,'' and the end of program.

While each APT word has a specific meaning, the meaning is also related to parts of the statement that follow. For instance, **FEDRAT** is *feed rate*. However, the word is followed by a slash (/) and a number like **FEDRAT/24.** This means the feed rate in *X, Y,* and *Z* directions will be 24 ipm (inches per minute). **GØFWD** is to *go forward* but **GØFWD/LIN2** provides direction of movement in relation to *line 1.* **ØUTTØL,** followed by /**.0005** directs the computer to stay within a 0.0005″ machining tolerance on the inside of curves.

A sample part is used to illustrate the major elements in APT programming. The four types of statements that are programmed are illustrated in figures 93-3, 93-4, and 93-5.

① Postprocessor Statements

```
        PART NØ 169 R
        MACHIN/ BGPT 12NC
        INTØL/.001
        ØUTØL/.001
        CUTTER/1.000
        CØØLNT/ØN
        CLPRNT
        FEEDRAT/2
        SPINDL/750
```

FIGURE 93-3 APT program postprocessor statements

② Geometry Statements

```
STPE  = PØINT/0, 0, .750
PT 1  = PØINT/.500, .500
PT 2  = PØINT/5.500,0
PT 3  = PØINT/0, 3.500
PT 4  = PØINT/4.500, 4.500
PT 5  = PØINT/1.500, 3.000
L1    = LINE/P1, P2
C1    = CIRC1/5.500, 3.500
L2    = LINE/P2, P3, TANTØ C1
L3    = LINE/P4, TANTØ C2
C2    = CIRC2/1.500, 3.000
L4    = LINE/P5, P6, TANTØ C1
```

FIGURE 93-4 APT program geometry statements

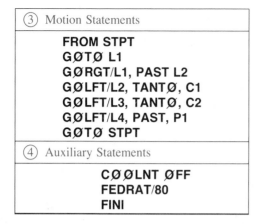

③ Motion Statements

FROM STPT
GØTØ L1
GØRGT/L1, PAST L2
GØLFT/L2, TANTØ, C1
GØLFT/L3, TANTØ, C2
GØLFT/L4, PAST, P1
GØTØ STPT

④ Auxiliary Statements

CØØLNT ØFF
FEDRAT/80
FINI

FIGURE 93-5 APT program motion and auxiliary statements

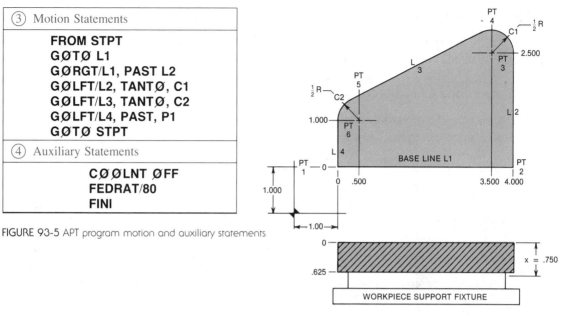

FIGURE 93-6 Sample part for APT programming example

Review and Self-Test

ASSIGNMENT UNIT 93

A. Transferring Binary-coded Information

1. a. Identify the channels on the EIA tape that are used for binary notation.
   b. Give the binary notation symbols for each column value.
   c. State the decimal equivalent value for each binary channel.
2. (a) Determine binary notation and (b) the decimal numeric value of the following binary numbers.

	Binary Numbers	Binary Notation (*a*)	Decimal Equivalent (*b*)
A	1 0 1 0		
B	1 0 1 0 1		
C	1 1 0 0 0 0		
D	1 0 1 0 1 0 1		

3. State three basic functions of a numerical control system.
4. Describe briefly what a block means in relation to programming for NC.

B. Nondimension Numerical Control (NC) Words and Functions

1. Give three examples of nondimensional words and characters used in programming each particular machine tool setup functions.

C. CNC, CAP, and CAM Systems

1. Identify two CNC tape-controlled programming features.
2. Cite a main difference between the following computer systems.
   a. Computer-aided design (CAD) and computer-aided manufacturing (CAM).
   b. Manually programmed numerical control (NC) and computerized numerical control (CNC).

D. Computer-aided Programming

1. a. Read the information provided on the accompanying section of a program sheet for machining a specific part.
   b. Mark the tape to enter the program information for blocks 000 and 001.
   c. Alongside the tape, add notes and brackets to explain each function and activity in each block.

Station Number	Block ID	Preparatory Function	X Function	Y Function	Res. No	Tool No.	Miscellaneous Function
Loading	000	5	00000	00000	00	4	02
1	001	5	67500	03750	00	1	12
2	002	6	87500	10250	00	1	80

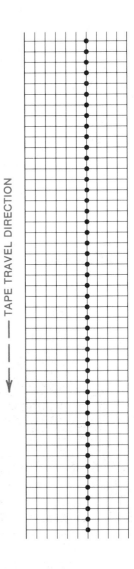

TAPE TRAVEL DIRECTION

2. a. State the *X* and *Y*-axis positions for holes 1, 2, and 3 and the locations for the elongated slot at 4, 5, 6, and 7.
   b. Give the *Z*-axis dimension for the slot depth.

Feature	Dimensions		
	X	*Y*	*Z*
1			
2			
3			
4			
5			
6			
7			

Unit 94 Achievement Review on High Technology Applications

OBJECTIVES OF THE UNIT

This achievement review serves as an overall test for Section 20. The unit is designed to measure the student's/trainee's ability to

- *Relate numerical control principles of programming to actual work situations.*
- *Deal with binary notation, binary numeric values, and decimal equivalent values; convert quantities to meet different problem requirements for NC programming.*
- *Identify common characteristics of NC programming manuscripts, CNC functions, and applications of the APT system.*
- *Read an industrial part drawing and prepare a table of rectangular coordinates of X, Y, and Z-axis movements.*
- *Translate NC information on a program sheet into a word address format tape for controlling basic machine tool processes.*

A. NUMERICAL CONTROL (NC) PROGRAMMING PRINCIPLES

1. Identify two conditions that determine whether the use of a fixed reference point away from a workpiece or a floating zero is to be used.
2. Cite one advantage of using rectangular (Cartesian) coordinates for establishing dimensions and positions of any element or feature of a part.

B. BINARY NOTATION, NUMERIC VALUES, AND NUMERICAL CONTROL PROGRAMMING

1. Explain briefly the numerical control of all machine, cutting tool, and processing movements by coded information provided by the binary decimal system.

2. Add the following binary numbers and then convert the answer to decimal number equivalents.

   a. 1 0 1 0
       1 1 0 1 1

   b. 1 1 1 0 1 0
        1 1 1 1 1

   c. 1 1 0 0 1 1 1
      1 1 0 1 1 0 0 1

   d. 1 1 0 1
        1 1 1
       1 0 1 1 0

   e.   1 1 1 0 0 1
        1 0 1 0 1 1 1 1
        1 0 1 0 1 0 1 1 1
      1 1 1 1 0 1 0 1 1 0

3. Record the following binary notations as binary numbers.
   a. $2^5 + 2^4 + 2^3$
   b. $2^6 + 2^5 + 2^3 + 2^1$
   c. $2^8 + 2^7 + 2^6 + 2^1 + 2^0$
   d. $2^8 + 2^7 + 2^6 + 2^5 + 2^4 + 2^2 + 2^1$

4. a. Convert the following decimal numbers into equivalent binary numbers.
      a. 36   b. 106   c. 516   d. 1,011
   b. Add the column of decimal number values and check the answer against the addition of the binary number values.

C. FUNCTIONS OF NC AND CNC SYSTEMS

1. Identify four basic types of statements that control all NC programming.

2. Name two characteristics of all programming manuscripts.

3. State three unique CNC functions.

4. Give two major characteristics of the automatically programmed tools (APT) system.

D. NC PROGRAMMING: CARTESIAN COORDINATES, BINARY-CODED INFORMATION, AND WORD ADDRESS FORMATS

1. a. Read the drawing of the Slotted Plate for features, exact dimensions and specifications, and manufacturing processes.
   b. Relist the sequence of operations on the cutter path drawing. Relocate the zero reference point 6.000″ from the left side and 4.000″ down from the bottom side of the Slotted Plate.

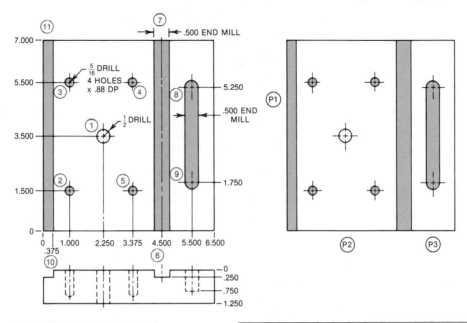

Tool List		
1	$\frac{1}{2}''$ Drill, stub	
2	$\frac{5}{16}''$ Drill, stub	
3	.500″ End Mill	
Locating Pins		
P1		
P2	.500″ Diam	
P3		

Position Number	Coordinates		
	X	Y	Z
1			
2			
3			
4			
5			
6			
7			
8			
9			
10			
11			

c. Determine the *X, Y,* and *Z* machine table positions (coordinates) for drilling holes 1, 2, 3, 4, and 5, and milling slots 6 through 11. *Note the location* of the fixed reference point off the workpiece.

Sample Section of Program Sheet for Slotted Plate

Sequence Number H or N	Preparatory Functions G	Coordinates		Miscellaneous Functions	Position Number	Tool	Head Pos.	Spindle Speed (ipm)	Speed (RPM)	Depth of Cut (Z Axis)	Table Feed (ipm)	Tool Remarks
		X	Y									
H001	G81	08250	07500	M51	1	1	1	6	1400	01250		$\frac{1}{2}$" drill, stub
N001	G80			M06	1							Tool change
H002	G81	07000	05500	M53	2	2	1	5	2200	00875		$\frac{5}{16}$" drill, stub
H003	G81		09375	M26	3							
H004	G81	09375		M26	4							
H005	G81		05500	M06	5							Tool Change

d. Read the *program sheet* for the examples of work processes that are given.

e. Mark up a section of tape for drilling hole #1, the tool change M06, and sequence #H002.

f. Along the side of the tape, cross-reference each line of characters with notes to identify the functions within the block for H001, N001, and H002.

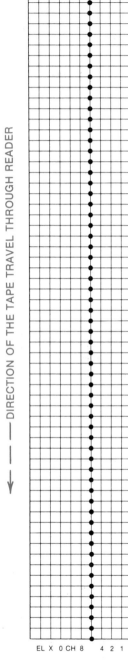

DIRECTION OF THE TAPE TRAVEL THROUGH READER

EL X 0 CH 8   4 2 1

SECTION OF TAPE
TO BE CODED

APPENDIX

TABLE 1 Standard Tables of Customary Units of Measure

Linear Measure		
12 inches (in.)	=	1 foot (ft).
3 ft.	=	1 yard (yd.)
16 1/2 ft.	=	1 rod (rd.)
5 1/2 yd.	=	1 rd.
320 rd.	=	1 mile
1760 yd.	=	1 mile
5280 ft.	=	1 mile

Surface Measure		
144 sq. in.	=	1 sq. ft.
9 sq. ft.	=	1 sq. yd.
30 1/4 sq. yd.	=	1 sq. rd.
160 sq. rd.	=	1 acre
640 acres	=	1 sq. mile
43,560 sq. ft.	=	1 acre

Cubic Measure (Volume)		
1728 cu. in.	=	1 cu. ft.
27 cu. ft.	=	1 cu. yd.
128 cu. ft.	=	1 cord

Angular (Circular) Measure		
60 sec. (")	=	1 min. (')
60'	=	1 degree (°)
90°	=	1 quadrant
360°	=	1 circle

Time Measure		
60 seconds (sec.)	=	1 minute (min.)
60 min.	=	1 hour (hr.)
24 hr.	=	1 day
7 days	=	1 week
52 weeks	=	1 year
365 days	=	1 year
10 years	=	1 decade

Liquid Measure		
4 gills	=	1 pint (pt.)
2 pt.	=	1 quart (qt.)
4 qt.	=	1 gallon (gal.)
231 cu. in.	=	1 gal.
31.5 gal.	=	1 barrel (bbl.)
42 gal.	=	1 bbl. of oil
8 1/2 lb.	=	1 gal. water
7 1/2 gal.	=	1 cu. ft.

Weights of Materials		
0.096 lb.	=	1 cu. in. aluminum
0.260 lb.	=	1 cu. in. cast iron
0.283 lb.	=	1 cu. in. mild steel
0.321 lb.	=	1 cu. in. copper
0.41 lb.	=	1 cu. in. lead
112 lb.	=	1 cu. ft. Dowmetal
167 lb.	=	1 cu. ft. aluminum
464 lb.	=	1 cu. ft. cast iron
490 lb.	=	1 cu. ft. mild steel
555.6 lb.	=	1 cu. ft. copper
710 lb.	=	1 cu. ft. lead

Avoirdupois Weight		
16 ounces (oz.)	=	1 pound (lb.)
100 lb.	=	1 hundredweight (cwt.)
20 cwt.	=	1 ton
2000 lb.	=	1 ton
8 1/2 lb.	=	1 gal. of water
62.4 lb.	=	1 cu. ft. of water
112 lb.	=	1 long cwt.
2240 lb.	=	1 long ton

Dry Measure		
2 cups	=	1 pt.
2 pt.	=	1 qt.
4 qt.	=	1 gal.
8 qt.	=	1 peck (pk.)
4 pk.	=	1 bushel (bu.)

Miscellaneous		
12 units	=	1 dozen (doz.)
12 doz.	=	1 gross
144 units	=	1 gross
24 sheets	=	1 quire
20 quires	=	1 ream
20 units	=	1 score
6 ft.	=	1 fathom

TABLE 2 Standard Tables of Metric Units of Measure

Linear Measure		
Unit	**Value in Meters**	**Symbol or Abbreviation**
micron	0.000 001	μ
millimeter	0.001	mm
centimeter	0.01	cm
decimeter	0.1	dm
meter (unit)	1.0	m
dekameter	10.0	dam
hectometer	100.0	hm
kilometer	1 000.00	km
myriameter	10 000.00	Mm
megameter	1 000 000.00	

Surface Measure		
Unit	**Value in Square Meters**	**Symbol or Abbreviation**
square millimeter	0.000 001	mm^2
square centimeter	0.000 1	cm^2
square decimeter	0.01	dm^2
square meter (centiare)	1.0	m^2
square dekameter (are)	100.0	a^2
hectare	10 000.0	ha^2
square kilometer	1 000 000.0	km^2

Volume		
Unit	**Value in Liters**	**Symbol or Abbreviation**
milliliter	0.001	mL
centiliter	0.01	cL
deciliter	0.1	dL
liter (unit)	1.0	L
dekaliter	10.0	daL
hectoliter	100.0	hL
kiloliter	1 000.0	kL

Mass		
Unit	**Value in Grams**	**Symbol or Abbreviation**
microgram	0.000 001	μg
milligram	0.001	mg
centigram	0.01	cg
decigram	0.1	dg
gram (unit)	1.0	g
dekagram	10.0	dag
hectogram	100.0	hg
kilogram	1 000.0	kg
myriagram	10 000.0	Mg
quintal	100 000.0	q
ton	1 000 000.0	t

Cubic Measure		
Unit	**Value in Cubic Meters**	**Symbol or Abbreviation**
cubic micron	10^{-18}	μ^3
cubic millimeter	10^{-9}	mm^3
cubic centimeter	10^{-6}	cm^3
cubic decimeter	10^{-3}	dm^3
cubic meter	1	m^3
cubic dekameter	10^3	dam^3
cubic hectometer	10^6	hm^3
cubic kilometer	10^9	km^3

TABLE 3 Metric and Customary Decimal Equivalents for Fractional Parts of an Inch

DECIMAL EQUIVALENTS						
Fraction		Decimal Equivalent		Fraction		Decimal Equivalent
		Customary (in.)	Metric (mm)			Customary (in.)	Metric (mm)

Fraction	Customary (in.)	Metric (mm)	Fraction	Customary (in.)	Metric (mm)
1/64	.015625	0.3969	33/64	.515625	13.0969
1/32	.03125	0.7938	17/32	.53125	13.4938
3/64	.046875	1.1906	35/64	.546875	13.8906
1/16	.0625	1.5875	9/16	.5625	14.2875
5/64	.078125	1.9844	37/64	.578125	14.6844
3/32	.09375	2.3813	19/32	.59375	15.0813
7/64	.109375	2.7781	39/64	.609375	15.4781
1/8	.1250	3.1750	5/8	.6250	15.8750
9/64	.140625	3.5719	41/64	.640625	16.2719
5/32	.15625	3.9688	21/32	.65625	16.6688
11/64	.171875	4.3656	43/64	.671875	17.0656
3/16	.1875	4.7625	11/16	.6875	17.4625
13/64	.203125	5.1594	45/64	.703125	17.8594
7/32	.21875	5.5563	23/32	.71875	18.2563
15/64	.234375	5.9531	47/64	.734375	18.6531
1/4	.250	6.3500	3/4	.750	19.0500
17/64	.265625	6.7469	49/64	.765625	19.4469
9/32	.28125	7.1438	25/32	.78125	19.8438
19/64	.296875	7.5406	51/64	.796875	20.2406
5/16	.3125	7.9375	13/16	.8125	20.6375
21/64	.328125	8.3384	53/64	.828125	21.0344
11/32	.34375	8.7313	27/32	.84375	21.4313
23/64	.359375	9.1281	55/64	.859375	21.8281
3/8	.3750	9.5250	7/8	.8750	22.2250
25/64	.390625	9.9219	57/64	.890625	22.6219
13/32	.40625	10.3188	29/32	.90625	23.0188
27/64	.421875	10.7156	59/64	.921875	23.4156
7/16	.4375	11.1125	15/16	.9375	23.8125
29/64	.453125	11.5094	61/64	.953125	24.2094
15/32	.46875	11.9063	31/32	.96875	24.6063
31/64	.484375	12.3031	63/64	.984375	25.0031
1/2	.500	12.7000	1	1.000	25.4000

TABLE 4 Decimal and Millimeter Equivalents

Millimeter Equivalents of Decimals (0.01″ to 0.99″)									
Dec.	0	1	2	3	4	5	6	7	8	9
0.0	. . . .	0.254	0.508	0.762	1.016	1.270	1.524	1.778	2.032	2.286
0.1	2.540	2.794	3.048	3.302	3.556	3.810	4.064	4.318	4.572	4.826
0.2	5.080	5.334	5.588	5.842	6.096	6.350	6.604	6.858	7.112	7.366
0.3	7.620	7.874	8.128	8.392	8.636	8.890	9.144	9.398	9.652	9.906
0.4	10.160	10.414	10.688	10.922	11.176	11.430	11.684	11.938	12.192	12.446
0.5	12.700	12.954	13.208	13.462	13.716	13.970	14.224	14.478	14.732	14.986
0.6	15.240	15.494	15.748	16.022	16.256	16.510	16.764	17.018	17.272	17.526
0.7	17.780	18.034	18.288	18.542	18.796	19.050	19.304	19.558	19.812	20.066
0.8	20.320	20.574	20.828	21.082	21.336	21.590	21.844	22.098	22.352	22.606
0.9	22.860	23.114	23.368	23.622	23.876	24.130	24.384	24.638	24.892	25.146

Example   0.1″ = 2.540 mm, 0.75″ = 19.050 mm

Decimal Equivalents of Millimeters (1 mm to 99 mm)									
mm	0	1	2	3	4	5	6	7	8	9
0	. . . .	0.0394	0.0787	0.1181	0.1575	0.1968	0.2362	0.2756	0.3150	0.3543
1	0.3937	0.4331	0.4724	0.5118	0.5512	0.5906	0.6299	0.6693	0.7087	0.7480
2	0.7874	0.8268	0.8661	0.9055	0.9449	0.9842	1.0236	1.0630	1.1024	1.1417
3	1.1811	1.2205	1.2598	1.2992	1.3386	1.3780	1.4173	1.4567	1.4961	1.5354
4	1.5748	1.6142	1.6535	1.6929	1.7323	1.7716	1.8110	1.8504	1.8898	1.9291
5	1.9685	2.0079	2.0472	2.0866	2.1260	2.1654	2.2047	2.2441	2.2835	2.3228
6	2.3622	2.4016	2.4409	2.4803	2.5197	2.5590	2.5984	2.6378	2.6772	2.7165
7	2.7559	2.7953	2.8346	2.8740	2.9134	2.9528	2.9921	3.0315	3.0709	3.1102
8	3.1496	3.1890	3.2283	3.2677	3.3071	3.3464	3.3858	3.4252	3.4646	3.5039
9	3.5433	3.5827	3.6220	3.6614	3.7008	3.7402	3.7795	3.8189	3.8583	3.8976

Example   10 mm = 0.3937″, 57 mm = 2.2441″

TABLE 5 Conversion Factors for SI, Conventional Metric, and Customary Units Used in Physical Science Applications

Category	Conversion of Customary and Conventional Metric Units to SI Metrics			Conversion of SI Metrics to Customary and Conventional Metric Units		
	From Customary or Conventional Metric Unit	To SI Metric	Factor (A) (Multiply by)	From SI Metric	To Customary or Conventional Metric Unit	Factor
acceleration	ft/s²	m/s²	$0.304\ 8$*	m/s²	ft/s²	Multiply by the
	in/s²		$2.540\ 0\ \times 10^{-2}$*		in/s²	reciprocal of the
area	ft²	m²	$9.290\ 3 \times 10^{-2}$	m²	ft²	multiplier (Factor A)
	in²		$6.451\ 6 \times 10^{-4}$*		in²	
density	g/cm³	kg/m³	$1.000\ 0 \times 10^{3}$*	kg/m³	g/cm³	which is used in
	lb (mass)/ft³		$16.018\ 5$		lb (mass)/ft³	conversion to the SI
	lb (mass)/in³		$2.768\ 0 \times 10^{4}$		lb (mass)/in³	metric unit
energy	Btu (thermochemical)		$1.054\ 3 \times 10^{3}$		Btu (thermochemical)	
	cal (thermochemical)		$4.184\ 0$*		cal (thermochemical)	
	eV		$1.602\ 1 \times 10^{-19}$		eV	
	erg	J	$1.000\ 0 \times 10^{-7}$*	J	erg	
	ft • lb (force)		$1.355\ 8$		ft • lb (force)	
	kW•h		$3.600\ 0 \times 10^{6}$*		kW•h	
	Wh		$3.600\ 0 \times 10^{3}$*		Wh	
flow (liquid and solid)	ft³/min		$4.719\ 5 \times 10^{-4}$		ft³/min.	
	ft³/s	m³/s	$2.831\ 7 \times 10^{-2}$	m³/s	ft³/s	
	in³/min		$2.731\ 2 \times 10^{-7}$		in³/min	
	lb (mass)/s		$0.453\ 6$		lb (mass)/s	
	lb (mass)/min	kg/s	$7.559\ 9 \times 10^{-3}$	kg/s	lb (mass)/min	
	tons (short, mass)/h		$0.252\ 0$		tons (short, mass)/h	
force	dyne		$1.000\ 0 \times 10^{-5}$*		dyne	
	kg (force)	N	$9.806\ 6$	N	kg (force)	
	lb (force)		$4.448\ 2$		lb (force)	
heat	Btu (thermochemical)/ft²		$1.134\ 9 \times 10^{4}$		Btu (thermochemical)/ft²	
	cal (thermochemical) /cm²	J/m²	$4.184\ 0 \times 10^{4}$*	J/m²	cal (thermochemical)/cm²	
	ft²/h	m²/s	$2.580\ 6 \times 10^{-5}$	m²/s	ft²/h	
length	yd		$0.914\ 4$		yd	
	ft		$0.304\ 8$*		ft	
	in	m	$2.540\ 0 \times 10^{-2}$*	m	in	
	μ (micron)		$1.000\ 0 \times 10^{-6}$*		μ (micron)	
	mil		$2.540\ 0 \times 10^{-5}$*		mil	

(continued)

* Denotes exact value.

TABLE 5 Conversion Factors for SI, Conventional Metric, and Customary Units Used in Physical Science Applications (continued)

Category	Conversion of Customary and Conventional Metric Units to SI Metrics			Conversion of SI Metrics to Customary and Conventional Metric Units		
	From Customary or Conventional Metric Unit	To SI Metric	Factor (A) (Multiply by)	From SI Metric	To Customary or Conventional Metric Unit	Factor
mass	lb (mass avoirdupois)	kg	0.453 6	kg	lb (mass, avoirdupois)	Multiply by the reciprocal of the multiplier (Factor A) which is used in conversion to the SI metric unit
	oz (mass, avoirdupois)		2.835 0 x 10^{-2}		oz (mass, avoirdupois)	
	ton, long = 2 240 lb (mass)		1.016 0 x 10^3		ton, long = 2 240 lb (mass)	
	ton, metric		1.000 0 x 10^3 *		ton, metric	
	ton, short = 2 000 lb (mass)		0.907 2 x 10^3		ton, short = 2 000 lb (mass)	
power	Btu (thermochemical)/min	W	17.572 5	W	Btu (thermochemical)/min	
	cal (thermochemical)/min		6.973 3 x 10^{-2}		cal (thermochemical)/min	
	erg/s		1.000 0 x 10^{-7} *		erg/s	
	ft · lb (force)/min		2.259 7 x 10^{-2}		ft · lb (force)/min	
	hp (550 ft · lb/s)		7.457 0 x 10^2		hp (550 ft · lb/s)	
pressure (stress)	atm (760 torr)	N/m^2	1.013 2 x 10^5	N/m^2	atm (760 torr)	
	dyne/cm^2		0.100 0*		dyne/cm^2	
	g (force)/cm^2		98.066 5*		g (force)/cm^2	
	kg (force)/cm^2		9.806 6 x 10^4		kg (force)/cm^2	
	lb (force)/in^2 (or psi)		6.894 8 x 10^3		lb (force)/in^2 (or psi)	
	lb (force)/in^2 (or psi)	kg (force)/mm^2	7.030 7 x 10^{-4}	kg (force)/mm^2		
	torr (mm mercury at 0°C)	N/m^2	1.333 2 x 10^2	N/m^2	torr (mm mercury at 0°C)	
velocity	ft/min	m/s	5.080 0 x 10^{-3} *	m/s	ft/min	
	in/s		2.540 0 x 10^{-2} *		in/s	
	mph	km/h	0.447 0	km/h	mph	
	mph		1.609 3			
volume	ft^3	m^3	2.831 7 x 10^{-2}	m^3	ft^3	
	in^3		1.638 7 x 10^{-5}		in^3	
	liter		1.000 0 x 10^{-3} *		liter	
temperature	deg C	K	$t_K = t_C + 273.15$	K	deg C	

*Denotes exact value

TABLE 6 Conversion of English and Metric Units of Measure

Linear Measure

Unit	Inches to milli-metres	Milli-metres to inches	Feet to metres	Metres to feet	Yards to metres	Metres to yards	Miles to kilo-metres	Kilo-metres to miles
1	25.40	0.03937	0.3048	3.281	0.9144	1.094	1.609	0.6214
2	50.80	0.07874	0.6096	6.562	1.829	2.187	3.219	1.243
3	76.20	0.1181	0.9144	9.842	2.743	3.281	4.828	1.864
4	101.60	0.1575	1.219	13.12	3.658	4.374	6.437	2.485
5	127.00	0.1968	1.524	16.40	4.572	5.468	8.047	3.107
6	152.40	0.2362	1.829	19.68	5.486	6.562	9.656	3.728
7	177.80	0.2756	2.134	22.97	6.401	7.655	11.27	4.350
8	203.20	0.3150	2.438	26.25	7.315	8.749	12.87	4.971
9	228.60	0.3543	2.743	29.53	8.230	9.842	14.48	5.592

Example   1 in. = 25.40 mm,   1 m = 3.281 ft.,   1 km = 0.6214 mi.

Surface Measure

Unit	Square inches to square centi-metres	Square centi-metres to square inches	Square feet to square metres	Square metres to square feet	Square yards to square metres	Square metres to square yards	Acres to hec-tares	Hec-tares to acres	Square miles to square kilo-metres	Square kilo-metres to square miles
1	6.452	0.1550	0.0929	10.76	0.8361	1.196	0.4047	2.471	2.59	0.3861
2	12.90	0.31	0.1859	21.53	1.672	2.392	0.8094	4.942	5.18	0.7722
3	19.356	0.465	0.2787	32.29	2.508	3.588	1.214	7.413	7.77	1.158
4	25.81	0.62	0.3716	43.06	3.345	4.784	1.619	9.884	10.36	1.544
5	32.26	0.775	0.4645	53.82	4.181	5.98	2.023	12.355	12.95	1.931
6	38.71	0.93	0.5574	64.58	5.017	7.176	2.428	14.826	15.54	2.317
7	45.16	1.085	0.6503	75.35	5.853	8.372	2.833	17.297	18.13	2.703
8	51.61	1.24	0.7432	86.11	6.689	9.568	3.237	19.768	20.72	3.089
9	58.08	1.395	0.8361	96.87	7.525	10.764	3.642	22.239	23.31	3.475

Example   1 sq. in. = 6.452 cm^2 1 m^2 = 1.196 sq. yd.,   1 sq. mi. = 2.59 km^2

Cubic Measure

Unit	Cubic inches to cubic centi-metres	Cubic centi-metres to cubic inches	Cubic feet to cubic metres	Cubic metres to cubic feet	Cubic yards to cubic metres	Cubic metres to cubic yards	Gallons to cubic feet	Cubic feet to gallons
1	16.39	0.06102	0.02832	35.31	0.7646	1.308	0.1337	7.481
2	32.77	0.1220	0.05663	70.63	1.529	2.616	0.2674	14.96
3	49.16	0.1831	0.08495	105.9	2.294	3.924	0.4010	22.44
4	65.55	0.2441	0.1133	141.3	3.058	5.232	0.5347	29.92
5	81.94	0.3051	0.1416	176.6	3.823	6.540	0.6684	37.40
6	98.32	0.3661	0.1699	211.9	4.587	7.848	0.8021	44.88
7	114.7	0.4272	0.1982	247.2	5.352	9.156	0.9358	52.36
8	131.1	0.4882	0.2265	282.5	6.116	10.46	1.069	59.84
9	147.5	0.5492	0.2549	371.8	6.881	11.77	1.203	67.32

Example   1 cm^3 = 0.06102 cu. in.,   1 gal. = 0.1337 cu. ft.

Volume or Capacity Measure

Unit	Liquid ounces to cubic centi-metres	Cubic centi-metres to liquid ounces	Pints to litres	Litres to pints	Quarts to litres	Litres to quarts	Gallons to litres	Litres to gallons	Bushels to hecto-litres	Hecto-litres to bushels
1	29.57	0.03381	0.4732	2.113	0.9463	1.057	3.785	0.2642	0.3524	2.838
2	59.15	0.06763	0.9463	4.227	1.893	2.113	7.571	0.5284	0.7048	5.676
3	88.72	0.1014	1.420	6.340	2.839	3.785	11.36	0.7925	1.057	8.513
4	118.3	0.1353	1.893	8.454	3.170	4.227	15.14	1.057	1.410	11.35
5	147.9	0.1691	2.366	10.57	4.732	5.284	18.93	1.321	1.762	14.19
6	177.4	0.2029	2.839	12.68	5.678	6.340	22.71	1.585	2.114	17.03
7	207.0	0.2367	3.312	14.79	6.624	7.397	26.50	1.849	2.467	19.86
8	236.6	0.2705	3.785	16.91	7.571	8.454	30.28	2.113	2.819	22.70
9	266.2	0.3043	4.259	19.02	8.517	9.510	34.07	2.378	3.171	25.54

Example   1 L = 2.113 pt.,   1 gal. = 3.785 L

TABLE 7 Numerical Values of Trigonometric Functions for Decimal-Degree (Metric) Angles

0.0° to 8.0

Dec/Deg	Sin	Cos	Tan	Cot	Dec/Deg
0.0	0.000 00	1.000 0	0.000 00	Infinite	90.0
.1	0.001 75	1.000 0	0.001 75	572.957	.9
.2	0.003 49	1.000 0	0.003 49	286.477	.8
.3	0.005 24	1.000 0	0.005 24	190.984	.7
.4	0.006 98	1.000 0	0.006 98	143.237	.6
.5	0.008 73	1.000 0	0.008 73	114.589	.5
.6	0.010 47	0.999 9	0.010 47	95.489	.4
.7	0.012 22	0.999 9	0.012 22	81.847	.3
.8	0.013 96	0.999 9	0.013 96	71.615	.2
.9	0.015 71	0.999 9	0.015 71	63.657	.1
1.0	0.017 45	0.999 8	0.017 46	57.290	89.0
.1	0.019 20	0.999 8	0.019 20	52.081	.9
.2	0.020 94	0.999 8	0.020 95	47.740	.8
.3	0.022 69	0.999 7	0.022 69	44.066	.7
.4	0.024 43	0.999 7	0.024 44	40.917	.6
.5	0.026 18	0.999 7	0.026 19	38.188	.5
.6	0.027 92	0.999 6	0.027 93	35.801	.4
.7	0.029 67	0.999 6	0.029 68	33.694	.3
.8	0.031 41	0.999 5	0.031 43	31.821	.2
.9	0.033 16	0.999 5	0.033 17	30.143	.1
2.0	0.034 90	0.999 4	0.034 92	28.636	88.0
.1	0.036 64	0.999 3	0.036 67	27.271	.9
.2	0.038 39	0.999 3	0.038 42	26.031	.8
.3	0.040 13	0.999 2	0.040 16	24.898	.7
.4	0.041 88	0.999 1	0.041 91	23.859	.6
.5	0.043 62	0.999 0	0.043 66	22.904	.5
.6	0.045 36	0.999 0	0.045 41	22.022	.4
.7	0.047 11	0.998 9	0.047 16	21.205	.3
.8	0.048 85	0.998 8	0.048 91	20.446	.2
.9	0.050 59	0.998 7	0.050 66	19.740	.1
3.0	0.052 34	0.998 6	0.052 41	19.081	87.0
.1	0.054 08	0.998 5	0.054 16	18.464	.9
.2	0.055 82	0.998 4	0.055 91	17.886	.8
.3	0.057 56	0.998 3	0.057 66	17.343	.7
.4	0.059 31	0.998 2	0.059 41	16.832	.6
.5	0.061 05	0.998 1	0.061 16	16.350	.5
.6	0.062 79	0.998 0	0.062 91	15.895	.4
.7	0.064 53	0.997 9	0.064 67	15.464	.3
.8	0.066 27	0.997 8	0.066 42	15.056	.2
.9	0.068 02	0.997 7	0.068 17	14.669	.1
4.0	0.069 76	0.997 6	0.069 93	14.301	86.0
.1	0.071 50	0.997 4	0.071 68	13.951	.9
.2	0.073 24	0.997 3	0.073 44	13.617	.8
.3	0.074 98	0.997 2	0.075 19	13.300	.7
.4	0.076 27	0.997 1	0.076 95	12.996	.6
.5	0.078 46	0.996 9	0.078 70	12.706	.5
.6	0.080 20	0.996 8	0.080 46	12.429	.4
.7	0.081 94	0.996 6	0.082 21	12.163	.3
.8	0.083 68	0.996 5	0.083 97	11.909	.2
.9	0.085 42	0.996 3	0.085 73	11.664	.1
5.0	0.087 16	0.996 2	0.087 49	11.430	85.0
.1	0.088 89	0.996 0	0.089 25	11.205	.9
.2	0.090 63	0.995 9	0.091 01	10.988	.8
.3	0.092 37	0.995 7	0.092 77	10.780	.7
.4	0.094 11	0.995 6	0.094 53	10.579	.6
.5	0.095 85	0.995 4	0.096 29	10.385	.5
.6	0.097 58	0.995 2	0.098 05	10.199	.4
.7	0.099 32	0.995 1	0.099 81	10.019	.3
.8	0.101 06	0.994 9	0.101 58	9.845	.2
.9	0.102 79	0.994 7	0.103 34	9.677	.1
6.0	0.104 53	0.994 5	0.105 10	9.514	84.0
.1	0.106 26	0.994 3	0.106 87	9.357	.9
.2	0.108 00	0.994 2	0.108 63	9.205	.8
.3	0.109 73	0.994 0	0.110 40	9.058	.7
.4	0.111 47	0.993 8	0.112 17	8.915	.6
.5	0.113 20	0.993 6	0.113 94	8.777	.5
.6	0.114 94	0.993 4	0.115 70	8.643	.4
.7	0.116 67	0.993 2	0.117 47	8.513	.3
.8	0.118 40	0.993 0	0.119 24	8.386	.2
.9	0.120 14	0.992 8	0.121 01	8.264	.1
7.0	0.121 87	0.992 5	0.122 78	8.144	83.0
.1	0.123 60	0.992 3	0.124 56	8.028	.9
.2	0.125 33	0.992 1	0.126 33	7.916	.8
.3	0.127 06	0.991 9	0.128 10	7.806	.7
.4	0.128 80	0.991 7	0.129 88	7.700	.6
.5	0.130 53	0.991 4	0.131 65	7.596	.5
.6	0.132 26	0.991 2	0.133 43	7.495	.4
.7	0.133 99	0.991 0	0.135 21	7.396	.3
.8	0.135 72	0.990 7	0.136 98	7.300	.2
.9	0.137 44	0.990 5	0.138 76	7.207	.1
8.0	0.139 17	0.990 3	0.140 54	7.1154	82.0
Dec/Deg	**Cos**	**Sin**	**Cot**	**Tan**	**Dec/Deg**

82.0 to 90.0

8.0 to 15.0

Dec/Deg	Sin	Cos	Tan	Cot	Dec/Deg
8.0	0.139 17	0.990 3	0.140 54	7.1154	82.0
.1	0.140 90	0.990 0	0.142 32	7.0264	.9
.2	0.142 63	0.989 8	0.144 10	6.9395	.8
.3	0.144 36	0.989 5	0.145 88	6.8547	.7
.4	0.146 08	0.989 3	0.147 67	6.7720	.6
.5	0.147 81	0.989 0	0.149 45	6.6912	.5
.6	0.149 54	0.988 8	0.151 24	6.6122	.4
.7	0.151 26	0.988 5	0.153 02	6.5350	.3
.8	0.152 99	0.988 2	0.154 81	6.4596	.2
.9	0.154 71	0.988 0	0.156 60	6.3859	.1
9.0	0.156 43	0.987 7	0.158 38	6.3138	81.0
.1	0.158 16	0.987 4	0.160 17	6.2432	.9
.2	0.159 88	0.987 1	0.161 96	6.1742	.8
.3	0.161 60	0.986 9	0.163 76	6.1066	.7
.4	0.163 33	0.986 6	0.165 55	6.0405	.6
.5	0.165 05	0.986 3	0.167 34	5.9758	.5
.6	0.166 77	0.986 0	0.169 14	5.9124	.4
.7	0.168 49	0.985 7	0.170 93	5.8502	.3
.8	0.170 21	0.985 4	0.172 73	5.7894	.2
.9	0.171 93	0.985 1	0.174 53	5.7297	.1
10.0	0.173 6	0.984 8	0.176 3	5.6713	80.0
.1	0.175 4	0.984 5	0.178 1	5.6140	.9
.2	0.177 1	0.984 2	0.179 9	5.5578	.8
.3	0.178 8	0.983 9	0.181 7	5.5026	.7
.4	0.180 5	0.983 6	0.183 5	5.4486	.6
.5	0.182 2	0.983 3	0.185 3	5.3955	.5
.6	0.184 0	0.982 9	0.187 1	5.3435	.4
.7	0.185 7	0.982 6	0.189 0	5.2924	.3
.8	0.187 4	0.982 3	0.190 8	5.2422	.2
.9	0.189 1	0.982 0	0.192 6	5.1929	.1
11.0	0.190 8	0.981 6	0.194 4	5.1446	79.0
.1	0.192 5	0.981 3	0.196 2	5.0970	.9
.2	0.194 2	0.981 0	0.198 0	5.0504	.8
.3	0.195 9	0.980 6	0.199 8	5.0045	.7
.4	0.197 7	0.980 3	0.201 6	4.9594	.6
.5	0.199 4	0.979 9	0.203 5	4.9152	.5
.6	0.201 1	0.979 6	0.205 3	4.8716	.4
.7	0.202 8	0.979 2	0.207 1	4.8288	.3
.8	0.204 5	0.978 9	0.208 9	4.7867	.2
.9	0.206 2	0.978 5	0.210 7	4.7453	.1
12.0	0.207 9	0.978 1	0.212 6	4.7046	78.0
.1	0.209 6	0.977 8	0.214 4	4.6646	.9
.2	0.211 3	0.977 4	0.216 2	4.6252	.8
.3	0.213 0	0.977 0	0.218 0	4.5864	.7
.4	0.214 7	0.976 7	0.219 9	4.5483	.6
.5	0.216 4	0.976 3	0.221 7	4.5107	.5
.6	0.218 1	0.975 9	0.223 5	4.4737	.4
.7	0.219 8	0.975 5	0.225 4	4.4373	.3
.8	0.221 5	0.975 1	0.227 2	4.4015	.2
.9	0.223 3	0.974 8	0.229 0	4.3662	.1
13.0	0.225 0	0.974 4	0.230 9	4.3315	77.0
.1	0.226 7	0.974 0	0.232 7	4.2972	.9
.2	0.228 4	0.973 6	0.234 5	4.2635	.8
.3	0.230 0	0.973 2	0.236 4	4.2303	.7
.4	0.231 7	0.972 8	0.238 2	4.1976	.6
.5	0.233 4	0.972 4	0.240 1	4.1653	.5
.6	0.235 1	0.972 0	0.241 9	4.1335	.4
.7	0.236 8	0.971 5	0.243 8	4.1022	.3
.8	0.238 5	0.971 1	0.245 6	4.0713	.2
.9	0.240 2	0.970 7	0.247 5	4.0408	.1
14.0	0.241 9	0.970 3	0.249 3	4.0108	76.0
.1	0.243 6	0.969 9	0.251 2	3.9812	.9
.2	0.245 3	0.969 4	0.253 0	3.9520	.8
.3	0.247 0	0.969 0	0.254 9	3.9232	.7
.4	0.248 7	0.968 6	0.256 8	3.8947	.6
.5	0.250 4	0.968 1	0.258 6	3.8667	.5
.6	0.252 1	0.967 7	0.260 5	3.8391	.4
.7	0.253 8	0.967 3	0.262 3	3.8118	.3
.8	0.255 4	0.966 8	0.264 2	3.7843	.2
.9	0.257 1	0.966 4	0.266 1	3.7583	.1
15.0	0.258 8	0.965 9	0.267 9	3.7321	75.0
.1	0.260 5	0.965 5	0.269 8	3.7062	.9
.2	0.262 2	0.965 0	0.271 7	3.6806	.8
.3	0.263 9	0.964 6	0.273 6	3.6554	.7
.4	0.265 6	0.964 1	0.275 4	3.6305	.6
.5	0.267 2	0.963 6	0.277 3	3.6059	.5
.6	0.268 9	0.963 2	0.279 2	3.5816	.4
.7	0.270 6	0.962 7	0.281 1	3.5576	.3
.8	0.272 3	0.962 2	0.283 0	3.5339	.2
.9	0.274 0	0.961 7	0.284 9	3.5105	.1
16.0	0.275 6	0.961 3	0.286 7	3.4874	74.0
Dec/Deg	**Cos**	**Sin**	**Cot**	**Tan**	**Dec/Deg**

74.0 to 82.0

(continued)

Note. Numerical values for secant and cosecant functions may be determined by calculator as reciprocals for cosine and sine functions, respectively.

TABLE 7 Numerical Values of Trigonometric Functions for Decimal-Degree (Metric) Angles *(continued)*

16.0 to 24.0

Dec/Deg	Sin	Cos	Tan	Cot	Dec/Deg
16.0	0.275 6	0.961 3	0.286 7	3.4874	74.0
.1	0.277 3	0.960 8	0.288 6	3.4646	.9
.2	0.279 0	0.960 3	0.290 5	3.4420	.8
.3	0.280 7	0.959 8	0.292 4	3.4197	.7
.4	0.282 3	0.959 3	0.294 3	3.3977	.6
.5	0.284 0	0.958 8	0.296 2	3.3759	.5
.6	0.285 7	0.958 3	0.298 1	3.3544	.4
.7	0.287 4	0.957 8	0.300 0	3.3332	.3
.8	0.289 0	0.957 3	0.301 9	3.3122	.2
.9	0.290 7	0.956 8	0.303 8	3.2914	.1
17.0	0.292 4	0.956 3	0.305 7	3.2709	73.0
.1	0.294 0	0.955 8	0.307 6	3.2506	.9
.2	0.295 7	0.955 3	0.309 6	3.2305	.8
.3	0.297 4	0.954 8	0.311 5	3.2106	.7
.4	0.299 0	0.954 2	0.313 4	3.1910	.6
.5	0.300 7	0.953 7	0.315 3	3.1716	.5
.6	0.302 4	0.953 2	0.317 2	3.1524	.4
.7	0.304 0	0.952 7	0.319 1	3.1334	.3
.8	0.305 7	0.952 1	0.321 1	3.1146	.2
.9	0.307 4	0.951 6	0.323 0	3.0961	.1
18.0	0.309 0	0.951 1	0.324 9	3.0777	72.0
.1	0.310 7	0.950 5	0.326 9	3.0595	.9
.2	0.312 3	0.950 0	0.328 8	3.0415	.8
.3	0.314 0	0.949 4	0.330 7	3.0237	.7
.4	0.315 6	0.948 9	0.332 7	3.0061	.6
.5	0.317 3	0.948 3	0.334 6	2.9887	.5
.6	0.319 0	0.947 8	0.336 5	2.9714	.4
.7	0.320 6	0.947 2	0.338 5	2.9544	.3
.8	0.322 3	0.946 6	0.340 4	2.9375	.2
.9	0.323 9	0.946 1	0.342 4	2.9208	.1
19.0	0.325 6	0.945 5	0.344 3	2.9042	71.0
.1	0.327 2	0.944 9	0.346 3	2.8878	.9
.2	0.328 9	0.944 4	0.348 2	2.8716	.8
.3	0.330 5	0.943 8	0.350 2	2.8556	.7
.4	0.332 2	0.943 2	0.352 2	2.8397	.6
.5	0.333 8	0.942 6	0.354 1	2.8239	.5
.6	0.335 5	0.942 1	0.356 1	2.8083	.4
.7	0.337 1	0.941 5	0.358 1	2.7929	.3
.8	0.338 7	0.940 9	0.360 0	2.7776	.2
.9	0.340 4	0.940 3	0.362 0	2.7625	.1
20.0	0.342 0	0.939 7	0.364 0	2.7475	70.0
.1	0.343 7	0.939 1	0.365 9	2.7326	.9
.2	0.345 3	0.938 5	0.367 9	2.7179	.8
.3	0.346 9	0.937 9	0.369 9	2.7034	.7
.4	0.348 6	0.937 3	0.371 9	2.6889	.6
.5	0.350 2	0.936 7	0.373 9	2.6746	.5
.6	0.351 8	0.936 1	0.375 9	2.6605	.4
.7	0.353 5	0.935 4	0.377 9	2.6464	.3
.8	0.355 1	0.934 8	0.379 9	2.6325	.2
.9	0.356 7	0.934 2	0.381 9	2.6187	.1
21.0	0.358 4	0.933 6	0.383 9	2.6051	69.0
.1	0.360 0	0.933 0	0.385 9	2.5916	.9
.2	0.361 6	0.932 3	0.387 9	2.5782	.8
.3	0.363 3	0.931 7	0.389 9	2.5649	.7
.4	0.364 9	0.931 1	0.391 9	2.5517	.6
.5	0.366 5	0.930 4	0.393 9	2.5386	.5
.6	0.368 1	0.929 8	0.395 9	2.5257	.4
.7	0.369 7	0.929 1	0.397 9	2.5129	.3
.8	0.371 4	0.928 5	0.400 0	2.5002	.2
.9	0.373 0	0.927 8	0.402 0	2.4876	.1
22.0	0.374 6	0.927 2	0.404 0	2.4751	68.0
.1	0.376 5	0.926 5	0.406 1	2.4627	.9
.2	0.377 8	0.925 9	0.408 1	2.4504	.8
.3	0.379 5	0.925 2	0.410 1	2.4383	.7
.4	0.381 1	0.924 5	0.412 2	2.4262	.6
.5	0.382 7	0.923 9	0.414 2	2.4142	.5
.6	0.384 3	0.923 2	0.416 3	2.4021	.4
.7	0.385 9	0.922 5	0.418 3	2.3906	.3
.8	0.387 5	0.921 9	0.420 4	2.3789	.2
.9	0.389 1	0.921 2	0.422 4	2.3673	.1
23.0	0.390 7	0.920 5	0.424 5	2.3559	67.0
.1	0.392 3	0.919 8	0.426 5	2.3445	.9
.2	0.393 9	0.919 1	0.428 6	2.3332	.8
.3	0.395 5	0.918 4	0.430 7	2.3220	.7
.4	0.397 1	0.917 8	0.432 7	2.3109	.6
.5	0.398 7	0.917 1	0.434 8	2.2998	.5
.6	0.400 3	0.916 4	0.436 9	2.2889	.4
.7	0.401 9	0.915 7	0.439 0	2.2781	.3
.8	0.403 5	0.915 0	0.441 1	2.2673	.2
.9	0.405 1	0.914 3	0.443 1	2.2566	.1
24.0	0.406 7	0.913 5	0.445 2	2.2460	66.0

Dec/Deg	Cos	Sin	Cot	Tan	Dec/Deg

66.0 to 74.0

24.0° to 32.0°

Dec/Deg	Sin	Cos	Tan	Cot	Dec/Deg
24.0	0.406 7	0.913 5	0.445 2	2.2460	66.0
.1	0.408 3	0.912 8	0.447 3	2.2355	.9
.2	0.409 9	0.912 1	0.449 4	2.2251	.8
.3	0.411 5	0.911 4	0.451 5	2.2148	.7
.4	0.413 1	0.910 7	0.453 6	2.2045	.6
.5	0.414 7	0.910 0	0.455 7	2.1943	.5
.6	0.416 3	0.909 2	0.457 8	2.1842	.4
.7	0.417 9	0.908 5	0.459 9	2.1742	.3
.8	0.419 5	0.907 8	0.462 1	2.1642	.2
.9	0.421 0	0.907 0	0.464 2	2.1543	.1
25.0	0.422 6	0.906 3	0.466 3	2.1445	65.0
.1	0.424 2	0.905 6	0.468 4	2.1348	.9
.2	0.425 8	0.904 8	0.470 6	2.1251	.8
.3	0.427 4	0.904 1	0.472 7	2.1155	.7
.4	0.428 9	0.903 3	0.474 8	2.1060	.6
.5	0.430 5	0.902 6	0.477 0	2.0965	.5
.6	0.432 1	0.901 8	0.479 1	2.0872	.4
.7	0.433 7	0.901 1	0.481 3	2.0778	.3
.8	0.435 2	0.900 3	0.483 4	2.0686	.2
.9	0.436 8	0.899 6	0.485 6	2.0594	.1
26.0	0.438 4	0.898 8	0.487 7	2.0503	64.0
.1	0.439 9	0.898 0	0.489 9	2.0413	.9
.2	0.441 5	0.897 3	0.492 1	2.0323	.8
.3	0.443 1	0.896 5	0.494 2	2.0233	.7
.4	0.444 6	0.895 7	0.496 4	2.0145	.6
.5	0.446 2	0.894 9	0.498 6	2.0057	.5
.6	0.447 8	0.894 2	0.500 8	1.9970	.4
.7	0.449 3	0.893 4	0.502 9	1.9883	.3
.8	0.450 9	0.892 6	0.505 1	1.9797	.2
.9	0.452 4	0.891 8	0.507 3	1.9711	.1
27.0	0.454 0	0.891 0	0.509 5	1.9626	63.0
.1	0.455 5	0.890 2	0.511 7	1.9542	.9
.2	0.457 1	0.889 4	0.513 9	1.9458	.8
.3	0.458 6	0.888 6	0.516 1	1.9375	.7
.4	0.460 2	0.887 8	0.518 4	1.9292	.6
.5	0.461 7	0.887 0	0.520 6	1.9210	.5
.6	0.463 3	0.886 2	0.522 8	1.9128	.4
.7	0.464 8	0.885 4	0.525 0	1.9047	.3
.8	0.466 4	0.884 6	0.527 2	1.8967	.2
.9	0.467 9	0.883 8	0.529 5	1.8887	.1
28.0	0.469 5	0.882 9	0.531 7	1.8807	62.0
.1	0.471 0	0.882 1	0.534 0	1.8728	.9
.2	0.472 6	0.881 3	0.536 2	1.8650	.8
.3	0.474 1	0.880 5	0.538 4	1.8572	.7
.4	0.475 6	0.879 6	0.540 7	1.8495	.6
.5	0.477 2	0.878 8	0.543 0	1.8418	.5
.6	0.478 7	0.878 0	0.545 2	1.8341	.4
.7	0.480 2	0.877 1	0.547 5	1.8265	.3
.8	0.481 8	0.876 3	0.549 8	1.8190	.2
.9	0.483 3	0.875 5	0.552 0	1.8115	.1
29.0	0.484 8	0.874 6	0.554 3	1.8040	61.0
.1	0.486 3	0.873 8	0.556 6	1.7966	.9
.2	0.487 9	0.872 9	0.558 9	1.7893	.8
.3	0.489 4	0.872 1	0.561 2	1.7820	.7
.4	0.490 9	0.871 2	0.563 5	1.7747	.6
.5	0.492 4	0.870 4	0.565 8	1.7675	.5
.6	0.493 9	0.869 5	0.568 1	1.7603	.4
.7	0.495 5	0.868 6	0.570 4	1.7532	.3
.8	0.497 0	0.867 8	0.572 7	1.7461	.2
.9	0.498 5	0.866 9	0.575 0	1.7391	.1
30.0	0.500 0	0.866 0	0.577 4	1.7321	60.0
.1	0.501 5	0.865 2	0.579 7	1.7251	.9
.2	0.503 0	0.864 3	0.582 0	1.7182	.8
.3	0.504 5	0.863 4	0.584 4	1.7113	.7
.4	0.506 0	0.862 5	0.586 7	1.7045	.6
.5	0.507 5	0.861 6	0.589 0	1.6977	.5
.6	0.509 0	0.860 7	0.591 4	1.6909	.4
.7	0.510 5	0.859 9	0.593 8	1.6842	.3
.8	0.512 0	0.859 0	0.596 1	1.6775	.2
.9	0.513 5	0.858 1	0.598 5	1.6709	.1
31.0	0.515 0	0.857 2	0.600 9	1.6643	59.0
.1	0.516 5	0.856 3	0.603 2	1.6577	.9
.2	0.518 0	0.855 4	0.605 6	1.6512	.8
.3	0.519 5	0.854 5	0.608 0	1.6447	.7
.4	0.521 0	0.853 6	0.610 4	1.6383	.6
.5	0.522 5	0.852 6	0.612 8	1.6319	.5
.6	0.524 0	0.851 7	0.615 2	1.6255	.4
.7	0.525 5	0.850 8	0.617 6	1.6191	.3
.8	0.527 0	0.849 9	0.620 0	1.6128	.2
.9	0.528 4	0.849 0	0.622 4	1.6066	.1
32.0	0.529 9	0.848 0	0.624 9	1.6003	58.0

Dec/Deg	Cos	Sin	Cot	Tan	Dec/Deg

58.0° to 66.0°

(continued)

TABLE 7 Numerical Values of Trigonometric Functions for Decimal-Degree (Metric) Angles (continued)

32.0° to 39.0°

Dec/Deg	Sin	Cos	Tan	Cot	Dec/Deg
32.0	0.529 9	0.848 0	0.624 9	1.600 3	58.0
.1	0.531 4	0.847 1	0.627 3	1.594 1	.9
.2	0.532 9	0.846 2	0.629 7	1.588 0	.8
.3	0.534 4	0.845 3	0.632 2	1.581 8	.7
.4	0.535 8	0.844 3	0.634 6	1.575 7	.6
.5	0.537 3	0.843 4	0.637 1	1.569 7	.5
.6	0.538 8	0.842 5	0.639 5	1.563 7	.4
.7	0.540 2	0.841 5	0.642 0	1.557 7	.3
.8	0.541 7	0.840 6	0.644 5	1.551 7	.2
.9	0.543 2	0.839 6	0.646 9	1.545 8	.1
33.0	0.544 6	0.838 7	0.649 4	1.539 9	57.0
.1	0.546 1	0.837 7	0.651 9	1.534 0	.9
.2	0.547 6	0.836 8	0.654 4	1.528 2	.8
.3	0.549 0	0.835 8	0.656 9	1.522 4	.7
.4	0.550 5	0.834 8	0.659 4	1.516 6	.6
.5	0.551 9	0.833 9	0.661 9	1.510 8	.5
.6	0.553 4	0.832 9	0.664 4	1.505·1	.4
.7	0.554 8	0.832 0	0.666 9	1.499 4	.3
.8	0.556 3	0.831 0	0.669 4	1.493 8	.2
.9	0.557 7	0.830 0	0.672 0	1.488 2	.1
34.0	0.559 2	0.829 0	0.674 5	1.482 6	56.0
.1	0.560 6	0.828 1	0.677 1	1.477 0	.9
.2	0.562 1	0.827 1	0.679 6	1.471 5	.8
.3	0.563 5	0.826 1	0.682 2	1.465 9	.7
.4	0.565 0	0.825 1	0.684 7	1.460 5	.6
.5	0.566 4	0.824 1	0.687 3	1.455 0	.5
.6	0.567 8	0.823 1	0.689 9	1.449 6	.4
.7	0.569 3	0.822 1	0.692 4	1.444 2	.3
.8	0.570 7	0.821 1	0.695 0	1.438 8	.2
.9	0.572 1	0.820 2	0.697 6	1.433 5	.1
35.0	0..573 6	0.819 2	0.700 2	1.428 1	55.0
.1	0.575 0	0.818 1	0.702 8	1.422 9	.9
.2	0.576 4	0.817 1	0.705 4	1.417 6	.8
.3	0.577 9	0.816 1	0.708 0	1.412 4	.7
.4	0.579 3	0.815 1	0.710 7	1.407 1	.6
.5	0.580 7	0.814 1	0.713 3	1.401 9	.5
.6	0.582 1	0.813 1	0.715 9	1.396 8	.4
.7	0.583 5	0.812 1	0.718 6	1.391 6	.3
.8	0.585 0	0.811 1	0.721 2	1.386 5	.2
.9	0.586 4	0.810 0	0.723 9	1.381 4	.1
36.0	0.587 8	0.809 0	0.726 5	1.376 4	54.0
.1	0.589 2	0.808 0	0.729 2	1.371 3	.9
.2	0.590 6	0.807 0	0.731 9	1.366 3	.8
.3	0.592 0	0.805 9	0.734 6	1.361 3	.7
.4	0.593 4	0.804 9	0.737 3	1.356 4	.6
.5	0.594 8	0.803 9	0.740 0	1.351 4	.5
.6	0.596 2	0.802 8	0.742 7	1.346 5	.4
.7	0.597 6	0.801 8	0.745 4	1.341 6	.3
.8	0.599 0	0.800 7	0.748 1	1.336 7	.2
.9	0.600 4	0.799 7	0.750 8	1.331 9	.1
37.0	0.601 8	0.798 6	0.753 6	1.327 0	53.0
.1	0.603 2	0.797 6	0.756 3	1.322 2	.9
.2	0.604 6	0.796 5	0.759 0	1.317 5	.8
.3	0.606 0	0.795 5	0.761 8	1.312 7	.7
.4	0.607 4	0.794 4	0.764 6	1.307 9	.6
.5	0.608 8	0.793 4	0.767 3	1.303 2	.5
.6	0.610 1	0.792 3	0.770 1	1.298 5	.4
.7	0.611 5	0.791 2	0.772 9	1.293 8	.3
.8	0.612 9	0.790 2	0.775 7	1.289 2	.2
.9	0.614 3	0.789 1	0.778 5	1.284 6	.1
38.0	0.615 7	0.788 0	0.781 3	1.279 9	52.0
.1	0.617 0	0.786 9	0.784 1	1.275 3	.9
.2	0.618 4	0.785 9	0.786 9	1.270 8	.8
.3	0.619 8	0.784 8	0.789 8	1.266 2	.7
.4	0.621 1	0.783 7	0.792 6	1.261 7	.6
.5	0.622 5	0.782 6	0.795 4	1.257 2	.5
.6	0.623 9	0.781 5	0.798 3	1.252 7	.4
.7	0.625 2	0.780 4	0.801 2	1.248 2	.3
.8	0.626 6	0.779 3	0.804 0	1.243 7	.2
.9	0.628 0	0.778 2	0.806 9	1.239 3	.1
39.0	0.629 3	0.777 1	0.809 8	1.234 9	51.0
Dec/Deg	**Cos**	**Sin**	**Cot**	**Tan**	**Dec/Deg**

51.0° to 58.0°

39.0° to 45.0°

Dec/Deg	Sin	Cos	Tan	Cit	Dec/Deg
39.0	0.629 3	0.777 1	0.809 8	1.234 9	51.0
.1	0.630 7	0.776 0	0.812 7	1.230 5	.9
.2	0.632 0	0.774 9	0.815 6	1.226 1	.8
.3	0.633 4	0.773 8	0.818 5	1.221 8	.7
.4	0.634 7	0.772 7	0.821 4	1.217 4	.6
.5	0.636 1	0.771 6	0.824 3	1.213 1	.5
.6	0.637 4	0.770 5	0.827 3	1.208 8	.4
.7	0.638 8	0.769 4	0.830 2	1.204 5	.3
.8	0.640 1	0.768 3	0.833 2	1.200 2	.2
.9	0.641 4	0.767 2	0.836 1	1.196 0	.1
40.0	0.642 8	0.766 0	0.839 1	1.191 8	50.0
.1	0.644 1	0.764 9	0.842 1	1.187 5	.9
.2	0.645 5	0.763 8	0.845 1	1.183 3	.8
.3	0.646 8	0.762 7	0.848 1	1.179 2	.7
.4	0.648 1	0.761 5	0.851 1	1.175 0	.6
.5	0.649 4	0.760 4	0.854 1	1.170 8	.5
.6	0.650 8	0.759 3	0.857 1	1.166 7	.4
.7	0.652 1	0.758 1	0.860 1	1.162 6	.3
.8	0.653 4	0.757 0	0.863 2	1.158 5	.2
.9	0.654.7	0.755 9	0.866 2	1.154 4	.1
41.0	0.656 1	0.754 7	0.869 3	1.150 4	49.0
.1	0.657 4	0.753 6	0.872 4	1.146 3	.9
.2	0.658 7	0.752 4	0.875 4	1.142 3	.8
.3	0.660 0	0.751 3	0.878 5	1.138 3	.7
.4	0.661 3	0.750 1	0.881 6	1.134 3	.6
.5	0.662 6	0.749 0	0.884 7	1.130 3	.5
.6	0.663 9	0.747 8	0.887 8	1.126 3	.4
.7	0.665 2	0.746 6	0.891 0	1.122 4	.3
.8	0.666 5	0.745 5	0.894 1	1.118 5	.2
.9	0.667 8	0.744 3	0.897 2	1.114 5	.1
42.0	0.669 1	0.743 1	0.900 4	1.110 6	48.0
.1	0.670 4	0.742 0	0.903 6	1.106 7	.9
.2	0.671 7	0.740 8	0.906 7	1.102 8	.8
.3	0.673 0	0.739 6	0.909 9	1.099 0	.7
.4	0.674 3	0.738 5	0.913 1	1.095 1	.6
.5	0.675 6	0.737 3	0.916 3	1.091 3	.5
.6	0.676 9	0.736 1	0.919 5	1.087 5	.4
.7	0.678 2	0.734 9	0.922 8	1.083 7	.3
.8	0.679 4	0.733 7	0.926 0	1.079 9	.2
.9	0.680 7	0.732 5	0.929 3	1.076 1	.1
43.0	0.682 0	0.731 4	0.932 5	1.072 4	47.0
.1	0.683 3	0.730 2	0.935 8	1.068 6	.9
.2	0.684 5	0.729 0	0.939 1	1.064 9	.8
.3	0.685 8	0.727 8	0.942 4	1.061 2	.7
.4	0.687 1	0.726 6	0.945 7	1.057 5	.6
.5	0.688 4	0.725 4	0.949 0	1.053 8	.5
.6	0.689 6	0.724 2	0.952 3	1.050 1	.4
.7	0.690 9	0.723 0	0.955 6	1.046 4	.3
.8	0.692 1	0.721 8	0.959 0	1.042 8	.2
.9	0.693 4	0.720 6	0.962 3	1.039 2	.1
44.0	0.694 7	0.719 3	0.965 7	1.035 5	46.0
.1	0.695 9	0.718 1	0.969 1	1.031 9	.9
.2	0.697 2	0.716 9	0.972 5	1.028 3	.8
.3	0.698 4	0.715 7	0.975 9	1.024 7	.7
.4	0.699 7	0.714 5	0.979 3	1.021 1	.6
.5	0.700 9	0.713 3	0.982 7	1.017 6	.5
.6	0.702 2	0.712 0	0.986 1	1.014 1	.4
.7	0.703 4	0.710 8	0.989 6	1.010 5	.3
.8	0.704 6	0.709 6	0.993 0	1.007 0	.2
.9	0.705 9	0.708 3	0.996 5	1.003 5	.1
45.0	0.707 1	0.707 1	1.000 0	1.000 0	45.0
Dec/Deg	**Cos**	**Sin**	**Cot**	**Tan**	**Dec/Deg**

45.0° to 51.0°

TABLE 8 Powers and Roots of Numbers (1 through 100)

Num-ber	Powers Square	Powers Cube	Roots Square	Roots Cube	Num-ber	Powers Square	Powers Cube	Roots Square	Roots Cube
1	1	1	1.000	1.000	51	2,601	132,651	7.141	3.708
2	4	8	1.414	1.260	52	2,704	140,608	7.211	3.733
3	9	27	1.732	1.442	53	2,809	148,877	7.280	3.756
4	16	64	2.000	1.587	54	2,916	157,464	7.348	3.780
5	25	125	2.236	1.710	55	3,025	166,375	7.416	3.803
6	36	216	2.449	1.817	56	3,136	175,616	7.483	3.826
7	49	343	2.646	1.913	57	3,249	185,193	7.550	3.849
8	64	512	2.828	2.000	58	3,364	195,112	7.616	3.871
9	81	729	3.000	2.080	59	3,481	205,379	7.681	3.893
10	100	1,000	3.162	2.154	60	3,600	216,000	7.746	3.915
11	121	1,331	3.317	2.224	61	3,721	226,981	7.810	3.936
12	144	1,728	3.464	2.289	62	3,844	238,328	7.874	3.958
13	169	2,197	3.606	2.351	63	3,969	250,047	7.937	3.979
14	196	2,744	3.742	2.410	64	4,096	262,144	8.000	4.000
15	225	3,375	3.873	2.466	65	4,225	274,625	8.062	4.021
16	256	4,096	4.000	2.520	66	4,356	287,496	8.124	4.041
17	289	4,913	4.123	2.571	67	4,489	300,763	8.185	4.062
18	324	5,832	4.243	2.621	68	4,624	314,432	8.246	4.082
19	361	6,859	4.359	2.668	69	4,761	328,509	8.307	4.102
20	400	8,000	4.472	2.714	70	4,900	343,000	8.367	4.121
21	441	9,261	4.583	2.759	71	5,041	357,911	8.426	4.141
22	484	10,648	4.690	2.802	72	5,184	373,248	8.485	4.160
23	529	12,167	4.796	2.844	73	5,329	389,017	8.544	4.179
24	576	13,824	4.899	2.884	74	5,476	405,224	8.602	4.198
25	625	15,625	5.000	2.924	75	5,625	421,875	8.660	4.217
26	676	17,576	5.099	2.962	76	5,776	438,976	8.718	4.236
27	729	19,683	5.196	3.000	77	5,929	456,533	8.775	4.254
28	784	21,952	5.292	3.037	78	6,084	474,552	8.832	4.273
29	841	24,389	5.385	3.072	79	6,241	493,039	8.888	4.291
30	900	27,000	5.477	3.107	80	6,400	512,000	8.944	4.309
31	961	29,791	5.568	3.141	81	6,561	531,441	9.000	4.327
32	1,024	32,798	5.657	3.175	82	6,724	551,368	9.055	4.344
33	1,089	35,937	5.745	3.208	83	6,889	571,787	9.110	4.362
34	1,156	39,304	5.831	3.240	84	7,056	592,704	9.165	4.380
35	1,225	42,875	5.916	3.271	85	7,225	614,125	9.220	4.397
36	1,296	46,656	6.000	3.302	86	7,396	636,056	9.274	4.414
37	1,369	50,653	6.083	3.332	87	7,569	658,503	9.327	4.481
38	1,444	54,872	6.164	3.362	88	7,744	681,472	9.381	4.448
39	1,521	59,319	6.245	3.391	89	7,921	704,969	9.434	4.465
40	1,600	64,000	6.325	3.420	90	8,100	729,000	9.487	4.481
41	1,681	68,921	6.403	3.448	91	8,281	753,571	9.539	4.498
42	1,764	74,088	6.481	3.476	92	8,464	778,688	9.592	4.514
43	1,849	79,507	6.557	3.503	93	8,649	804,357	9.644	4.531
44	1,936	85,184	6.633	3.530	94	8,836	830,584	9.695	4.547
45	2,025	91,125	6.708	3.557	95	9,025	857,375	9.747	4.563
46	2,116	97,336	6.782	3.583	96	9,216	884,736	9.798	4.579
47	2,209	103,823	6.856	3.609	97	9,409	912,673	9.849	4.595
48	2,304	110,592	6.928	3.634	98	9,604	941,192	9.900	4.610
49	2,401	117,649	7.000	3.659	99	9,801	970,299	9.950	4.626
50	2,500	125,000	7.071	3.684	100	10,000	1,000,000	10.000	4.642

TABLE 9 Sizes of Tapers and Angles

Taper per foot	Included angle			Taper per inch
	Deg.	Min.	Sec.	
1/8	0	35	48	0.010416
3/16	0	53	44	0.015625
1/4	1	11	36	0.020833
5/16	1	29	30	0.026042
3/8	1	47	24	0.031250
7/16	2	5	18	0.036458
1/2	2	23	10	0.014667
9/16	2	41	4	0.046875
5/8	2	59	42	0.052084
11/16	3	16	54	0.057292
3/4	3	34	44	0.06250
13/16	3	52	38	0.067708
7/8	4	10	32	0.072917
15/16	4	28	24	0.078125
1	4	46	18	0.083330

Taper per foot	Included angle			Taper per inch
	Deg.	Min.	Sec.	
1 1/4	5	57	48	0.104666
1 1/2	7	9	10	0.125000
1 3/4	8	20	26	0.145833
2	9	31	36	0.166666
2 1/2	11	53	36	0.208333
3	14	15	0	0.250000
3 1/2	16	35	40	0.291666
4	18	55	28	0.333333
4 1/2	21	14	2	0.375000
5	23	32	12	0.416666
6	28	4	2	0.500000

TABLE 10 Calculator Processes for Common Algebraic, Geometric, and Trigonometric Applications

Geometric Calculation		Formula	Calculator Processes	Measurement Unit Value
Areas (A)	Square	$A = s^2$	(s) x (s)	square units
	Rectangle	$A = (\ell)(w)$	(ℓ) x (w)	" "
	Circle	$A = \pi r^2$	π to required accuracy (π) x (r) x (r)	" "
	Triangle	$A = \frac{(b \times h)}{2}$	(b x h) ÷ 2	" "
Perimeter (P)	Square	$P = 4(s)$	(s) x (4)	base unit
	Rectangle	$P = 2(\ell \times w)$	$(\ell$ x w) x 2	" "
	Triangle	$P = s_1 + s_2 + s_3$	$s_1 + (s_2) + (s_3)$	" "
	Cirumference	$C = \pi d$	(π) x (d)	" "
Degrees (°) and Radians (r)	Angle (as a radian value)	$\angle A = \frac{n°}{57.295\,78°}$	n° ÷ 57.295 78	radian
	Radian angle (r) (in degrees)	$d = r \times 57.295\,78$	(r) x (57.295 78)	degrees
Volume (V)	Cube	$V = s^3$	(s) x (s) x (s)	cubic units
	Rectangular solid	$V = (\ell) \times (w) \times (h)$	(ℓ) x (w) x (h)	" "
	Cylindrical solid	$V = \pi r^2 h$	(π) x (r) x (r) x (h) (π to required accuracy)	" "
	Sphere	$V = \frac{4\pi r^3}{3}$	(4) x (π) x (r) x (r) x (r) ÷ 3	" "
	Cone	$V = 0.2618h (D^2 + Dd + d^2)$	(D x D+ (D x d) + (d x d) x (h) x (0.2618)	" "

ANSWERS TO ODD-NUMBERED PROBLEMS

All answers To Achievement Review units are included in the Instructor's Guide.

PART ONE FUNDAMENTALS OF BASIC MATHEMATICS

Section 1 Whole Numbers

Unit 1 Addition of Whole Numbers

A. 1. a. 59   b. 120   c. 1,008   d. 12,987

    3. a. 57   b. 99   c. 105   d. 457   e. 1,650   f. 8,057

    5. A. VI   B. XXIV   C. XLIX   D. XCIV   E. CDXLIV   F. MCMLXXVI

B. 1. A. 37   C. 91   E. 965   G. 1,021   I. 219   K. 880

    B. 70   D. 144   F. 823   H. 1,362   J. 229   L. 24,134

C. 1. 12 inches   5. A. 84″   7. 67″   13. 94,468 kW

    3. 65 yards   B. 120″   9. 504,861 motors   15. 699 sq ft

                 C. 156″   11. 437′

Unit 2 Subtraction of Whole Numbers

A. 1. A. 44   C. 16   E. 144   G. 528   I. 184   K. 1,186

    B. 61   D. 39   F. 206   H. 778   J. 509   L. 878

    3. a. 8,108   b. 1,778   c. 1,018 feet   d. 6,638 miles

B. 1. 2,875 board feet

    3. Week 1: 780 miles   Week 3: 1,988 miles   Week 5: 1,936 miles

       Week 2: 1,089 miles   Week 4: 1,089 miles

    5. a. 149 millihenries   b. 213 watts   c. 24 750 ohms

7.

	Production for Processes				
	A	B	C	D	E
a. Fastest	2	2	2	3	1
b. Second Fastest	14	42	1,133	889	436
Slowest	21	126	1,222	2,199	878

    9. a. 1. 182   b. A. 19,557   c. 2,136 calories

         2. 598   B. 17,421

         3. 1,888

         4. 168

   11. 3875 Ω

Unit 3 Multiplication of Whole Numbers

A. 1. A. 84   B. 154   C. 702   D. 1,920   E. 847

B. 1. 20,808 work hours   5. (A) = 36″ (B) = 58″   7. 94,500 shingles

    3. $9,594.00   (C) = 25 ″ (D) = 36″   9. $162.16

                                                   11. 15,042 miles

13.

Part A			
Machine	(a)	(b)	(c)
1	32	288	864
2	48	432	1,296
3	72	648	1,944

Part B			
Machine	(a)	(b)	(c)
1	80	720	3,600
2	88	792	3,960
3	136	1,224	6,120

Part C			
Machine	(a)	(b)	(c)
1	800	7,200	86,400
2	960	8,640	103,680
3	1,056	9,504	114,048

Part D			
Machine	(a)	(b)	(c)
1	1,872	16,848	454,896
2	1,976	17,784	480,168
3	3,032	27,288	736,776

15. 222 in.^3 (222 cu in.)

Unit 4 Division of Whole Numbers

A. 1. A. 21   C. 21   E. 26   G. 11   I. 14
      B. 30   D. 45   F. 10   H. 30   J. 22

B. 1. 85 square yards   7. 8 inches   13. 1,475 calories
   3. 3 pounds   9. 1,292 watts   15. 182 milliamperes
   5. 21 gallons   11. $6.00

Section 2 Common Fractions

Unit 6 The Concept of Common Fractions

A. 1. a. $\frac{1}{8}$ f. $\frac{5}{6}$ k. $\frac{11}{16}$ p. $\frac{20}{32}$ u. $\frac{31}{64}$

b. $\frac{3}{8}$ g. $\frac{3}{16}$ l. $\frac{15}{16}$ q. $\frac{25}{32}$ v. $\frac{57}{64}$

c. $\frac{5}{8}$ h. $\frac{1}{4}$ m. $\frac{5}{32}$ r. $\frac{30}{32}$ w. $\frac{1}{2}$

d. $\frac{1}{6}$ i. $\frac{5}{16}$ n. $\frac{10}{32}$ s. $\frac{11}{64}$ x. $\frac{3}{4}$

e. $\frac{1}{2}$ j. $\frac{7}{16}$ o. $\frac{15}{32}$ t. $\frac{21}{64}$ y. $\frac{5}{6}$

z. $\frac{7}{8}$

B. 1. a. $\frac{1}{64}$ c. $\frac{3}{32}$ e. $\frac{1}{4}$ g. $\frac{9}{32}$ i. $\frac{1}{2}$ k. $\frac{3}{4}$

b. $\frac{1}{16}$ d. $\frac{7}{32}$ f. $\frac{17}{64}$ h. $\frac{5}{16}$ j. $\frac{5}{8}$ l. $\frac{7}{8}$

3. a. $\frac{1''}{8}$ b. $\frac{3''}{8}$ c. $\frac{5''}{8}$ d. $\frac{7''}{8}$

C. 1. a. $\frac{14''}{32} = \frac{7''}{16}$ b. $\frac{48''}{64} = \frac{3''}{4}$

c. $\frac{10''}{16} = \frac{5''}{8}$

e. $\frac{10''}{64} = \frac{5''}{32}$

d. $\frac{44''}{64} = \frac{11''}{16}$

f. $\frac{18''}{128} = \frac{9''}{64}$

3.

5. Check lines drawn to the lengths indicated.

7. A. $\frac{1''}{2}$ B. $1\frac{5''}{8}$ C. $1\frac{5''}{16}$ D. $2\frac{1''}{8}$ E. $6\frac{19''}{32}$ F. $1\frac{3''}{4}$ G. $1\frac{5''}{8}$ H. $\frac{11''}{16}$ I. $1\frac{1''}{4}$ J. $2\frac{9''}{16}$

Unit 7 Addition of Fractions

A. 1. 1   3. $\frac{3}{4}$ 5. $\frac{7}{8}$ 7. $1\frac{3}{4}$ 9. $2\frac{3}{4}$

B. 1. $128\frac{5}{12}$ 3. $34\frac{1}{4}$ 5. $15\frac{59}{64}$

C. 1. A. $5\frac{5''}{8}$ B. $7\frac{9''}{32}$ C. $7\frac{21''}{64}$ D. $7\frac{21''}{32}$ E. $10\frac{15''}{32}$

   3. A. $33\frac{1''}{4}$ B. $32\frac{5''}{8}$ C. $33\frac{7''}{32}$ D. $36\frac{13''}{32}$

   5. $102\frac{3''}{32}$

Unit 8 Subtraction of Fractions

A. 1. $\frac{1}{8}$ 3. $\frac{1}{2}$ 5. $\frac{1}{4}$ 7. $\frac{13}{32}$ 9. $\frac{1}{16}$

B. 1. $3\frac{1}{4}$ 3. $31\frac{19}{32}$ 5. $71\frac{3}{64}$

C. 1. $\frac{2}{3}$ 3. $25\frac{11}{16}$ 5. $350\frac{59}{64}$ 7. $\frac{7}{64}$ 9. $\frac{29}{32}$

D. 1. $\frac{2}{5}$ 3. $9\frac{1}{2}$ 5. $150\frac{11}{64}$ 7. $2\frac{1}{2}$ 9. $2\frac{7}{64}$

E. 1. $\frac{7''}{8}$ B. $1\frac{3''}{4}$ C. $2\frac{3''}{16}$ D. $2\frac{5''}{64}$ E. $\frac{37''}{64}$ 3. 7 inches   5. $\frac{5''}{8}$

   7. $1'-11\frac{7''}{8}$ 9. A. $11\frac{5''}{16}$ B. $6\frac{14''}{16}$ or $6\frac{7''}{8}$ C. $16\frac{25''}{32}$ D. $8\frac{57''}{64}$ 11. $\frac{13''}{16}$ 13. $9'-6\frac{3''}{4}$

Unit 9 Multiplication of Fractions

A. 1. $\frac{1}{8}$ 3. $\frac{7}{72}$ 5. $\frac{25}{72}$ 7. $\frac{13}{32}$ 9. $\frac{195}{512}$

B. 1. $\frac{7}{36}$ 3. $\frac{19}{32}$ 5. $2\frac{71}{72}$ 7. $\frac{255}{256}$ 9. $1\frac{169}{256}$
(Note: All multiplication and division steps precede all addition and subtraction steps unless otherwise shown by parentheses.)

C. 1. $2\frac{8}{9}$ 3. $14\frac{179}{256}$ 5. $14\frac{257}{512}$

D. 1. $\frac{1}{8}$ 3. $1\frac{3}{64}$

E. 1. A. $5\frac{9}{16}''$ B. $14\frac{29}{32}''$ C. $156\frac{7}{8}''$ D. $120\frac{13}{64}''$ E. $94\frac{51}{64}''$

3. A. $3\frac{9}{32}''$ B. $4\frac{3}{8}''$ C. $2\frac{3}{16}''$ D. $4\frac{3}{8}''$

5.

Method	Part A		Part B	
	Daily Production	Unit Cost	Daily Production	Unit Cost
1	75	150	$157\frac{1}{2}$	441
2	105	$262\frac{1}{2}$	$202\frac{1}{2}$	$632\frac{13}{16}$
3	$112\frac{1}{2}$	$309\frac{3}{8}$	$221\frac{1}{4}$	$912\frac{21}{32}$

Unit 10 Division of Fractions

A. 1. 3   3. $\frac{2}{3}$ 5. $\frac{3}{4}$ 7. $\frac{7}{10}$ 9. $\frac{39}{424}$

B. 1. 9   3. 8   5. $\frac{1}{64}$ 7. $26\frac{2}{3}$ 9. 3

C. 1. 1   3. $1\frac{17}{26}$ 5. $1\frac{95}{101}$ 7. $\frac{47}{102}$

D. 1. $\frac{1}{4}$ 3. $87\frac{3}{7}$ 5. $4\frac{49}{152}$ 7. $\frac{225}{1552}$

E. 1. 8 leads   3. 13 pieces   5. 386 pieces   7. 7.5 lbs/ft
9. a. A. 5 minutes 20 seconds B. 8 minutes C. 4 minutes D. 3 minutes 37 seconds E. 45 seconds
b. 21 minutes 42 seconds   11. $1\frac{1}{2}$ amperes (rounded to nearest $\frac{1}{2}$ ampere)

Section 3 Decimal Fractions

Unit 12 The Concept of Decimal Fractions

A. 1. A. .5   B. .3   C. .6   D. .3   E. .9   F. .1   G. .5   H. .77   I. .62
3. a. .7   b. .16   c. .015   d. .0011   e. .2152   f. 3.1875
5. a. 1.1   b. 3.09   c. 25.91   d. 272.067   e. 2525.0021   f. 362.2007
B. 1. Dimensions located on the tenth scale.

3. A. 1.007   B. 2.03   C. 1.15   D. .803   E. .90
5. a. .76   b. 1.95   c. 7.32   d. 29.41   e. 2.56   f. 18.27   g. 221.76   h. .90   i. 21.00
7. a. 9.094 ohms   b. 6.375 inches   c. 4.333 hours   d. 3.938 watts

Unit 13 Addition of Decimals

A. 1. a. 9.4   c. 220.72   e. 1.5   g. 31.52   i. 354.4096
b. 99.4   d. 287.726   f. 470.4   h. 727.505   j. 3226.08606
B. 1. A. 3.4375   B. 3.4688   C. 5.0001   D. 6.5313   E. 10.0001
3. A. 5.125   5. 1.7875″ thick   7. 9.39 amperes

Unit 14 Subtraction of Decimals

A. 1. A. .250   B. .2565   C. .8442   D. 1.8438   E. 1.2969
   3. A. 2.972   B. 2.097   C. 1.847   D. .500   E. 5.127
   5. A. .906   B. 1.625   C. 4.406   D. .370   E. .344
   7. 0.00125″ shim required   9. $I_C = 1.062$ A (amperes)

Unit 15 Multiplication of Decimals

A. 1. a. 7.2   c. 8.64   e. 127.4   g. 21.294   i. 680.58144   k. 267.50428
   b. 24   d. 37   f. 7.22   h. 69.4434   j. 3.0289281   l. 20.42789
B. 1. A. 64.395 pounds   B. 5.1675 pounds   C. 5.76375 pounds   D. 4.17375 pounds
C. 1. A. $6.56   B. $10.63   C. $9.01   D. $107.21   E. $0.56 Total Cost. $133.97
   3. A. 6.8 inches   B. 7.6 inches   C. 37.3 inches   D. 119.8 inches   E. 200.4 inches   Total = 371.9 inches
   5. A. 41.48 hr   B. 46.75 hr   C. 60.76 hr
   7. a. $61.35   b. $2.56   c. $0.57   d. $59.36   9. Power = 3.49 watts

Unit 16 Division of Decimals

A. 1. a. .118   b. .0237   c. 4   d. 4.01   e. .333   f. 2.540
B. 1. a. .750   b. .625   c. .563   d. .167   e. .097   f. .406   g. .453
C. 1. a. .500   b. .750   c. .375   d. .3125   e. .21875   f. .296875
D. 1. A. .010   B. .003   C. .0017   D. .0021   E. .0022
   3. A. 16   B. 37   C. 49   D. 30   E. 17
   5. A. $14\frac{7}{16}$ oz   B. $25\frac{2}{5}$ gr   C. $175\frac{3}{8}$ qt   D. $9\frac{9}{16}$ lb

Section 4 Measurement: Directed and Computed (Customary Units)

Unit 18 Principles of Linear Measure

A. 1. a. A. $\frac{1}{2}''$ B. $\frac{1}{4}''$ C. $41\frac{3}{4}''$ D. $42\frac{5}{8}''$ E. $63\frac{7}{8}''$
   b. A. $\frac{1}{8}$ C. $\frac{3}{4}$ E. $2\frac{5}{8}$ G. $\frac{5}{8}$ I. $1\frac{15}{16}$ K. $\frac{21}{32}$ M. $1\frac{13}{32}$ O. $2\frac{29}{32}$ Q. $\frac{47}{64}$ S. $1\frac{19}{32}$
      B. $\frac{1}{2}$ D. $1\frac{7}{8}$ F. $\frac{5}{16}$ H. $1\frac{1}{2}$ J. $2\frac{7}{16}$ L. $1\frac{1}{4}$ N. $2\frac{5}{32}$ P. $\frac{15}{64}$ R. $1\frac{1}{16}$ T. $2\frac{37}{64}$
   3. a. $3\frac{3}{4}$ b. $1\frac{3}{4}$ c. $3\frac{1}{4}$ d. $3\frac{5}{8}$ e. $3\frac{11}{16}$ f. $2\frac{5}{16}$ g. $2\frac{31}{32}$ h. $3\frac{11}{16}$ i. $1\frac{3}{64}$ j. $3\frac{7}{16}$
   5. A. $\frac{15}{32}$ B. $\frac{1}{2}$ C. $\frac{27}{32}$ D. $\frac{1}{2}$ E. $2\frac{1}{16}$ F. $\frac{11}{32}$ G. 1   H. $5\frac{23}{32}$
   7. a. $3\frac{7}{10}$ b. 1   c. $3\frac{17}{50}$ d. $2\frac{33}{50}$
B. 1. A. 1.5002   B. 2.7502   C. 1.3333   D. 4.0835   E. 5.4168   F. 6.7501
   3. A. 1.660   C. 1.073   E. .391   G. 1.266   I. .953
      B. 2.385   D. 1.510   F. .500   H. 2.609   J. 5.328
C. 1. A. .012   B. .075   C. .200   D. .250   E. .562   F. .9375
D. 1. A. .5003   B. .3902   C. .2553
E. 1. A. .500   B. .525   C. 2.463   D. 9.906
F. 1. A. .500 (one block) or .300 + .200 or .350 + .150 or .400 + .100
      B. .750 (one block) or .500 + .250 or .400 + .350 or .450 + .300
      C. .750 + .125
      D. .150 + .115 or .130 + .135 or .140 + .125 or .120 + .145
      E. .1001 + .500
      F. .1007 + .650
      G. .1003 + .149 + 1.000

H. .1008 + .100 + 2.000

I. .1009 + .104 + .800 + 4.000 + 3.000 + 2.000 + 1.000

J. .135 + .750 + .600 + .400 + 4.000 + 3.000 + 2.000 + 1.000

(*Note.* Many other combinations are possible and acceptable.)

Unit 19 Principles of Angular and Circular Measure

A. 1. a. 60° b. 80° c. 135° d. 50° e. 105°

3. a. $\frac{1}{2}°$ b. $\frac{3}{4}°$ c. $1\frac{1}{4}°$ d. 300′ e. 450′ f. 84′ g. 300″ h. 330″

B. 1. `a. 30° b. 255° c. 100° d. 1°16′30″ e. 82°48′ f. 93°5′20″

C. 1. A. 80° B. 150° C. 32° D. 170° E. 105° F. 160° G. 85° H. 20°

3. A semicircular protractor is to be used to lay out the following angles: A. 10° B. 25° C. 100° D. 120°

5. A. 237° B. 134° C. 48° D. 217°

D. 1. A. 2.094 B. 7.854 C. 20.617 D. 26.495

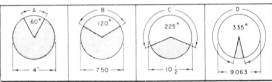

A. $\frac{60}{360} \times 4 \times 3.1416 = 2.0944″$ B. $\frac{120}{360} \times 7.5 \times 3.1416 = 7.854″$

C. $\frac{225}{360} \times 10.5 \times 3.1416 = 20.61675$ or $20.617″$ D. $\frac{335}{360} \times 9.063 \times 3.1416 = 26.4950763$ or $26.495″$

3. A. 4′–9″   B. 6′–10″   C. 8′–3″

Unit 20 Principles of Surface Measure

A. 1. A. $(8)^2 = 64$ sq in.

B. $(10.50)^2 = 110.25$ sq in.

C. $(76)^2 = 5776$ sq in. or 40.111 sq ft

D. $20.6 \times 12.4 = 255.44$ sq in.

E. $60 \times 16 = 960$ sq in. or 6.667 sq ft

F. $4.75 \times 3.10 = 14.725$ sq ft

3. 1,904 sq ft

B. 1. A. $10.25 \times 8.50 = 87.125$ sq in. or $87\frac{1}{8}$ sq in.

B. $(3.4 \times 1.2) - .4(3.4 - 2.2) = 3.6$ sq in.

C. $(30.25 \times 16.75) - (14.5 \times 8.5) = 383.4375$ or $383\frac{7}{16}$ sq in.

C. 1. A. $.5 \times .6 \times (1.4 + .9) = .69$ sq in.

B. $.5 \times 9.5 \times (12.5 + 10.25) = 108.0625$ or $108\frac{1}{16}$ sq in.

C. $.5 \times 5.8 \times (13.4 + 17.6) - (2.4)^2 = 84.14$ sq. in.

3. Lot A is 12,348 sq ft   Lot B is 12,299 sq ft

D. 1. A. $.5 \times 12 \times 8.5 = 51$ sq in.

B. $.5 \times 3.2 \times 5.8 = 9.28$ sq in.

C. $.5 \times 1.02 \times 15.8 = 8.058$ sq in.

D. $.8 \times 2.6 + .5 \times .8 \times 2.6 - (4)^2 = 2.96$ sq in.

E. 1. A. $.7854 \times (.8)^2 = .50$ sq in.

B. $.7854 (6)^2 = 28.27$ sq in.

C. $.7854 \times (5.25)^2 = 21.65$ sq in.

D. $3.1416 \times (.5)^2 = .79$ sq in.

E. $3.1416 \times (2)^2 = 12.57$ sq in.

F. $3.1416 \times (3.8)^2 = 45.36$ sq in.

3. 79.97 sq in.

F. 1. A. $\frac{1}{3} \times (12)^2 \times .7854 = 37.699$ sq in.

B. $\frac{1}{5} \times (2.125)^2 \times 3.1416 = 2.837$ sq in.

C. $\frac{1}{12} \times (4.8^2 - 1.92^2) \times .7854 = 1.267$ sq in.

G. 1. A. $7.75 \times 3.1416 \times 8.5 = 206.95$ sq in. lateral surface

      B. $5 \times 3.1416 \times 7.5 + 2 \times (5)^2 \times .7854 = 157.08$ sq in. total surface

Unit 21 Principles of Volume Measure

A. 1. a. $2 \times 1728 = 3456$ cu in.

    b. $1.5 \times 1728 = 2592$ cu in.

    c. 3.625 or $\frac{29}{9} \times 1728 = 6264$ cu in.

    d. $10 \times 1728 + 19 = 17,299$ cu in.

    e. $3456 \div 1728 = 2$ cu ft

    f. $18.144 \div 1728 = .0105$ cu ft

    g. $8640 \div 1728 = 5$ cu ft

    h. $1944 \div 1728 = 1$ cu ft 216 cu in.

    i. $3 \times 27 = 81$ cu ft

    j. $\frac{13}{3} \times 27 = 117$ cu ft.

    k. $5 \times 27 + 7 = 142$ cu ft

    l. $7 \times 27 + 19 = 208$ cu ft

B. 1. A. $6 \times 6 \times 6 = 216$ cu. in.   B. $(8.5)^3 = 614.125$ cu. in.   C. $(1.5)^3 = 3.375$ cu. ft.

    3. A. $\frac{1}{2}$ B. $\frac{1}{4}$ C. $2\frac{1}{4}$

C. 1. 512 cu yd

    3. A. $(6 \times 4 - 2 \times 3) \times 12 = 216$ cu in.

       B. $(20.2 \times 6.8 - 4.2 \times 4.2) \times 17 + 7.6 \times 12.6 \times 3.4 = 2360.8$ cu in.

D. 1. A. $(4)^2 \times .7854 \times 10 = 125.66$ cu in.

       B. $(12.5)^2 \times .7854 \times 24.5 = 3006.61$ cu in.

       C. $(1.6)^2 \times 3.1416 \times 6.4 = 51.47$ cu in.

    3. A. $(5^2 - 2^2) \times .7854 \times 10 = 164.9$ cu in.   B. $(4.25^2 - 1.5^2) \times .7854 \times 12 = 149$ cu in.

    5. $(3.625 \times 250 + 107) \times .75^2 \times .7854 \times .28 \times .42 = \$52.64$

E. 1. $\{12 \times 6 - (3 \times 2 \times 2 \times .7854)\} \times 16 \times 20 \times .26 = 5206.3$ lb

F. 1. a. 16 qt   c. 15 qt   e. $10\frac{1}{2}$ g. 7 pt 3 gills   i. 9 gal 1 qt   k. 3 bbl 2 gal   m. 2 gal 2 qt

    b. 25 qt   d. 13 pt   f. 17 pt 2 gills   h. 4 gal 1qt   j. 2 bbl   l. 3 gal   n. 1,155 cu in.

                                                                     o. 1,097.25 cu in.

Section 5 Percentage and Averages

Unit 23 The Concepts of Percent and Percentage

A. 1. 25%   51% $37\frac{1}{2}$%

    3. A. 65%   B. $6\frac{1}{2}$%   C. .6%   D. $6\frac{2}{3}$%   E. $12\frac{1}{4}$%

    5. A. 150%   B. 75%   C. $12\frac{1}{2}$%   D. 205%   E. $2\frac{1}{2}$%   F. .4%

B. 1. A. 200   B. 325   C. 62.5'   D. 1.25 tons   E. .563 lb

C. 1. A. \$20.00   B. 9.6 sheets   C. 12.6 acres   D. .65 cu yd   E. 26 kW   F. 364.5 lb

    3. A. 106 lb tin   B. 43.2 lb tin   C. 75 lb tin

         106 lb lead   4.8 lb lead   50 lb lead

    5. 53.592 hp

Unit 24 Application of Percentage, Base, and Rate

A. 1. A. 1920 tons   B. 843.75 gal   C. 5.4188 in.   D. 147.0825 sq ft   E. 18.531 sheets

    3. (a) The amount of profit is \$456.   (b) The percent of profit is 12%.

    5. a. $1290 \times (100 - 12.5) = 1,128$ (whole) castings   b. $1290 \times .975 \div 1,128 = \$1.12$

B. 1. A. $120 \div .90 = 133\frac{1}{3}$ hp   C. $137.8 \div .072 = 1913.89$ lb   E. $126.5 \div .0325 = 3892.3$ in.

       B. $2016 \div 54 = 3733\frac{1}{3}$ cables   D. $78.5 \div .0625 = 1207.69$ bars

    3. $\frac{2.4}{120} = .02 = 2\%$

C. 1. $\frac{1}{5} = .20 = 20\%$ acid $100 - 20 = 80\%$ water   3. $346 \times .82 = 283.7$ rpm

5.

	Copper	Tin	Zinc	Phosphorus	Lead	Iron	
A	320	44	32.8	1.6	1.2	.4	pounds
B	435.7	45.7	38.8	1.8	2.7	.3	pounds

Unit 25 Averages and Estimates

A. 1. 11.675″   3. 1.2498″

B. 1. $1235 - (212 + 224 + 232 + 275) = 292$ units

Section 6 Finance

Unit 27 Money and Time Calculations

A. 1. A. One dollar and twelve cents

   B. Twenty dollars and seven cents

   C. Two thousand six hundred ninety-two dollars and seventy-five cents

   D. Two cents

   E. Thirty-seven cents

   F. Two dollars and thirty-seven and one-half cents

   G. Forty-six and one-quarter cents

B. 1. $3 \times \frac{14}{12} \times 2 \times .886 = \$6.20$ 3. $\frac{3}{4} \times 24 \times 36 \times 96 \div 231 \times 5.50 = \$1481.14$

   $\frac{3}{4} \times 24 \times 36 \times 96 \div 231 = 269.3$ gal

C. 1. A. 360 min   B. $3\frac{1}{6}$ days   C. 720 sec   D. 510 min   E. $1\frac{2}{3}$ min   F. 23 hr 20 min

Unit 28 Manufacturing Costs and Discounts

A. 1. A. $381.00   B. $623.00   C. $595.93   D. $723.90   E. $930.61

   3. a. $277.20   d. $304.92 materials   e. $845.46

   b. $491.04 $515.59 labor

   c. $23.76 $24.95 overhead

B. 1. $19.50 \div 12 \times .78 = \$1.27$

   3. a. Distributor B gives the best discount ($103.68). Distributor A gives $103.49 discount; distributor C, $101.89.

   b. The net cost of the fabric from distributor B is $220.32.

   5. (a) (1)   (a) (2)   (b) (1)   (b) (2)

   $2152.50 $2006.74 $4957.50 $5103.26

Unit 29 Payrolls and Taxes

A. 1. A. 49 hr $995.10   C. 46 hr, $533.60   E. 51 hr, $268.38

   B. 46 hr $700.70   D. 36 hr, $315.40   F. 40 hr, $174.00

B. 1. A. $40 + (4 \times 1.5) \times 11.44 \times .07 = 36.84$ D. $40 \times 4.60 \times .07 = \$12.88$

   B. $40 + (2 \times 1.5) \times 8.28 \times .07 = \$24.92$ E. $32 \times 3.82 \times .07 = \$\,8.56$

   C. $40 + (8 \times 1.5) \times 5.00 \times .07 = \$18.20$

C. 1.

Worker	Regular Hours Worked	Overtime		Weekly Earnings	Exemption Allowance	Tax A 20%	Tax B 2%	Take-Home Pay
		Hours Worked	Hours Pay					
A	40	12	21	$762.50	$72.00	$138.10	$13.81	$610.59
B	40	9	$13\frac{1}{2}$	465.45	54.00	82.29	8.23	374.93
C	34			332.20	54.00	55.64	5.56	271.00
D	40	17	$27\frac{1}{2}$	353.70	72.00	36.34	5.63	291.50
E	37			147.26	36.00	22.25	2.23	122.70

Section 7 Graphs and Statistical Measurements

Unit 31 Development and Interpretation of Bar Graphs

A. 1. The vertical scale of the graph shows percents from 0% to 50%. The vertical bars represent: fuel (15%); products (9%); miscellaneous (15%).

B. 1. a. 1984   b. 1979   c. Using 610 in 1981 and 720 in 1984, percent of increase is 18%.

Unit 32 Development and Interpretation of Line Graphs

A. 1.

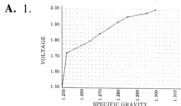

B. 1.

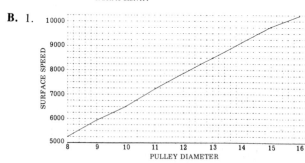

3.

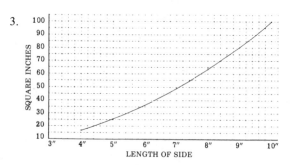

5. a. (1) 275'/min,   (2) 550'/min,   (3) 840'/min surface speed
   b. (1) 6"   (2) $9\frac{1}{2}''$ (3) 11"

C. 1. a. 42.5 to 60.0 hours of production time

   b. System A. 600 units   System B. 500 units   System C. 740 units
   c. System A. 600 to 880 units   System B. 500 to 1030 units   System C. 740 to 1300 units
   d. System A has the greatest variation in hourly production.
   e. System A production gradually increases from 600 to 630 to 675 units.
      System B drastically increases from 500 to 660 units and then at a constant rate to 720 units at 47.5 hours.
      System C has a constant rate rise from 740 to 820 to 900 units at 47.5 hours.

Unit 33 Development and Application of Circle Graphs
A. 1. The circle is divided into two areas. The larger area $(81\frac{1}{4}\%)$ represents the 1,300 men. The smaller area $(18\frac{3}{4}\%)$ represents the 300 women.
B. 1. a. 35,280   c. Men: 18–35   10,080   Women: 18–35   3,024
      b. 71% men   29% women   36–55   12,600   36–55   6,250

Unit 34 Statistical Measurements
A. 1. a. (1) 34.13% in the + range; 34.13% in the − range
        (2) 3,413 bearings in the + range; 3,413 bearings in the − range.
      b. 2,718 second quality bearings are acceptable   c. 456 total number of rejects
B. 1. a. Single sample size of 75 is required
      b. The lot is accepted if no more than four parts in the sampling are defective. The lot is rejected when five or more parts are defective.
      c. 42 parts are required in the first sample in the double sampling plan.
      d. If there are three or four rejects in the first batch of 42 parts, a second sampling of 84 parts is required, totaling 126 parts.
      e. (1) Eight sampling sequences are used.   (2) Total number of parts in the eight samples is 128.
C. 1. a. The tensile strength range is from 20,000 to 90,000 lb/sq in.
      b. (1) The average (mean) tensile strength is 39,429 lb/sq in. correct to the nearest pound.
        (2) The median tensile strength is 26,000 lb/sq in.
        (3) The tensile strength mode is 20,000 lb/sq in.

PART TWO FUNDAMENTALS OF SI METRIC MEASUREMENTS

Section 8 Metrication: Systems, Instruments, and Measurement Conversions

Unit 36 SI Metric Units of Measurement
A. 1.1. (f)   2. (a)   3. (i)   4. (h)   5. (d)   6. (e)
B. 1. A. 10 mm   C. 232 mm   E. 288 mm   G. 3.5 cm   I. 27.6 cm
      B. 55 mm   D. 261 mm   F. 2 cm   H. 5.6 cm   J. 29.4 cm
    3. A. 31 mm   B. 15 mm   C. 38 mm   D. 30 mm   E. 126 mm or 12.6 cm
    5. (a) A. 4.1 cm   B. 2.3 cm   C. 3.8 cm   D. 2.6 cm
      (b) E. 12.8 cm
C. 1. A. 3.96 cm   B. 7.16 cm   C. 3.4 cm   D. 10.56 cm   E. 13.96 cm   F. 17.36 cm
    3. A. 4.24 cm   C. 2.62 cm   E. 1.116 cm   G. 3.216 cm   I. 2.316 cm
      B. 6.14 cm   D. 3.84 cm   F. 1.316 cm   H. 6.716 cm   J. 13.516 cm
      (*Note*. In industrial practice, metric dimensions E through I are rounded to two decimal places for machining as indicated.)
D. 1. A. 900 cm^2 B. 225 dm^2
    3. A. $60/360 \times (2.6) \times 0.785\ 4 = 88\ 488$ dm^2
      B. $240/360 \times (8.4^2 - 2.2^2) \times .785\ 4 = 34.410\ 992$ cm^2
    5. (a) 20.45 square meters   (b) 179 boards
E. 1. A. $(6.2)^3 = 238.328$ dm^3 B. $12 \times 1.8 \times .94 = 2.030\ 4$ m^3 (all changed to meters)
    3. A. $(8.2)^2 \times 0.785\ 4 \times 24.6 = 1\ 299.13$ cm^3
      B. $(2.4)^2 \times 3.141\ 6 \times 3.2 = 57.91$ dm^3
      C. $(6.4)^2 \times 0.785\ 4 \times 44.8 = 1\ 441.22$ cm^3
F. 1. $7.4 \times 3.2 \times 2.6 \times 1.057 = 61.5$ liters
    3. A. 19 305.60 cm^3 B. 2.09 liters   C. 1070.12 qt   D. 827.51 gal

G. 1.

		Larger Units		Smaller Units	
a.	**Designation**	kilogram	hectogram	milligram	centigram
b.	**Unit Symbol**	kg	hg	mg	cg
c.	**Value in Relation to One Gram**	1000 g	100 g	$\frac{1}{1000}$ g or 0.001 g	$\frac{1}{100}$ g or 0.01 g

H. 1. Annealing 1,400°F; tempering 449.6°F.

Unit 37 Precision Measurement: Metric Measuring Instruments

A. 1. (A). 3.00 mm   (B). 4.50 mm   (C). 6.75 mm   (D). 7.373 mm   (E). 8.755 mm

B. 1. (A). 5.00 mm   (B). 10.52 mm   (C). 19.83 mm   (D). 8.504 mm   (E). 16.874   (F). 16.925 mm

C. 1. (A). 0.537″   (B). 1.884″

D. 1. Hole A, 179.40 mm; hole B, 107.46 mm; hole C, 128.60 mm.

E. 1. (A). 71.24 mm   (B). 233.58 mm   (C). 380.94 mm

F. 1. (A). 30° 0′   (B). 5° 15′   (C). 62° 50′

G. 1.

Dimension		Series of Gage Blocks (mm)	Sizes of Blocks in Each Series (mm)
A	58.555 (mm)	0.001	1.005
		0.01	1.05
		0.1	1.5
		1.0	5.0
		10.0	50.0
B	155.863 (mm)	0.001	1.003
		0.01	1.06
		0.1	1.8
		1.0	2.0
		25.0	100.0 and 50.0

Unit 38 Base, Supplementary, and Derived SI Metric Units

A. 1. *Examples.* Range of measurement; measurements in two systems; formulas, or conversion factors, or mathematical processes; equivalent values in the second system.

   3. a. Supplementary units: radian, steradian; b. Measurement symbol: rad, sr.

   c. The radian measures plane angles; steradian, solid angles (or the area of a spherical surface).

   d. 1 radian (rad) = 57.295 78°, 1 steradian (sr) = radius (r) of circle squared

B. 1. The addition of nonsignificant zeros helps to identify and control the precise degree of accuracy required for one or more dimensions on a drawing.

   3. **Examples.**   SI metric   125.72 mm ± 0.1 mm

                Customary units   7.8125″ ± 0.0001″

C. 1. The controlling dimension identifies the prime system of measurement in which an object and its features are designed.

Unit 39 Conversion: Factors and Processes (SI and Customary Units)

A. 1. *Note*. The fractional inch values are first changed to decimal inch values (column b) and then equivalent mm values correct to three decimal places (column c).

Part	Linear Measurement	(b) Decimal Value (.001″)	(c) Equivalent mm Value (.001 mm)	(d) Equivalent (.01 mm) Measurements to (.001″) Precision
A	$\begin{smallmatrix}x\\x\end{smallmatrix}\ 7^7$	7.875″	200.025	200.03
B	$4\frac{13″}{64}$	4.203″	106.759	106.76
C	$\frac{31″}{32}$	0.969″	24.606	24.61

3. Example. Hard conversion requires that standards of accuracy and dimensional measurements be in one measurement system without regard to the availability of tooling or the capacity to produce the part to be interchangeable in the second measurement system.

5. a. 2 mm   b. 57.5 dm   c. 13.06 cm   d. 5.94 m   e. 602.36 in.   f. 5 in   g. 2 in.   h. 22.97 ft

7. a. 0.000 75 m³   c. 5.89 yd³   e. 1,700 ℓ   g. 3 ℓ
   b. 353.14 ft³   d. 0.102 m³   f. 2 ℓ   h. 119.23 ℓ

B. 1. Conversion tables usually contain such information as a. identification of the known unit of measure (given value); b. units in the new required unit of measure; c. conversion factors; or categories under which units are grouped; notation on the precision of the factors.

3. A. $\frac{1}{25.4}$ B. $\frac{0.007\ 560}{1}$ C. $\frac{1}{6\ 894.757}$ D. $\frac{15\ 432.9}{1}$

5.
	(a)	(b)
A.	68 947.57	689 450 Pa
B.	627.63	7 657 kg (force)

PART THREE FUNDAMENTALS OF ELECTRONIC CALCULATORS

Section 9 Calculators: Basic and Advanced Mathematical Processes

Unit 41 Four-Function Calculators: Basic Mathematical Processes

A. 1. (1) c   (3) a   (5) b

3. (a) The C key may be pressed to reset the contents of the accumulator display and the arithmetic register to zero. (*Note*. On some machines the C key is pressed twice.)

(b) The CE or CD key is pressed to erase an error, followed by the entry of the correct new value.

B. 1.

	Required Significant Digits	Quantity Examples	
		Whole Number Values	Mixed Number Values
A	1	7	0.000 07
B	3	295	0.002 95
C	5	17 346	17.034
D	7	2 528 946	250.096 4

C. 1.

	29 375.	Kilometers (km)
Problem	7 645.2	km
(add)	987.96	km
	1 969.7	km

Step (a)	Keyboard Entry (b)	Display Reads (c)
1	CC	0
2	29 375	29 375
3	+	29 375
4	7 645.2	7 645.2
5	+	37 020.2
6	987.96	987.96
7	+	38 008.16
8	1 969.7	1 969.7
9	=	39 977.86 km **Ans**

3. Problem Example: Divide 27 three times by 2.62 as a constant.

Step	Keyboard Input	Display Readout
1	CC	0
2	27	27
3	÷	27
4	2.62	2.62
5	=	10.305 343
6	=	3.933 337
7	=	1.501 273 6 **Ans**

Unit 42 Four-Function Calculators: Advanced Mathematical Processes

A. 1. *Overflow.* (a) A condition where the numerical capacity of the calculator is exceeded. Condition may be cleared by dividing by a multiple of ten. **Example.** 10 787 932.873

Underflow. (b) A condition where a significant number of nonsignificant numbers are lost. Again, the condition may be cleared by using submultiples of ten. Example. .000 000 004 98 displays as: .000 000 00

3. 1 161.0 mm

B. 1. The computed values shown in the table are rounded off to four places.

Kind of Measurement	Equivalent Measurements					
	A	B	C	D	E	F
Fractional	$\frac{1''}{16}$	$\frac{1''}{64}$	$\frac{1''}{8}$	$\frac{5''}{16\bullet}$	$1\frac{31''}{64}$	$4\frac{2,656''}{10,000}$
Decimal	0.0625"	0.0156"	0.125"	0.3125"	1.4843"	4.2656"
Metric (mm)	1.5875	0.3969	3.1750	7.9375	37.7	108.346

C. 1. 801.018 1 MHz

D. 1. An algorithm is a set of mathematical rules, procedures and logical decisions that are fed into a calculator. The algorithm is used to translate advanced mathematical processes into the four basic processes that are within the capacity of the calculator.

3.

	Quantity	Power Value	
		Display Readout	**Measurement**
A	8^6	262144	262 144
B	7.2^3	373.248	373.25
C	$8.4^2 \times 10.6^2$	7928.1216	7928.12
D	$10.04^3 \times 16.2^2$	265 601.87	265 601.87
E	2.2^{12}	12855.001	12 855.00
F	0.46^{15}	0.0000087	0.000 0087
G	π^4 Use $\pi = 3.1416$	97.4100	(4 places) 97.4100
H	α^6 Use $\alpha = 2.7183$	403.44494	(3 places) 403.445

E.

Display Readout (a)	Required Dimension (b)
A. 9.2736	9.273 6
B. 5.6629	5.662 9
C. 7.461457	7.461 5
D. 2.9916291	2.991 6

(*Note*. The algorithm for problem D. is used as an example to show how the dimension was calculated.)

$\sqrt[6]{716.88}$ First estimate is 3.

Substituting this value,

$R = [716.88 \div (3)^5 + (5)(3)] \div 6$

$= 2.99169$ Second estimate $= \sqrt[6]{716.968}$

$R = [716.88 \div (2.9917)^5 + (5)(2.9917)] \div 6$

$= 2.9916291$ Third estimate $= \sqrt[6]{716.880}$ **Ans**

The 6th power of 2.991 6 (rounded to four places) is within the required precision.

Unit 43 Scientific Calculators: Operation, Functions, and Programming

A. 1. **Examples.** · The scientific calculator is internally programmed to perform direct operations in statistical measurement, algebra, geometry, and trigonometry.

· The scientific calculator is designed to store, recall, exchange, and sum data.

3. a. Press the ON/C clear entry/clear key before any function or operation key is pressed to remove an incorrect reading.

b. Press the ON/C key twice.

B. 1.

Numerical Entry (a)	Function Key (b)	Display (c)	Notes (d)
	ON/C	0	Clear display and register
3.1416	× ((	3.1416	3.1416 stored pending evaluation of ( )
8.2	+	8.2	(8.2 + stored
5.73	)	13.93	(8.2 + 5.73) evaluated
	÷	43.762488	3.1416 × 13.93 evaluated
	((	43.762488	43.762488 stored pending evaluation of next ( ).
9.84	−	9.84	(9.84 − stored
3.652	)	6.192	(9.84 − 3.652 evaluated)
	=	7.0675852 Ans	6.192 divided into 3.1416 × (8.2 + 5.73)

   3. A. 440 lb   B. 428 lb   C. 3,129 lb

C. 1.

Numerical Entry (a)	Key Function and Symbol (b)	Display (c)
25.4	1/x Reciprocal Key	25.4

   3. The problem requires the listing of (a) each numerical entry, (b) name and sequence of keys, and (c) display for finding the mass (volume) of a cored bronze casting. The rounded-off final answer to two decimal places is 7 413.33 cm³.

D. 1.

	Circuit A	Circuit B
Total Resistance (R_T) =	87.08 Ω (ohms)	139.72 Ω (ohms)
Mean resistance =	14.5133 Ω	29.953 Ω
Standard Deviation =	9.069	13.252
Variance =	82.239	175.615

PART FOUR FUNDAMENTALS OF APPLIED ALGEBRA (WITH CALCULATOR APPLICATIONS)

Section 10 Symbols, Terms, and Signed Numbers

Unit 45 The Concepts of Symbols and Terms

A. 1. Area of Rectangle $(A) = (x + 2) \times (x + 7) = x^2 + 9x + 14$

   3. The center distances are

Ⓐ. $\dfrac{(L) - \left(\frac{1}{6}L + \frac{1}{6}L\right)}{3}$ or $\dfrac{\frac{2}{3}L}{3}$ or $\frac{2}{9}L$ Ⓑ. $\dfrac{(L) - \left(\frac{1}{6}L + \frac{1}{6}L\right)}{1\frac{1}{2}}$ or $\dfrac{\frac{2}{3}L}{1\frac{1}{2}}$ or $\frac{4}{9}L$ Ⓒ. $(L) - \left(\frac{1}{6}L + \frac{1}{6}L\right)$ or $\frac{2}{3}L$

B. 1.

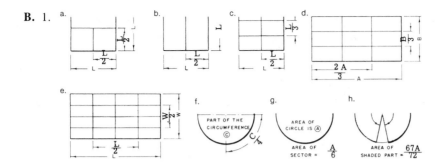

3. 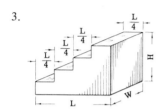　　5.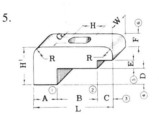

Unit 46 The Concept of Signed Numbers

A. 1. a. Define direction
　　b. Indicate quantities (or value)

B. 1.

Size	Quantity	Direction
a. -6 mm	6 mm	$(-)$ Left of zero reference point
b. $+27''$	$27''$	$(+)$ Right of zero
c. 1500 V	1500 V	$(+)$ Right of zero
d. -0.006 s	0.006 s	$(-)$ Left of zero
e. $+27.5$ N/m^2	27.5 N/m^2	$(+)$ Right of zero
f. 0	0	Reference point
g. -2.75 km	2.75 km	$(-)$ Left of zero
h. $57\frac{5''}{8}$	$57\frac{5''}{8}$	$(+)$ Right of zero

3. a. -14　　c. $\frac{1}{3}$ (or $+\frac{1}{3}$)
　 b. 37 (or $+37$)　d. -0.008

C. 1. a. $|3|$　b. $|15|$　c. $|439.75|$　d. $|0.5869|$　e. $|0|$　f. $|\frac{1}{2}|$　g. $|12\frac{9}{16}|$

D. 1. a. The ampere readings that correspond to the test items follow.
　　(1) Ⓐ $= -3$　　(4) Ⓡ $= 0$
　　(2) Ⓜ $= -3.5$　(5) Ⓚ $= -1$
　　(3) Ⓘ $= +7$　　(6) Ⓢ $= +5$
　 b. The following letters correspond to the ampere readings.
　　(1) $-3 =$ Ⓐ　(3) $-6 =$ Ⓝ　(5) $+4 =$ Ⓛ　(7) $+8 =$ Ⓠ
　　(2) $+1 =$ Ⓔ　(4) $-9 =$ Ⓑ　(6) $-7 =$ Ⓙ　(8) $-2 =$ Ⓟ

c. (1) Ⓠ   (2) Ⓓ   (3) Ⓖ   (4) Ⓙ

E. 1. $+23$ mph   3. $-\frac{1}{4}^{\circ}$ 5. $-\$75.00$ (or $-\$75$)

7. The contraction of $0.000\,09''$, expressed as a signed number, is $-0.000\,09''$.

Unit 47 Application of Positive and Negative Signed Numbers

A. 1. a. $+9$ e. $+13X$ i. $+15\frac{3}{4}L$ m. $+18$

  b. $+5$ f. $+7Y$ j. $+2.5A$ n. $+15\frac{1}{2}$

  c. -2 g. $-3AB$ k. $-13\frac{3}{8}L$ o. -16

  d. -14 h. $-16XYZ$ l. $-11.938B$ p. $+3.625A$

3. $+\$13,473$

B. 1. a. $+2$ c. -11 e. $-2a$ g. $-51A$

  b. $+13$ d. $+2$ f. $+21b$ h. $+16XY$

3. a. A b. $13x+y$ c. $6AB+3LM$ d. $9M+6B$

C. 1. a. 12 b. -11 c. -21.76 d. 50

3. a. 77 b. -91 c. 149 d. -42

D. 1. a. 2 b. -14 c. -4 d. 5

3. a. 36.67 b. $.179$ c. -137.5 d. -36.25

E. 1. a. $37,358.4$ b. -7976.9 c. 1138.7

Unit 48 Graphic Solutions to Signed Number Problems

A. 1. a. A number line is to be drawn to accommodate positive and negative signed numbers from $+6.5$ to -3.5 in increments of 0.5.

  b. Vectors are to be used to graphically add three sets of signed numbers as shown on the number line.

  (1) -1 mm and $+3$ mm
  (2) -3.5 mm and $+2.5$ mm
  (3) The reading on the number line is -5.5 mm. (-1.5 mm and -4.0 mm)

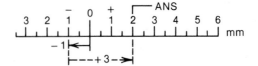

3. a.

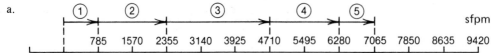

  b. The converted values read on the number line are
  ① 750 RPM $=$ 785 sfpm   ③ $4,500$ RPM $=4,710$ sfpm   ⑤ $6,750$ RPM $=7,065$ sfpm
  ② $2,250$ RPM $=2,355$ sfpm   ④ $6,000$ RPM $=6,280$ sfpm
  c. The RPM at maximum speed (at the end of six seconds) is $6,750$.

B. 1. A number line is to be used to graphically solve the five problems involving the subtraction of signed numbers. The answers are
  a. $+2$ b. -2.5 c. $+6.0$ (or 6.0)   d. $+1.75$ V (or 1.75 V)   e. $14.375''$ (or $+14.375''$)

C. 1. a.

Problem	Unit of Measurement	Signed Number Groupings, Quantities, and Processes			
A	Liters	$(10 \div 2) + (6.5 \times 4) - (2.8 \div 0.7) - (-3 + 7)$			
		* 12 26 -4 -4			
B	Milliamperes (mA)	$(-2.5 + 4.5) - (1.0 \times 3.5) + (-6.6 \div 2.2) + (7.4 - 9.9)$			
		* 2.0 -3.5 -3 -2.5			
C	Force (newton N)	$(10.5 - 7.5) + (-20 \times .5) + [-(77 \div 14) \quad - \quad (47 - 4.5 - 68.5)]$			
		* 3.0 -10 $[$ -5.5 $+26$ $]$			
		* 3.0 -10 $+20.5$			

    b. Each * vector is to be plotted graphically on a number line on which the measurement scale is identified.

    c. The readings on the number lines are problem A, 30 liters; problem B -7.0 mA (or -7.0 milliamperes); problem C $+13.5$ N (or 13.5 newtons).

Unit 49 Application of Symbols in Addition of Terms

A. 1. A. $25X$ **B.** $18\frac{1}{4}L$ **C.** $26.6W$ **D.** $30.63H$ **E.** $33.13A$ **F.** $34.2AB$

  3. a. $13l + 5w$ **c.** $26\frac{3}{4}cd + \left(-9\frac{1}{6}d\right)$ **e.** $+21\frac{1}{2}lw + 12\frac{3}{8}lh + \left(-12\frac{3}{4}l\right)$

      b. $-8ab + 5b$ **d.** $32.00xy + (-17.45y) + (-32.5z)$

B. 1. B. $2\frac{1}{2}A$ **C.** $4A$ **D.** $2\frac{1}{8}A$ **E.** $2\frac{7}{8}A$ **F.** $4\frac{5}{8}A$

  3. A. $3L$ **B.** $4\frac{1}{2}L$ **C.** $6L$ **D.** $6\frac{7}{8}L$ **E.** $7\frac{5}{8}L$ **F.** $2\frac{1}{2}W$ **G.** $3\frac{3}{8}W$

  5. a. $R_T = (R_1 + R_4 + R_7) + (R_2 + R_3) + (R_5 + R_6)$

      b. $R_T = (12.5 + 10.8 + 6.6) + (4.70 + 9.25) + (8.25 + 9.75)$

                   X Group   Y Group   Z Group

      c. (1) R_X group $= 29.9$ ohms   (4) R_X and R_Y groups $= 43.85$ ohms

           (2) R_Y group $= 13.95$ ohms R_Y and R_Z groups $= 31.95$ ohms

           (3) R_Z group $= 18.00$ ohms   (5) R_T for the series circuit $= 61.85$ ohms

Unit 50 Application of Symbols in Subtraction of Terms

A. 1. A. $13X$ **B.** $9\frac{1}{4}W$ **C.** $8\frac{13}{16}L$ **D.** $6.6 N$ **E.** $5.52YZ$ **F.** $51.4XYZ$

  3. A. 3.4 **B.** 1.2 **C.** 3.6 **D.** 2.025 **E.** 7.65 **F.** 8.712

B. 1. a. $11P + 9L$ **c.** $8\frac{1}{4}X + 7Y$ **e.** $36.75HD - 17.50CD$

      b. $9ab - 12bc$ **d.** $54.75lw + 22.87wd$ **f.** $1\frac{11}{16}l + 12\frac{1}{8}wd$

C. 1. A. $2\frac{1}{2}L$ **B.** $\frac{3}{4}L$ **C.** $2\frac{3}{16}L$ **D.** $1\frac{15}{16}L$ **E.** $2\frac{1}{16}L$ **F.** $1\frac{1}{4}L$ **G.** $\frac{1}{2}L$

  3. a. The employees net wage (N_w) as an algebraic expression follows.

$$N_W = G_W - [0.15G_W + 0.07G_W + 0.036G_W + 0.045G_W + 0.06G_W + 0.010G_W]$$
$$= G_W - [0.375G_W]$$

      b. Net wage (N_W) $= 0.625G$

Unit 51 Application of Symbols in Multiplication of Terms

A. 1. A. $50XY$ **B.** $46AB$ **C.** $57.12LM$ **D.** $39.375ABC$ **E.** $102XY$ **F.** $21\frac{7}{8}ADC$

  3. A. $12(7.5 + 6.5) = 168$ **C.** $12(4.25 + 6.5) = 129$ **E.** $12(7.75 + 6.5) = 171$

      B. $12(5 + 6.5) = 138$ **D.** $12(6.5 + 6.5) = 156$ **F.** $12(5.063 + 6.5) = 138.756$

B. 1. A. $4L \times 3W \times 2H = 24LWH$

    B. $6.2L \times 8.4W \times 4.4H = 229.15LWH$

    C. $(7\frac{1}{2}L \times 5\frac{3}{8}H) - (4\frac{1}{2}L \times 2\frac{1}{2}H) \times 2\frac{1}{4}W = 65\frac{25}{64}LWH$

    D. $6.28D \times 4.2L = 26.38DL$

    E. $.7854 + 3.5D^2 \times 12.5H = 120.26\ D^2H$

    F. $(3.5D^2 - 2.25D^2) \times .7854 \times 1.5W = 8.47D^2W$

    3. a. The algebraic expressions for determining values (1) through (6) are given in the chart.

    b. The numerical quantities, obtained by substituting known values with each algebraic expression, are listed in the chart as *computed volumes and weights.* Computed volumes and weights (V_A through V_E) and weights (W_A through W_E) are correct to three decimal places. V_T, W_T, V_M, and W_M are accurate to two decimal places.

    (*Note.* The standard volume (S) = 0.314 lb.)

Electrical Conductor	Algebraic Expressions (a)		Computed (b)	
	Finding Volumes	**Finding Weights**	**Volumes**	**Weights (lb)**
A	$V_A = [(S + .10) \cdot (10)]$	$W_A = (V_A) \cdot (0.314)$	11.000S	3.454
B	$V_B = [(S + .15) \cdot (5)]$	$W_B = (V_B) \cdot (0.314)$	5.750S	1.806
C	$V_C = [(S + .06) \cdot (10)]$	$W_C = (V_C) \cdot (0.314)$	9.400S	2.952
D	$V_D = [(S + .074) \cdot (6)]$	$W_D = (V_D) \cdot (0.314)$	5.556S	1.745
E	$V_E = [(S + .0825) \cdot (10)]$	$W_E = (V_E) \cdot (0.314)$	10.825S	3.399
	$V_T = V_A + V_B + V_C + V_D + V_E$	$W_T = (V_T) \cdot (0.314)$	42.53S	13.35
	$V_M = \frac{V_T}{5}$	$W_M = \frac{W_T}{5}$	8.51S	2.67

Unit 52 Application of Symbols in Division of Terms

A. 1. a. L c. $4\frac{1}{8}B$ e. 4   g. 6   i. $15 + 30y$

     b. $2A$ d. $1.603X$ f. $3a$ h. $40l$

B. 1. $M = \left(12 - 2 \times \frac{1}{2} - 4 \times \frac{1}{8}\right) \div 5 = 2.1$ m

3.

	(1) $L = 254.0$ mm	(2) $L = 381.0$ mm	(3) $L = 317.5$ mm	(4) $L = 479.4$ mm	(5) $L = 622.3$ mm
A	.635	.953	.794	1.199	1.556
B	.826	1.238	1.032	1.558	2.022
C	1.524	2.286	1.905	2.877	3.734
D	1.778	2.667	2.223	3.356	4.356
E	2.223	3.334	2.778	4.195	5.445

5.
Formula	Test Bar #1	Test Bar #2	Test Bar #3	Test Bar #4
A. $L - 2M$	3"	4.5"	6.25"	7.375"
B. $\frac{A}{4}$	.75"	1.125"	1.5625"	1.8438"
C. $2B$	1.5"	2.25"	3.125"	3.6875"
D. $3B$	2.25"	3.375"	4.6875"	5.5313"
E. $4B$	3"	4.5"	6.25"	7.375"

7. a. $R_T = \dfrac{R_1 \cdot R_2 \cdot R_3}{R_1 R_2 + R_1 R_3 + R_2 R_3}$

b. Circuit A. $R_T = \frac{(20) \cdot (80) \cdot (120)}{(20 \cdot 80) + (20 \cdot 120) + (80 \cdot 120)} = 14.12\ \Omega$ **Ans** (c)

Circuit B.   *Note*. Change the 4200 Ω to 4.2 kΩ first.

$$R_T = \frac{(4.2) \cdot (2.3) \cdot (0.5)}{(4.2 \cdot 2.3) + (4.2 \cdot 0.5) + (2.3 \cdot 0.5)} = 0.37413\ \text{k}\Omega.$$

$0.37413\ \text{k}\Omega = 374.13\ \Omega$ **Ans** (c)

Section 11 Equations

Units 54 A Concept of Equations

A. 1. $X = C - M = 14 - 12.6$

3. $C = (D + 4D) \div 2$ $C = \frac{20}{2} = 10$

5. $P = \frac{1}{N}$

7. A. $10A = 16$ A. 1.6 m
   B. $4\frac{1}{2}A$ B. $4.5 \times 1.6 = 7.2$ m
   C. $B + 2A$ C. $7.2 + 2 \times 1.6 = 10.4$ m
   D. $C + 2\frac{1}{2}A$ D. $10.4 \times 2.5 \times 1.6 = 14.4$ m
   E. $16'' - 5\frac{1}{2}A$ E. $5.5 \times 1.6 = 7.2$ m

B. 1. a. $2X + Y$ e. $2\pi(R_1 - R_2)$ h. $1\frac{1}{2}(l + 2n)$

b. $2n(2 - l)$ f. $3\frac{1}{7}(D_1 - D_2)$ i. $6(a + 3b - 5c)$
c. $3(XY + 2LM)$ g. $.64(2P_4 - P_5)$ j. $1.57B(A + C - 2D)$
d. $\pi(D_1 + D_2)$

Unit 55 Solving Equations by Addition and Subtraction

A. 1. $3A + 2A = 10.5$ or $A = 2.1$ mm $B + 2B + B = 4.8$; $B = 1.2$ mm

3. Outside Diameter
   A. 44.45 mm   B. 25.40 mm   C. 20.64 mm   D. 36.51 mm   E. 35.71 mm

5. $L = 2M - 2H$
   A. $2 \times 8 - 2 \times 2 = 12''$ D. $2 \times 2.5 - 2 \times .5 = 4''$ 7. $V_t = 6.63$ volts
   B. $2 \times 10 - 2 \times 3 = 14''$ E. $2 \times 1\frac{3}{4} - 2 \times \frac{1}{4} = 3''$ $R_t = 0.232$ ohms
   C. $2 \times 6\frac{1}{4} - 2 \times 2\frac{1}{4} = 8''$ F. $2 \times 1.25 - 2 \times .375 = 1.75''$

B. 1. $L = O - (R_1 + R_2)$
   A. $12 - 3\frac{1}{2} = 8\frac{1''}{2}$ C. $10 - 2\frac{5''}{16} = 7\frac{11''}{16}$ E . $8.25 - 2 = 6.25''$
   B. $11 - 2\frac{7}{8} = 8\frac{1''}{8}$ D. $9\frac{1}{2} - 2\frac{1}{16} = 7\frac{7''}{16}$ F. $7.625 - 1.562 = 6.063''$

3. $N - 1 + 4.6 + 5.3 + 2.8 = 16.2$ $N = 4.5$ mm
5. A. 0.55   B. 2.475   C. 4.779
7. $.120''$ B. $1\frac{11''}{64}$ C. $1.17X$ D. $.935A$ E. 1.43 cm

Unit 56 Analyzing, Expressing, and Solving Equations by Multiplication and Division

A. 1. a. $X = 48$ c. $U = 11$ e. $c = 6.28$ g. $c = 7.0686$
   b. $M = 25$ d. $n = 30$ f. $a = 5.084$ h. $B = .0315$
B. 1. a. $X = 3$ b. $Y = 3$ c. $A = 18$ d. $B = -.9$ e. $L = 14.875$
C. 1. $2{,}799.36\ W$
3. $N = .6495 \div D$ A. $N = 10$ C. $N = 32$ E. $N = 64$
   B. $N = 5$ D. $N = 18$ F. $N = 28$

D. 1. a. $X = 2$ b. $Y = 6.4$ c. $Z = 5.25$ d. $N = 20$ e. $L = 8.16$ f. $A = 20.5$
3. $D = C/\pi$
   A. $10''$ B. $18.183''$ C. $1.5''$ D. $1.799''$

5. A. 350°   B. 399.2°   C. 804.8°   D. 161.6°   E. 207.14°

7. A. $3 = 6B$ $B = \frac{1}{2}$

    D. $1.5 \times 4.6 = 1.25F \div .75$ $F = 4.14$

    B. $40 = 2n/25$ $n = 500$

    E. $2.2C = 2200 \times 1.8 \div 25$ $C = 72$

    C. $4 \times 1.25 = (4.75 - 2.13) \div 3P$ $P = .1747$

9. $B = \dfrac{2\left(\frac{22}{7}\right)(L)(W)(R)}{33,000}$

    A. $B = \left(2 \times \frac{22}{7} \times 14 \times 15 \times 550\right) \div 33,000 = 726,000 \div 33,000 = 22$

    B. $B = \left(2 \times \frac{22}{7} \times 17\frac{1}{2} \times 30\frac{1}{2} \times 750\right) \div 33,000 = 2,516,250 \div 33,000 = 76\frac{1}{4}$

    C. $B = \left(2 \times \frac{22}{7} \times 36\frac{3}{4} \times 75\frac{1}{4} \times 66\right) \div 33,000 = 1,147,261\frac{1}{2} \div 33,000 = 34\frac{3}{4}$

Section 12 Ratio and Proportion

Unit 58 The Concept of Ratio

A. 1. Measurements   a. $2\frac{7}{8}''$ b. $1\frac{31}{32}''$ c. $3''$ d. $1\frac{1}{8}''$ e. $2\frac{9}{16}''$

     Ratios   b:a = 63:92   a:c = 23:24   d:b = 36:63 or 4:7   c:e = 48:41

   3. A. 10:21   B. 31:17   C. 27:31   D. 21:37   E. 10:31 $A_1 = \frac{31}{32}$ $A_2 = \frac{5}{16}$

   5. $350:17\frac{1}{2} = 20:1$

   7. A. = 2:5   B. = 1:2   C. = 1:3   D. = 39:125   E. = 1:3

   9. a. A. 1:1 or 1/1   B. 1:2 or 1/2   C. $1:2\frac{1}{2}$ or $1/2\frac{1}{2}$ D. 8:9 or 8/9   E = 9:22 or 22/9

     b. A. 1:1 or 1/1   B. 2:1 or 2/1   C. $2\frac{1}{2}:1$ or $2\frac{1}{2}/1$ D. 9:8 or 9/8   E = 22:9 or 22/9

Unit 59 The Concept of Proportion

A. 1. $3.5:8.5 = X:450 = 185.3$ milliliters (ml)   **Ans**

   3. a. $.58 \times 7.82 \times (1728 + 864) = 11,756.275$ oz or 734.77 lb

     b. $734.77 \times 0.4536 = 339.29$ kg

   5. $10,900:560 = x:2.5$ $x = 48.7$ minutes

B. 1. $4 \times 1700 \div 3200 = 2.125$ or 2″ to nearest $\frac{1}{2}''$

   3. Step 1. $20.32 \times 300 \div 8.89 = 685.6$ or 686 RPM

     Step 2. $16.51 \times 300 \div 12.7 = 390$ RPM

     Step 3. $12.7 \times 300 \div 16.51 = 230.8$ or 231 RPM

     Step 4. $8.89 \times 300 \div 20.32 = 131.25$ or 131 RPM

   5. A. 240 kg   B. 222 kg   C. 80 kg   D. 2,173 kg

Unit 60 Application of Relationships to Proportion

A. 1. A. $3 \times 7.8 \div 8.3 = 2.819$ oz   D. $8.25 \times 7.8 \div 8.3 = 7.753$ lb

     B. $1 \times 7.8 \div 8.3 = .94$ lb   E. $12.125 \times 7.8 \div 8.3 = 11.395$ lb

     C. $2.5 \times 7.8 \div 8.3 = 2.349$ lb

   3. (These figures can only be approximate.)

     A. $1''$ C. $\frac{11}{16}''$ E. $4\frac{1}{2}''$ G. $1''$ I. $\frac{3}{8}''$ K. $3\frac{1}{4}''$ M. $\frac{9}{16}''$ O. $\frac{15}{64}''$

     B. $2\frac{1}{2}''$ D. $\frac{5}{16}''$ F. $\frac{3}{4}''$ H. $\frac{9}{16}''$ J. $1\frac{7}{8}''$ L. $\frac{3}{4}''$ N. $\frac{27}{32}''$ P. $\frac{1}{16}''$

Section 13 Exponents

Unit 62 The Concept of Exponents

A. 1. a. 2^2 b. 4^4 c. 10^7 d. 5^4 e. $.4^2$ f. 1.2^2 g. $.5^3$ h. 6.25^2

B. 1.

Factors	Exponential Form
A. 2×2	2^2
B. $2 \times 2 \times 2 \times 2$	2^4
C. 5×5	5^2
D. $2 \times 2 \times 2 \times 2 \times 2 \times 2$	2^6
E. $.3 \times .3$	$.3^2$
F. $2 \times .3 \times 2 \times .3$	$(2 \times .3)^2$ or $.6^2$
G. $3 \times 3 \times .3 \times .3$	$(3 \times .3)^2$ or $.9^2$
H. $\frac{1}{2} \times \frac{1}{2} \times \frac{1}{2} \times \frac{1}{2}$	$\left(\frac{1}{2}\right)^4$
I. $\frac{1}{2} \times \frac{1}{2} \times \frac{1}{2} \times \frac{1}{2} \times \frac{1}{2} \times \frac{1}{2}$	$\left(\frac{1}{2}\right)^6$
J. $5 \times 5 \times 5$	5^3
K. $2 \times 2 \times 2 \times 5 \times 5 \times 5$	$(2 \times 5)^3$ or 10^3
L. $2 \times 2 \times 2 \times 2 \times 2 \times 2 \times 2 \times 2 \times 2$	$(2'')^9$
M. $3 \times 3 \times 3 \times .3 \times .3 \times .3$	$(3'' \times .3)^3$ or $(.9'')^3$
N. $2 \times 2 \times 2 \times 2 \times 2 \times 2 \times .3 \times .3 \times .3$	$(2 \times 2 \times .3)^3$ or $(1.2)^3$

C. 1. $A = (D)^2 \times .7854$ $\qquad$ $A = (r)^2 \times \pi$
A. $(22.5)^2 \times .7854 = 39.609 \text{ cm}^3$ $\qquad$ C. $(2.4)^2 \times 3.1416 = 18.096 \text{ cm}^2$
B. $(18.75)^2 \times .7854 = 276.117 \text{ cm}^2$ $\qquad$ D. $(10.32)^2 \times 3.1416 = 334.588 \text{ mm}^2$
3. $P = I^2(R)$ $\qquad$ a. 1440 watts $\qquad$ b. 1224 watts $\qquad$ c. 2537.28 watts

Unit 63 Scientific Notation and Calculator Functions

A. 1. A. 3.24×10^{-5} in. to 1.44×10^{-5} in. $\qquad$ E. 1.6×10^{-5} in.
B. 8.1×10^{-4} mm to 3.6×10^{-4} mm $\qquad$ F. 4×10^3 Å (Angstrom)
C. 2×10^{-5} in. $\qquad$ G. 2×10^3 vps (vibrations/sec)
D. 5×10^{-4} mm $\qquad$ H. 4×10^7 cps (cycles/sec)

B. 1. A. $0.000\ 002\ 655\ \Omega/\text{cm}^3$ $\qquad$ D. $0.000\ 0013''/°\text{F}$
B. $0.000\ 001\ 673\ \Omega/\text{cm}^3$ $\qquad$ E. $2300\ \text{kg/m}^3$
C. $0.000\ 0024''/°\text{F}$ $\qquad$ F. $550\ \text{lb/ft}^3$

C. 1. a. Ⓐ Exponent $\qquad$ Ⓑ Mantissa $\qquad$ Ⓒ Integer $\qquad$ Ⓓ Decimal Value
Ⓔ Floating Negative Sign $\qquad$ Ⓕ Negative Exponent Sign
b. (1) 7.34251×10^{16} $\qquad$ (2) -5.2503×10^{-8}

D. 1. a. $11.667\ \text{ft}^3/\text{min}$ (or $1.1167^{10^1}\ \text{ft}^3/\text{min}$) $\qquad$ 3. $1\ 867\ 008 \times 10^{18}\text{eV}$ (electron/volts)
b. $0.316\ \text{m}^3/\text{min}$ (or $3.16^{10^{-1}}\ \text{m}^3/\text{min}$)

Unit 64 Algebraic Multiplication of Numbers and Letters with Exponents

A. 1. a. 2^4 b. 3^6 c. 10^{12} d. 9^{16} e. $\left(\frac{1}{2}\right)^5$ f. $\left(\frac{1}{4}\right)^8$ g. 1.2^{13} h. 3.02^{14}

B. 1. a. $16.92(10^6)$ volts $\qquad$ b. $229.376(10^9)$ grams $\qquad$ c. $396.39(10^{-12})$ hertz $\qquad$ d. $2\ 747.55(10^{-6})$ meters

C. 1. a. $A^2 + 4A + 4$ $\quad$ d. $y^2 + 6y + 9$ $\qquad$ g. $n^3 + 3n^2 + 3n + 1$ $\quad$ j. $m^3 + 21m^2 + 147m + 343$
b. $C^2 + 10C + 25$ $\quad$ e. $X^2 + 8X + 16$ $\qquad$ h. $Z^2 + 12Z + 36$ $\quad$ k. $R^2 - 2R + 1$
c. $B^2 + 16B + 64$ $\quad$ f. $d^3 + 9d^2 + 27d + 27$ $\quad$ i. $g^3 + 6g^2 + 12g + 8$ $\quad$ l. $S^3 - 3S^2 + 3S - 1$

D. 1. a. $(3.12^{10^{-6}}) \times (0.425^{10^{-3}}) \times (105^{10^{-2}})$ $\qquad$ c. and d. $139.23^{10^{-11}}$ Hz
3. a. $1.5^{10^2} \times 3.14159 \times 8.5 \times 2.857^{10^{-1}}$ $\qquad$ c. and d. 1.14^{10^3} '/min

Unit 65 Algebraic Division of Numbers and Letters with Exponents

A. 1. a. $4^2 = 16$ c. $12^2 = 144$ e. $2.5^2 = 6.25$ g. $\left(\frac{1}{2}\right)^2 = \frac{1}{4}$

   b. $3^3 = 27$ d. $132^2 = 17{,}424$ f. $.63^1 = .63$ h. $\left(\frac{5}{8}\right)^2 = \frac{25}{64}$

B. 1. a. A^1 or A b. B^3 c. X^3 d. $A^{(c-d)}$ e. Y^{2m} f. C^{-4a} or $\frac{1}{C_{4a}}$

C. 1. a. $3.0(10^{-2})$ meters   c. $0.75(10^3)$ meters
   b. $4.5(10^{-3})$ liters   d. $1.2(10^9)$ volts

D. 1.

	Required		(a) **Programming Information Exponential Form**	Computed Quantities	
	Quantity	**Electrical Unit of Measure**		(b) **Scientific Notation Format**	(c) **Standard Notation**
A	Total Resistance (R_t)	kΩ (kiloohms)	1.6 kΩ + 6.4 kΩ + 272 Ω + 2.6 kΩ $1.6^{10^3} + 6.4^{10^3} + .272^{10^3} + 2.6^{10^3}$	10.872^{10^3}	10.872 kΩ
B	Current (I)	mA (milliamperes)	$\dfrac{27.4}{10.872\ \text{K}^{10^3}}$	2.520^{10^3} A	2 520 000 mA
C	Source Voltage (E)	V (volts)	$27.4 \times \dfrac{1.6^{10^3}}{10.872^{10^3}}$	4.032 V	
			$27.4 \times \dfrac{6.4^{10^3}}{10.872^{10^3}}$	16.130 V	
			$27.4 \times \dfrac{.272^{10^3}}{10.872^{10^3}}$	.686 V	
			$27.4 \times \dfrac{2.6^{10^3}}{10.872^{10^3}}$	6.553 V	
D	Check on Total Source Voltage	V (volts)		27.4 V	
E	Total Circuit Power (P_t)	mW (milliwatts)	$2.52^{10^{-3}} \times 27.4$	$69.048^{10^{-3}}$ W	69.048 mW

Section 14 Radicals

Unit 67 Square Root of Numbers

A. 1. a. 6   c. 10   e. 12   g. .5   i. 1.2   k. 60
   b. 8   d. 25   f. 16   h. 1.1   j. 1.5   l. 1.7

B. 1. a. 1.41   e. $\sqrt{15.\ 00\ 00\ 00}$ $\begin{array}{r} 3.\ \ 8\ \ 7\ \ 2\ \ \textbf{Ans}\ 3.87 \end{array}$ h. $\sqrt{3\ 20.\ 00\ 00\ 00}$ $\begin{array}{r} 1\ \ 7.\ \ 8\ \ 8\ \ 8\ \ \textbf{Ans}\ 17.89 \end{array}$

$\begin{array}{r} 1.\ \ 7\ \ 3\ \ 2\ \ \textbf{Ans}\ 1.73 \end{array}$
b. $\sqrt{3.\ 00\ 00\ 00}$

$$\begin{array}{rr} & 1 \\ 27 & 2\ 00 \\ & 1\ 89 \\ 343 & 11\ 00 \\ & 10\ 29 \\ 3462 & 71\ 00 \\ & 69\ 24 \end{array}$$

$$\begin{array}{rr} & 9 \\ 68 & 6\ 00 \\ & 5\ 44 \\ 767 & 56\ 00 \\ & 53\ 69 \\ 7742 & 2\ 31\ 00 \\ & 1\ 54\ 84 \end{array}$$

$$\begin{array}{rr} & 1 \\ 27 & 2\ 20 \\ & 1\ 89 \\ 348 & 31\ 00 \\ & 27\ 84 \\ 3568 & 3\ 16\ 00 \\ & 2\ 85\ 44 \\ 35768 & 30\ 56\ 00 \\ & 28\ 61\ 44 \end{array}$$

c. $\sqrt{7} = 2.65$ f. $\sqrt{17} = 4.12$ i. $\sqrt{431} = 20.76$
d. $\sqrt{5} = 2.24$ g. $\sqrt{24} = 4.90$

C. 1. a. $\frac{2}{5}$ c. $\frac{3}{4}$ e. $\frac{6}{7}$ g. $\frac{8}{9}$ i. $\frac{.4}{.5}$ k. $\frac{1}{1.1}$

   b. $\frac{4}{5}$ d. $\frac{5}{6}$ f. $\frac{3}{5}$ h. $\frac{10}{11}$ j. $\frac{.5}{.6}$ l. $\frac{.8}{1.2}$

3. a. $\frac{8}{9}$ b. $\frac{21}{26}$ c. $\frac{3.3}{6.5}$ or $\frac{33}{65}$ d. $\frac{10.488}{11.384}$

D. 1. a. $\frac{2.24}{2.65}$ b. $\frac{2.24}{3.46}$ c. $\frac{6.63}{5.92}$ d. $\frac{1.73}{5.66}$ e. $\frac{2.24}{3.53}$ f. $\frac{.35}{.49}$

E. 1. A. $\sqrt{16 + 9} = 5''$ D. $\sqrt{100 + 256} = 18.868''$
   B. $\sqrt{43.56 + 77.44} = 11''$ E. $\sqrt{31.36 + 60.84} = 9.602''$
   C. $\sqrt{1 + 4} = 2.236''$

3. A. 60.95 cm   B. 88.89 cm   C. .964 m   D. 1.58 m

F. 1. Diameter of wire strands in cable A = 6.58 circular mils;
   cable B = 10.64 circular mils.

3. The lengths of the sides of the integrated circuit chips are
   A. 0.125''   B. 3.200 mm   C. 4.126 mm

Unit 68 Roots of Algebraic Numbers

A. 1. a. $2a$ c. $6c$ e. $1.2Y$ g. $\frac{4A}{5B}$ i. $\frac{1.565X}{2.683Y}$

   b. $3b$ d. $.5X$ f. $1.6Z$ h. $\frac{5.196b}{5.657c}$

B. 1. $\sqrt{7}$ 3. $(2 + x)\sqrt{Y}$ 5. 0

C. 1. $zy\sqrt{ab}$ 3. $.30\sqrt[3]{48a^2}$ 5. $2\sqrt{3b}$

D. 1. Crankshaft diameter = 4.359''.

3. The diameter of each concrete column is $1' - 4\frac{3''}{4}$

Unit 69 Application of Square Root Tables

A. 1. a. 1.41   b. 1.73   c. 2.24   d. 2.45   e. 2.65   f. 3.16
B. 1. a. $\sqrt{32} = 4\sqrt{2} = 5.66$ d. $\sqrt{324} = 18.00$ g. $4\sqrt{.2} = 1.79$
   b. $\sqrt{80} = 4\sqrt{5} = 8.94$ e. $\sqrt{1694} = 41.12$ h. $\sqrt{2501.7} = 50.02$
   c. $\sqrt{63} = 3\sqrt{7} = 7.94$ f. $5\sqrt{1.1} = 5.24$
C. 1. A. $Z = \sqrt{9 + 16} = 5''$ B. $Z = \sqrt{36 + 49} = 9.220''$

    C. $Z = \sqrt{156.25 + 225} = 19.525''$ D. $Z = \sqrt{73.96 + 110.25} = 13.572''$

3. $\sqrt{440 \div 200} = \sqrt{2.2} = 1.483$ amperes

5. $c = \sqrt{96.4 \times 2.5 \div 8} = \sqrt{30.125} = 5.488$ inches

7. a. 13.65 ft   b. $13' - 7\frac{3''}{4}$ (to nearest $\frac{1''}{4}$)

Section 15 Formulas and Complex Equations

Unit 71 The Concept and Use of Algebraic Formulas

A. 1. a. $D = 2R$ b. $A = S^2$ c. $L = 1\frac{1}{4}C$ d. $A = \sqrt{B^2 + C^2}$ e. $CS(\text{in inches}) = \pi D(\text{RPM})$

    3. Answers depend on formulas used in each case.

B. 1. $H_o = H_1 - [(E) \cdot (H_1)]$

    $H_o = 4,212$ calories

    3. $d = \sqrt{\frac{A}{.7854}}$ $d = 0.125''$ 5. a. $R_1 = \frac{R_t \cdot R_2}{R_2 - R_t}$ b. $R_1 = 19.84\ \Omega$

C. 1. a. $R = o - (2d + 2c)$ b. $L = L_1 + L_2 + \left(\frac{t}{2}\right)$ c. $L = 2C + 3.25\left(\frac{D_1 + D_2}{2}\right)$

Unit 72 Complex Equations

Computed Values	Check	Method
A. 1. a. $X = 5, Y = 4$	$X + 4Y = 21$	
	$(5) + (4) \cdot (4) = 21$	Substitution
	$21 = 21$	
b. $A = 11\frac{1}{7}, B = 32\frac{4}{7}$	$4A - B = 12$	
	$(4) \cdot \left(11\frac{1}{7}\right) - \left(32\frac{4}{7}\right) = 12$	Substitution
	$12 = 12$	
c. $z = \frac{1}{2}, q = \frac{4}{5}$	$6z - 5q = -1$	
	$6\left(\frac{1}{2}\right) - 5\left(\frac{4}{5}\right) = -1$	Addition
	$3 - 4 = -1$	
	$-1 = -1$	
d. $a = -\frac{20}{23}, b = -1\frac{19}{23}$	$5a + 2b + 8 = 0$	
	$5\left(-\frac{20}{23}\right) + 2\left(-1\frac{19}{23}\right) + 8 = 0$ $\left(-\frac{100}{23}\right) + \left(-\frac{84}{23}\right) + 8 = 0$	Addition
	$-\frac{184}{23} + 8 = 0$ $0 = 0$	

Equations	Computed Values	Check
B. 1. $X + Y = 2.50$	$X = 2.00''$	$X + Y = 2.50$
$X + X/4 = 2.50$	$Y = 0.50''$	$(2.00) + (0.50) = 2.50$
		$2.50 = 2.50$
3. $X + Y = 32$	Cylinder $X = 8$ ft^3	$X + Y = 32$
$3Y + Y = 32$	Cylinder $Y = 24$ ft^3	$(8) + (24) = 32$
		$32 = 32$

Unit 73 Application of Special Formulas and Handbook Data

A. 1. Examples of two major functions served by formulas.

    a. To state precise relationships among quantities so that unknown values may be accurately determined.

    b. To provide a simplified form for establishing required measurements that are interrelated in an equality.

    3. The strain is $0.000\ 071\ ''/''$.

    5. The change in length is $4.212''$.

    7. The impedance is 6.9 ohms.

B. 1.

Gear	Number of Teeth	Diametral Pitch (P)*	Addendum (a)	Dedendum (d)	Whole Depth (WD)	Working Depth (W_d)	Circular Pitch (P_c)
A	36	12	0.083	0.096	0.179	0.167	0.262
B	24	6	0.167	0.193	0.360	0.333	0.524
C	48	18	0.056	0.064	0.120	0.111	0.175

*Rounded to whole number value

PART FIVE FUNDAMENTALS OF APPLIED GEOMETRY WITH CALCULATOR APPLICATIONS

Section 16 Geometric Lines and Shapes

Unit 75 The Concepts of Lines, Angles, and Circles

A. 1.

Line	Direction and Length		
	Horizontal	**Vertical**	**Inclined 45°**
AB	3″		
AC		4″	
AD		$3\frac{1}{2}″$	
XY			152 mm
XZ			130 mm

    3. *a* and a_1 = 8.123 V   5. *A* = 32.522 V

B. 1. a. 69° 46′   b. 114° 46′   c. 141° 54′ 36″   d. 290° 36′ 52″

    3. a. ∠A = 81° 45′   b. ∠B = 141° 32′   c. ∠C = 90° 2′ 45″

C. 1. Angles 1 and 6–obtuse angles
      Angles 2 and 4–right angles
      Angles 3 and 5–acute angles

    3. ∠A = 145°   ∠D = 90°
       ∠B = 60°   ∠E = 160°
       ∠C = 150°

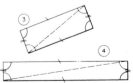

D. 1. A. radius = 0.8125 m   B. diameter = 1.75 m   C. chord = 0.438 m   D. arc = $\frac{1}{8}$ × 12.688 = 1.586 m

Unit 76 Basic Flat Shapes

A. 1. a. Equilateral   b. Obtuse scalene   c. Acute isosceles   d. Right isosceles

B. 1.   3.

C. 1.

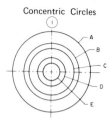

Concentric Circles ①

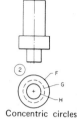

② Concentric circles

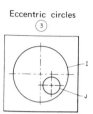

Eccentric circles ③

D. 1. Trapezoid   Parallelogram   Hexagon   Trapezoid
   3. A. Requires the use of a protractor to draw a square with 2-inch sides.
      B. A parallelogram is to be drawn with a protractor. The inclined sides at 60° to the horizontal are 2 inches long; the horizontal, 3 inches long.
      C. A hexagon is to be drawn with a protractor. The distance across flats is $2\frac{1}{4}$ inches.
      D. An octagon is required to be drawn with the aid of a protractor; distance across corners is $2\frac{1}{2}$ inches.

Unit 77 Basic Solid Shapes

A. 1. A. Cylinder   E. Cylinder Hole   H. Cylinder
       B. Cone   F. Square or Rectangular Hole   I. Truncated Square Prism (Duct)
       C. Hemisphere   G. Triangular Prism   J. Truncated Cylinder (Pipe)
       D. Rectangular Solid

  3.

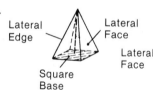

a.   b.   c.   d.

B. 1. a. Cone   b. Sphere   c. Hemisphere

C. 1.

Lateral Edge   Lateral Face   Lateral Face   Square Base

Unit 78 Congruent and Symmetrical Plane and Solid Geometric

A. 1. a.

Side	Included Angle	Side
AB	Angle A	AC
AC	Angle C	CB
BC	Angle B	BA
BA	Angle B	BC
AB	Angle A	AC
BC	Angle C	CA

B. 1. Yes. All corresponding parts are equal.
   3. No. Difference in one set of corresponding parts.
C. 1. Balance   Smooth operation   Simplified production   Appearance   Replacement

D. 1. a. b.

Section 17 Formulas Applied to Geometric Forms

Unit 80 Applications of Formulas to Plane and Solid Geometric Forms (Regular Polygons)

A. 1. $\left(48 \times 48 \times \frac{4}{12}\right) \div 27 = 28.44$ cu yds

3. Formula: $A = \left(4S + \frac{1}{2}\right)H$

    A. $\left(4 \times 8 + \frac{1}{2}\right) \times 8 = 260$ sq in.   C. $\left(4 \times 39 + \frac{1}{2}\right) \times 39 = 6103.5$ sq in. or 42.4 sq ft

    B. $\left(4 \times 9.5 + \frac{1}{2}\right) \times 9.5 = 365.75$ D. $\left(4 \times 64 + \frac{1}{2}\right) \times 64 = 16,416$ sq in. or 114 sq ft
    sq in.   E. $\left(4 \times 57.5 + \frac{1}{2}\right) \times 57.5 = 13,253.75$ sq in. or 92.04 sq ft

B. 1. $\left[\left(32 \times 28\right) + \left(24\frac{2}{3} \times 30\right)\right] \div 9 = 181.78$ sq yd
  3. A. $(24 \times 36 \times 30) \div 231 = 112.2$ gallons   D. $\left(5\frac{1}{2} \times 6\frac{1}{3} \times 10\frac{1}{2} \times 7.48\right) \div 31\frac{1}{2} = 86.851$ barrels
    B. $(12.5 \times 25 \times 28) \div 231 = 37.875$ gallons   E. $\left(20\frac{2}{3} \times 15\frac{1}{2} \times 12\frac{1}{2} \times 7.48\right) \div 31\frac{1}{2} = 950.828$ barrels
    C. $\left(2\frac{1}{3} \times 3\frac{1}{4} \times 2\right) \times 7.48 = 113.45$ gallons

C. 1. A. 3.03 m   B. 2.36 m   C. 3.88 m
D. 1. A. $1 \times 1 = 1$ B. $2 \times 2 = 4$ C. $2 \times 2 = 4$ D. $2 \times 1 = 2$ E. $2 \times 1 = 2$
  3. a. $4(\sqrt{930}) + 1.3 = 123.3$ cm   b. $4(\sqrt{1650}) + 1.3 = 163.7$ cm   c. $4(\sqrt{3721}) + 1.3 = 245.3$ cm

Unit 81 Application of Formulas to Plane and Solid Circular Forms

A. 1. A. $14 \times \frac{22}{7} = 44''$ C. $3\frac{1}{2} \times \frac{22}{7} = 11'$ E. $9 \times \frac{22}{7} = 28.286''$

    B. $70 \times \frac{22}{7} = 220''$ D. $10.5 \times \frac{22}{7} = 33'$

  3. $C = \pi D$ $D = \frac{C}{\pi}$
    A. 20.321 cm   B. 11.434 cm   C. 28.298 mm   D. 45.480 mm   E. 0.485 m or 485.103 mm

B. 1. A. $2 \times 1.75 \times 3.1416 \times \frac{90}{360} = 2.7489''$ C. $2 \times 4.25 \times 3.1416 \times \frac{120}{360} = 8.9012''$
    B. $2 \times 6.8 \times 3.1416 \times \frac{90}{360} = 10.6814''$

  3. **Customary (U.S.) Unit Values**   **Metric Unit Values**

    A. $\frac{180}{360} \times 10 \times 3.142 = 15.710''$ $\frac{180}{360} \times 25.4 \times 3.142 = 39.903$ cm

    B. $\frac{90}{360} \times 12.25 \times 3.142 = 9.622''$ $\frac{90}{360} \times 31.12 \times 3.142 = 24.445$ cm

    C. $\frac{60}{360} \times 12.6 \times 3.142 = 6.598''$ $\frac{60}{360} \times .320 \times 3.142 = 0.168$ m

D. $\frac{135}{360} \times 2 \times 8 \times 3.142 = 18.852''$ $\frac{135}{360} \times 2 \times 20.32 \times 3.142 = 47.884$ cm

E. $\frac{72}{360} \times 2 \times 6.25 \times 3.142 = 7.855''$ $\frac{72}{360} \times 2 \times 158.75 \times 3.142 = 199.517$ mm

C. 1. $V = \pi r^2 H$

A. $\frac{22}{7} \times 7^2 \times 10 = 1\ 540$ cu in.

D. $\frac{22}{7} \times 14^2 \times 15 \times 7.48 \div 31.5 = 2194.133$ barrels

B. $\frac{22}{7} \times 10^2 \times 32 = 10\ 057$ cu in.

E. $\frac{22}{7} \times 10.25^2 \times 12.5 \times 7.48 \div 31.5 = 980.108$ barrels

C. $\frac{22}{7} \times 7.25^2 \times 18 = 2\ 973.54$ cu in.

3. $[A_1$ (in meters) $- A_2$ (in meters)$] \times H =$ Quantity required

A. $A_1 = (0.9906)^2 \times 0.7854 = .7707$
   $A_2 = (0.7112)^2 \times 0.7854 = \underline{.3973}$
                                    $.37344 \times 1.8 = 0.67$ m^3 **Ans**

B. $A_1 = (1.2246)^2 \times 0.7854 = 1.2166$
   $A_2 = (0.889)^2 \times 0.7854\ \ = \underline{0.62072}$
                                    $0.59588 \times 2.4 = 1.43$ m^3 **Ans**

Section 18 Common Geometric Constructions

Unit 83 Basic Geometric Constructions

A. 1. Problems require construction for parallel lines.
B. 1. Problems require construction for bisecting lines.
C. 1. Problems involve construction for erecting a perpendicular to a line at a given point.
D. 1. Problems require construction for bisecting an angle.

E. 1.

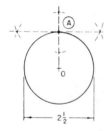 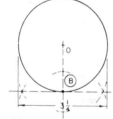 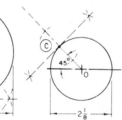

F. 1. a. A full-size layout of the Slotted Plate is required. The basic geometric constructions are to be shown for parallel and perpendicular lines, bisecting and other dividing of angles, and constructing tangent lines, circles, and arcs.
     b. Dimensions are to be included.

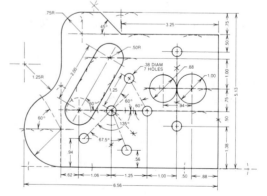

Unit 84 Constructions Applied to Geometric Shapes

A. 1. a. Connect points on circumference located by two diameters at right angles.

    b. Connect alternate points on circumference located by striking off six radius length chords on circle.

    c. Connect points on circumference located by striking off six radius length chords on circle.

  3. $2H = 5.196''$

B. 1. a. Construct parallel tangents at ends of two perpendicular diameters.

    b. Construct tangents at points on circumference located by two consecutive chords of radius length.

    c. Construct tangents at points on circumference located by six chords of radius length.

  3. a. The dimensioned layout of the Sheet Metal Template shows the dotted geometric construction lines (to a 2:1 Scale).

    b. The drawing is to be dimensioned.

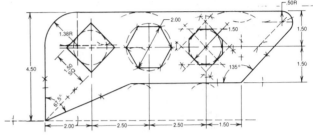

PART SIX FUNDAMENTALS OF APPLIED TRIGONOMETRY (WITH CALCULATOR APPLICATIONS)

Section 19 Right, Acute, and Oblique Triangles

Unit 86 The Concept of Right Triangle Trigonometry

A. 1. ∡A

    a = side opposite

    b = side adjacent

    c = hypotenuse

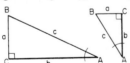

B. 1.

Function	Abbreviation	Trigonometry Ratios	
		Expressed as Sides	Sides Represented by Letters
Sine	sin ∠A	$\dfrac{\text{opposite}}{\text{hypotenuse}}$	$\dfrac{a}{c}$
Cosine	cos ∠A	$\dfrac{\text{adjacent}}{\text{hypotenuse}}$	$\dfrac{b}{c}$
Tangent	tan ∠A	$\dfrac{\text{opposite}}{\text{adjacent}}$	$\dfrac{a}{b}$
Cotangent	cot ∠A	$\dfrac{\text{adjacent}}{\text{opposite}}$	$\dfrac{b}{a}$
Secant	sec ∠A	$\dfrac{\text{hypotenuse}}{\text{adjacent}}$	$\dfrac{c}{b}$
Cosecant	csc ∠A	$\dfrac{\text{hypotenuse}}{\text{opposite}}$	$\dfrac{c}{a}$

3. Sin angle $D = \frac{4.8}{8.0} = .600$ Cot angle $D = \frac{6.4}{4.8} = 1.333$

Cos angle $D = \frac{6.4}{8.0} = .800$ Sec angle $D = \frac{8.0}{6.4} = 1.25$

Tan angle $D = \frac{4.8}{6.4} = .750$ Csc angle $D = \frac{8.0}{4.8} = 1.667$

5. The length of the guy wire is 19.25 meters.

C. 1.

Function	Cofunction
sin 45°	cos 45°
tan 20°	cot 70°
sec $29\frac{1}{2}°$	csc $60\frac{1}{2}°$

D. 1. Using the Pythagorean theorem of $a^2 = b^2 + c^2$ and substituting the values of 14 feet and 7 feet, the length of the rafters is 15'-8".

3. The accuracy of the 90° center lines of the holes if found by substituting the value of the hypotenuse (60.00 cm, the 54.54 cm horizontal center distance, and the 25.01 cm height in the Pythagorean formula. The measurement check is 3600.1, which is within the $\pm 5'$ accuracy required. Thus, the 90° angle is correct.

E. 1. a. The impedance is found by formula.

$$Z = \sqrt{R^2 + (X_L - X_C)^2} = \sqrt{30^2 + (25 - 40)^2} = \sqrt{30^2 + (-15)^2} = 33.54 \ \Omega$$

b. The phase angle $\angle\theta = \tan\theta = \frac{X_L - X_C}{R} = -0.5000$

The tan θ of 0.5000 indicates a negative angle in the fourth quadrant of $-26° \ 34'$.

F. 1. a. The 30° 54' 30" angle setup requires the following (or an equivalent) selection of angle gage blocks. 30° + (1' + 3' + 20' + 30') + 30".

b. The 15° 27' 15" angle setup requires the following (or an equivalent) selection of angle gage blocks. 15° + (30' − 3') + (20" − 5").

Unit 87 Applications of Tables of Trigonometric Functions

A. 1. Since these are construction problems, measurements will only approximate exact computation.

Opposite Side	Angle A	Sine Angle A
2"	30°	214 = .5
2.165"	60°	$\frac{2.165}{2.5} = .86603$
1.6"	$29\frac{1}{2}°$	$\frac{1.6}{3.25} = .49242$

B. 1. The straight-line symbols indicate whether the numerical function values are read downward or upward in a trigonometric function table and whether the left or right column minutes are used.

(1) sin 73° 10' ⌐   (3) tan 163° 39' ⌐   (5) sec 253° 24' ⌐   (7) tan 343° 22' ⌐
(2) cos 106° 42' ⌐   (4) cot 196° 58' ⌐   (6) csc 286° 17' ⌐

C. 1. a. and b.

Given Angle		Required Trigonometric Function Value (a)	Name and Angle Measurement of the Complementary Angle (b)
A	1° 30'	sin 0.0262	cos 88° 30'
B	10° 30'	cos 0.9833	sin 79° 30'
C	25° 30'	tan 0.4770	cot 64° 30'
D	50° 40'	sin 0.7735	cos 39° 20'
E	60° 36'	cos 0.4909	sin 29° 24'
F	75° 25'	tan 3.8436	cot 14° 35'

3.

	Trigonometric Function Value	Quadrant (a)	Angle Measurement (b)
A	sin 0.30043	I	17° 29′
B	cos 0.30071	I	72° 30′
C	cos −0.95372	II	162° 30′
D	tan 0.51283	III	207° 9′
E	sec −1.29160	III	219° 16′
F	cot −0.60921	IV	301° 21′
G	sin −0.63383	IV	320° 40′

D. 1.

	Trigonometric Function Value	Quadrant	Angle Measurement
A	sin A = 0.30071	I	17.5°
B	cos A = 0.25038	I	75.5°
C	tan A = −0.98270	II	135.5°
D	cot A = −0.04891	II	92.8°
E	sec A = −1.00490	III	185.7°
F	csc A = −1.72628	III	215.4°
G	cos A = 0.18567	IV	280.7°
H	cot A = −4.27374	IV	347.4°

3. a. 0.38671   c. −1.04853
   b. −0.92407   d. −1.00486

E. 1. A. = cos 30° × hyp. = .866 × 2.25 = 1.949″
       B. = sin 30° × hyp. = .5 × 2.25 = 1.125″

5. tan A = $\frac{305 \text{ mm}}{1200 \text{ mm}}$ = 0.2500
   A = 14° 2′

3. tan $\frac{1}{2}$ angle A = $\frac{(4.00 - 2.5) \div 2}{4}$ = $\frac{.75}{4}$ = .1875

   tan $\frac{1}{2}$ angle A = 10° 37′ 10″   angle A = 21° 14′

Unit 88 Four-Function and Scientific Calculator Programming: Trigonometric Applications

A. 1. Algorithms provide a sequential series of steps that are used with a four-function calculator to increase the capacity of the instrument in solving trigonometry problems.

3. With the possible introduction of error in trigonometric calculations by using a calculator alone, more accurate results are produced by using conventional methods supplemented by the calculator to perform those operations where there is no question about the accuracy of the result.

5. Sin θ = $1.745\alpha[1 + 0.508\alpha^2(0.15\alpha^2 - 1)]$; $\alpha = \theta \div 100$

B. 1. The natural trigonometric functions for angles A, B, and C using a scientific calculator are given in the table.

	Angles	sin	cos	tan	csc
A	45°	0.7071	0.7071	1.0000	1.4142
B	32° 30′ (32.5°)	0.5373	0.8434	0.6371	1.8612
C	85° 49′ 12″ (85.82°)	0.9973	0.0729	13.6828	1.0027

3.

	Given Measurement	Numerical Entry (a)	Trigonometric Function Value(b)	Reciprocal Function and Value (c)
A	sin 60°	60	sin 0.8660	csc 1.1547
B	tan 85° 43'	85.72	tan 13.352	cotan 0.0748
C	sec 58° 35' 30"	58.59	sec 1.9188	cos 0.5216

C. 1. a. $\overline{1}$ b. $\overline{2}$ c. $\overline{4}$ d. 1   e. 2   f. 4

3. a. 3.8820   b. 0.3514   c. $\overline{3}$.6645

5. Common logarithms: (a) 1.8061 (b) 0.2367, rounded to four decimal places.
Natural logarithms: (c) 0.7307 (d) −6.5293, rounded to four decimal places.

D. 1. a.

Angle	Degrees	Radians	Grads
A	45	0.7854	50
B	67.5	1.178	75
C	135.56	2.366	150.6
D	270	4.712	300

b. The equivalent of 135.56° is 135° 33' 36".

Unit 89 Acute Triangles

A. 1. Csc 30° = $O \div 1.375$ $O = 2 \times 1.375$ $O = 2.75''$
    Cot 30° = $A \div 1.375$ $A = 1.732 \times 1.375$ $A = 2.3815''$

3. a. $h = 1.9(\tan 60)$ b. $V_c = 1.05(3.61)((3.29)1/3\pi r^2 h$
     $= 1.9(1.7321)$ $= 12.47 \text{ m}^3$ volume of cone **Ans**
     $= 3.29$ meters high **Ans**   c. $15.24 + 3.29 = 18.53$ m overall length **Ans**

B. 1. Angle $A = \frac{1}{2} \times \frac{360}{14} = 12° 51' 25''$ Sin angle $A = xy \div 2.1875$
   $xy = .22252 \times 2.1875 = .48676$ $xz = 2xy = .9735''$

Unit 90 Oblique Triangles

A. 1. A. $n^2 = 4^2 + 3^2 - 2 \times 4 \times 3 \times .17635$ $n^2 = 29.2324$ $n = 5.4''$
   (*Note.* Use law of supplements. Note that the cosine of an angle of more than 90° is a negative.) Dimensional values are rounded to one decimal place.
   $\frac{\text{Sin } 100}{\text{Sin M}} = \frac{5.41}{4}$ $\frac{.98481}{\text{Sin M}} = \frac{5.41}{4}$ Sin M = .7281   M = 46° 44'
   Angle O = 180° − (100° + 46° 44') = 33° 16'

   B. $m^2 = 3.5^2 + 2^2 - 2 \times 2 \times 3.5 \times .57358$ $m^2 = 8.21988$ m = 2.9''
   $\frac{\text{Sin } 55° (.81915)}{\text{Sin O}} = \frac{2.867}{2}$ Sin O = .57143   Angle O = 34° 51'

   C. $O^2 = 7.8^2 + 5.6^2 - 2 \times 7.8 \times 5.6 \times .48481$ $O^2 = 49.871844$ O = 7.062''
   $\frac{\text{Sin O} (.87462)}{\text{Sin M}} = \frac{7.062}{5.6}$ Sin M = .69452   M = 43° 59'
   Angle N = 180° − (61° + 43° 59') = 75° 1'

   D. $n^2 = 4.2^2 + 1.8^2 - 2 \times 4.2 \times 1.8 \times (-.08715)$ $n^2 = 22.1977$ n = 4.7''
   $\frac{\text{Sin } 95° (.99619)}{\text{Sin O}} = \frac{4.711}{1.8}$ Sin O = .3806   Angle O = 22° 22'
   Angle M = 180° − (95° + 22° 22') = 62° 38'

   E. $m^2 = 2.5^2 + 4.6^2 - 2 \times 2.5 \times 4.6 \times .54464$ $m^2 = 14.88328$ m = 3.9''
   $\frac{\text{Sin } 57 (.83867)}{\text{Sin O}} = \frac{3.858}{2.5}$ Sin O = .54345   Angle O = 32° 55'
   Angle N = 180° − (57° + 32° 55') = 90° 5'

3. Use law of sines   X = 32° 20'   y = 21'7" or 21.6'

5. $\cos X = \dfrac{(18.415)^2 + (27.6225)^2 - (13.6525)^2}{2(18.415 \times 27.6225)} = .89914 = 25° 57' 36''$

$\cos Z = \dfrac{(27.6225)^2 + (13.6525)^2 - (18.415)^2}{2(27.6225 \times 13.6525)} = .80914 = 35° 59' 12''$

$\cos Y = 180° - (25° 57' 32'' + 35° 59' 12'') = 118° 3' 16''$

PART SEVEN HIGH TECHNOLOGY APPLICATIONS OF MATHEMATICS

Section 20 Systems and Programming for Advanced Applications

Unit 92 Numerical Control: Cartesian Coordinate and Binary Systems

A. 1. a. Point-to-point programming provides direct commands for a particular process before moving on to the next event. Continuous-path programming blends one contour or movement in a continuing cycle.

    b. Absolute dimensioning relates each dimension to a fixed reference point. Incremental dimensioning provides a build-on moving from one dimension to the next.

B. 1. a. The drawing shows the reference point floated to the center of the Mounting Plate.

    b. The X, Y, and Z rectangular coordinate dimensions relate to the floated zero.

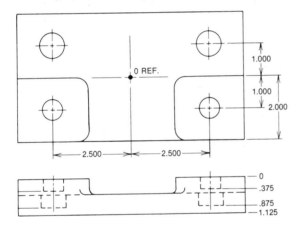

Hole	Rectangular Coordinate Dimensions (0.001″)		
	X	Y	Z
1	−2.500	−1.000	0.375
2	2.500	−1.000	0.375
3	2.500	1.000	0.875
4	−2.500	1.000	0.875

C. 1. **Examples.**

    a. The binary system provides stable $+$ and $-$, on/off, conductor/nonconductor, and charged/discharged conditions.

    b. Binary numbers are easily added, subtracted, multiplied, and divided.

    c. Binary numbers respond to instantaneous processing of data for signals that control processes and products.

  3. a. $12_{10} = 1\,1\,0\,0_2$ c. $155_{10} = 1\,0\,0\,1\,1\,0\,1\,1_2$ e. $1\,298_{10} = 1\,0\,1\,0\,0\,0\,1\,0\,0\,1\,0_2$

     b. $36_{10} = 1\,0\,0\,1\,0\,0_2$ d. $477_{10} = 1\,1\,1\,0\,1\,1\,1\,0\,1_2$

Unit 93 Numerical Control Applications: Computer Numerical Control (CNC), Computer-Assisted Design (CAD), and Computer-Assisted Manufacturing (CAM)

A. 1. a. The channels on EIA tape are

    b. The binary notation symbols are

    c. The decimal equivalents are

1	2	3	and	4
2^0	2^1	2^2		2^3
1	2	4		8

3. **Examples.**
   a. Controlling table and workpiece positioning movements.
   b. Controlling tool positioning and rotation movements.
   c. Establishing speed and feed rates, machining, and operational sequences.

B. 1.

Examples.	Nondimensional Programming Words	Character
a. Preparatory function	_____	G
b. Tool selection	_____	T
c. Miscellaneous function	_____	M

C. 1. *Examples* of two CNC-tape controlled program features.
   a. Direct feed rates in sfpm (mm/min) or speeds in RPM.
   b. Simultaneous rotary table movement with *XY*- or *Z*-axis movements.

D. 1. a. Dots on the tape identify the alphanumeric values and the word statements.
   b. The tape is marked with program information for blocks 000 and 001.
   c. The notes and brackets along the side of the tape identify each function and activity in each block.

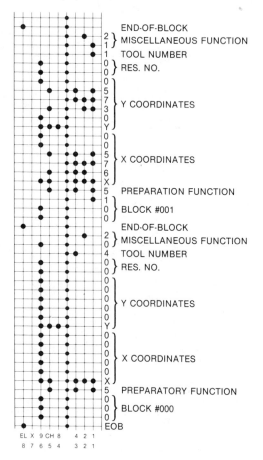

GLOSSARY

Algebra A branch of mathematics requiring symbols, terms, and signed numbers in problems that cannot be solved by the four basic mathematical processes alone.

Algorithm A set of mathematical rules, numerical procedures, or logical decisions that also may be used with an electronic calculator to solve advanced mathematical problems.

Alphanumeric Code Instructions (primarily related to numerical control) that consist of numbers, letters, and other symbols. Letter-number symbols that form the foundation for input-output signals into computerized equipment.

Angular Measure The size of an opening formed by two intersecting lines or surfaces. Angles are measured in radians or steradians and in degrees (°), minutes ('), and seconds ("). Angular measure involves the four basic arithmetical processes with angles and parts of angles.

Area of a Circle $A = \pi$ (radius)^2 *or* $A = 0.7854$ (diameter)^2

Area of a Triangle The surface area of a triangle equals the base $\times \frac{1}{2}$ altitude.

Axiom A self-evident mathematical truth. For example, the value of an equation is not changed if the same process and quantity are applied to both sides of an equation.

Balanced Equation The values on both sides of an equation are equal.

Bar Graphs Heavy lines or bars of a definite length to represent specific quantities on a grid having vertical and horizontal lines. A scale indicates the value between each line.

Base Units in SI Metrics Seven base units circumscribe all physical science measurements. The base units include meter, kilogram, second, ampere, candela, kelvin, and mole.

Basic Arithmetical Processes The four processes of adding, subtracting, multiplying, and dividing. Included, also, is any combination of the four processes.

Binary Numbers A numbering system comprising only two numeric symbols: 0 and 1. A base-two numerical notation system on which computers depend. Two conditions of operation of electronic circuits: on-off, plus or minus, charge or discharge, and conductor or nonconductor.

Cartesian Coordinate System A graphical system used to describe in mathematical terms any absolute point from any other point in space. Precise positioning of a feature or dimension of a part along three coordinate axes that are mutually perpendicular.

Celsius (°C) A derived unit of temperature measurement in SI metrics, expressed as degrees Celsius (°C). Celsius temperatures replace centigrade temperatures.

Circle A closed curved line on a flat surface. Every point on the closed curved line is the same distance from a fixed given point called a center. *Concentric circles* have a common center. *Eccentric circles* are "off-center" and do not originate from the same center.

Circle Graph The division of a circle into a specific number of parts, each sector denoting a fixed relationship to all other data presented.

Circular Measure The measurement of circles, curved surfaces, cylinders, and other circular shapes.

Circumference The distance around the periphery of a circle, measured either in standard units of linear measure or in degrees.

Collecting Terms The process of combining letters, symbols, or numbers within grouping symbols.

Complex Equation A formula or equation with two unknown variables. Complex equations are solved by either substitution or addition.

Computer-aided Design (CAD) Generating design features, dimensioning, and other information from data and specifications that are programmed, stored, retrieved, and modified from computer memory and other computer functions.

Computer-aided Manufacturing (CAM) A complete computerized system that combines the capacity of microprocessors, minicomputers, and main-line computers to control multiple machine and instrument functions in manufacturing.

Computer-aided Programming (CAP) Programming machine tools, instruments, and other equipment by computers rather than by manual programming.

Computer Numerical Control (CNC) The automated control of all functions of a machine tool, instrument, or other equipment. Input, output, and storage of command signals that are generated by computer rather than by manual programming.

Congruence A geometric term indicating that all the physical properties of size and shape in one part are identical in another part. Two figures are congruent if they differ only in location. Two line segments of equal length or two circles of equal diameter are congruent.

Constant A fixed mathematical relationship for which a specific symbol is used universally.

Conventional Units French and European metric systems using the meter as the basic unit of measure. Unit designations, standards, values, and derivation of base and other units differ from SI metrics.

Conversion of Quantities The changing of a quantity from a unit of measure to an equivalent value in another unit of measure. The process of multiplying or dividing (as may be required) a specified quantity by a conversion factor.

Cubical Solid A figure formed by extending an original square surface so that all corners are square and the edges are the same length.

Cubic Measure (also Volume Measure) Measurement of the space occupied by a body that has three linear dimensions of length, height, and depth. Cubic measure is the product of these three dimensions, expressed in cubic units.

Customary Units United States units of measure based upon the inch, yard, and pound, similar to the British Standard units.

Cylinder of Revolution (Right Cylinder) Two equal and perpendicular bases with a lateral (vertical outside) surface around the bases. The area of the lateral surface of a cylinder of revolution equals the circumference × height.

Decimal-Degree (Trigonometric Function Table) A table of numerical trigonometric function values for metric angle measurements. Metric angle measurements that are expressed in whole and/or decimal parts of a degree.

Decimal Equivalent A conversion of a customary unit of measure to its equivalent value as a decimal.

Decimal Fraction A fraction whose denominator is a multiple of ten: for example, 100; 1,000; and 10,000. Decimal fractions are written on one line with a decimal point in front of the numerical value.

Decimal Places The number of digits to the right of the decimal point. The number of places is usually related to the degree of accuracy required in the answer.

Depreciation The loss in value of a process, material, or product.

Derived Units in SI Metrics Units that are derived from the base units in SI metrics to satisfy varying computational needs. The complementary derived units relate to quantities of length, mass, time, electric current, thermodynamic temperature, luminous intensity, and amount of substance.

Division The process of determining how many times one value is contained in another. Division is a simplified method of subtraction.

Equation A mathematical statement indicating that the quantities or expressions on both sides of an equality sign (=) are equal.

Estimating A shortcut mathematical process of determining a range against which an actual answer may be checked for accuracy.

Exponent A simple mathematical statement that indicates a quantity is to be multiplied by itself a number of times. The exponent is a small number (superscript) written to the right and slightly above a quantity. The exponent indicates the power to which the number is to be *raised* (multiplied by itself).

Expressing Quantities Stating a quantity either in terms of numbers, or a combination of numbers and symbols. *Numerical terms* signifies that numbers are used to express a quantity. *Literal terms* relates to the use of numbers and symbols to express a quantity.

Expression The statement of a problem in abbreviated form using numerical and literal terms.

Extracting the Square Root The process of determining the equal factors that, when multiplied together, give the original value.

Factoring The steps used in determining the series of smaller numbers that, when multiplied by each other, produces the original number.

Formula A shortened method of expressing relationships and a combination of mathematical processes that consistently give the same solution.

Formula Evaluation The process of (1) substituting all known numerical and literal factors in a formula and (2) solving for a particular variable by performing the indicated operations.

Gage Block Measurement A system for establishing a precise measurement by ''wringing'' together hardened and precisely finished steel blocks against which measurements may be established. Precision blocks used to establish precise linear and angular dimensions.

Geometry A branch of mathematics relating to points, lines, planes, closed flat shapes, and solids. These elements may be used alone or in combination to describe, design, construct, and test every visible object.

Grouping Symbols Symbols that are used to group quantities together to simplify reading and mathematical processes. The four common grouping symbols are parentheses, brackets, braces, and a bar to cover a quantity.

Hertz The frequency of one cycle per second.

Improper Fraction A fraction with a numerator greater than the denominator.

Indirect (Inverse) Proportion Ratios that vary inversely. The two terms in the first or second ratio are reversed.

Interpolation The process of calculating an intermediate value of a function that lies between two known values.

Interpolation (Machining) The method of advancing a workpiece or a tool from one point in a program to the next point. Producing a contoured feature in a workpiece. Continuous path machining controlled by interpolation.

Joule The work done when a force of one newton is moved a distance of one meter in the direction of the force.

Kilogram A base unit of mass in SI metrics. A kilogram is equal to 2.2 pounds or 1,000 grams.

Law of Cosines The square of any side of a triangle is equal to the sum of the squares of the other sides minus twice the product of the two sides and the cosine of their included angle.

Law of Sines An abbreviated method of expressing that the sides of a triangle are proportional to the sines of the opposite angles.

Line Graph A grid of lines at fixed values on which straight, curved, or broken lines are plotted to present information visually.

Line Segment The working length of a line defined by end points.

Literal Equation A specific type of equation in which some or all of the quantities are represented by letters instead of numbers.

Logarithm of a Number The exponent indicating the power to which it is necessary to raise a number to produce a given number.

Lowest (Least) Common Denominator The smallest number into which each number in a set of denominators will divide exactly.

Magnitude A specific quantity, size, or amount of a force.

Mantissa The decimal part of a logarithm.

Members (Equation) The expressions that appear on either side of the equality sign in an equation. The quantity on the left side is the *first member*. The *second member* appears on the right side.

Meter A base unit of length in SI metrics and conventional metric measurements. For most practical purposes the meter equals 39.37 inches.

Metricize The process of converting any other unit of measure to its SI metric equivalent.

Micrometer A precision instrument for taking linear measurements in any one of the measurement systems, depending on the calibrations on the micrometer. The distance between the anvil and a spindle is accurately read from a calibrated barrel and thimble. Vernier micrometers have an additional set of graduated lines. These make it possible to read to a more precise degree (0.002 mm and 0.0001″) than regular micrometers.

Multiplication A simplified method of addition.

Multiplying and Dividing Quantities The product of multiplying two like negative terms or any number of positive terms is $(+)$. The product is $(-)$ when an odd number of negative terms is multiplied. These rules for the use of signs apply also to division.

Newton A force that, when applied to a body of one kilogram mass, produces an acceleration of the body of one meter per second per second.

Normal Frequency Distribution A bell-shaped curve formed by plotting actual production sizes in a random distribution against a design dimension for a part. Distribution of part sizes as contrasted with a required dimension.

Numerical Control (NC) A system of coded instructions that controls processes and movements of machine tools and instruments.

Parallelogram A figure having two pairs of parallel sides. The area = length × height.

Percent A short way of relating a given number of parts to a whole, which is equal to 100 (percent). One percent is $\frac{1}{100}$ of the whole.

Percentage The product of the base times the rate.

Plane Geometry That phase of geometry that relates to objects having two dimensions that lie within a plane.

Plane (Plane Surface) A flat surface on which a straight line connecting two points lies.

Positive and Negative Quantities *Positive numbers* refers to all numbers that are greater than zero. *Negative numbers* relates to values that are less than zero.

Prefixes and Symbols (SI Metrics) Multiple and submultiple values of the base-ten notation stem are identified by a system of prefixes. Each prefix is identified by an SI symbol. SI metric quantities are usually defined by the appropriate prefix followed by the unit of measure.

Probability The number of ways an event can occur divided by the total number of ways the event can occur.

Proper Fraction A fraction with a numerator smaller than the denominator.

Proportion Two ratios, equal in value, that are placed on opposite sides of an equality sign.

Protractor A measuring tool for angles. Protractors vary from simple flat tools graduated in degrees to more precise instruments. A movable blade protractor and the universal-bevel protractor are graduated to permit readings in degrees and minutes. The vernier protractor with vernier scales is used for still greater degrees of accuracy.

Pyramid A solid geometric object formed by connecting each corner of a flat shape with a point (*vertex*) outside the base.

Pythagorean Theorem A formula expressing the relationship among the three sides of any right triangle. $A^2 = B^2 + C^2$.

Radical Sign A mathematical shorthand way of indicating that the equal factors of the value under the radical symbol $(\sqrt{})$ are to be determined.

Ratio A comparison of one quantity with another like quantity or value.

Rectangle A figure whose opposite sides are parallel and whose adjacent sides are at right angles to each other. The area = length × height.

Resultant A single vector that represents the sum of a given set of vectors.

Right Prism A solid formed by a flat form as it moves perpendicular to its base to a specified altitude.

The shape of the base determines the name of the right prism.

Rounding (Rounding Off) Reducing the number of digits to a lesser number than the total number available. A quantity is reduced to the number of significant digits that produces a desired degree of accuracy in the result.

Scientific Notation System A mathematical system where quantities are written and computed in a simplified form using power of ten multiples.

Sector of a Circle The surface or area between the center and the circumference of a circle that is included within a given angle. The area of a sector equals the area of the circle multiplied by the fractional part that the sector occupies.

Signed or Directed Numbers Positive ($+$) and negative ($-$) numbers that indicate either direction from a fixed reference point or that identify the mathematical operation.

Significant Digits Any digit that is needed to define a specific value or quantity. The number of significant digits is based on the implied or required precision associated with the problem.

SI Metrics (SI) An up-to-date International System of Units of measurement established cooperatively among most industrialized nations. SI measurements depend on base, supplementary, and derived units, and on combinations of these units to accurately measure and quantitatively define all measurable objects. SI is the standard abbreviation for *Le Système International d'Unités*.

Soft Conversion Refers to measurements that are computed directly without consideration for product design. Direct conversion of measurement values from either SI metric or customary units of measure.

Solid Angle Represents the enclosed area that is equal to a square whose sides are equal in length to the radius of a sphere. Measured in steradian units of measure.

Solid Geometry That phase of geometry that relates to objects formed and measured by three or more dimensions. Features that lie in two or more planes.

Solids of Revolution The shape taken by outside lines or lines of a flat form when revolved around an axis.

Statistical Measurement Manipulating data and large numbers of measurements by mathematical computations.

Stretchout Shop term applied to a two-dimensional layout that becomes a three-dimensional object when rolled or formed into a closed figure.

Supplementary Units in SI Metrics The radian and steradian constitute the two supplementary units for measuring angles in SI metrics. The radian relates to a plane angle; the steradian to a solid angle.

Surface Measure The measurement of a part, object, mechanism, or other physical mass that has length and height. A surface is "measured" when its length and height (in the same unit of measure) are multiplied. The product is the *area* in square units.

Symbols A simplified way of identifying and working with quantities, units of measure, and mathematical processes and of communicating information.

Symmetrical The corresponding points of an object are equidistant from an axis.

Terms Parts in a mathematical expression that are separated by such signs as ($+$) and ($-$). *Like terms* relates to the different terms in an expression that have the same literal factor. When the literal factors are different or unlike, they are called *unlike terms*.

Terms (Proportion) The two outside terms of a proportion are the *extremes*. The two inner terms of a proportion are the *means*.

Theorem A mathematical truth that can be proven.

Transposing Terms A method of moving all known terms to one side of an equation and all unknown terms to the other side. The sign of each transposed quantity is changed.

Transversal A line that intersects two or more lines.

Trapezoid A four-sided figure in which two of the sides (called *bases*) are parallel. The area $=$ the sum of the two bases $\times \frac{1}{2}$ altitude.

Triangle Three straight line segments joined at the ends to form a closed flat shape.

Trigonometric Functions An equation expressing the ratio between two sides. The six common ratios are sine (sin), cosine (cos), tangent (tan), cotangent (cot), secant (sec), and cosecant (csc).

Trigonometry A branch of mathematics that deals with the measurement of angles, triangles, and distances.

Variable A letter or other symbol in an equation that represents a quantity that may have more than one value.

Vector A line segment that represents direction ($+$ or $-$) and magnitude (size or quantity).

Vernier Caliper Micrometer An instrument on which linear measurements are determined by adding the reading on a graduated beam and a graduation on the vernier scale of a movable leg.

Volt The difference of electrical potential between two joints of a conductor carrying a constant current of one ampere. The power dissipated between these two points equals one watt.